T0141932

Advances in Intelligent Systems and Computing

Volume 764

The series "Advances in Intelligent Systems and Computing" contains publications on theory, applications, and design methods of Intelligent Systems and Intelligent Computing. Virtually all disciplines such as engineering, natural sciences, computer and information science, ICT, economics, business, e-commerce, environment, healthcare, life science are covered. The list of topics spans all the areas of modern intelligent systems and computing such as: computational intelligence, soft computing including neural networks, fuzzy systems, evolutionary computing and the fusion of these paradigms, social intelligence, ambient intelligence, computational neuroscience, artificial life, virtual worlds and society, cognitive science and systems, Perception and Vision, DNA and immune based systems, self-organizing and adaptive systems, e-Learning and teaching, human-centered and human-centric computing, recommender systems, intelligent control, robotics and mechatronics including human-machine teaming, knowledge-based paradigms, learning paradigms, machine ethics, intelligent data analysis, knowledge management, intelligent agents, intelligent decision making and support, intelligent network security, trust management, interactive entertainment, Web intelligence and multimedia.

The publications within "Advances in Intelligent Systems and Computing" are primarily proceedings of important conferences, symposia and congresses. They cover significant recent developments in the field, both of a foundational and applicable character. An important characteristic feature of the series is the short publication time and world-wide distribution. This permits a rapid and broad dissemination of research results.

More information about this series at http://www.springer.com/series/11156

Radek Silhavy
Editor

Artificial Intelligence and Algorithms in Intelligent Systems

Proceedings of 7th Computer Science On-line Conference 2018, Volume 2

Editor
Radek Silhavy
Faculty of Applied Informatics
Tomas Bata University in Zlín
Zlín
Czech Republic

ISSN 2194-5357 ISSN 2194-5365 (electronic)
Advances in Intelligent Systems and Computing
ISBN 978-3-319-91188-5 ISBN 978-3-319-91189-2 (eBook)
https://doi.org/10.1007/978-3-319-91189-2

Library of Congress Control Number: 2018943432

Printed on acid-free paper

This Springer imprint is published by the registered company Springer International Publishing AG part of Springer Nature
The registered company address is: Gewerbestrasse 11, 6330 Cham, Switzerland

Preface

This book constitutes the refereed proceedings of the modern trends and approaches of artificial intelligence research and its application to intelligent systems. The paper discusses hybridization of algorithms, new trends in neural networks, optimization algorithms, and real-life issues related to artificial methods' application.

This book constitutes the refereed proceedings of the Artificial Intelligence and Algorithms in Intelligent Systems of the 7th Computer Science On-line Conference 2018 (CSOC 2018), held online in April 2018.

CSOC 2018 has received (all sections) 265 submissions, 141 of them were accepted for publication. More than 60% of accepted submissions were received from Europe, 30% from Asia, 5% from Africa, and 5% from America. Researchers from 30 countries participated in CSOC 2018 conference.

CSOC 2018 conference intends to provide an international forum for the discussion of the latest high-quality research results in all areas related to Computer Science. The addressed topics are the theoretical aspects and applications of Computer Science, Artificial Intelligences, Cybernetics, Automation Control Theory and Software Engineering.

Computer Science On-line Conference is held online and modern communication technology, which is broadly used to improve the traditional concept of scientific conferences. It brings equal opportunity to participate to all researchers around the world.

I believe that you will find following proceedings interesting and useful for your own research work.

March 2018

Radek Silhavy
Editor

Organization

Program Committee

Program Committee Chairs

Petr Silhavy	Tomas Bata University in Zlin, Faculty of Applied Informatics
Radek Silhavy	Tomas Bata University in Zlin, Faculty of Applied Informatics
Zdenka Prokopova	Tomas Bata University in Zlin, Faculty of Applied Informatics
Roman Senkerik	Tomas Bata University in Zlin, Faculty of Applied Informatics
Roman Prokop	Tomas Bata University in Zlin, Faculty of Applied Informatics
Viacheslav Zelentsov	Doctor of Engineering Sciences, Chief Researcher of St. Petersburg Institute for Informatics and Automation of Russian Academy of Sciences (SPIIRAS)

Program Committee Members

Boguslaw Cyganek	Department of Computer Science, University of Science and Technology, Krakow, Poland
Krzysztof Okarma	Faculty of Electrical Engineering, West Pomeranian University of Technology, Szczecin, Poland
Monika Bakosova	Institute of Information Engineering, Automation and Mathematics, Slovak University of Technology, Bratislava, Slovak Republic

Pavel Vaclavek	Faculty of Electrical Engineering and Communication, Brno University of Technology, Brno, Czech Republic
Miroslaw Ochodek	Faculty of Computing, Poznan University of Technology, Poznan, Poland
Olga Brovkina	Global Change Research Centre Academy of Science of the Czech Republic, Brno, Czech Republic and Mendel University of Brno, Czech Republic
Elarbi Badidi	College of Information Technology, United Arab Emirates University, Al Ain, United Arab Emirates
Luis Alberto Morales Rosales	Head of the Master Program in Computer Science, Superior Technological Institute of Misantla, Mexico
Mariana Lobato Baes	Superior Technological of Libres, Mexico
Abdessattar Chaâri	Laboratory of Sciences and Techniques of Automatic control and Computer engineering, University of Sfax, Tunisian Republic
Gopal Sakarkar	Shri. Ramdeobaba College of Engineering and Management, Republic of India
V. V. Krishna Maddinala	GD Rungta College of Engineering & Technology, Republic of India
Anand N. Khobragade	Maharashtra Remote Sensing Applications Centre, Republic of India
Abdallah Handoura	Computer and Communication Laboratory, Telecom Bretagne, France

Technical Program Committee Members

Ivo Bukovsky
Maciej Majewski
Miroslaw Ochodek
Bronislav Chramcov
Eric Afful Dazie
Michal Bliznak
Donald Davendra
Radim Farana
Martin Kotyrba
Erik Kral
David Malanik
Michal Pluhacek
Zdenka Prokopova
Martin Sysel
Roman Senkerik
Petr Silhavy
Radek Silhavy
Jiri Vojtesek
Eva Volna
Janez Brest
Ales Zamuda
Roman Prokop
Boguslaw Cyganek
Krzysztof Okarma
Monika Bakosova
Pavel Vaclavek
Olga Brovkina
Elarbi Badidi

Organizing Committee Chair

Radek Silhavy — Tomas Bata University in Zlin, Faculty of Applied Informatics

Conference Organizer (Production)

OpenPublish.eu s.r.o.
Web: http://www.openpublish.eu
Email: csoc@openpublish.eu

Conference Website, Call for Papers

http://www.openpublish.eu

Contents

Analysis of Affective and Gender Factors in Image Comprehension of Visual Advertisement

Gabrielė Liaudanskaitė[1], Gabrielė Saulytė[1], Julijus Jakutavičius[1], Eglė Vaičiukynaitė[2], Ligita Zailskaitė-Jakštė[1], and Robertas Damaševičius[1](✉)

[1] Department of Multimedia Engineering, Kaunas University of Technology, Kaunas, Lithuania
robertas.damasevicius@ktu.lt
[2] Department of Marketing, Kaunas University of Technology, Kaunas, Lithuania

Abstract. Advertisement is an integral part of the daily life in modern society. Nevertheless, the advertising industry often ignores the fact that people might understand ads differently and behave accordingly. Eye movements, fixation times, gaze landing sites and areas of interest in visual images are known as good predictors of customer behavior. Here we perform eye tracking of several versions (neutral, positively and negatively valence) of popular ads and analyze the dependency of eye fixation times on affective and gender factors. The results show that there are statistically significant differences on how affectively charged images are perceived, but no statistically significant differences between genders have been found when shorter fixations representing unconscious processing of visual information are analyzed.

Keywords: Image comprehension · Gaze tracking · Visual advertisement Affective neuromarketing · Gender studies

1 Introduction

Marketing professionals know that a well-designed visual ad should have a catchy visual story that forces the viewer to stop. Specific features of images such as colors can stimulate, excite, and form different emotions and result in different reaction. The need to execute effective advertising campaigns to promote products and services motivates the research in this area of science [1]. The sensomotoric, affective and cognitive response of consumers to marketing stimuli is analyzed by neuromarketing, a field of science that applies the methods of neuroscience and physiological computing to evaluate and predict consumer behavior [2].

To explain the role of emotions in visual understanding, several theoretical models of emotion have been proposed. Some models propose a small number of universal emotional states, i.e., discrete emotions such as joy, fear, etc. (see [3], for a review). The appraisal models claim that specific emotions are influenced by appraisal processes

R. Silhavy (Ed.): CSOC 2018, AISC 764, pp. 1–11, 2019.
https://doi.org/10.1007/978-3-319-91189-2_1

which integrate the situational context of an event (see [4]). The dimensional models suggest that emotion can be explained by a dimensional space, such as a 2D space spanning valence and arousal. Valence describes the extent to which an emotion is positive or negative, whereas arousal refers to emotional intensity, i.e., the strength of the associated emotional state.

Since advertising seeks to influence the user's mental functions - first of all, perception, behavior, thinking and emotions, the interest in the role of affective factors on the perception of advertising is steadily growing, and so there is a growing number of studies that help to understand how advertising is most affected by emotion [5, 6].

The task is made more difficult by known fact that the content can not be applied indiscriminately, e.g., if one ad engages children, it will not necessarily be suitable for adults. Moreover, there is a dual perception of the emotions visible in the image [7] as follows:

(1) Syncretic cognition, which is occurring instinctively, spontaneously and broadly subjectively. The image is perceived by what it is visually recognizable and known from before, it causes relevant emotions (e.g., joy, sorrow, fear, etc.); and
(2) Analytical cognition, which is slower. The image is perceived by thinking about the visible characters, analyzing their meanings. The analytical perception is less subjective, since other well-known descriptions of the image are used.

While advertisers are often trying to disturb people emotions and thus force them to engage in an advertised product or product, a user who first encounters a positive impression on a product becomes more influenced by emotions (syncretic perception) than rational decisions (analytical perception).

Recently, the researchers began evaluating the neurological correlates of consumer behavior using the electroencephalography (EEG), electromyography (EMG), galvanic skin response (GSR) and gaze tracking technology in order to understand marketing-relevant human behavior [8]. The quantitative data collected and properly analyzed provides an advantage over the commonly used surveys and questionnaires, which rely on the ability and willingness of the participants to honestly and accurately report their attitudes and behaviors.

Eye movements, gaze landing sites and areas of interest are known to be good predictors of customer behavior. People spend more time for examining (i.e., fixating their gaze on) options that they will eventually choose [9]. For example, consumers spent 54% more time looking at the ads of businesses that they ended up choosing [10]. Unusual brand extension cause disruption of visual processing and result in longer visual fixation times as evidenced in [11]. The emotional state can be recognized using gaze tracking technology [12], while different factors such as the emotional state and gender have influence on visual processing and understanding [13].

The aim of the paper is to analyze gender and affect related differences in the perception of visual advertising. The results of experiments using gaze tracking are presented and analyzed, and the implications for neuromarketing are discussed.

The structure of the remaining parts of the paper is as follows. Section 2 describes the experiments. Section 3 evaluates the results and concludes.

2 Experiments

2.1 Subjects

The experiment was conducted at Kaunas University of Technology (Kaunas, Lithuania). Due to the specifics of the hardware requirements, the place of execution and the limited access to the gaze tracking device, only students of 18–25 years old were invited to participate. As this social group of 18–25 age range is often a target audience in commercial advertising, we consider the use of this social group in research as appropriate [14]. Twenty people - 10 women (50%) and 10 men (50%) have participated. 45% of respondents were 18–21 years old, while 55% were 22–25 years old. All subjects reported to have normal vision. Participants signed a written consent form. The participants also received written task instructions on the procedure.

2.2 Dataset

The analyzed data included three original ads and their modified versions with emotions introduced (changed) using image processing software as follows (also see Fig. 1): "Trump for President 2016" (source: http://adsoftheworld.com/), "Charming Chocolate", "Charming Chocolate" ad created by the students of KTU and "I am Muslim" ad by People Against Suffering Oppression and Poverty (PASSOP).

Fig. 1. Analyzed images: original (top) and modified (bottom): "Trump for President 2016" (left), "Charming Chocolate" (center), "I am Muslim" (right)

2.3 Methodology

The experiment was performed on the basis of a mixed quantitative and qualitative research methodology. To achieve the objective, an eye tracking device and special software were used. Research was conducted using the methodology presented in [15], which sought to find out how people react to smiling faces in promotional imagery. The aim is to find out whether two groups of subjects of similar composition, which are shown the same advertisements in which only the expression of a person's face differs, can notice the difference in the intensity of the reaction.

For each participant the gaze-tracking device was adjusted and calibrated. Each experiment consisted of 3 stages, while calibration was repeated at the beginning of each stage. Following the experiments, the raw data was pre-processed and eye fixations shorter than 60 ms were discarded because of contamination from equipment measurement noise and blinking effects [16]. The upper threshold of fixations has been set to 600 ms as recommended in [17].

2.4 Procedure

Before the experiment, the intensity of human emotions seen in each advertisement was measured using FaceReader software (Noldus, the Netherlands). The program can recognize the basic facial expressions: happiness, surprise, sadness, anger, fear, contempt, disgust, and neutrality. The intensity of each emotion is evaluated on a scale from 0 to 1, where zero means that emotion is not recognized, and one - the emotion is very strong. The results of emotions measurements are shown in Table 1.

Table 1. Face emotion measurements with FaceReader

No.	Ad	Variant	Recognized emotion	Value	Assigned emotion
1	"Trump for President 2016"	Original	Neutral	0.4716	Happy
		Modified	Anger	0.2055	Sad
			Sad	0.2124	
			Neutral	0.5408	
2	"Charming Chocolate"	Original	Joy	0.9965	Happy
		Modified	Joy	0.1504	Sad
			Neutral	0.7603	
3	"I am Muslim"	Original	Neutral	0.9996	Sad
		Modified	Joy	0.9999	Happy

The participants were divided into two equal groups with an equal number of women and men of the same age, emotional advertisements were displayed, or modified with the use of additional photos for retouching. One ad was displayed for a maximum of 30 s, but each respondent was able to switch the ad on when she did not want to watch it for longer. Numeric data collected during the experiment was the timeframe for fixation of sight of the software (start time and duration in ms).

2.5 Statistical Analysis

Usually, power laws are used to model human visual perception. The distribution of eye fixation times can be adequately characterized using a two-parameter Weibull probability distribution [18, 19], and its performance on the eye-gaze data is shown to be better than other power law based approaches. The probability density function of a Weibull random variable is:

$$pdf(x) = \frac{k}{\lambda}\left(\frac{x}{\lambda}\right)^{k-1} e^{-(x/\lambda)^k} \quad (1)$$

where $k > 0$ is the shape parameter and $\lambda > 0$ is the scale parameter of the distribution. A value of $k > 1$ indicates that the failure rate increases with time. This happens if there is an aging or fatigue [20] process.

Weibull distribution parameters, gamma and beta, highly correlate (between 84 and 93%) with neural responses in the early visual system [21]. Sometimes none of the considered models represents the data appropriately. Thus, it should be tested how well the integrated Weibull distribution fits the data. The goodness of fit between experimentally established distribution and real data is evaluated using R^2 value, which is calculated as follows:

$$R^2 = \frac{\sum_{i=1}^{n} (\hat{x}_i - \bar{x})^2}{\sum_{i=1}^{n} (x_i - \bar{x})^2} \quad (2)$$

here $\bar{x}$ is the mean, and $\hat{x}$ is the fitted value.

Having the parameters of Weibull distribution, the mean and standard deviation of a Weibull random variable can be expressed as follows:

$$E(X) = \lambda\Gamma\left(1 + \frac{1}{k}\right) \quad (3)$$

and

$$std(X) = \lambda\sqrt{\Gamma\left(1 + \frac{2}{k}\right) - \left(\Gamma\left(1 + \frac{1}{k}\right)\right)^2} \quad (4)$$

here $\Gamma(\cdot)$ is the Gamma function:

$$\Gamma(z) = \int_0^\infty x^{z-1} e^{-x} dx \quad (5)$$

To evaluate if two data samples belong to the same statistical distribution, a number of statistical tests can be used.

The Kruskal-Wallis method tests if two or more classes have equal median and gives the value of p. If the value of p is close to 0 it means that the feature contains discriminative information.

The Wilcoxon rank sum test is used to establish if there is a significant difference between two sample groups using their ranks. If the ranks of the sample groups are significantly ($\alpha = 0.05$) separated, the test statistic identifies significant difference.

The third test statistic is obtained by bootstrapping, a way of estimating statistical parameters from the sample by performing resampling with replacement. For a two-sample bootstrap, we independently draw bootstrap samples with replacement from each sample, and compute the Fisher–Pitman test [22] statistic that compares the samples. The bootstrap distribution is used for confidence intervals and standard errors. Given the observed value of the difference (Δ) between the means of the samples $x = (x_1, x_2, \ldots, x_n)$ and $y = (y_1, y_2, \ldots, y_n)$ drawn from two probability distributions X and Y, the p-value is calculated as follows:

$$p_{value} = \frac{1 + \sum_{i=1}^{B} [\Delta_i \geq |\Delta|]}{B+1} \tag{6}$$

here $[\cdot]$ is the Iverson bracket operator, and B is the number of bootstrap resamples, here we set it to 1000 as suggested in [23].

For calculation of confidence, the percentile intervals are used. The sample values are arranged in ascending order and numbered sequentially with the smallest value receiving the number 1, obtaining the ranks of the sample values. For 95% confidence interval, the 2.5 and 97.5 percentiles are used, which corresponds to $0.025n + 0.5$ and $0.975n + 0.5$ ordered values, where n is the sample size [24].

2.6 Hypotheses

H1. There is no difference between genders in eye fixation times. Hypotheses are:

$$H_0^{(1)} : \mu_{male} = \mu_{female}; H_1^{(1)} : \mu_{male} \neq \mu_{female}$$

H2. There is no difference between affect of images in eye fixation times. Hypotheses are:

$$H_0^{(2)} : \mu_{happy} = \mu_{sad}; H_1^{(2)} : \mu_{happy} \neq \mu_{sad}$$

To analyze the interaction effects in population groups, the following sub-hypotheses are formulated:

H1a. There is no difference between genders in eye fixation times of the positively valenced images. Hypotheses are:

$$H_0^{(1a)} : \mu_{happy,male} = \mu_{happy,female}; H_1^{(1a)} : \mu_{happy,male} \neq \mu_{happy,female}$$

H1b. There is no difference between genders in eye fixation times of the negatively valenced images. Hypotheses are:

$$H_0^{(1b)} : \mu_{sad,male} = \mu_{sad,female}; H_1^{(1b)} : \mu_{sad,male} \neq \mu_{sad,female}$$

H2a. There is no difference between affect of images in the eye fixation times in males. Hypotheses are:

$$H_0^{(2a)} : \mu_{happy,male} = \mu_{sad,male}; H_1^{(2a)} : \mu_{happy,male} \neq \mu_{sad,male}$$

H2b. There is no difference between affect of images in the eye fixation times in females. Hypotheses are:

$$H_0^{(2b)} : \mu_{happy,female} = \mu_{sad,female}; H_1^{(2b)} : \mu_{happy,female} \neq \mu_{sad,female}$$

2.7 Results

All data were analyzed with custom written scripts and executed in MATLAB (Mathworks, Inc., Natick, Massachusetts). First, we determine the statistical distribution of data by fitting. Following [25], Weibull distribution can be used as a simple predictor of fixation location data, which can be explained by the image local contrast statistics. In our case, the goodness-of-fit between real data and fitted Weibull distribution data is 0.96.

When analyzing the eye fixation time data in the Weibull distribution parameter space (scale parameter λ vs shape parameter k, see Eq. 1), we can notice a difference both in gender and affective factors (see Fig. 2). Here individual parameter values were generated from initial population of data by bootstrapping. For males, the probability of having a larger value of scale parameter λ than for females is $p = 0.6274$, while for the shape parameter k, the difference is not significant ($p = 0.6191$). In the affective dimension, the probability of a larger value of the scale parameter λ for happy vs sad samples of data is $p = 0.8493$, while the probability of a larger value of the shape parameter k is $p = 0.9308$. This means that positive emotions lead to longer mean fixation times and more outliers in the right tail of the value distribution.

Following [26], we differentiate between short (correcting and ambient) fixations, and long (focal) fixations. Long fixations last longer than 300 ms and are attributed to conscious visual processing process, while shorter fixations are related to unconscious processing of information.

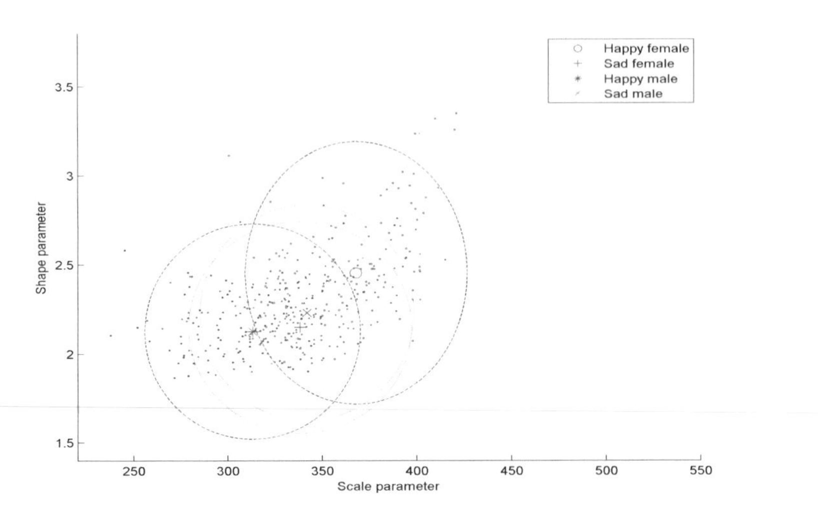

Fig. 2. Gaze data in the Weibull distribution parameter space (with 0.95 confidence ellipses)

We used a repeated-measures ANOVA test to examine the differences between fixation times in each of the experimental groups. A two-way factorial ANOVA was performed to examine the interaction effect between the fixation times.in the affective dimension (happy vs sad) and gender dimension (male vs female). For short fixations (<300 ms), the ANOVA revealed a significant effect between affect (happy vs sad) and fixation times ($F(2, 49) = 4.07$, $p = 0.0493$). The effect was significant in male subjects ($F(2, 24) = 4.75$, $p = 0.0398$), but not significant in female subjects ($F(2, 23) = 0.61$, $p = 0.4422$). In the gender dimension (male vs female), the effect was insignificant ($F(2, 48) = 0.15$, $p = 0.7016$). The effect also was not significant both in images labeled as happy ($F(2, 23) = 0.05$, $p = 0.8305$) and as sad ($F(2, 24) = 0.68$, $p = 0.4178$).

The results were confirmed by the Kruskal-Wallis test, which revealed a significant difference between fixation times in affective dimension (happy vs sad) ($p = 0.0176$), but failed to reject the null hypothesis of equal medians for gender dimension male vs female ($p = 0.8396$), as well as in different subgroups: happy male vs happy female ($p = 0.8597$), sad male vs sad female ($p = 0.4419$), happy male vs sad male ($p = 0.0517$), and happy female vs sad female ($p = 0.3143$). Note that although rejected, the affective effect is stronger for males than for females.

The results were also confirmed by the ranksum test, which revealed a significant difference between fixation times in the affective dimension (happy vs sad) ($p = 0.0496$), but failed to reject the null hypothesis of equal medians for the gender dimension ($p = 0.9200$), as well as in different subgroups: happy male vs female ($p = 0.8712$), sad male vs female ($p = 0.5388$), happy male vs sad male ($p = 0.1048$), and happy female vs sad female ($p = 0.2712$). Note that although rejected, the affective effect is stronger for males than for females.

The statistical testing results are summarized in Table 2.

Table 2. Summary of statistical test results

Hypothesis	Kruskal-Wallis test outcome (*p*-value)	Ranksum test outcome (*p*-value)	ANOVA1 test outcome (*p*-value)	Bootstrap test outcome (*p*-value)
H1	Fail to reject (0.8396)	Fail to reject (0.9200)	Fail to reject (0.6747)	0.6847
H1a	Fail to reject (0.8597)	Fail to reject (0.8712)	Fail to reject (0.8121)	0.8601
H1b	Fail to reject (0.4419)	Fail to reject (0.5388)	Fail to reject (0.3958)	0.3841
H2	**Reject (0.0176)**	**Reject (0.0496)**	**Reject (0.0378)**	**0.9214**
H2a	Fail to reject (0.0517)	Fail to reject (0.1048)	Fail to reject (0.0589)	0.8917
H2b	Fail to reject (0.3143)	Fail to reject (0.2712)	Fail to reject (0.4816)	0.5244

3 Evaluation and Conclusion

In this paper, we have analyzed the affect and gender related differences of how subjects react to emotionally charged advertisements. Four different statistical tests (Kruskal-Wallis, ranksum, ANOVA, bootstrapping) confirmed the significant difference ($p < 0.05$) between short eye fixation times (<300 ms) in the affective dimension, however we failed to discover any gender related differences of how the images are processed by human visual system.

This observation agrees well with earlier findings in adult subjects, suggesting that there is a direct link between eye movement behavior characterized by fixation duration and distinct modes of visual processing for different visual stimuli (such as photographs and emotional stimuli [27] as well as for different tasks (localization vs. identification [28]). Specifically, the ambient mode characterized by short fixation times has been related to pre-attentive scanning associated with the visual exploration of the spatial layout of an image, which is characteristic to the early stages of viewing [28]. Our findings are in line with these previous observations.

However, the findings of this study are preliminary and limited in nature. We need to repeat the study with a larger dataset and a larger population of subjects to confirm or reject the present results. Despite these potent limits, the results of the current study are promising in terms of demonstrating a new modality, which can serve as an assistive tool for evaluation of visual ads in the context of neuromarketing.

Acknowledgement. We acknowledge V. Stankevicius, Š. Baltramonaityte, L. Balockaite and "Choco Group" for the use of the "Charming Chocolate" photo.

References

1. Clay, R.A.: Advertising as science. Am. Psychol. Assoc. **33**(9), 38 (2002)
2. Senior, C., Lee, N.: A manifesto for neuromarketing science. J. Consum. Behav. **7**(4–5), 263–271 (2008)
3. Levenson, R.W.: Basic emotion questions. Emotion Rev. **3**, 379–386 (2011)
4. Ellsworth, P.C., Scherer, K.R.: Appraisal processes in emotion. In: Davidson, R.J., Goldsmith, H., Scherer, K.R. (eds.) Handbook of Affective Sciences, pp. 572–595. Oxford University Press, New York and Oxford (2003)
5. Yadati, K., Katti, H., Kankanhalli, M.S.: CAVVA: computational affective video-in-video advertising. IEEE Trans. Multimedia **16**(1), 15–23 (2014)
6. Kwong, C.K., Jiang, H., Luo, X.G.: AI-based methodology of integrating affective design, engineering, and marketing for defining design specifications of new products. Eng. Appl. AI **47**, 49–60 (2016)
7. Chaudhuri, A., Buck, R.: Media differences in rational and emotional responses to advertising. J. Broadcast. Electron. Media **39**(1), 109–125 (2009)
8. Lee, N., Broderick, A.J., Chamberlain, L.: What is 'neuromarketing'? A discussion and agenda for future research. Int. J. Psychophysiol. **63**(2), 199–204 (2007)
9. Krajbich, I., Armel, C., Rangel, A.: Visual fixations and the computation and comparison of value in simple choice. Nat. Neurosci. **13**(10), 1292–1298 (2010)
10. Lohse, G.L.: Consumer eye movement patterns on yellow page advertising. J. Advertising **26**, 61–73 (1997)
11. Stewart, A.J., Pickering, M.J., Sturt, P.: Using eye movements during reading as an implicit measure of acceptability of brand extensions. Appl. Cognit. Psychol. **18**, 697–709 (2004)
12. Maskeliunas, R., Raudonis, V.: Are you ashamed? Can a gaze tracker tell? PeerJ. Comput. Sci. **2**, e75 (2016)
13. Raudonis, V., Maskeliunas, R., Stankevicius, K., Damasevicius, R.: Gender, age, colour, position and stress: how they influence attention at workplace? In: 17th International Conference on Computational Science and Its Applications (ICCSA) vol. 5, pp. 248–264 (2017)
14. Coursaris, C.K., Sung, J., Swierenga. S.J.: Effects of message characteristics, age, and gender on perceptions of mobile advertising – an empirical investigation among college students. In: Ninth International Conference on Mobile Business and 2010 Ninth Global Mobility Roundtable (ICMB-GMR), pp. 198–205 (2010)
15. Berg, H., Söderlund, M., Lindström, A.: Spreading joy: examining the effects of smiling models on consumer joy and attitudes. J. Consum. Market. **32**(6), 459–469 (2015)
16. Helo, A., Pannasch, S., Sirri, L., Rämä, P.: The maturation of eye movement behavior: scene viewing characteristics in children and adults. Vis. Res. **103**, 83–91 (2014)
17. Jacob, R.J.K.: Eye tracking in advanced interface design. In: Barfield, W., Furness, T. (eds.) Virtual Environments and Advanced Interface Design. Oxford University Press, Oxford (1994)
18. Bosworth, R.G., Petrich, J.A.F., Dobkins, K.R.: Effects of attention and laterality on motion and orientation discrimination in deaf signers. Brain Cognit. **82**, 117–126 (2013)
19. Yanulevskaya, V., Geusebroek, J.M.: Significance of the Weibull distribution and its sub-models in natural image statistics. In: Proceedings of the International Conference on Computer Vision Theory and Applications, pp. 355–362 (2009)

20. Vasiljevas, M., Gedminas, T., Sevcenko, A., Janciukas, M., Blazauskas, T., Damasevicius R.: Modelling eye fatigue in gaze spelling task. In: 2016 IEEE 12th International Conference on Intelligent Computer Communication and Processing (ICCP), pp. 95–102 (2016)
21. Scholte, H.S., Ghebreab, S., Waldorp, L., Smeulders, A.W.M., Lamme, V.A.F.: Brain responses strongly correlate with Weibull image statistics when processing natural images. J. Vis. **9**, 1–9 (2009)
22. Fisher, R.A.: The Design of Experiments, 8th edn., pp. 1–248. Hafner Publishing Company, New York (1971)
23. Briggs, A.H., Wonderling, D.E., Mooney, C.Z.: Pulling cost-effectiveness analysis up by its bootstraps: a non-parametric approach to confidence interval estimation. Health Econ. **6**, 327–340 (1997)
24. Diem, K., Seldrup, J., Lentner, C. (eds.): Geigy scientific tables. In: Introduction to Statistics, Statistical Tables and Mathematical Formulae—Percentiles, vol. 2, 8th edn. Ciba-Geigy Limited, Basel (1982)
25. Yanulevskaya, V., Marsman, J.B., Cornelissen, F., Geusebroek, J.-M.: An image statistics-based model for fixation prediction. Cognit. Comput. **3**(1), 94–104 (2011)
26. Velichkovsky, B.B., Rumyantsev, M.A., Morozov, M.A.: New solution to the midas touch problem: identification of visual commands via extraction of focal fixations. Procedia Comput. Sci. **39**, 75–82 (2014)
27. Pannasch, S., Helmert, J.R., Roth, K., Walter, H.: Visual fixation durations and saccade amplitudes: shifting relationship in a variety of conditions. J. Eye Mov. Res. **2**(2), 1–19 (2008)
28. Unema, P., Pannasch, S., Joos, M., Velichkovsky, B.: Time course of information processing during scene perception: The relationship between saccade amplitude and fixation duration. Visual Cognit. **12**(3), 473–494 (2005)

FARIP: Framework for Artifact Removal for Image Processing Using JPEG

T. M. Shashidhar[1(✉)] and K. B. Ramesh[2]

[1] Visvesvaraya Technological University, Belagavi, Karnataka, India
shashidhartm2014@gmail.com
[2] Department of Electronics and Instrumentation Engineering, RV College of Engineering, Bengaluru, Karnataka, India

Abstract. Irrespective of a significant advancement of the compression technique in digital image processing, still the presence of artifacts or fingerprints do exists, even in a smaller scale. Such presence of artifacts is mis-utilized by the miscreants by invoking their attacks where it is quite hard to differentiate tampered image due to normal problems or malicious attack. Therefore, we present a very simple modeling of a system called as FARIP i.e. Framework of Artifact Removal in Image Processing that utilize the quantization process present in JPEG-based compression and results in perfect removal of the traces from a given image. This also acts as a solution towards the image that has been generated by the compression technique performed from JPEG standard. The comparative analysis shows that proposed system offers better signal quality as compared to the existing standards of compression.

Keywords: Compression · JPEG · JEG2000 · Artifacts · Traces
Digital image processing · Deblocking · Quantization

1 Introduction

Image forensic does increasingly gaining attention owe to its explicit capability to disclosing the underlying fact [1]. The prime assumption in this concept is any operations on the top of the image will statistically alter the significant metadata information of an image [2]. Such statistical unique characteristics are called as fingerprint. The validation process includes identification of such digital fingerprints also called as traces or artifacts [3]. At present, there has been various research work being carried out towards identification/localization of such traces [4, 5]. In majority of the existing cases of research, there is a dependency on a large section of the signal in order to carry out the analysis. Image compression plays a significant role in this regards. There is a higher significance level for the fingerprints generated as majority of the images are subjected to the compression by their capturing device or during the process of storing it in memory [6]. When an attacker tampers the image, they make use of various forgery tools in order to invoke an attack [7] by ensuring that both the tamper image and original image doesn't bear a strong perceptual difference. The adversary has multiple ways to compute such difference in order to avoid getting caught and hence it is essential that a unique compression technique be explored in order to ensure proper

R. Silhavy (Ed.): CSOC 2018, AISC 764, pp. 12–20, 2019.
https://doi.org/10.1007/978-3-319-91189-2_2

identification of such traces. Irrespective of various existing compression techniques [8, 9], we have not come across simpler technique for elimination of the traces due to the compression process in image processing.

Therefore, this paper introduces a very simpler technique where a JPEG-based compression has been introduced with an emphasis on quantization process in order to yield a slightly iterative process for generating high quality images. Section 2 discusses about the existing research work followed by problem identification in Sect. 3. Section 4 discusses about proposed solution followed by elaborated discussion of research methodology implementation in Sect. 5. Comparative analysis of proposed system with existing system of compression of accomplished result is discussed under Sect. 6 followed by conclusion in Sect. 7.

2 Related Work

This section upgrades the information about research contribution following our prior review work [10]. Identification of traces in images has been investigated by Pasquini et al. [11] using statistical approach. Lipman et al. [12] have addressed the reflection artifacts for enhancing image quality on two dimensional shear waves. Adoption of sparse representation along with structural quantization approach was studied by Zhao et al. [13]. Liu et al. [14] improved the performing of deblocking technique using knowledge-based approach. Study towards elimination of artifacts using filtering mechanism has been seen in the work carried out by Gong et al. [15]. Fan et al. [16] have performed an investigation towards spatial and transformed domain and its effect on artifacts removal in images. Ebrahimi et al. [17] have presented a passive technique for identification of blocking artifacts using experimental approach. Min et al. [18] have studied the possible impact of artifacts arise from compression on perceptual quality. Study towards nano-scale imaging along with its artifacts minimization was carried out by Fairbairn and Moheimani [19]. Adoption of predictive approach is being carried out by Bannan et al. [20]. Li et al. [21] have introduced a solution towards artifacts removal from compression process to be used in steganalysis process. Study using case study of medical image and its associated artifacts effect has been investigated by Mauldin et al. [22] by analyzing geometric factors of the image. Sharma et al. [23] have presented a study where both image quality and compression performance is equally emphasized. Study towards localizing the image forgery has been carried out by Bianchi et al. [24] considering the artifacts caused due to JPEG compression. Fairbairn [25] has implemented a controller design for reducing the image artifacts. Ferrara et al. [26] have carried out the study to localize the region of image forgery considering probability map of artifacts. Goto et al. [27] have presented a study for minimizing the compression artifacts using total variation technique. Johnson et al. [28] have presented visual analysis technique for investigating the compression artifacts on pathological images. Similar technique has been also carried out by Kim and Sim [29] using adaptive approach based on signal weight. Tang et al. [30] have presented Markov model for eliminating blocking artifacts using indexing mechanism. The next section outlines the research problems.

3 Problem Description

A closer look into the existing techniques on artifact removal shows that they are quite capable of addressing the visible traces associated with blocking. Such models are restricted to images of minimal to medium scale quality. However, for better performance, there is a need for eliminating both visual as well as traces from statistical approaches. Such process must not result in an image that could be detected forensically incorporated fingerprints. Moreover, a very simple technique is also demanded using JPEG-based compression approach that could offer better computational time and works on multiple forms of images. The next section introduces the proposed solution to address this problem.

4 Proposed Solution

The proposed system of FARIP uses a very simple modeling of compression which also performs removal of the traces. The proposed solution of addressing the problems associated with the blocking artifacts is based on the fact that JPEG compression could be suitably modified in such a way that all the traces could be significantly eliminated precisely. The scheme of proposed system is showcased in Fig. 1.

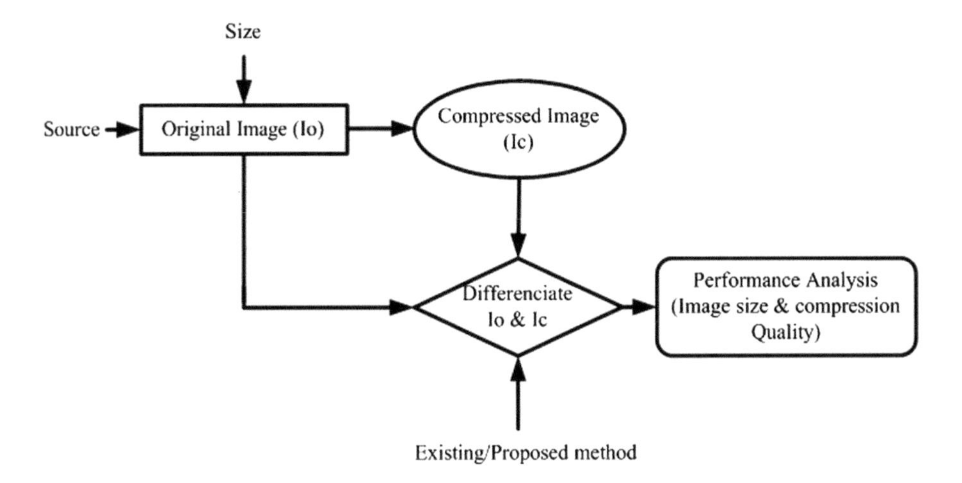

Fig. 1. Flow of artifacts removal by FARIP

Figure 1 highlights how a normal image is inflicted with blocking artifacts and slowly the proposed FARIP performs removal of any form of traces during the blocking process resulting in final decompressed image with good signal quality. The complete process of artifacts removal is carried out using JPEG standard emphasizing on the quantization steps. The next section outlines adopted research methodology. In the above Fig. 1, the original image (Io) will be selected from the desired folder and its size is computed. Later, a compressed image (Ic) will be obtained from Io. The difference among both the Io and Ic with respect to PSNR and MSE and finally, the performance analysis is performed on the results obtained from existing method and proposed method. The performance analysis is done by considering the image size and compression quality parameters to explore the significance of proposed method.

5 FARIP System Design

The proposed FARIP is subjected to the artifact removal using the JPEG standard itself. For this purpose, we apply Laplace distribution for initiating the system design considering the distribution of the coefficients over a range of specific bands as follows,

$$\alpha(A = a) = c.e^{d} \tag{1}$$

In the above Eq. (1), α represents parametric model and where A is transform coefficient. The constant c represents $\gamma/2$, where γ represents unique value of the sub-band. The exponential power variable d will represent – $\gamma|a|$ The distribution of the priorly compressed JPEG image is carried out using discrete approach of Laplace distribution as follows,

$$\alpha(B = b) = \left\{ \begin{array}{ll} 1 - e^{d_1} & b = 0 \\ e^{-d_2} & b = k\beta_{i,j} \\ 0 & otherwise \end{array} \right\} \tag{2}$$

The exponential term d_1 and d_2 will mean $\gamma.\beta_{i,j}/2$ and $\gamma|b|$ respectively. In order to perform any form of modification of the priorly compressed image, the initial operations of JPEG compression steps were carried out in order to acquire transformed coefficients. It is due to the fact that the ultimate phase of the decompression using JPEG consists of projecting the decompressed values of the pixels back to coefficient values acquired from the quantized image. Hence, it is feasible to obtain the coefficient values by iterating the process of quantization on the scale of new coefficient B' in such a way that $B = \beta_{i,j}\ (B'/\beta_{i,j})$. The similar quantized coefficients were obtained using a highest probability value of model attribute. With an aid of this process, it is now feasible to compute the distribution of sub-band coefficient prior to performing JPEG compression. We consider that η represents cumulative observations from all the sub-bands where η is sum of number η_o of sub-band coefficient (that is usually valued as zeros) and all the non-zero coefficient η_1. Therefore, the empirical expression will be,

$$\delta = \sum_{k=1}^{\eta} |b_k| \tag{3}$$

Once the γ is computed, the anti-forensic dither is added for the entire coefficient in the sub-band. In majority of the records of quantization, the sub-bands with higher coefficient is used as it doesn't offer higher significant perceptual changes. It has also been seen that sub-bands and their respective variance minimizes when there is a sub-band transition from lower to higher frequency. Owing to this process, the different forms of sub-band that has higher value of coefficient is then subjected to quantization to the scale of zero while performing JPEG compression. This will also mean that the computed value of γ is available for the explored sub-bands that makes impossible to perform any form of modification of such values of coefficient. Owing to the involvement of the coefficient value obtained in final stages of decompression, there is no dependency of amending any coefficient of such sub-bands. Normally, the dc sub-band of the coefficients has greater variation on different forms of images and hence there is no reported parametric framework for such distribution. The distribution of the dc sub-band associated with the non-quantized coefficient is actually distributed uniformly that falls under the range of interval quantization. Therefore, it is feasible to construct coefficient using anti-forensic approach whose distribute is somewhat equivalent to the non-quantized distribution. This is carried out by incorporating uniformly distributed dither to the dc sub-bands of quantized coefficient. It is also seen that the dynamic range of the transformed values is quite higher in contrast to the quantized range where there is less number of coefficient that are possibly quantized over a specific constant. Hence, there are very less number of the coefficient values that are present over a specific interval.

Once the modified transformed coefficient is acquired anti-forensically, the system that performs inverse transforms operation of such coefficient block. The yielding transformed blocks are then integrated in the enhanced image using anti-forensic image. If the colored image is down-sampled at the time of the JPEG compression, the system than constructs an amended version of the transformed image that significant assists in removing the residual traces for a given image due to the JPEG compression. The chrominance layer that is recently down-sampled is then subjected to the interleaving operation in order to construct an image free from traces. Hence, a successful and yet a simple trace-removal procedure is presented for image processing.

6 Results Discussion

The implementation of the proposed system is carried out on colored image of *.tif* extension using MATLAB. The complete execution of the proposed scheme is carried out on 32-bit machine on core-*i*3 processor. Figure 2 shows the visual outcome of the compression based artifact removal procedure using existing JP2 compression and JPEG compression standard. The colored image is initially converted to grayscale image which when implemented by the proposed system incorporates the blocking artifacts on it. Finally, the decompressed image is obtained where the blocking artifacts are removed using both the compression scheme. One of the interesting facts to observe is that spatial factor of the fingerprints is used for identifying the type and form of attack. When such fingerprints are found to vary significant on the local scale that such condition can be actually used for identifying the tampered image as well as it also significantly assists in detection of the tampered region within an image. Another significant observation is that if the blocking artifacts are found not to perform appropriate alignment with the residual image than the yielding mismatch regions can be used to identify the traces of image tampering. If the proposed technique is implemented with the software of the camera in order to perform quantization than it will be impossible for the image forger to remove any form of traces by their technique of tampering the metadata of the image. Therefore, the proposed system offers very simple scheme for artifact removal.

Type of image/method	Existing method (JP2 compression)	Proposed method (JPEG compression)
Original image		
Grayscale image		
Artifact image	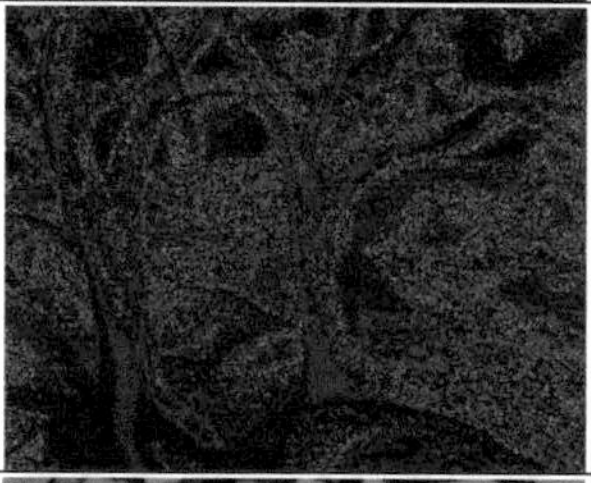	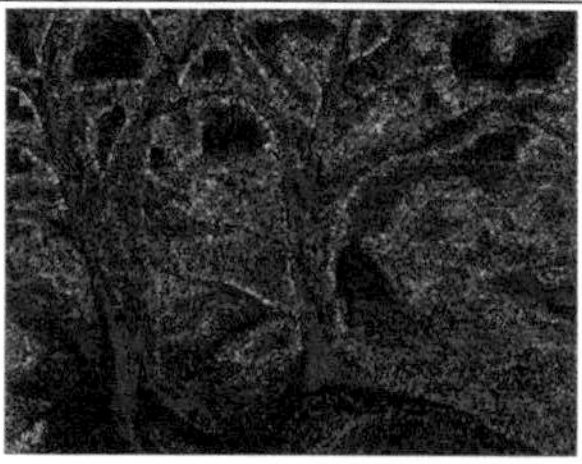
Image after removing artifact		
Original Size	1776950 byte	1776950 byte
Compressed Size	7999 byte	9014byte
PSNR	29.3953	28.4217
MSE	23.5042	31.9305

Fig. 2. Existing JPEG compression

We also compared our study outcome with that of existing system with respect to image size and peak signal-to-noise ratio as shown in Figs. 3 and 4. The outcome shows JP2 and JPEG compression and is found that JPEG compression offers better signal quality as noticed from the decompressed image quality. The complete execution time of proposed system is found to be 0.02663 s.

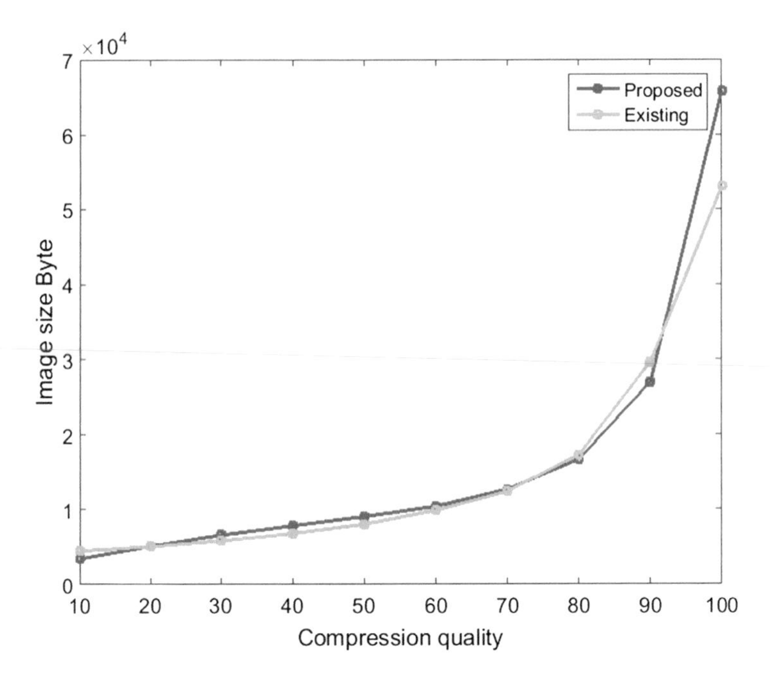

Fig. 3. Analysis of image size

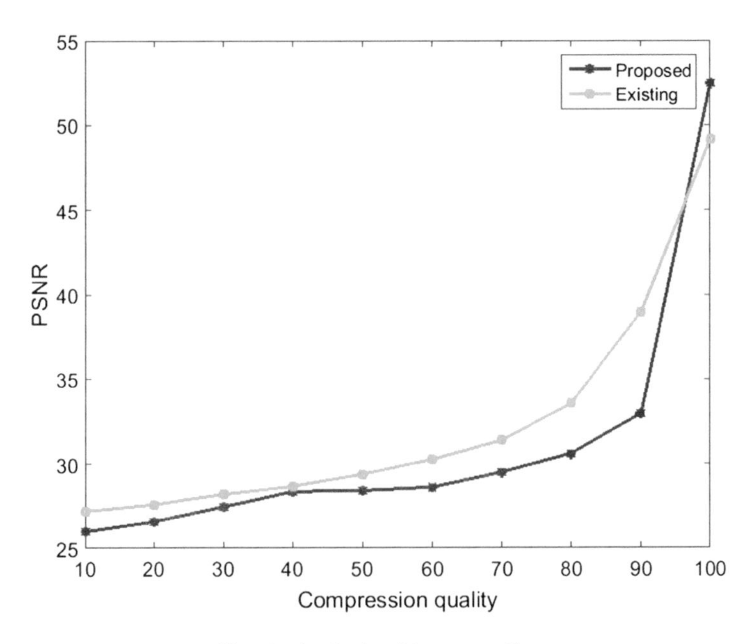

Fig. 4. Analysis of image quality

7 Conclusion

At present there are multiple approaches for existing forensic methodologies that are capable of identifying the different forms of malicious image tampering; however, they are not powered with any sorts of capability to hide the fingerprints of such manipulations. This operating of hiding traces or fingerprint is quite essential as presence of any forms of traces makes the image quite perceptible which makes the task of catching hold of attacks on image quite challenging. Therefore, the proposed system presents a very simple technique where JPEG based compression methodology has been used for removing the traces or fingerprint. This is one of the most cost effective approaches owing to less inclusion of sophisticated computational steps with faster response.

References

1. Ho, A.T.S., Li, S.: Handbook of Digital Forensics of Multimedia Data and Devices. Wiley, Hoboken (2015)
2. Sencar, H.T., Memon, N.: Digital Image Forensics: There is More to a Picture than Meets the Eye. Springer, New York (2012)
3. Pal, R.: Innovative Research in Attention Modeling and Computer Vision Applications. IGI Global, Hershey (2015)
4. Haritha, C., Ganesan, M., Sumesh, E.P.: A survey on modern trends in ECG noise removal techniques. In: 2016 International Conference on Circuit, Power and Computing Technologies (ICCPCT), Nagercoil, pp. 1–7 (2016)
5. Zou, Y., Nathan, V., Jafari, R.: Automatic identification of artifact-related independent components for artifact removal in EEG recordings. IEEE J. Biomed. Health Inform. **20**(1), 73–81 (2016)
6. Seadle, M.: Quantifying Research Integrity. Morgan & Claypool Publishers, San Rafael (2016)
7. Elwin, J.G.R., Aditya, T.S., Shankar, S.M.: Survey on passive methods of image tampering detection. In: 2010 International Conference on Communication and Computational Intelligence (INCOCCI), Erode, pp. 431–436 (2010)
8. Zhao, Z.-H., Wang, W.-Y.: A lossless compression method of JPEG file based on shuffle algorithm. In: 2010 2nd International Conference on Advanced Computer Control, Shenyang, pp. 160–162 (2010)
9. Bhagat, A.P., Atique, M.: Medical images: formats, compression techniques and DICOM image retrieval a survey. In: 2012 International Conference on Devices, Circuits and Systems (ICDCS), Coimbatore, pp. 172–176 (2012)
10. Shashidhar, T.M., Ramesh, K.B.: Reviewing the effectivity factor in existing techniques of image forensics. Int. J. Electr. Comput. Eng. **7**(6), 3558–3569 (2017)
11. Pasquini, C., Boato, G., Pérez-González, F.: Statistical detection of JPEG traces in digital images in uncompressed formats. IEEE Trans. Inf. Forensics Secur. **12**(12), 2890–2905 (2017)
12. Lipman, S.L., Rouze, N.C., Palmeri, M.L., Nightingale, K.R.: Evaluating the improvement in shear wave speed image quality using multidimensional directional filters in the presence of reflection artifacts. IEEE Trans. Ultrason. Ferroelectr. Freq. Control **63**(8), 1049–1063 (2016)

13. Zhao, C., Zhang, J., Ma, S., Fan, X., Zhang, Y., Gao, W.: Reducing image compression artifacts by structural sparse representation and quantization constraint prior. IEEE Trans. Circuits Syst. Video Technol. **27**(10), 2057–2071 (2017)
14. Liu, Y., Li, X., Kong, A.W.K.: Speeding up the knowledge-based deblocking method for efficient forensic analysis. In: 2014 IEEE Symposium on Computational Intelligence in Biometrics and Identity Management (CIBIM), Orlando, FL, pp. 185–194 (2014)
15. Gong, Y., Yu, T., Chen, B., He, M., Li, Y.: Removal of cardiopulmonary resuscitation artifacts with an enhanced adaptive filtering method: an experimental trial. BioMed Res. Int. (2014)
16. Fan, W., Wang, K., Cayre, F., Xiong, Z.: JPEG anti-forensics with improved tradeoff between forensic undetectability and image quality. IEEE Trans. Inf. Forensics Secur. **9**(8), 1211–1226 (2014)
17. Ebrahimi, A., Ibrahim, S., Ghazizadeh, E., Alizadeh, M.: Paint-doctored JPEG image forensics based on blocking artifacts. In: 2015 International Conference and Workshop on Computing and Communication (IEMCON), Vancouver, BC, pp. 1–5 (2015)
18. Min, X., Zhai, G., Gao, Z., Hu, C.: Influence of compression artifacts on visual attention. In: 2014 IEEE International Conference on Multimedia and Expo (ICME), Chengdu, pp. 1–6 (2014)
19. Fairbairn, M.W., Moheimani, S.O.R.: Control techniques for increasing the scan speed and minimizing image artifacts in tapping-mode atomic force microscopy: toward video-rate nanoscale imaging. IEEE Control Syst. **33**(6), 46–67 (2013)
20. Bannan, K.E., Handler, W.B., Wyenberg, C., Chronik, B.A., Salisbury, S.P.: Prediction of force and image artifacts under MRI for metals used in medical devices. IEEE/ASME Trans. Mechatron. **18**(3), 954–962 (2013)
21. Li, H., Luo, W., Huang, J.: Countering anti-JPEG compression forensics. In: 2012 19th IEEE International Conference on Image Processing, Orlando, FL, pp. 241–244 (2012)
22. Mauldin, F.W., Owen, K., Tiouririne, M., Hossack, J.A.: The effects of transducer geometry on artifacts common to diagnostic bone imaging with conventional medical ultrasound. IEEE Trans. Ultrason. Ferroelectr. Freq. Control **59**(6), 1101–1114 (2012)
23. Sharmaa, A., Bautistab, P., Yagib, Y.: Balancing image quality and compression factor for special stains whole slide images. Anal. Cell. Pathol. **35**, 101–106 (2012)
24. Bianchi, T., Piva, A.: Image forgery localization via block-grained analysis of JPEG artifacts. IEEE Trans. Inf. Forensics Secur. **7**(3), 1003–1017 (2012)
25. Fairbairn, M.W., Moheimani, S.O.R.: A switched gain resonant controller to minimize image artifacts in intermittent contact mode atomic force microscopy. IEEE Trans. Nanotechnol. **11**(6), 1126–1134 (2012)
26. Ferrara, P., Bianchi, T., De Rosa, A., Piva, A.: Image forgery localization via fine-grained analysis of CFA artifacts. IEEE Trans. Inf. Forensics Secur. **7**(5), 1566–1577 (2012)
27. Goto, T., Kato, Y., Hirano, S., Sakurai, M., Nguyen, T.Q.: Compression artifact reduction based on total variation regularization method for MPEG-2. IEEE Trans. Consum. Electron. **57**(1), 253–259 (2011)
28. Johnson, J.P., Krupinski, E.A., Yan, M., Roehrig, H., Graham, A.R., Weinstein, R.S.: Using a visual discrimination model for the detection of compression artifacts in virtual pathology images. IEEE Trans. Med. Imaging **30**(2), 306–314 (2011)
29. Kim, J., Sim, C.B.: Compression artifacts removal by signal adaptive weighted sum technique. IEEE Trans. Consum. Electron. **57**(4), 1944–1952 (2011)
30. Tang, C., Kong, A.W.K., Craft, N.: Using a knowledge-based approach to remove blocking artifacts in skin images for forensic analysis. IEEE Trans. Inf. Forensics Secur. **6**(3), 1038–1049 (2011)

SOPA: Search Optimization Based Predictive Approach for Design Optimization in FinFET/SRAM

H. Girish[1(✉)] and D. R. Shashikumar[2]

[1] J C Bose Centre for Research and Development, Department of ECE, Cambridge Institute of Technology, K.R. Puram, Bangalore, India
hgirishphd@gmail.com

[2] Department of Computer Science Engineering, Cambridge Institute of Technology, K.R. Puram, Bangalore, India

Abstract. FinFET/SRAM has been contributing to the new evolution of modern-day memory units that are used over broader scale of computing units and other sophisticated devices. A review analysis is performed over existing system to find that existing approaches are more inclined towards improvement in performance parameters and very less towards design optimization. Hence, a novel approach is introduced and is named as Search Optimization based Predictive Approach (SOPA) for optimizing the design structure of FinFET/SRAM so that it can ensure highest degree of fault tolerance when used in broader scale of dynamic applications and modern computing devices. In this, analytical methodology used where the proposed computational model is found to offer reduced computational time and more yield in increasing simulation iteration. The study contributes to progressive convergence of elite design of FinFET/SRAM rather than recursive design and hence cost effective.

Keywords: FinFET · Static RAM · Memory units · Design structure Optimization · Yield

1 Introduction

With the growing usage of the Nano scale circuits in different forms of electrical and electronic device, there is a growing concern about the memory utilization exclusively related to SRAM [1]. Although SRAM offers multiple advantageous factor e.g. simplicity in operation, higher speed, and multiple forms of accessibility, but still it is shrouded by various problems which is cost and size [2, 3]. The usage cost of SRAM is practically more than DRAM whereas it also occupies more space than DRAM. In order to rescue this, the transistor structure of FinFET is always studied with respect to SRAM [4]. However, there are still more problems associated with FinFET/SRAM e.g. area, stability, yield, access time, power consumption, etc. [5, 6]. various conventional modeling e.g. double gated, back gated design, designs with dynamic feedback, etc. are presented in past for FinFET/SRAM design [7, 8]. However, still some problems exist. Currently it is been found that, majority of the design methodologies have been carried out considering hardware-based approach as well as prototyping but there some

R. Silhavy (Ed.): CSOC 2018, AISC 764, pp. 21–29, 2019.
https://doi.org/10.1007/978-3-319-91189-2_3

open-end problems still are not solved e.g. (i) predictive design structure of FinFET/SRAM to offer better performance with accuracy, (ii) existing studies have narrowed scale of implementation of system error that doesn't seem to offer fault tolerance to the SRAM design, (iii) less studies are carried out to jointly study the impact of design on yield, energy, and response of the memory units. Hence, a broader look towards design optimization is missing from the literature that motivates the research community to further carry out design optimization of FinFET/SRAM.

Hence, this paper introduces a novel search-based optimization technique towards enhancing the design of FinFET/SRAM. Section 2 discusses about the existing research work followed by problem identification in Sect. 3. Section 4 discusses about proposed methodology followed by elaborated discussion of algorithm implementation in Sect. 5. Comparative analysis of accomplished result is discussed under Sect. 6 followed by conclusion in Sect. 7.

2 Related Work

A discussion on existing system is addressed in [9] over SRAM and is identified certain set of problems of cost optimization that was addressed in our most recent implementation [10]. This section will update more information about approaches towards enhancing performance of FinFET/SRAM. The work carried out by Bhattacharya and Jha [11] addressed the problem of stabilization of its reading operation using monolithic integration approach. Study toward identification of errors in SRAM was carried out by Fang and Oates [12]. Chen et al. [13] have introduced a technique for facilitating an effective writing operation for reducing the supply voltage. Yang et al. [14] addresses the problem associated with near-threshold based operation improving the reading performance. Karl et al. [15] has developed a prototype design of SRAM emphasizing on collapsing voltage for minimizing energy consumption during write operation. Zhang et al. [16] have performed optimization using dopant over their prototype of SRAM targeting as scaling the voltage factor. Kulkarni et al. [17] have adopted an amplification technique for enhance sensing performance in FinFET. Hu et al. [18] have presented SRAM design with better improvement over noise for improved performance of reading and writing operation. Bhattacharya and Jha [19] have presented a synthesis-based approach for accelerating the process of extracting the capacitance factor for a given layout arrays of SRAM. The technique also introduces an exclusive partitioning process for better efficiency. Song et al. [20] have presented a scheme for addressing the power issue in 128 mb of SRAM focusing on multiple density factors. Mishra and Mahapatra [21] have carried out a simulation-based approach for investigating the stability problems associate with SRAM. This work is reported to offer fault tolerance operation however their approach was limited to only normal operating conditions of circuits. Asenov et al. [22] have presented a unique approach of design optimization considering variability factor in the simulation study. The work carried out by Joshi et al. [23] has investigated variability of devices with respect to the SRAM using statistical approach. All the above existing approaches do not address the design optimization problem of FinFET/SRAM and hence offers a broader scope of carrying out a research work. The next section outlines the problems understood by reviewing all the above studies.

3 Problem Description

After reviewing the existing approaches towards FinFET/SRAM design and is observed that there has been good amount of research using hardware platform, however, none of these works has ever claimed of optimal throughput outcome or any other form of broader gain in performance. Also, it is found that existing approaches couldn't offer fault tolerance as they were not much standard computational model with an extensive analysis to prove it. Moreover, the memory units and their respective communication with other connected memory units have not been investigated in past with respect to memory and power that is significant degraded in the upcoming design of chip multiprocessor. Therefore, the problems statement is "*developing a system that balances multiple aspects e.g. lower scale of power consumption, increased yield, and faster processing is all that is demanded.*"

4 Proposed Methodology

The design of the proposed system introduces a predictive approach where a search-optimization based methodology is applied in order to enhance the throughput performance of FinFET/SRAM. Figure 1 highlights the proposed scheme.

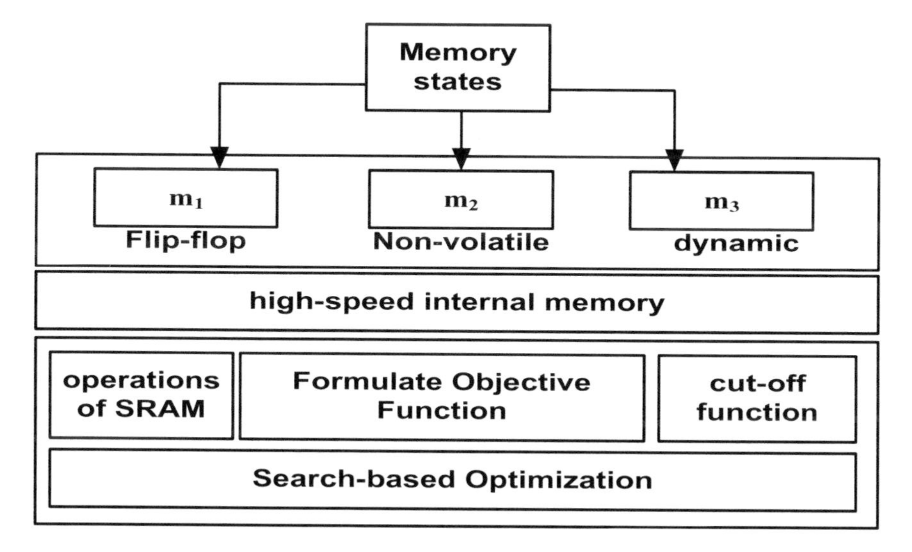

Fig. 1. Implemented system scheme

The proposed system introduces three different functionalities of FinFET/SRAM which when combined with internal memory should offer maximum throughput without any dependencies of increasing resources. A novel mechanism of search-based optimization scheme is applied that considers various operations of SRAM, formulates an objective function, whose process is controlled by a novel cut-off function in order to achieve convergence. The next section briefs about algorithm implementation.

5 Algorithm Design

For a memory unit to be fault tolerant, it is necessary to ensure that it can cater up all the situation of resource utilization and querying process. It is also necessary to ensure how each memory units communicates with each other in order to accomplish the processing task. Therefore, the formulation of the proposed system is carried out for multiple states of memory corresponding to FinFET/SRAM. The proposed algorithm introduces a novel search-optimization-based predictive approach where the input to the algorithm are m_1, m_2, m_3 (different memory), α (operations of SRAM), γ (high speed memory (internal)), σ (memory units), τ (cut-off function), p_{def} (defined parameter) and final outcome is accomplishment of converging point. The incorporated steps of the algorithm are:

Algorithm for design optimization of FinFET SRAM
Input: m_1, m_2, m_3, α, γ, σ, τ, p_{def}
Output:converging point
Start
1. *init*m_1, m_2, m_3
2. Set $f_{obj}(x) \rightarrow \sum_{i=1}^{5} \alpha$
3. form Matrix β=[m_{prime}, M_{onchip}]
4. arrange $\beta=\gamma(m_1, m_2, m_3)$ in σ
5. compute τ=abs{H}
6. **While** τ is not achieved do
7.   **For**p=0 to μ
8.   randSelect(p_{def}) from μ
9. $r_{adap} \rightarrow g(\tau)$
10. **End**
11. Assess all outcomes and shortlist best
12. **End**
13. Convergence achieved
End

The first part of the algorithm focuses on coupling different forms of memory states of SRAM considering the adoption of high speed internal memory for speeding up the computation process. The variables m_1, m_2, and m_3 correspond to higher capability of memory (Line-1). The next part of the algorithm is to formulate an objective function to ensure design optimization. The proposed algorithm formulates an objective function $f_{obj}(x)$ which has dependency of 5 parameters of operations of SRAM i.e. $\alpha 1$, $\alpha 2$, $\alpha 3$, $\alpha 4$, and $\alpha 5$ corresponds to (i) reading operation in local memory along with its cost, (ii) writing operation in local memory with its cost, (iii) reading operation of remote memory along with its cost, (iv) writing operation of remote memory along with its cost, and (v) cost of making the data go mobile from one to another memory segment (Line-2). The matrix β is created which maintains two forms of memory i.e. prime memory (M_{prime}) and on chip memory (M_{onchip}). The next step is to define a function γ (high speed memory) where the elements of memory are going to be structured for direct accessibility when queried for allocation (Line-4). This step has direct impact on positive energy saving as well as balanced allocation of necessary resources. The algorithm than initiates begin its search optimization for obtaining better convergence point of design. For this, a novel formulation of a cut-off function τ is applied to compute instead of initializing like in majority of studies in existing system. The computation of the cut-off function is carried out by computing a matrix H which depends on cumulative state of memory and cost function (Line-5). The complete evaluation is also dependent on cardinality of data items as well as cost of data as well. Line-6 to Line-12 corresponds to proposed search optimization scheme with an intention to obtain better convergence point of design of SRAM. The initial step in this search-based optimization is to perform random selection of population, where population is a set of all best possible outcomes of resource allocation policies of FinFET/SRAM. The proposed system will apply a greedy approach in order to perform random selection of population μ (Line-8). The next step is to apply a function g on cut-off function (Line-9), which corresponds to replicate behaviour with adaptive approach r_{adap}. This process is enhancement on the prior population extraction where it offers formulating more elite outcome from a set of one of more elite population. The computation of the function $g(\tau)$ mainly depends on maximum value τ_{max}, best value τ_{best}, and average value of cut-off function τ_{avg} and is empirically represented as,

$$g(\tau) = c \cdot \frac{\tau^p}{\tau^q}$$

Where the notation τ_p and τ_q corresponds to $(\tau_{max} - \tau_{best})$ and $(\tau_{max} - \tau_{avg})$ respectively. The variable c represents system matrix whose values depends on final application to be designed and hence could be configured likewise. The prime idea of this search optimization is to assess the presence of sufficient resource for allocation of particular memory units in FinFET/SRAM. It also has the capability to assess its adjacent cores of memory units and easily localize certain units. Such approaches are not seen in any existing approach for design enhancement of SRAM. The final step of search optimization includes resisting of the accomplishing the convergence point. A convergence point is defined as an instance of an outcome where minimal resource is used to obtained better yield of SRAM that is more capable of making it fault tolerant. Therefore, the final step of proposed algorithm is to performing sorting and to check for the best values of the outcome. The process is continued until and unless all the populations μ are checked (Line-11 and Line-13). An interesting fact to see for is the process of swapping the memory unit with each other in order to cater up the dynamic job requirements. This process results in building more capability to the FinFET/SRAM for exhibiting better convergence performance with less inclusion of memory parameters. Hence, this proposed optimization technique will also ensure better energy conservation scheme as well as it also has the capability to minimize and control of leakage power. At the same time, the proposed algorithm encourages progressive steps and not iterative steps, for which reason computational burden is kept within a specific limit. The next section outlines the results being obtained.

6 Results Discussion

A simulation study is carried out in MATLAB on 6 bit machine equipped with windows platform. Later, a hypothetical task is generated to assess the memory factor with respect to throughput and response time. The minimum configuration of the test platform uses core i3 process of 2 GHz frequency. The baseline memory states m_1, m_2, and m_3 are configured with multiple energies of access operation (0.2 nJ–20 nJ), delay (0.5 ns–20 ns), and leakage power (2 mW–100 mW). The study outcome of proposed system using Search Optimization (SO) is compared with the standard Particle Swarm Optimization (PSO).

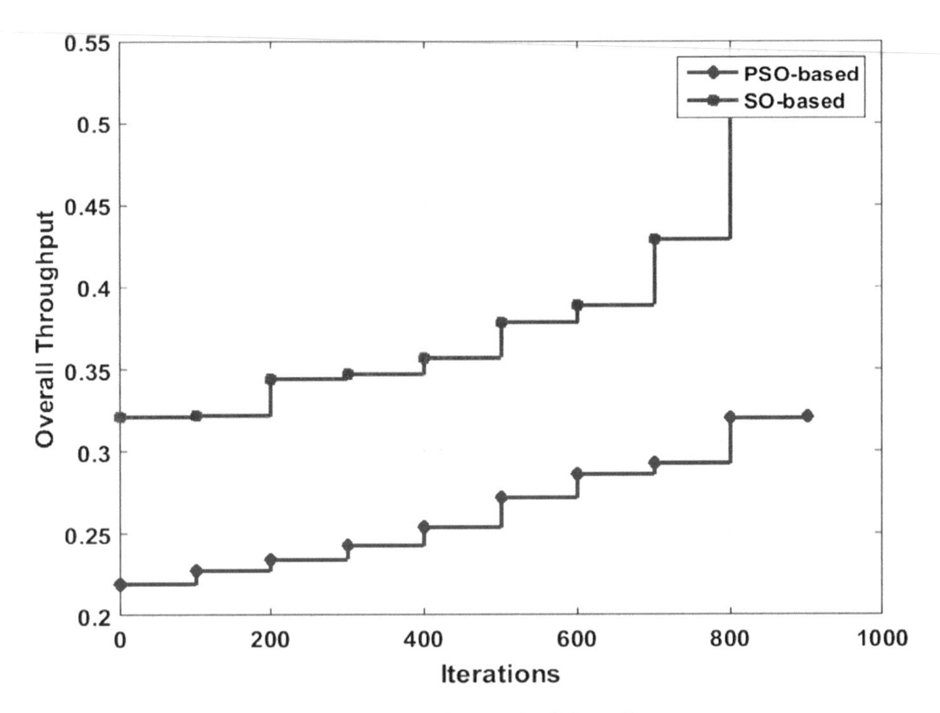

Fig. 2. Comparative result of throughput

The outcome shown in Fig. 2 highlights that proposed system of SO offers better throughput performance in contrast to PSO based approach. The prime reason for this trend is – PSO based design optimization scheme offers iterative process to obtain the elite outcome of convergence. This causes increase in latency owing to excessive read-write operation resulting in maximum resource allocation. Hence, PSO based schemes offers fault tolerance for only lesser iteration. In the meanwhile, proposed SO-based scheme offers a progressive exploration process where the all the cumulative outcomes are already defined and process of optimization narrows down the search process using a cut-off function and hence makes the process much progressive and less iterative. Normally, such optimization process is not witness in existing FinFET/SRAM

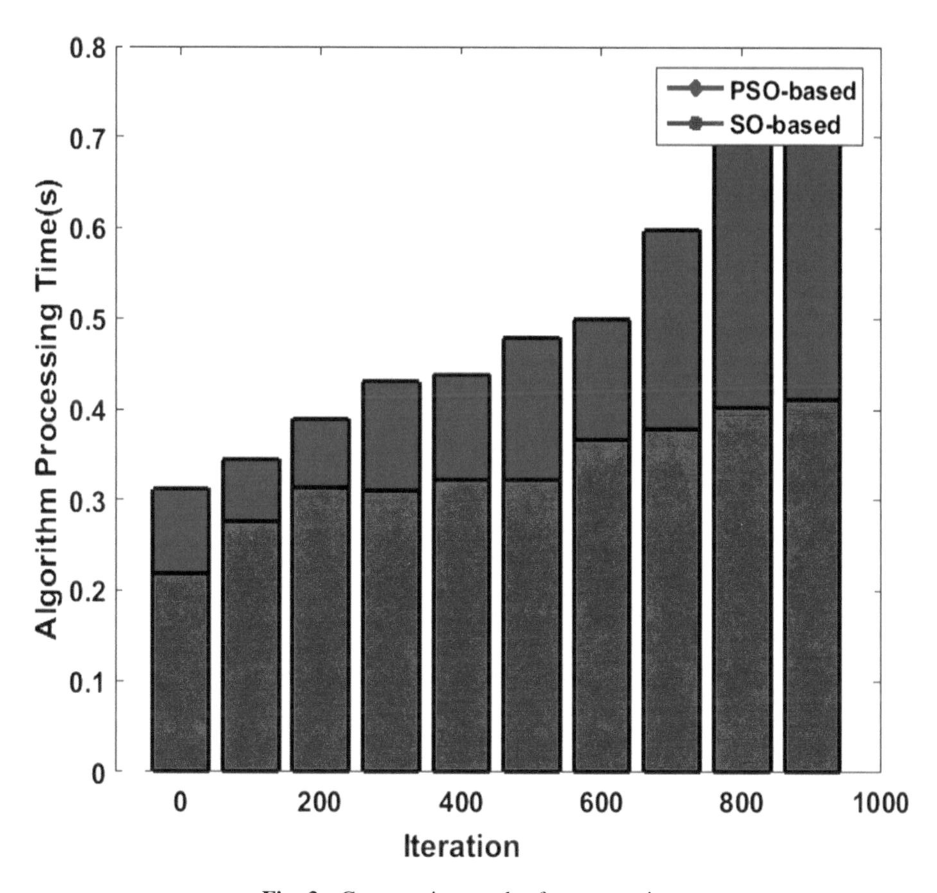

Fig. 3. Comparative result of response time

design in any of the existing system and hence the proposed system can be said to offer better balance between the energy being depleted during read-write operation (Fig. 3).

A closer look into the proposed technique shows that it offers significant reduction of the write operation up to 35% and minimizes energy consumption to 20% approximately. Owing to adoption of randomized process of initiating a search, proposed algorithm performs faster operation which contributes to faster algorithm execution time. The second reason for faster response time is that proposed algorithm performs parallel search process of obtaining elite outcome of convergence point. This cut downs the computational time to 50% and is quite useful for handling massive and dynamic task. At the same time, greedy mechanism further simplifies the process of initiating and managing the search depending on the best value of the cut-off function that consistently updates it. The computational complexity of the proposed system is therefore very less and it has nearly uniform trend, which makes the performance quite predictive for the proposed system. The outcome of graphical trend therefore exhibits that proposed system offers better throughput performance measured with respect to probability that could suit well to any form of applications using FinFET/SRAM with less energy consumption, more yield, and faster computational time. It is best suitable for upcoming designs of registers with high speed operation, sophisticated smart phones with advanced mobile networks, buffer management in n-based routers, etc.

7 Conclusion

The novelty of the proposed system is that it introduces three different roles of SRAM in the form of memory units with an objective to minimize the energy depletion as well as any other issues concerning with the latency. The proposed technique introduces a novel search-based optimization approach that uses a highly progressive mechanism in order to achieve its target. The implementation of the proposed system is basically a computational model that suits well with any computational environment. The study outcome shows that proposed system offers better response and higher throughput.

References

1. Wicht, B.: Current Sense Amplifiers for Embedded SRAM in High-Performance System-on-a-Chip Designs. Springer, Heidelberg (2013)
2. Alioto, M.: Enabling the Internet of Things: From Integrated Circuits to Integrated Systems. Springer, Cham (2017)
3. Sun, G.: Exploring Memory Hierarchy Design with Emerging Memory Technologies. Springer, Cham (2013)
4. Dasgupta, S., Kaushik, B.K., Kumar Pal, P.: Spacer Engineered FinFET Architectures: High-Performance Digital Circuit Applications. CRC Press, Boca Raton (2017)
5. Reis, R., Cao, Y., Wirth, G.: Circuit Design for Reliability. Springer, New York (2015)
6. Gildenblat, G.: Compact Modeling: Principles, Techniques and Applications. Springer, Dordrecht (2010)
7. Chauhan, Y.S., Lu, D.D., Sriramkumar, V., Khandelwal, S.: FinFET Modeling for IC Simulation and Design: Using the BSIM-CMG Standard. Academic Press, London (2015)
8. Weste, N.H.E., Harris, D.: CMOS VLSI Design: A Circuits and Systems Perspective. Pearson Education India (2015)
9. Girish, H., Shashikumar, D.R.: Insights of performance enhancement techniques on FinFET-based SRAM cells. Commun. Appl. Electron. (CAE) **5**(6), 20–26 (2016)
10. Girish, H., Shashikumar, D.R.: Cost-effective computational modeling of fault tolerant optimization of FinFET-Based SRAM cells. In: Computer Science On-line Conference, pp. 1–12. Springer (2017)
11. Bhattacharya, D., Jha, N.K.: Ultra-high density monolithic 3-D FinFET SRAM with enhanced read stability. IEEE Trans. Circ. Syst. I Regul. Pap. **63**(8), 1176–1187 (2016)
12. Fang, Y.P., Oates, A.S.: Characterization of single bit and multiple cell soft error events in planar and FinFET SRAMs. IEEE Trans. Device Mater. Reliab. **16**(2), 132–137 (2016)
13. Chen, Y.H., et al.: A 16 nm 128 Mb SRAM in High-κ metal-gate FinFET technology with write-assist circuitry for low-VMIN applications. IEEE J. Solid State Circ. **50**(1), 170–177 (2015)
14. Yang, Y., Park, J., Song, S.C., Wang, J., Yeap, G., Jung, S.O.: Single-ended 9T SRAM cell for near-threshold voltage operation with enhanced read performance in 22-nm FinFET technology. IEEE Trans. Very Large Scale Integr. (VLSI) Syst. **23**(11), 2748–2752 (2015)

15. Karl, E., et al.: A 0.6 V, 1.5 GHz 84 Mb SRAM in 14 nm FinFET CMOS technology with capacitive charge-sharing write assist circuitry. IEEE J. Solid State Circ. **51**(1), 222–229 (2016)
16. Zhang, X., Connelly, D., Takeuchi, H., Hytha, M., Mears, R.J., Liu, T.J.K.: Comparison of SOI versus bulk FinFET technologies for 6T-SRAM voltage scaling at the 7-/8-nm node. IEEE Trans. Electron Devices **64**(1), 329–332 (2017)
17. Kulkarni, J.P., et al.: 5.6 Mb/mm^2 1R1 W 8T SRAM arrays operating down to 560 mV utilizing small-signal sensing with charge shared bitline and asymmetric sense amplifier in 14 nm FinFET CMOS technology. IEEE J. Solid State Circ. **52**(1), 229–239 (2017)
18. Hu, V.P.H., Fan, M.L., Su, P., Chuang, C.T.: Analysis of GeOI FinFET 6T SRAM cells with variation-tolerant WLUD read-assist and TVC write-assist. IEEE Trans. Electron Devices **62**(6), 1710–1715 (2015)
19. Bhattacharya, D., Jha, N.K.: TCAD-assisted capacitance extraction of FinFET SRAM and logic arrays. IEEE Trans. Very Large Scale Integr. VLSI Syst. **24**(1), 329–333 (2016)
20. Song, T., et al.: A 10 nm FinFET 128 Mb SRAM With assist adjustment system for power, performance, and area optimization. IEEE J. Solid State Circ. **52**(1), 240–249 (2017)
21. Mishra, S., Mahapatra, S.: On the impact of time-zero variability, variable NBTI, and stochastic TDDB on SRAM cells. IEEE Trans. Electron Devices **63**(7), 2764–2770 (2016)
22. Asenov, A., et al.: Variability aware simulation based Design-Technology Cooptimization (DTCO) Flow in 14 nm FinFET/SRAM cooptimization. IEEE Trans. Electron Devices **62**(6), 1682–1690 (2015)
23. Joshi, R., et al.: A universal hardware-driven PVT and layout-aware predictive failure analytics for SRAM. IEEE Trans. Very Large Scale Integr. VLSI Syst. **24**(3), 968–978 (2016)

Analysis of the Quality of the Painting Process Using Preprocessing Techniques of Text Mining

Veronika Simoncicova[1(✉)], Pavol Tanuska[1], Hans-Christian Heidecke[1], and Stefan Rydzi[2]

[1] Faculty of Materials Science and Technology in Trnava, Slovak University of Technology, Bratislava, Slovakia
{veronika.simoncicova,pavol.tanuska}@stuba.sk, christian.heidecke@gmx.de

[2] PredictiveDataScience, s. r. o., Klzava 31, Bratislava, Slovakia
stefan.rydzi@predictivedatascience.sk

Abstract. Text mining is a relatively new area of computer science, and its use has grown immensely lately. The aim is to join two dataset from different data sources and to acquire information about percentage defects from the painting process, which are transmitted from the manufacturing to the end customers. The data sets are totally different and for their joining using text attributes, preprocessing are needed.

Keywords: Text mining · Data set · Data · Defect

1 Introduction

Real-word data are often incomplete, noisy, uncertain, and unreliable. Information redundancy may exist among the multiple pieces of data that are interconnected in a large network. Information redundancy can be explored in such networks to perform quality data cleaning, data integration, information validation, and trustability analysis by network analysis [1].

There are many other kinds of semi-structured or unstructured data, such as spatiotemporal, multimedia, and hypertext data, which have interesting applications. Such data carry various kinds of semantics, are either stored in or dynamically streamed through a system and call for specialized data mining methodologies. Thus, mining multiple kinds of data, including spatial data, spatiotemporal data, cyber-physical system data, multimedia data, text data, web data, and data streams, are increasingly important tasks in data mining [1].

Text mining has recently come to the forefront, especially of researching text information from social networks, various text documents, but the spectrum of use is very widely.

Feldman [2] wrote, that text mining is a new and exciting area a of computer science research that tries to solve the crisis of information overload by combining techniques from data mining, machine learning, natural language processing,

R. Silhavy (Ed.): CSOC 2018, AISC 764, pp. 30–38, 2019.
https://doi.org/10.1007/978-3-319-91189-2_4

information retrieval, and knowledge management. Similarly, link detection – a rapidly evolving approach to the analysis of text that shares and builds on many of the key elements of text mining – also provides new tools for people to better leverage their burgeoning textual data resources. The main tasks of link detection are to extract, discover, and link together sparse evidence form vast amounts of data sources, to represent and evaluate the significance of the related evidence, and to learn patterns to guide to extraction, discovery, and linkage of entities.

Data preprocessing is used on the different kind of data for example data from social network [3], data obtained from a laser confocal microscope [4] or surveillance of healthcare data [5] and many various another data. We decided to apply it, especially several techniques for data pre-processing, on painting process data and data from authorised service with the aim of analysing the quality of the painting process in one of the automotive company.

2 Preprocesing Techniques of Text Mining

As the text data often contains different formats like number formats, date formats and the most common words unlikely to help Text mining such as prepositions, articles, and pro-nouns. These can be eliminated by the pre-processing techniques. These techniques eliminate noise from text data, later identify the root word for actual words and reduce the size of the text. We know several different pre-processing methods, for example, tokenization, stop word removal and stemming for the text. Methods of stemming and stop wording were used by the quality analysis, therefore, they will be described in more detail.

All changes will contribute to better, faster and more efficient analysis of data. Preprocessing the data is very important part of data pre-processing, as the quality of acquired results depends on the quality of the used data [6, 7].

2.1 Stemming

Stemming, this method is used to identify the root/stem of a word [8]. For example, the words argue, argued, argues, arguing, all can be stemmed to the word "argu" (Fig. 1).

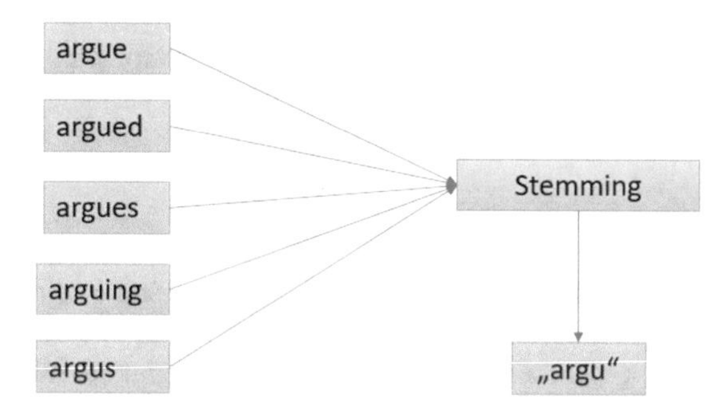

Fig. 1. Stemming process

There are mainly two defects in stemming [9]:

- over stemming,
- under stemming.

Over-stemming is when two words with different stems are stemmed to the same root. This is also known as a false positive. Under-stemming is when two words that should be stemmed to the same root are not. This is also known as a false negative.

2.2 Stop Word Removal

Many words in documents recur very frequently but are essentially meaningless as they are used to join words together in a sentence. It is commonly understood that stop words do not contribute to the context or content of textual documents. Due to their high frequency of occurrence, their presence in text mining presents an obstacle in understanding the content of the documents. Stop words are very frequently used common words like 'and', 'are', 'this' etc. They are not useful in classification of documents. So, they must be removed. However, the development of such stop words list is difficult and inconsistent between textual sources. This process also reduces the text data and improves the system performance. Every text document deals with these words which are not necessary for text mining applications [9].

3 Analysing the Dataset from the Manufacturing

In this section we will summarize the practices for collecting, integrating, interpreting, preprocessing and presenting data. The aim is to join two data sets from the different data sources and to acquire information concerning percentage of defects transferred from the manufacturing to the end customers. First dataset represents painting defects in manufacturing and the second dataset stands for records from the authorised services, where the claimed painting defects from the produced and sold products are stored.

These datasets contain description of defects, description of their location, but their quantity is very different, dataset from the service contains hundreds of records and dataset from the painting shop contains millions records about painting defects from the manufacturing.

Currently, these datasets cannot be compared by the Department of Quality. We work with useful software for analysis data, Rapidminer, to solve this issue, which will be described in the next chapter. Text information is the key attribute are, since we used several text mining techniques for the text preprocessing.

3.1 Rapidminer

Rapidminer is currently one of the most used an open source predictive analytics platform for data analysis. It is accessible as a stand-alone application for information investigation and as a data mining engine for the integration into own products. Rapidminer provides an integrated environment for data mining and machine learning procedures, including [10]:

- extracting the data from different source systems; transforming the data and loading into a data warehouse (DW) or data repository other applications,
- data pre-processing and visualization,
- predictive analytics and statistical modelling, evaluation, and deployment.

What makes it even more powerful is that it provides learning schemes, models and algorithms from WEKA and R scripts [10].

Rapidminer offers a large amount of different operators, which can be easily extended using existing extensions. There are packages for text processing, web mining, weka extensions, R scripting, series extension, python scripting, anomaly detection and more [11].

The Rapidminer text extension adds all operators necessary for statistical text analysis. Text from different data sources can be loaded and can be transformed by different filtering techniques, to analyses text data. The Rapidminer text extensions support several text formats including plain text, HTML, or PDF. It also provides standard filters for tokenization, stemming, stop word filtering, or n-gram generation. The Text Processing package, can be installed and updated through the Market Rapidminer menu item under the Help menu. The Text Mining extension uses a special class for handling documents, called the document class. This class stores the whole document in combination with additional meta information [11].

In an industrial company, there is large amount of data sources. In order to carry out our task, we will use two data sources, an internal data source for production data and an external source for field data respectively data from the market. All numeric values were changed in accordance with the company's security policy.

3.2 Dataset from Autorised Services

The system with field data is used to analyse the quality of the produced products. In that system, service data is stored, where the various attributes detail the reported claimed defects that have been reported to customers at authorised service centers. The unique object or product number (UVN), name of defect, the wrong part of the product, the exact location defects on the product, the type of product, production date, type of defect, the mileage and other attributes that accurately describe defect on the claimed

Tab. 1. The example set of service dataset

UVN	Wrong part	Location	Type	Production date	KM	Type of defect	Name of defect
11111	Front bonnet	Right	Sedan	12.10.2016	3215	Paint	Paint scratches
22222	Door	Left	SUV	16.06.2015	20085	Paint	Paint scratches
33333	Back bonnet	Front left	Kombi	07.11.2016	15	Paint	Skimmed place
44444	Door	Front right	SUV	08.07.2016	715	Paint	Spotting
55555	Door	Back left	Kombi	11.05.2015	0	Paint	Paint scratches
66666	Mirror cover	Right	SUV	06.03.2015	7	Paint	Paint scratches

products, are the most important attributes. Table 1 represents a sample of data from the services. For us, text is the key attribute. However, it requires to be preprocessed, as sometimes the same defect differs in name

Data analysis is currently performed manually and only on current data in a time interval of up to one month. The data, we obtained, concern products belonging to the year 2015 and 2016. There are exactly 274 records from this period. The number of products with defects is 174, containing different types of products. We analysed only painting defects from manufacturing.

3.3 Dataset from the Painting Shop

The production system database is used to write painting defects in production, but only a part of the data can be obtained from this database. A data sample of the painting defects from production is shown in Table 2, where we can see the attributes such as unique object number, defect description, equipment, object type, occurrence time of an defect, and number of repetitions, indicating how many times the object/product has returned to the repair.

Tab. 2. The example set of painting defects

UVIM	Error description	Equipment	Status	TN	Type	Ocurrence time	Repetiton
11111	Front bonnet right scratcher	HJ444	LQ00	92	Sedan	01.12.2016	0
10000	Door left front painting scratcher	HJ333	LQ00	92	Sedan	01.12.2016	1
23555	Back bonnet spotting	HJ444	LQ00	88	SUV	01.12.2016	1
44444	Door right front skimmmed places	HJ222	LQ00	88	SUV	01.12.2016	0
66666	Front bonnet right skimmed places	HJ333	LQ00	88	SUV	01.12.2016	0
88888	Door back left painting scratcher	HJ111	LQ00	72	Kombi	01.12.2016	1

The dataset contains painting defects directly from the paint shop as well as painting defects recorded in the assembly line. This dataset, we added based on painting experts' recommendations. The total number of records is 4.5 million for the period from January 2015 to March 2017. The number of objects with defects is 440 thousand.

3.4 The Process of Analysis Datasets

The first step was to select the test example set from the dataset from painting shop using "Sample" operator, since the amount of data has a great demand on computer memory. The two datasets were joined using a unique identifier, unique vehicle number, where, there is the example set of each kind of data and joining these two datasets new dataset was created, which is located on the right of the figure. We identified objects occurring in both datasets. The main issue stemmed from the incorrect description of a wrong part of the object and from non-exact location of the defects on object/product, not matching the defect description.

Dataset containing only those examples where the key attributes of both input example sets match is the result of the first phase. The total number of matching records is 1225. "Join" operator connects records to each other, i.e. if there are 2 records of the same in the service dataset and 4 records stemming from painting, 2 * 4 will result in 8 entries.

In the second phase, the problem of different defect names had to be solved and at the same time, only totally or partially identical ones were to be selected. Defect descriptions were different in each dataset, and there were various problems in comparing these attributes, for example, a different format of defect and placement names, different naming, diacritics, and others. These defects were removed using Text mining methods for text editing; stemming and stop word techniques were used. We worked with Rapidminer software, "Stem (Snowball)" operator was utilised. This operator stems words by applying stemming algorithms written for the Snowball language. Various stemming algorithms for different languages can be chosen. "Filter Stopword" was used to remove stopwords from a document.

After this adjustment, the second phase, shown in Fig. 2, was carried out. In the Figure, individual data sets already been linked based on the primary identifier are depicted. In this file, we modified text attributes using pre-processing techniques of Text mining, and by comparing text we determined the complete or partially consistent records.

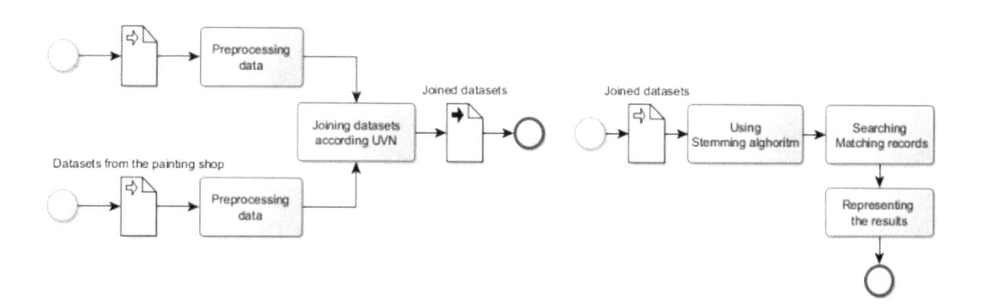

Fig. 2. The process of the modification of text attributes

As a result of these adjustments, we identified exactly 40 records from a total of 274 entries; all matches found being product A, which could be explained by having only 2% of product B defects in the painting data, and the rest 98% concerning the product A. The total number of products matched was 24, i.e., 8.8% of the total number of products recorded in the service (174 objects/products). These were the first relevant results. Consequently, they were checked by the quality department, and following the consultations and their recommendations, we adjusted the process and received a percentage of 10% of the total number of tests recorded in the service.

Our accuracy of identifying the same mistakes was about 80%, which a very impressive result, as the fact that not all matching records had to be real and the same has to be taken into account. It is necessary to amend that sometimes an defect occurs just as a coincidence and this deficiency cannot be effectively eliminated. The results were checked by the quality department using a program to analysis all complaints, including photos.

The entire process of editing and identifying the particular disorders transmitted to a particular consumer is shown in Fig. 3.

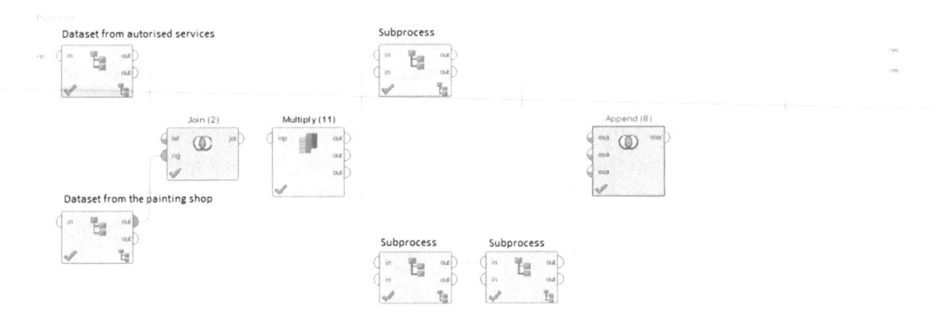

Fig. 3. The process of analysis of quality

The process consists of retrieving data that had to be partially edited, i.e., it was necessary to modify data types, create a new attribute, and edit the names of some attributes. The next step was to associate data sets with the join operator, and the last step was to define totally identical records using modified text attributes and their cross-referencing.

4 Conclusion

The main objective was to make efficient and to improve the analysis of the quality of product painting in the particular industrial company. The Department of Quality currently works only with service data that has been evaluated from a variety of perspectives, and when an increased incidence of a particular defect is recorded, the painting process is to be analysed and corrective actions can be taken. The dataset obtained from services is not possible to be compared with such a large amount of data from the painting shop. Text analysis and the proposed process helped expanding the view of the quality of the produced products and streamlined and accelerated work; the entire process of retrieving new data and comparing it takes about few minutes.

The all stages are depicted on the Fig. 4, where is:

- Identification the aim - clear and exactly defined aim is important for process of data analysis,
- Obtaining data sources - this process is very complicated in a large company and it is therefore necessary to identify and collect all necessary data for analysis,

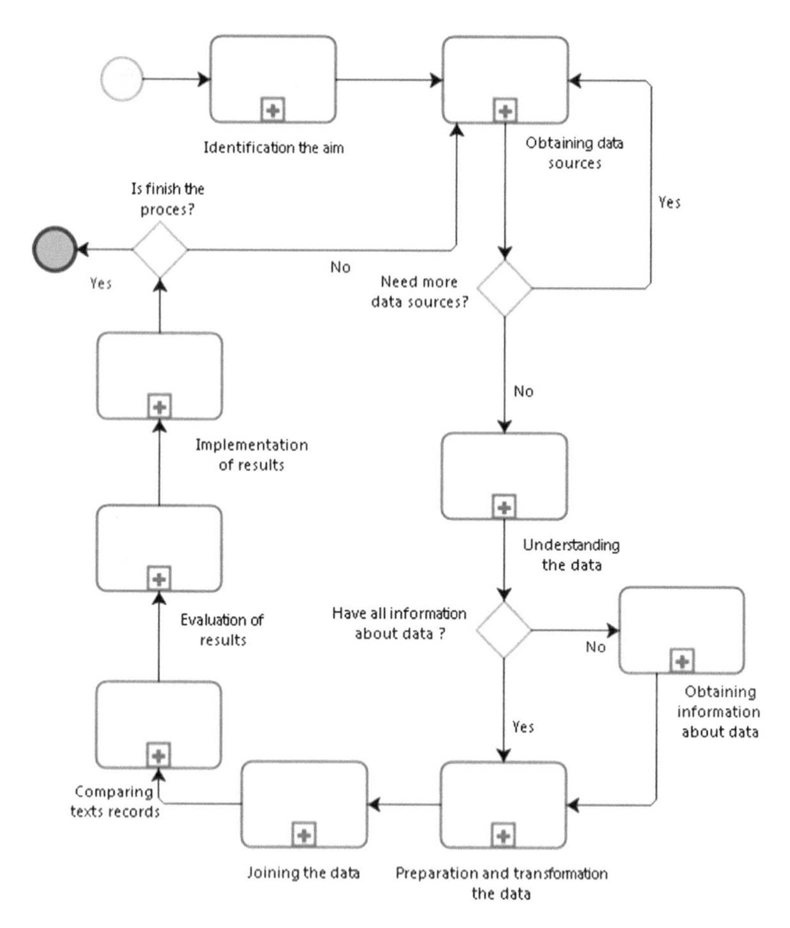

Fig. 4. The process of analysis of quality

- Understanding the data - find detailed information about the data
- Preparation and transformation the data the longest and most complicated stage of the whole process, this stage affects the final results,
- Joining the data - the basis is to find the primary key to be linked,
- Comparing text records - the main purpose is to find exactly identical or partially identical text records,
- Evaluation of results - a summary of the results achieved,
- Implementation of results.

These results will be further used to extend the process on the longer time frame of painting defects. We are currently working on data processing from 2011 to 2015, when changes were made to the names of defects and locations and the structure needed to be unified. We also plan to use some datamining methods and other text mining tools. Later, we plan to extend the dataset to other defects, stemming for example from engine, electrical issues and others.

Acknowledgment. This publication is the result of implementation of the project: "Increase of Power Safety of the Slovak Republic" (ITMS: 26220220077) supported by the Research & Development Operational Programme funded by the ERDF and project VEGA 1/0673/15: "Knowledge discovery for hierarchical control of technological and production processes" supported by the VEGA.

References

1. Han, J., Kamber, M., Pei, J.: Data Mining Concepts and Techniques, 14 February 2018. https://books.google.de/books?hl=sk&lr=&id=pQws07tdpjoC&oi=fnd&pg=PP1&dq=data+mining+text+mining&ots=tzEy0-pzX2&sig=3y8SbPuEoEeYkbE8A69jA2st890#v=onepage&q&f=false
2. Feldman, R., Sanger, J.: The text mining handbook, 14 February 2018. https://books.google.de/books?hl=sk&lr=&id=U3EA_zX3ZwEC&oi=fnd&pg=PR1&dq=feldman+text+mining&ots=2NxKMiDwOG&sig=hDTiHAMhaeJ83NzmtS8CME4PmZA#v=onepage&q=feldman%20text%20mining&f=false
3. Domingos, P.: Mining Social Networks for Viral Marketing, 14 February 2018. http://ncwebcenter.com/domingos05.pdf
4. Bezak, T., Elias, M., Spendla, L., Kebisek, M.: Complex roughness determination process of surfaces obtained by laser confocal microscope, 14 February 2018. http://sci-hub.hk/, http://ieeexplore.ieee.org/abstract/document/7555111/
5. Obenshain, M.K.: Application of Data Mining Techniques to Healthcare Data, 14 February 2018. http://sci-hub.hk/, https://www.cambridge.org/core/journals/infection-control-and-hospital-epidemiology/article/application-of-data-mining-techniques-to-healthcare-data/7EE5E7B1FA8B1C535FBC7A3881EC42
6. Simoncicova, V., Hrcka, L., Tadanai, O., Tanuska, P., Vazan, P.: Data Pre-processing from Production Processes for Analysis in Automotive Industry, 14 February 2018. http://archive.ceciis.foi.hr/app/public/conferences/1/ceciis2016/papers/DKB-3.pdf
7. Ramasubramanian, C., Ramya, R.: Effective preprocessing activities in text mining using improved porter's stemming algorithm. Int. J. Adv. Res. Comput. Commun. Eng. **2**(12), December 2013. ISSN (Online): 2278-1021
8. Gurusamy, V.: Preprocessing Techniques for Text Mining, 20 July 2017. https://www.researchgate.net/publication/273127322_Preprocessing_Techniques_for_Text_Mining
9. Gupta, G., Malhotra, S.: Text documents tokenization for word frequency count using rapid miner (taking resume as an example). Int. J. Comput. Appl. (0975-8887). International Conference on Advancement in Engineering and Technology (ICAET 2015) (2015)
10. Akthar, F., Hahne, C.: RapidMiner 5 Operator Reference, 20 July 2017. https://rapidminer.com/wp-content/uploads/2013/10/RapidMiner_OperatorReference_en.pdf
11. RapidMiner text mining extension, 20 July 2017. http://www.predictiveanalyticstoday.com/rapidminer-text-mining-extension/

Bioinspired Algorithm for 2D Packing Problem

Vladimir Kureichik, Liliya Kureichik, Vladimir Kureichik Jr., and Daria Zaruba(✉)

Southern Federal University, Rostov-on-Don, Russia
vkur@sfedu.ru, ilu@rambler.com, kureichik@yandex.ru, daria.zaruba@gmail.com

Abstract. The paper considers one of the most important problem of resource allocation – packing of units within 2D space. The problem is NP-hard. A problem formulation is made, as well as restrictions and boundary conditions are found out. To solve the considered problem the authors suggest to use firefly optimization algorithm on the basis of which there are developed a bioinspired algorithm. This algorithm allows to obtain sets of quazi-optimal solutions for the 2D packing problem within polynomial time. Also, there are suggested mechanisms for encoding and decoding of alternative solutions. and presented a scheme of firefly algorithm for 2D packing problem. On the basis of the suggested algorithm there are developed software for computational experiments on benchmarks. Experimental investigations were carried out taking into account time and quality of alternative solutions. As a result, experiments shows the effectiveness of the developed algorithm.

Keywords: Packing problem · Optimization · Bioinspired algorithm
Firefly algorithm

1 Introduction

Today, informatics and computer technology is promising, since it is directly related with advanced technologies. One of the components of computer science as an applied discipline is the computer-aided design. The development of computer-aided design systems is supported by a solid scientific and technical basis. Computer-aided design systems make it possible to develop and improve the design methodology on the basis of the advanced achievements of fundamental sciences, and stimulate the development of the theory of complex systems and objects design [1].

At present, resource allocation problem became more relevant and can be solved by optimization methods. A variety of resource allocation optimization problem combine in term "packing" such as material cutting, container packing, scheduling, partitioning of objects and etc. This problem is NP-hard and to solve it there are used different heuristics algorithms, which allow to obtain optimal and quazi-optimal solutions in appropriate time. One of the relevant approaches to solve 2D packing problem is the use of bioinspired algorithms [2–5]. The main goal of the work is developing and researching of the bioinspired algorithm for 2D packing problem.

R. Silhavy (Ed.): CSOC 2018, AISC 764, pp. 39–46, 2019.
https://doi.org/10.1007/978-3-319-91189-2_5

2 Formulation of 2D Packing Problem

For packing units within restricted 2D space input data are 2D dimensions and a list of units.

The 2D packing problem can be formulated in the following way. Let $Q = \{q_1, q_2, \ldots, q_n\}$ is a set of units and N is a set of containers in which need to place units. Each unit and container has geometrical dimensions. Units need to lace within containers in such way, that placement should correspond to the selected criterion and objective function value should tend to optimal. Output data are plan of packing, objective function value and computational time [4, 6, 7].

Let us describe the 2D placement problem within restricted area more formally. Given 2D area with width W and length L. Also, there are a set of units $Q = \{q_1\}$ in number N. The set Q splits into disjoint sets $Q_1, Q_2, \ldots, Q_t$ such that sum of dimensions in each set of Q_i is less than 1. It is generally believed that all units from Q_i are packed into the one container with dimensions 1. The given problem has following restrictions [9]:

1. All units cannot intersect the boundaries of a given area, i.e. satisfy the system of inequalities (1):

$$\begin{cases} x_{1i} \geq 0; \\ y_{1i} \geq 0; \\ x_{2i} \leq L_x; \\ y_{2i} \leq W_y; \end{cases} \tag{1}$$

2. Total area of all units cannot exceed packaging area (2):

$$\sum\nolimits_{i=1}^{n} (l_i \cdot w_i) \leq L_x \cdot W_y. \tag{2}$$

3. The absence of intersections, i.e. in one area of space, you can not put two objects:

$$\begin{gathered}\left(x_{2i} \cdot x_{1j} \& y_{2i} \geq y_{1j}\right) \vee \left(x_{2i} \cdot x_{1j} \& y_{2i} \geq y_{1j}\right) \vee \left(x_{2i} \cdot x_{1j} \& y_{2i} \geq y_{1j}\right) \vee \left(x_{2i} \cdot x_{1j} \& y_{2i} \geq y_{1j}\right) \\ \vee \left(x_{2i} \cdot x_{1j} \& y_{2i} \geq y_{1j}\right) \vee \left(x_{2i} \cdot x_{1j} \& y_{2i} \geq y_{1j}\right) \vee \left(x_{2i} \cdot x_{1j} \& y_{2i} \geq y_{1j}\right) = 1 \forall i \leq n, j \leq (i \neq j).\end{gathered} \tag{3}$$

4. All elements must be supported by their base on the surface.

The optimization criterion is a ratio of placeable units to the total number of units. The objective function is represented as

$$S = \frac{\sum N_P}{N_B} \rightarrow 1, \tag{4}$$

where N_P is a number of placeable units and N_B is a total number of units.

3 Firefly Algorithm

The firefly algorithm was proposed by Xin-She Yang in 2007. The main idea of firefly algorithm lies in the following statements: all fireflies can attract each others independent of its gender; attractiveness of a firefly for other individuals is proportional to its brightness; less attractive fireflies move towards a more attractive ones; the brightness of the firefly's illumination, seen by another firefly, decreases with increasing distance between the fireflies; if the firefly does not see a firefly brighter than himself, then it moves randomly [8–11].

On the basis of this model the authors suggested the algorithm for the 2D packing problem. The main aim is not complete copying of the fireflies' behavior, but concerning of base search principles. The scheme of the algorithm is shown on the Fig. 1.

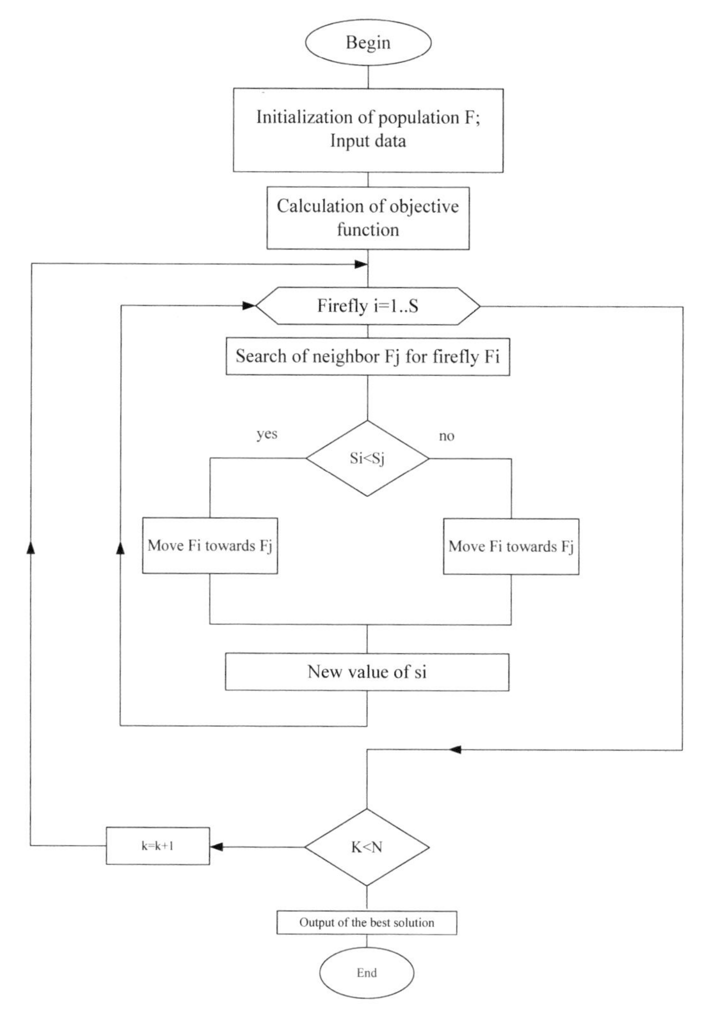

Fig. 1. The scheme of the 2D packing algorithm based on firefly algorithm

Let us describe the developed algorithm in detail.

1. Initially, input search parameters are set and an initial population F is generated. Input parameters are a size of population S, a number of iteration N, width and height of units.
2. Calculation of the objective function (OF) value for initial population. Each individual in population has a brightness of illumination. A value of brightness, which represented as $F_i \in F, i \in [1 : |F|]$, corresponds to OF value at this point of time.
3. Selection of firefly and search of more attractive individual taking into account neighborhood radius of the firefly.

Attractiveness of the individual F_i for the individual F_j is calculated in following way:

$$\beta_{i,j} = \frac{\beta_0}{1 + \gamma r_{i,j}^2}; \tag{5}$$

where $i, j \in F, i \in [1 : |F|], i \neq j, r_{i,j}$ is a distance between fireflies F_i and F_j, β_0 is a mutual attractiveness of fireflies at zero distance between them, γ is a coefficient of light absorption in environment.

4. If a more attractive individual is found, the firefly moves toward the more attractive neighbor. The firefly F_i is considered as neighbor of the firefly F_j, $i \in [1 : |F|], i \neq j$ under the following conditions: Euclidian distance between these individuals do not exceed a current radius r_i; current brightness of F_j exceeds brightness of F_i, i.e. $l_j > l_i$. Otherwise, the firefly moves randomly.
5. Calculation of new location of the firefly on the basis of its OF value. New radius of the firefly F_i is defined as follows:

$$r_i^{'} = min(r_{min}, max(0, (r_i + \xi(t - |T_i|)))), \tag{6}$$

where $i \in [1 : |F|], i \neq j$, T_i is a set of neighbors, r_i is a minimum permissible radius, t is a desired number of neighbors, ξ is a positive constant.

6. If a number of iteration k is less then predetermined N, than k = k + 1. Otherwise, the end of the algorithm.
7. Output the best solution.

4 Encoding of Solutions

During the algorithm implementation, it is necessary to represent solutions in appropriate form, i.e. encode solutions. The authors use a vector chromosome, which represent as a tuple $\langle b_i, l_i \rangle$, where b_i is a number of the i-th block, l_i is a option of rotation of the i-th block. In the chromosome the order of blocks is set in the process of decoding. Each block can take two possible positions (Fig. 2). Since the authors consider a packing of one-dimension blocks, which numbers are represented as non-negative integers, then it is necessary to represent orientation options as "–1" and "–2".

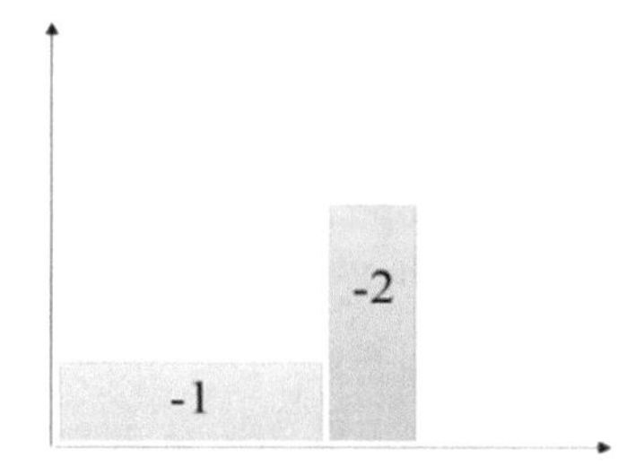

Fig. 2. Orientation options for the block

Figure 3 shows an example of encoding solutions for placement on plain, and Fig. 4 shows an example of placement on 2D space.

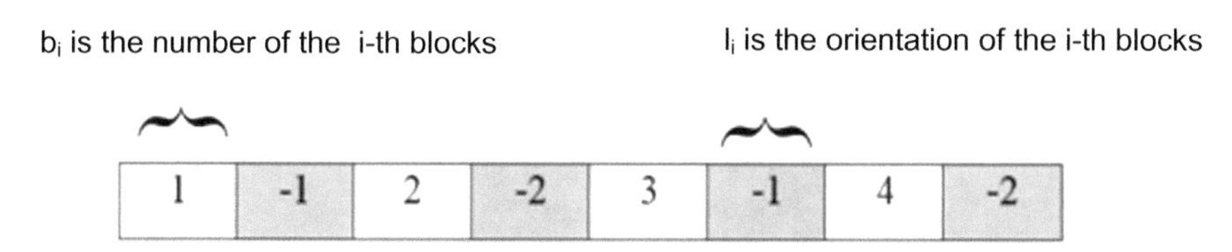

Fig. 3. Encoding solution

Decoding is a sequential process; each block is situated according to given position. Elements are situated from up to down and from left to down with regard to begin of container.

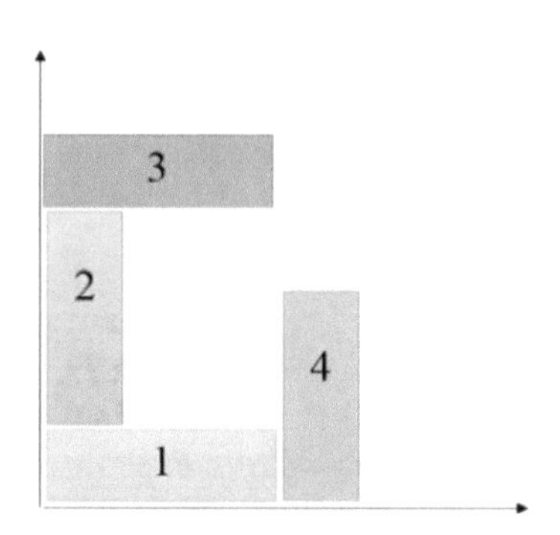

Fig. 4. Blocks placement according with encoded solution

During chromosome decoding each block has several possible positions. The first block is situated in (0;0), i.e. in origin of coordinates. Then, after the first block is placed in accordance with the chromosome, the coordinates of the newly placed element are deleted for the next block (for the first block - (0, 0)) and the set of coordinates of this block is added. With the placement of the next block, all possible points are ordered in increasing Y and X coordinates, and the first one is chosen which satisfies constraints mentioned above.

Let us demonstrate the mutation operator (MO) by chromosome transformation, which show some placement on blocks within container space (Fig. 5).

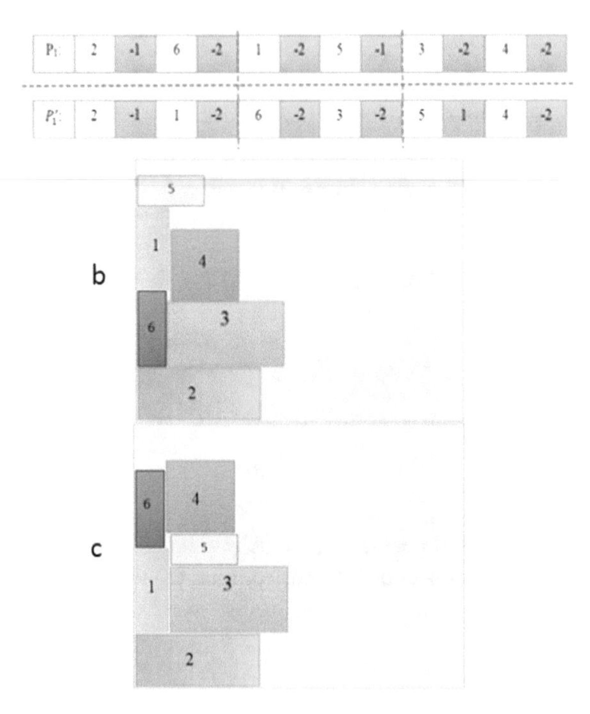

Fig. 5. Mutation of the parent chromosome; a – operator mutation application; b – blocks placement before mutation operator; c – blocks placement after mutation operator

5 Experiments

Objects of researching are genetic and firefly algorithms, and genetic operators block (selection, crossover, mutation, inversion and reduction) [12, 13]. For the developed algorithm there were carried out a computational experiment, results are represented on Figs. 6 and 7. Diagrams are shown a dependency of objective function value and running time on number of blocks.

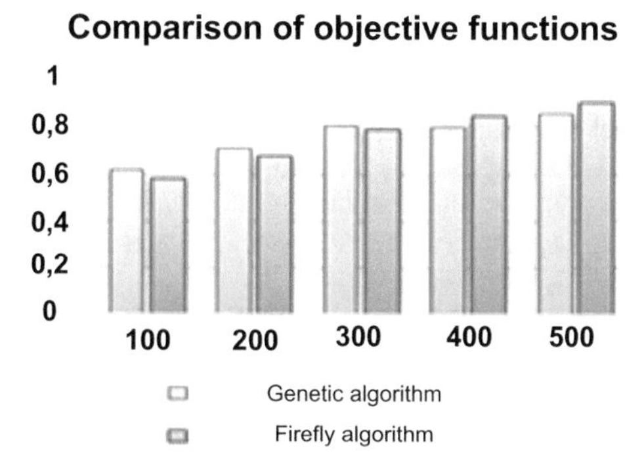

Fig. 6. Dependence of objective function value on blocks quantity

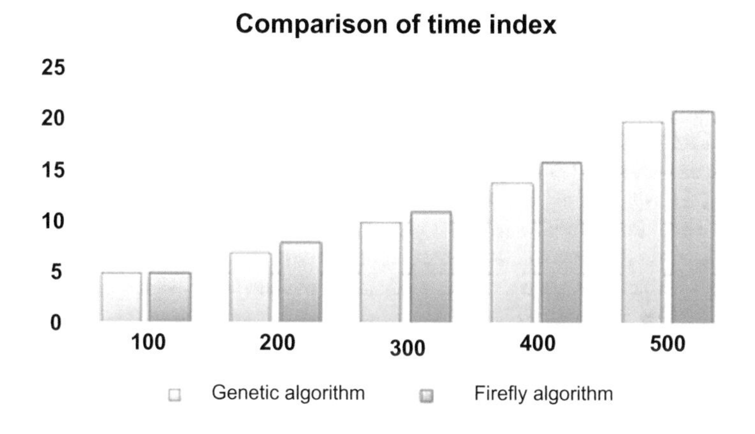

Fig. 7. Dependence of time on blocks

Obtained results shows that the firefly algorithm allows to obtain quality solutions within appropriate time.

6 Conclusion

The paper considers the 2D packing problem. To solve given problem authors developed the bioinspired algorithm based on the firefly optimization method, which can receive sets of quazi-optimal solutions within polynomial time. Also there were created a software to carry out sets of computational experiments on test examples (benchmarks). As a result, there were estimate the effectiveness of the developed algorithm, as well as its recommended settings for obtaining optimal solutions.

References

1. Alpert, C.J., Dinesh, P.M., Sachin, S.S.: Handbook of Algorithms for Physical design Automation. Auerbach Publications Taylor & Francis Group, USA (2009)
2. Zeng, J., Yu, H.: A study of graph partitioning schemes for parallel graph community detection. Parallel Comput. **58**, 131–139 (2016)
3. Kureichik, V., Zaruba, D., Kureichik Jr., V.: Hybrid approach for graph partitioning. Adv. Intell. Syst. Comput. **573**, 64–73 (2017)
4. Kureichik, V., Kureichik Jr., V., Zaruba, D.: Hybrid bioinspired search for schematic design. Adv. Intell. Syst. Comput. **451**, 249–255 (2016)
5. Kureichik, V.V., Zaporozhets, D.Y., Zaruba, D.V.: Partitioning of VLSI fragments based on the model of glowworm's behavior. In: Proceedings of the 19th International Conference on Soft Computing and Measurements, SCM 2016, pp. 268–272 (2016). art. no. 7519750
6. Zaporozhets, D.Y., Zaruba, D.V., Kureichik, V.V.: Hybrid bionic algorithms for solving problems of parametric optimization. World Appl. Sci. J. **23**, 1032–1036 (2013)
7. Kureichik, V., Glushchenko, A., Kureichik, V., Orlov, A.: Bioinspired approach for 3D packaging problem. In: Application of Information and Communication Technologies, AICT 2016 - Conference Proceedings (2017)
8. Yang, X.-S., Fong, S., He, X., Deb, S., Zhao, Y.: Swarm intelligence: today and tomorrow. In: Proceedings - 2016 3rd International Conference on Soft Computing and Machine Intelligence, ISCMI 2016, pp. 219–223 (2017). статья № 8057470
9. Qi, X., Zhu, S., Zhang, H.: A hybrid firefly algorithm. In: Proceedings of 2017 IEEE 2nd Advanced Information Technology, Electronic and Automation Control Conference, IAEAC 2017, pp. 287–291 (2017). статья № 8054023
10. Tong, N., Fu, Q., Zhong, C., Wang, P.: A multi-group firefly algorithm for numerical optimization. J. Phys. Conf. Ser. **887**(1) (2017)
11. Wang, H., Cui, Z., Sun, H., Rahnamayan, S., Yang, X.-S.: Randomly attracted firefly algorithm with neighborhood search and dynamic parameter adjustment mechanism. Soft. Comput. **21**(18), 5325–5339 (2017)
12. Kacprzyk, J., Kureichik, V.M., Malioukov, S.P., Kureichik, V.V., Malioukov, A.S.: Experimental investigation of algorithms developed. In: Studies in Computational Intelligence, vol. 212, pp. 211–223, 227–236 (2009)
13. http://vlsicad.ucsd.edu/UCLAWeb/cheese/ispd98.html

Authorship Identification Using Random Projections

Robertas Damaševičius[1](✉), Jurgita Kapočiūtė-Dzikienė[2], and Marcin Woźniak[3]

[1] Department of Software Engineering, Kaunas University of Technology, Kaunas, Lithuania
robertas.damasevicius@ktu.lt
[2] Faculty of Informatics, Vytautas Magnus University, Kaunas, Lithuania
[3] Institute of Mathematics, Silesian University of Technology, Gliwice, Poland

Abstract. The paper describes the results of experiments in applying the Random Projection (RP) method for authorship identification of online texts. We propose using RP for feature dimensionality reduction to low-dimensional feature subspace combined with probability density function (PDF) estimation for identification of the features of each author. In our experiments, we use the dataset of Internet comments posted on a web news site in Lithuanian language, and we have achieved 92% accuracy of author identification.

Keywords: Author identification · Text mining · Digital text forensics Random projections

1 Introduction

Internet is a conducive platform for spreading anonymous opinions, alternative facts and even fake news [1]. In many cases, it is impossible to trace the original author of such messages. Consequently, the authorship analysis of anonymous (or pseudonymous) online texts (such as SMSs, Facebook posts, Twitter tweets, emails and web forum comments) has attracted an increased interest in the cyber forensics and text mining research communities.

Digital forensic analysis of textual documents and messages to tackle the anonymity problem is called authorship analysis [2]. It requires performing the statistical analysis of syntactical and linguistic (stylometric) features of texts on order to assign them to suspected authors. The problem may be reduced to classification of texts into classes based on their authorship. The applications of authorship analysis are various and include fraud detection [2], spam email filtering [3], fight against cyber bullying [4], and plagiarism detection [5].

Authorship identification as a concept is related to the task of plagiarism detection [6], and is often used synonymously to author verification [7], authorship attribution [8], author unmasking [9], and author recognition [10]. The domain of author identification is based on the assumption that each author produces texts with his/her own unique style due to an unconscious use of words and language grammar. Therefore, by

R. Silhavy (Ed.): CSOC 2018, AISC 764, pp. 47–56, 2019.
https://doi.org/10.1007/978-3-319-91189-2_6

measuring quantitative characteristics of words and grammar use in the texts, the author of any text can be deduced explicitly.

The main challenge in authorship analysis and correct author identification is to learning the stylistic characteristics typical for different authors with a sufficiently high accuracy to make a valid decision [3]. Different indicators of the individual authorship have been identified in the literature, including individual peculiarities (lexical richness) of vocabulary and dictionary usage, the use of unusual words or syntactic language constructs and stylistic and sub-stylistic features. In many cases, such language features are used sub-consciously and can not be falsified at will [11].

The determination of authorship of short online texts is even more challenging as compared to the more longer and formal text documents such as novels or poems, because the effectiveness of some language based metrics very much depend upon the number of words in the text, there can be a disproportional number of grammatical errors, some Internet communities use their own dialects (jargon), and the style of the author can vary significantly depending upon the intended recipient of messages [12].

Text classification is the task to automatically classify text documents into a number of predefined classes. When a typical workflow of classification is used, first, text documents are represented as feature vectors where feature values may encode syntactical and stylistic characteristics of the documents such as the presence of words, their parts, word n-grams, etc. in the document. To make documents of different lengths comparable, each feature vector is typically normalized by dividing each feature value by text length or other normalizing quantity. The resulting feature vectors are high-dimensional but usually sparse thus raising the need for feature dimensionality reduction. A review of authorship attribution approaches specifically applied for texts written in the Lithuanian language has been presented in [13]. Next, the classifier is trained using a set of labelled training documents, and then the classifier is ready to be used for classifying new text documents with unknown labels. However, in case of supervised classification, a large number of labelled training examples is needed to build a sufficiently accurate classifier. Preparation of training dataset is usually performed manually, which is time consuming and labour intensive [14].

Other challenges of text classification include dataset imbalance issues such as feature space heterogeneity (uneven distributions of features for classes) [15], class overlapping [16], and disjoint data sets [17]. These difficulties reduce the effectiveness of training a conventional classifier model for authorship identification, because the trained classifiers are influenced negatively by complexity of feature distribution. The infamous ‘curse of dimensionality’ setts limits on the analysis of big and rich datasets that are generated by mining text documents. While classification in itself generally is fast, the training process is often painstakingly slow, as the complexity of training algorithms is high and their do not scale well with big data [18]. As cyber security of state often requires that internet is monitored live, the task may become infeasible. Therefore, in many cases a dimensionality reduction to lower-dimensional feature subspace is required as a pre-requisite before further analysis can be performed.

Authorship identification of texts written in a national (i.e., other than English) language can add additional problems (unique national letters, syntactical and grammatical features, etc.) to the authorship recognition tasks [19]. For example, Lithuanian language alphabet, additionally to common Latin alphabet letters, has 18 extra letters with diacritics (9 capital and 9 small), some of which are unique (e.g., there is no other alphabet with the Ė (ė) letter). These letters are included in several 8-bit single-byte coded character sets (such as ISO/IEC 8859-13). Some symbols have encoding issues due to different encoding standards used on different platforms. Text pre-processing and classification with incorrect encoding may lead to incorrect results.

The novelty of this paper is the adoption of the Random Projections (RP) method for feature selection for authorship attribution task. The organization of the paper is as follows. We analyse the application of the RP method for classification of texts in Sect. 2. The results of experiments are presented in Sect. 3. Finally, we formulate the conclusions in Sect. 4.

2 Method

Random Projection (RP) is a simple data dimensionality reduction technique that projects the original highly-dimensional data to a lower-dimensional data subspace using a specifically constructed random matrix [20]. It has been successfully applied to many machine learning tasks, including classification [21], clustering [22], regression [23], and information retrieval [24].

RP differs from other data dimension reduction methods, such as Principal Component Analysis (PCA), as it does not pre-define any criterion such as variance for deriving projections. A random projection from n dimensions to m dimensions is a linear transformation represented by a random $n \times m$ matrix R, which can be generated as proposed by Achlioptas [25] as follows:

$$r_{ij} = \begin{cases} +1, & probability 1/6 \\ 0, & probability 2/3 \\ -1, & probability 1/6 \end{cases} \tag{1}$$

Given a n-dimensional data set represented as an $p \times n$ matrix X, where p is the number of data points in X, the mapping $X \times R$ results in a reduced dimension dataset X' represented as an $p \times m$ matrix. If $m < n$, then the original feature dimensionality is reduced.

The theoretical justification of RP has been provided by the Johnson-Lindenstrauss Lemma [26], which guarantees that Euclidean inter-point distances in the reduced dimension random subspace, derived via linear transformation from the original higher-dimensional space, are approximately preserved. Each projection captures a linear yet randomly weighted combination of all original features and in this way preserves information present in the original data.

The third step in the application of RP for authorship attribution is feature selection. As the dimensionality of the projected space is low, we can consider the projected low-dimensional observations as a set of random and mutually independent variables for which we can estimate the probability density function (PDF) by using the kernel density estimation (KDE) method [27].

If x_1, x_2, …, x_N is a sample of some random variable, then its PDF can be approximated as follows:

$$\hat{f}_h(x) = \frac{1}{Nh} K\left(\frac{x - x_i}{h}\right) \tag{2}$$

here K is some kernel function, and h is its smoothing parameter. Here K is taken to be a standard Gaussian function of a feature with mean zero and variance 1 as follows:

$$K(x) = \frac{1}{\sqrt{2\pi}} e^{-\frac{1}{2}x^2} \tag{3}$$

For a two dimensional case, we can calculate the bi-variate PDF as a product of uni-variate PDFs as follows:

$$\hat{f}(x, y) = \hat{f}(x) \cdot \hat{f}(y) \tag{4}$$

here x and y are data in each dimension, respectively.

As RP essentially is a stochastic method, each time it produces a different projection of the original data points, therefore, revealing only a part of the higher-dimensional data manifold. So there is no a priori way of knowing what information will be captured by a particular random projection, but by taking several random projections and by defining some king of a metric for evaluating each projection, with high probability a 'better' projection can derived [28], which captures more information from the original high-dimensional data. In case of the classification problem, we are interested in a mapping that more efficiently separates data points belonging to different classes.

In this paper, as a criterion for estimating the 'quality' mapping, we use the Jaccard distance metric between the PDFs of data points representing each class (i.e., an author). Jaccard distance is easily adaptable to multidimensional feature spaces where compared points belong to different classes. Therefore, this type of methodology fits well to the multidimensional text features in the domain of authorship identification.

The Jaccard distance, which measures dissimilarity between sets, is complementary to the Jaccard similarity coefficient and can be obtained by subtracting the Jaccard coefficient from 1, or, equivalently, by dividing the difference of the sizes of the union and the intersection of two sets by the size of the their union as follows:

$$d_J(A, B) = 1 - J(A, B) = \frac{|A \cup B| - |A \cap B|}{|A \cup B|} \tag{5}$$

For feature selection, the χ^2 metric is used as a measure for feature selection. χ^2 measures the correlation between the feature and the authorship class (label). Only features whose χ^2 value is more than the pre-defined threshold value can be considered further. The relevance of feature t with the author set c is calculated as follows:

$$\chi^2(t, c) = \frac{N(AD - BC)^2}{(A + C)(B + D)(A + B)(C + D)} \tag{6}$$

here A is the number of times both feature t and author set c exists, B is the number of times feature t exists, but author set c does not exist, C is the number of times feature t does not exist, but author set c exists, D is the number of times both feature t and author set c does not exist, and N be the number of the training samples [29].

3 Experiments

3.1 Dataset

The datasets consists of the Internet comments collected from www.delfi.lt news website (topics "in Lithuania" and "Abroad") in January 2015–August 2015 [30]. These comments were uploaded by anonymous readers expressing their opinions about news articles. The composed corpus contains text messages of 1,000 authors.

When assigning authorship labels, we followed an assumption that identity of author can be revealed, if his/her texts are written under the same unique IP address and the same unique pseudonym (taking both together as a single unit). None of the authors were included: if the same pseudonym was used for writing from different IP addresses (dynamic IP problem), or if different pseudonyms were used while using the same computer connected into the same computer network (i.e., using the same IP address).

Dataset as cleaned using the following rules: Replies, meta-information and non-Lithuanian alphabetic letters (except punctuation marks) were filtered out; no texts shorter than 30 symbols (excluding white-spaces); and texts with out-of-vocabulary words, missing diacritics, etc. (spoken non-normative Lithuanian language). As some users may be writing texts omitting Lithuanian language letters (nasal letters and other letters with diacritics), such texts were also discarded from the text dataset.

The descriptive statistics of the text dataset is presented in Table 1.

Table 1. Descriptive statistics of text corpus

Number of authors	Number of texts	Number of words	Avg. text length, words	Random baseline	Majority baseline
10	14,443	289,462	20.042	0.001	0.018
100	63,131	1,511,823	23.947	0.002	0.018
1000	155,078	4,068,231	26.233	0.003	0.018

We compare our results with the random baseline and the majority baseline. The random baseline is calculated as follows:

$$\sum_j P^2(c_j) \tag{7}$$

The majority baseline is calculated as follows:

$$\max(P(c_j)) \tag{8}$$

here $P(c_j)$ is the probability of that some texts written by particular author c_j.

These baselines show the lowest accuracy threshold, which must be exceeded to claim that applied method is effective for the authorship identification task.

3.2 Features

We have analysed and performed experiments with the following Lithuanian-language specific lexical features: ratio of function words, cumulative ratio of all function words, ratio of stop word, cumulative ratio of all stop words, ratio of specific word ending, ratio of bigrams uncommon to Lithuanian language, cumulative ratio of all uncommon bigrams, ratio of words with Lithuanian prefix ‘ne’ (a Lithuanian prefix for negation), ratio of each Lithuanian specific letter, cumulative ratio of all Lithuanian specific letters, ratio of specific abbreviations, cumulative ratio of all specific abbreviation, ratio of each Lithuanian simile, and cumulative ratio of all similes [31].

The types of features used are summarized (and examples given) in Table 2.

Table 2. Lithuanian lexical features

No.	Feature types	Examples of features and description
1	Function words	Examples: ant, apie, ar, arba, aš, be Ratio of each Lithuanian function word
2	All function words	Cumulative ratio of all Lithuanian function words
3	Stop words	Examples: į, įkypai, įstrižai, šįjį, šalia, še, šiąją Ratio of each Lithuanian stop word
4	All stop words	Cumulative ratio of all Lithuanian stop words
5	Word endings	Examples: a, ai, ajam, ame, ams, ant Ratio of each Lithuanian specific word ending
6	Uncommon bigrams	Examples: qu, sh, zh, ch, ux, xu Ratio of each bigram uncommon to Lithuanian language
7	All uncommon bigrams	Cumulative ratio of all uncommon bigrams
8	Prefix ‘ne’	Ratio of words with prefix ‘ne’
9	Letters	Examples: ą, č, ę, ė, į, š Ratio of each Lithuanian language specific letter
10	All letters	Cumulative ratio of all Lithuanian language specific letters
11	Abbreviations	Examples: doc., dr., gyd., kun., tūkst., vyr. Ratio of each Lithuanian specific abbreviation
12	All abbreviations	Cumulative ratio of all Lithuanian specific abbreviation
13	Similes	Examples: pavyzdžiui, kaip, lyg, tarkim, tarytum Ratio of each Lithuanian simile
14	All similes	Cumulative ratio of all Lithuanian similes

3.3 Results

We selected the target number of dimensions as 2, and repeated random projection procedure 100 times during the "training stage" (we used 80% of texts). The best random projection was selected using the best separation criterion by Jaccard distance. The accuracy of the method was calculated during the "testing stage" (using 20% texts). The separation of texts into training and testing tests was not entirely random, as we had to ensure that for each author in the training stage there was a sample of text available in the testing stage and vice versa. The author assignment decision was made by selecting an author, for whom the PDF value was the largest at a given point in a feature space representing the particular text message.

To analyse the impact of the length of text message on the accuracy of author identification, we have introduced a threshold of minimal length of a text (in characters), which were used to construct the training and testing datasets. For example, the dataset with text messages of at least 2000 characters long consisted of 55 texts written by 8 different authors with an average length of 2942 characters.

For evaluation of the method, the 10-fold cross-validation was used. The results of our experiments are given in Table 3. We can see that the accuracy of author identification increases, if longer texts are available for analysis. The best results are achieved with the texts of at least 1800 characters long ($0.92 \div 0.03$), while longer texts show no considerable improvement in accuracy.

Table 3. Summary of results

Min. length, chars	Avg. length	No. of authors	No. of texts	Min. accuracy	Avg. accuracy	Std.	Max. accuracy
2000	2942	8	55	0.8545	0.8989	0.0251	0.9636
1800	2825	8	62	0.7903	**0.9229**	0.0364	**0.9677**
1600	2677	9	71	0.7446	0.8742	0.0440	0.9437
1400	2519	10	82	0.8049	0.8666	0.0327	0.9390
1200	2320	14	98	0.5306	0.7361	0.1034	0.8673
1000	2053	22	125	0.4080	0.5314	0.1040	0.7200

The results are also summarized graphically in Fig. 1.

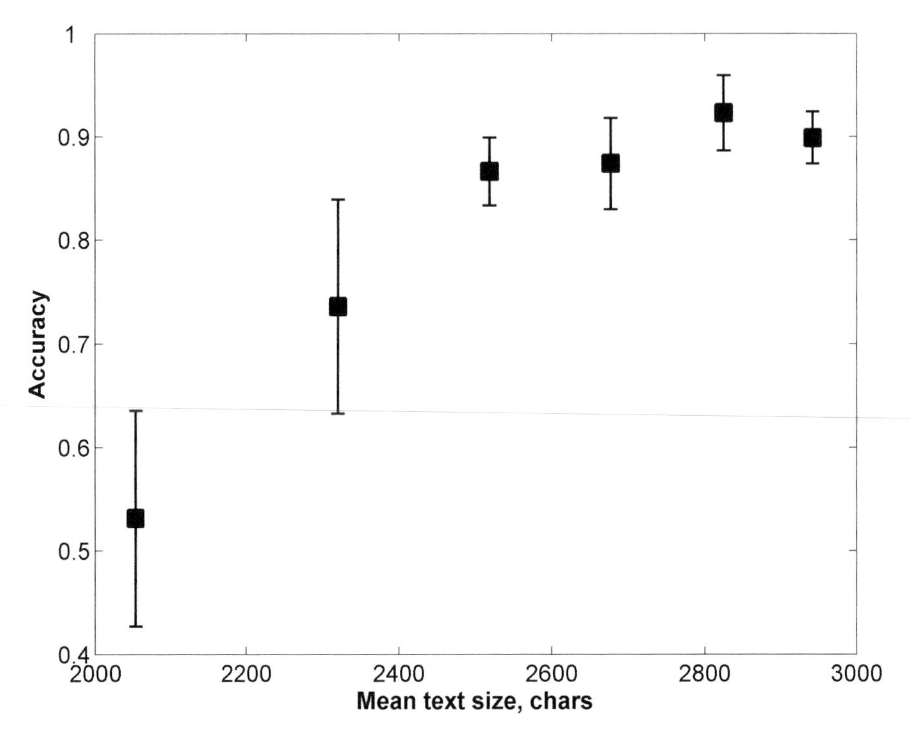

Fig. 1. Authorship identification results

4 Conclusions

We have demonstrated the use of Random Projections (RP) as a fast and effective method for feature dimensionality reduction in the text mining domain. In conjunction with classification methods, it can be used to enable fast and accurate authorship attribution.

In this paper, we report the authorship identification results on the dataset of online Lithuanian language texts (online news website comments) obtained using a set of Lithuanian-language specific features. The dimensionality of feature vectors was reduced using the RP method. Then, we have fitted the probability density function (PDF) of the low-dimensional feature values representing the texts of each author and selected the best random projection, which ensures the best separation of authors in the low dimensional feature subspace.

The results of experiments performed with the texts of different length showed that the best accuracy (92%) was achieved using the texts of at least 1800 characters of length (2825 characters on average).

The limitations of our study is as follows. While the intended use of our method is the identification of authorship of online posts or comments, usually these comments are short. However, we have achieved our top results with the comments, which are quite long (1800 of characters equals to about one (1) page of text typed on a standard A4 page layout).

In future research, we plan to use a large dataset with a larger number of authors available, to analyse different domains (e.g. blogs, tweets, etc.) of online texts as well as texts written in different languages.

References

1. Bean, J.: The medium is the fake news. Interactions **24**(3), 24–25 (2017)
2. Iqbal, F., Binsalleeh, H., Fung, B.C.M., Debbabi, M.: A unified data mining solution for authorship analysis in anonymous textual communications. Inf. Sci. **231**, 98–112 (2013)
3. de Vel, O., Anderson, A., Corney, M., Mohay, G.: Mining e-mail content for author identification forensics. SIGMOD Rec. **30**(4), 55–64 (2001)
4. Pillay, S.R., Solorio, T.: Authorship attribution of web forum posts. In: Proceedings of the eCrime Researchers Summit (eCrime), pp. 1–7 (2010)
5. Potthast, M., Stein, B., Barrón, A., Rosso, P.: An evaluation framework for plagiarism detection. In: Proceedings of the 23rd International Conference on Computational Linguistics (COLING 2010), pp. 997–1005 (2010)
6. Stein, B., Nedim Lipka, N., Prettenhofer, P.: Intrinsic plagiarism analysis. Lang. Resour. Eval. **45**(1), 63–82 (2011)
7. van Dam, M., Hauff, C.: Large-scale author verification: temporal and topical influences. In: 37th International ACM SIGIR Conference on Research & Development in Information Retrieval (SIGIR 2014), pp. 1039–1042 (2014)
8. Sanderson, C., Guenter, S.: Short text authorship attribution via sequence kernels, Markov chains and author unmasking: an investigation. In: Proceedings of the 2006 Conference on Empirical Methods in Natural Language Processing (EMNLP 2006), pp. 482–491 (2006)
9. Kestemont, M., Luyckx, K., Daelemans, W., Crombez, T.: Cross-genre authorship verification using unmasking. Engl. Stud. **93**(3), 340–356 (2012)
10. Clark, J.H., Hannon, C.J.: A classifier system for author recognition using synonym-based features. In: Mexican International Conference on Advances in Artificial Intelligence, MICAI 2007. LNCS, vol. 4827, pp. 839–849. Springer (2007)
11. Luyckx, K., Daelemans, W.: Authorship attribution and verification with many authors and limited data. In: 22nd International Conference on Computational Linguistics, COLING 2008, vol. 1, pp. 513–520 (2008)
12. Sukhoparov, M.E.: Mechanism of establishing authorship of short messages posted by users of internet portals by methods of mathematical linguistics. Aut. Control Comp. Sci. **49**, 813–819 (2015)
13. Kapociute-Dzikiene, J., Venckauskas, A., Damasevicius, R.: A comparison of authorship attribution approaches applied on the Lithuanian language. In: Federated Conference on Computer Science and Information Systems, FedCSIS 2017, pp. 347–351 (2017)
14. Nagy, T.I., Farkas, R., Csirik, J.: On positive and unlabeled learning for text classification. In: Proceedings of the 14th International Conference on Text, Speech and Dialogue (TSD 2011), pp. 2019–226 (2011)
15. Wang, Y.: An incremental classification algorithm for mining data with feature space heterogeneity. Math. Probl. Eng. **2014**, art. 327142, 9 p. (2014)
16. Prati, R.C., Batista, G.E.A.P.A., Monard, M.C.: Class imbalances versus class overlapping: an analysis of a learning system behavior. In: Mexican International Conference on Advances in Artificial Intelligence, MICAI 2004. LNCS, vol. 2972, pp. 312–321 (2004)
17. He, H., Garcia, E.A.: Learning from imbalanced data. IEEE Trans. Knowl. Data Eng. **21**(9), 1263–1284 (2009)

18. Lim, T.S., Loh, W.Y., Shih, Y.S.: A comparison of prediction accuracy, complexity, and training time of thirty-three old and new classification algorithms. Mach. Learn. **40**, 203 (2000)
19. Venckauskas, A., Damasevicius, R., Marcinkevicius, R., Karpavicius, A.: Problems of authorship identification of the national language electronic discourse. In: 21st International Conference on Information and Software Technologies - ICIST 2015. CCIS, vol. 538, pp. 415–432. Springer (2015)
20. Bingham, E., Mannila, H.: Random projection in dimensionality reduction: applications to image and text data. In: 7th ACM SIGKDD International Conference on Knowledge Discovery and Data Mining (KDD 2001), pp. 245–250 (2001)
21. Fradkin, D., Madigan, D.: Experiments with random projections for machine learning. In: 9th ACM SIGKDD International Conference on Knowledge Discovery and Data Mining, pp. 517–522 (2003)
22. Carraher, L.A., Wilsey, P.A., Moitra, A., Dey, S.: Random projection clustering on streaming data. In: 2016 IEEE 16th International Conference on Data Mining Workshops (ICDMW), pp. 708–715 (2016)
23. Thanei, G.A., Heinze, C., Meinshausen, N.: Random projections for large-scale regression. In: Ahmed, S. (ed.) Big and Complex Data Analysis. Contributions to Statistics, pp. 51–68. Springer, Cham (2017)
24. Oh'uchi, H., Miura, T., Shioya, I.: Retrieval for text stream by random projection. In: International Conference on Information Systems Technology and its Applications (ISTA), pp. 151–164 (2004)
25. Achlioptas, D.: Database-friendly random projections. In: 20th ACM SIGMOD-SIGACT-SIGART Symposium on Principles of Database Systems, PODS 2001, p. 274 (2001)
26. Matoušek, J.: On variants of the Johnson-Lindenstrauss lemma. Random Struct. Alg. **33**, 142–156 (2008)
27. Parzen, E.: On estimation of a probability density function and mode. Ann. Math. Stat. **33**(3), 1065 (1962)
28. Palmer, A.D., Bunch, J., Styles, I.B.: The use of random projections for the analysis of mass spectrometry imaging data. J. Am. Soc. Mass Spectrom. **26**, 315–322 (2015)
29. Naga Prasad, S., Narsimha, V.B., Vijayapal Reddy, P., Vinaya Babu, A.: Influence of lexical, syntactic and structural features and their combination on authorship attribution for Telugu text. Proced. Comput. Sci. **48**, 58–64 (2015). International Conference on Computer, Communication and Convergence (ICCC 2015)
30. Kapociute-Dzikiene, J., Utka, A., Sarkute, L.: Authorship attribution of Internet comments with thousand candidate authors. In: 21st International Conference on Information and Software Technologies, ICIST 2015. CCIS, vol. 538, pp. 433–448. Springer (2015)
31. Venckauskas, A., Karpavicius, A., Damasevicius, R., Marcinkevicius, R., Kapociute-Dzikiene, J., Napoli, C.: Open class authorship attribution of Lithuanian Internet comments using one-class classifier. In: Federated Conference on Computer Science and Information Systems, FedCSIS 2017, pp. 373–382 (2017)

A Method for Intelligent Quality Assessment of a Gearbox Using Antipatterns and Convolutional Neural Networks

Andrzej Tuchołka(✉), Maciej Majewski, Wojciech Kacalak, and Zbigniew Budniak

Faculty of Mechanical Engineering, Koszalin University of Technology, Raclawicka 15-17, 75-620 Koszalin, Poland
{andrzej.tucholka,maciej.majewski,wojciech.kacalak, zbigniew.budniak}@tu.koszalin.pl
http://wm.tu.koszalin.pl

Abstract. Taking gearbox as a reference structure, authors apply a method for grading the quality of mechanical structures using a convolutional neural network trained with antipatterns found in gearbox constructions. Antipatterns are used as a quality reference embodied in a neural network, which is used for classifying tested structures to match the antipatterns taught to it.

The measure of similarity to antipatterns (used for training and abstracted by the neural network) is interpreted as the quality measure and so the inversed sum of similarities to each of the antipattern classes used in training is considered a quantitative grade of quality.

Such grading enables automated cross-comparison of structures based on their quality (defined as differentiation from used antipatterns).

Keywords: Antipattern · Structure quality · Convnet

1 Introduction

Mechanical Engineering has been built upon mathematical models enabling design of technically advanced solutions. Recent emergence of novel algorithms (i.e. Convolutional Neural Network) created an opportunity to increase the speed with which the evaluation of mechanical structures can be performed. This is achievable by increasing the ability of computers to detect and analyze patterns in structured data.

Quality evaluation methods based on simulations require manual and detailed definition of the design of each of tested structures. Such approach, to be precise, requires detailed definition of the simulated construction and associated simulation models. Physics based simulations of such constructions are widespread, but even highly optimized ones, due to their inherent complexity, lack in performance and require manual scoping, configuration, and interpretation.

R. Silhavy (Ed.): CSOC 2018, AISC 764, pp. 57–68, 2019.
https://doi.org/10.1007/978-3-319-91189-2_7

To define a quality grade, simulation results have to be compared against a pre-defined quality reference (or at least each other) to define the best among tested ones. Even for trivial design errors, simulation based approach makes it difficult to timely evaluate the quality of a large population of elements (e.g. a population of computer generated mechanical designs of gearboxes).

Convolutional Neural Network (Convnet) is used in many object recognition and classification techniques, and is known for its applications in image recognition. Most notably it is used to explore the structured content of the image [1], but also in text and other structured data classification problems. Proposed method of intelligent quality assessment is aimed to aid machine designers in early detection of reproducible design errors. Still, to define a meaningful quality grade, our method requires a library of incorrect designs (antipatterns) that can be observed in tested constructions.

Comparing the traditional, simulation based approach with our method, we emphasize that our quality results cannot be considered exhaustive, as they do not attempt to predict the behavior of the construction. This means that this method cannot be used for reasoning about the performance of the construction.

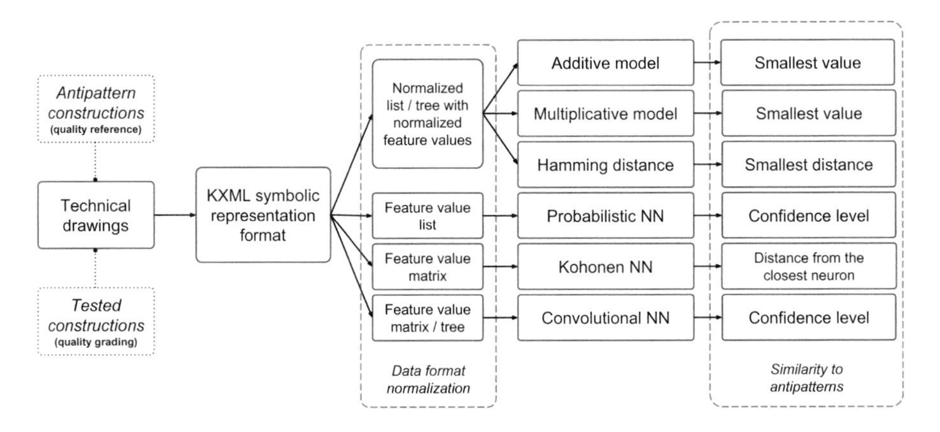

Fig. 1. Overview of the method and applied models for intelligent quality assessment

Testing different numerical models Fig. 1 for such pattern identification, we've found neural networks very useful for working with incomplete datasets (e.g. enabling suggestions during the work on the designs). Additionally, the low computation cost of classifying a structure by a pre-trained model is an attractive addition, compared to simpler iterative models (i.e. Antipattern Matching Factor [2]).

This paper presents a method for quality assessment [3] of a worm gearbox using Convolutional Neural Network as a classification mechanism. A worm gearbox is a construction characterized by multiple features meant to provide casing for the oil, protection, heat dispersion, worm gear positioning, and other. All these features can malfunction in multiple ways, mainly due to design errors. We present a method that can be used to automatically detect and prevent such errors from being added to mechanical designs.

1.1 Antipatterns as a Quality Reference

Antipatterns representing mechanical design knowledge prove themselves useful in enhancing quality related processes [5]. Usage of antipatterns as the training dataset for the network, provides the method with an ability to perform its calculations with regards to concrete quality references. In case of antipatterns, such references are examples of incorrect and repeatable solutions, that could be observed in the class of tested elements (i.e. worm gearbox).

The library of such antipatterns (i.e. gearbox design errors) is used, as the reference enabling quantification of the quality measurements. The similarity of the tested construction to the library of antipatterns can be measured and interpreted as a subjective, quantitative quality measure.

Antipatterns can be identified in features of mechanical constructions (Fig. 2) but also can be a function of the structure of the construction (Fig. 3). In both cases it is possible to provide a symbolic description of such constructions and calculate their similarity to a tested element.

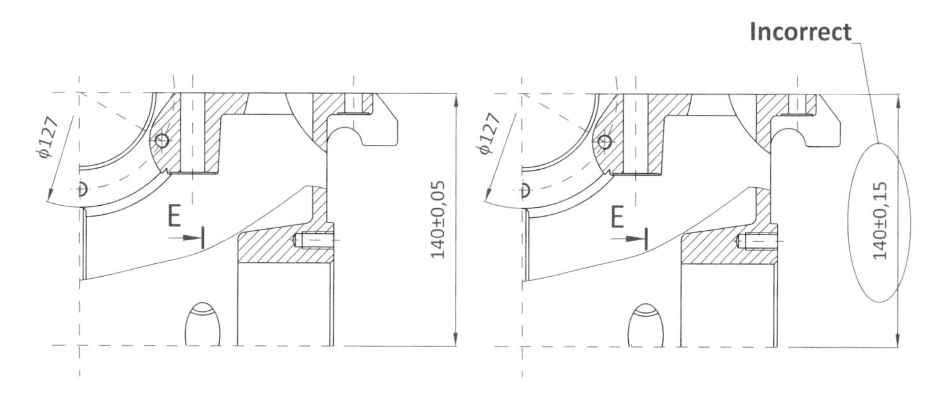

Fig. 2. Antipattern: precision (0,15 mm) of the spread of the axis of the worm and the worm-wheel can't be greater than the acceptable value (0,06 mm)

It is necessary to mention, that due to the nature of neural networks, the calculations done to classify the tested element cannot be directly analyzed for correctness. Furthermore, the classification done by neural networks result in a subjective confidence level rather then a precise quantitative value that can be reasoned upon. Still, upon converging on the training data, the neural network is able to detect structured data patterns.

2 Preparing the Training Dataset

The aim of applying convolutional neural network is to enable detecting similarities in design (i.e. repeating patterns). To allow the network to create an abstraction over the antipatterns used for training, we can either: transform the data into image format, or we can redesign the data transformations, and convolutions to match the symbolic data structure.

The scope of the symbolic representation of antipattern features, in training data, should allow to clearly define the pattern and the incorrect nature of the antipattern (e.g. see Fig. 2). Additionally, to allow the network to properly abstract over the antipattern, the error should be placed in different contexts (i.e. structures of symbolic nodes).

The antipattern visualized on Fig. 2, is referring directly to a feature of the construction being the precision of the spread between the main axis of a gearbox. As designed, the distance between the axis of the worm and the axis of the worm wheel, is 140 mm, but the precision of this dimension vary. In the incorrect case, it is twice the acceptable limit. Such design error in precision, can be easily overlooked, yet it will create an increased tip clearance of the gear and have a negative impact on the kinematic precision of the reduction gear caused by elevated kinematic deviations.

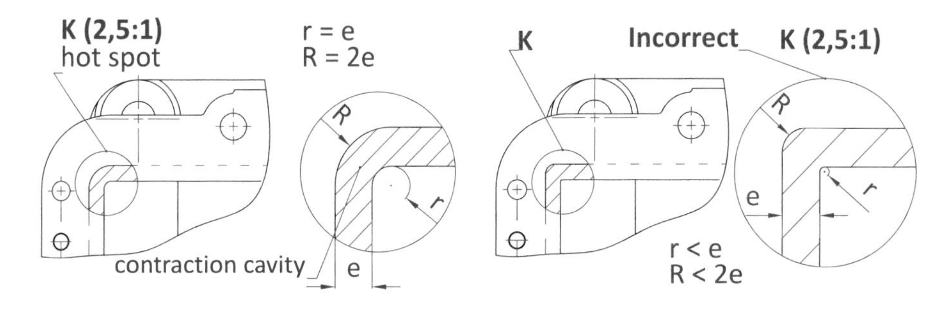

Fig. 3. Local shoulders forming an antipattern design - a hot spot

Another, more complex antipattern (Fig. 3) is reflecting an incorrect relation between curvature of the shoulder and the thickness of the wall on which this shoulder exists.

Having the technical drawings as input can be very convenient, as there are several neural network designs ready for image processing and detecting similarities in embedded objects. Still, the nature of technical drawings adds an additional set of challenges: rotations, varying notation styles and other often implied character of features of designed elements. Furthermore, the technical drawings themselves require additional notation to define the logic of the antipattern (i.e. mathematical on Fig. 3).

2.1 Symbolic Representation of the Antipattern Features of the Worm Gearbox

To represent common antipatterns found in worm gearboxes, we use a symbolic representation [4]. Challenges in normalization of data describing the constructions, come mainly from the character of technical drawings. Such difficulties come from their varying style, visual irregularities that are not the data (e.g. rotation, visibility, or positioning of the dimensions).

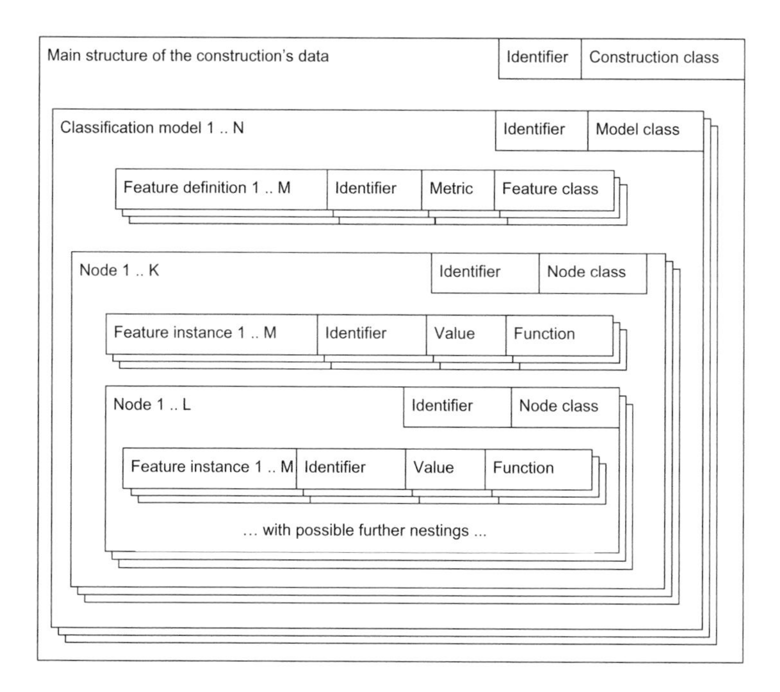

Fig. 4. The structure of KXML symbolic language

Another approach could be to select a specific computer interpretable data format. A notable ones to use could be the Standard for the Exchange of Product model data (ISO 10303) and ECMA-363 Universal 3D File Format. We have found these to be designed mainly for manufacturing and visualization purposes making the data normalization process more complex. It is worth to note, that provided adequate parsing models the method of comparing structures with antipatterns could work with any data format.

To avoid pitfalls of normalizing the data describing constructions, and complexity of existing computer interpretable data formats, we have created a symbolic language (KXML [4] Fig. 4) capable of representing the symbolical structure of a mechanical construction. In essence, provided a technical drawing, we manually decompose the construction into a symbolic, object-oriented, tree-like structure of nodes. Such data format simplifies further processing of the structured data. We format this symbolic representation in XML compatible notation to ease data imports, but the format itself is notation independent.

None of used numerical models allows for object oriented input, hence a natural loss occurring during the data transformation. Using the symbolic language, we can describe complex structures in many ways: using an XML or JSON compatible notations, a specialized declarative language similar to HTML, or in a minimized or symbolical notation. We have defined assumptions for creation of a language, whereby a structural object can be easily divided into constituent objects, and modifiers can be distinguished as elements for altering dimensions, shape, structure, or other properties of processed objects.

To represent the data describing the construction we use four main elements: (1) a classification model - providing the context for the analysis and interpretation of the features and structural nodes; (2) definitions of features - structuring the data used for calculating the properties of the construction; (3) instances of nodes and their structure - allowing for representing the construction decomposed explicitly into a node structure; (4) instances of features - containing values or math functions allowing to use them in the context of the concrete classification model and its position in the structure of nodes.

```
<struct class="gearbox" id="g001">
  <model class="mechanical">
    <feature id="r" unit="mm" />
    <feature id="R" unit="mm" />
    <feature id="e" unit="mm" />
    <node class="case">
      <node class="wall">
        <feature id="e" value="5" />
        <node class="thermal center">
          <node class="contraction cavity">
            <feature id="r" value="1" />
            <feature id="R" value="2" />
          </node>
        </node>
      </node>
    </node>
  </model>
</struct>
```

Listing 1. KXML representation of the antipattern (Fig. 3.)

Relations emerging from the structure of the language, through the properties of those four main elements, allow to represent contextual information between: (A) elementary elements of the construction - through the structure of classification models, nodes, feature definitions and instances; (B) class definitions for features and nodes enabling common reasoning on values enclosed in feature instances through class definition reuse; (C) inherited node and feature definitions allowing for common analysis of represented elements based on their ancestry and common feature scopes; (D) feature instances in different classification models - through the node identity mechanism; (E) indirect numerical relations included in mathematical functions relating feature values to each other.

Besides increased verbosity that simplifies the integration of the data, proposed KXML language, (in general a computer readable format makes) makes it easier to generate large populations of examples of a specific antipattern. Larger amounts of samples are required for neural networks to be able to properly abstract over the features they will classify.

For the neural network to converge on the antipattern it is necessary to generate multiple example structures to be used during the training process. Proposed symbolic description makes it also easier to generate large populations of examples of a specific antipattern that is taught to the network. In the antipattern visualized on Fig. 3 and describe on Listing 1, the incorrect nature of the design comes from the fact that both: the inner (r) and outer (R) radius' of the contraction cavity are smaller then the thickness (e) of the gearbox's wall.

2.2 Data Format Normalization

When it comes to creation of technological and organizational processes, object-based description is a natural mapping of such structures. Attempting to process as much information from the KXML symbolic description, we have to include the feature value and class information in the context and structural position of the node class and its nesting. Aiming to avoid the complexity brought by multi-dimensionality of the symbolic description, where each feature class would form one.

The requirements of the numerical models are much more rigid, mostly requiring a low-dimensional input. The additive (i.e. Antipattern matching factor) and multiplicative models we've tested require one-dimensional input, and providing a very basic ability to process tree-like datasets. Such ability is not allowing for differentiation between tree's nodes and hence becomes useless if the input data's structure is not static (impractical for real life structures).

The neural networks allow for more complex input structure, still due to computational complexity preferring a low number of dimensions. Introducing any dimension reducing model (e.g. Principal Component Analysis) would blur the relationship between the data even further then a normal neural network design. Aiming to preserve as much data as possible, we're using data matrices with lexically normalized structure.

3 Convnet Based Neural Network Design

Aligning the design of the algorithm used for antipattern detection, we take advantage of two main features of convolutional networks: the location invariance and compositionality. Location invariance enables detection of patterns in any location (position) of the symbolic data representation, while compositionality enables combining data patterns and non-direct data relations. Availability of such features in the numerical method used for data classification is a clear advantage compared to simple additive, multiplicative or deterministic models proposed earlier [3].

Application of the Convnet based network design (for detection of antipatterns in a symbolic representation of a mechanical construction), rather then analyzing image pixel or text input, takes the matrix of feature data created from the KXML symbolic description of the mechanical construction. This approach builds upon the Convnet designs for text classification, but requires defining a normalization technique that will preserve at least some of the structural information contained in KXML.

Flattening out internal tree data structure, the two dimensional input matrix contains a map of feature values of each node in the antipattern structure (Listing 1).

$$Antipattern1_{m,n} = \begin{bmatrix} 0\,0\,0 \\ 5\,0\,0 \\ 0\,0\,0 \\ 0\,1\,2 \end{bmatrix}$$

Such input representation of the data, given adequate kernel size and normalized columns, could enable detecting patterns in feature values. The incorrect nature of the antipattern (the values of r and R being too small compared to thickness of the wall e. Compared to the symbolic representation, the above matrix representation is omitting the dependencies between parts of the construction. Processing larger matrices (real sized designs), this pattern would be visible to the network due to proximity of the feature values (within kernel size).

Attempting to include more information from the KXML symbolic description, we have to include the feature value and class information in the context of the node class and its nesting. To avoid creation of additional dimensions, we can embed such information in a more complex matrix. Here, in addition to the feature values (columns 5–7), the matrix is expanded with class of the node (1st column) and node's nesting level (columns 2–4).

$$Antipattern1_{m,n} = \begin{bmatrix} 1\,0\,0\,0\,0\,0\,0 \\ 0\,2\,0\,0\,5\,0\,0 \\ 0\,0\,3\,0\,0\,0\,0 \\ 0\,0\,0\,4\,0\,1\,2 \end{bmatrix}$$

Compared, to rather minimal data matrix above, the full construction contains more nesting levels, and feature definitions are increasing the m (horizontal) dimension of the matrix. Its n (vertical) dimension will be directly related to the amount of nodes in the description. Each row of the above matrix is a node, and having the antipatterns being formed by relations between features of such nodes, we have a one dimensional dataset that can be processed as input.

3.1 Kernel and Filters

Our approach when designing the neural network, visualized on Fig. 5, processes a one-dimensional dataset containing construction's feature data. Due to the changes in the input structure, it is required to define and fine-tune the dimensions of the kernel along with the size of the step and stride with which it will be applied to the dataset. Additionally, data transformations (filters) can be used to highlight related feature columns for easier detection of patterns occurring in the data.

To enable the network to detect relations between them, the dimensions of the kernel have to span across multiple neighboring nodes. Due to the structure of proposed above data matrices, the kernel should contain an arbitrary amount of rows. Its amount should be synchronized with the complexity and abstraction level of antipatterns.

To keep the calculations simple, we have flattened out the data structure, which removes the need to create tree-like kernels. It is possible that such approach could yield better results, as the structural information would be more accessible for the network.

The filters that can be applied on the data, again need to reflect the modified input data structure. In the example above, the first four columns contain structure-related data and the last three - feature data. Hence the structures of the kernel, and the network will work with multi-dimensional data entities. This complexity should be reduced by applying filters. Namely, inspecting relations between features and forming feature-sets, highlighting specific relations between data items. Such filters make it easier for the network to notice and learn the patterns in input data and could be loosely compared to edge detection algorithms used in image recognition.

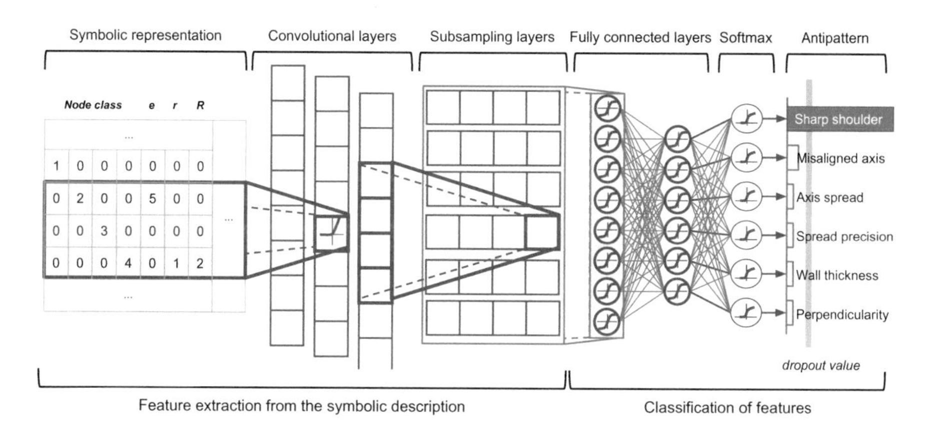

Fig. 5. A conceptual mapping of one-dimensional dataset and the Convnet design

3.2 Quality Assessment with Distance from Antipatterns

The essence of our method of intelligent quality assessment lies in detecting similarity of the tested element to a set of antipatterns defining the quality reference (Fig. 6). In this context, for the design of the classification algorithm to be useful in quality grading, it has to provide a quantitative value representing similarity to the library of the antipatterns rather then a single one.

Detecting just one, most similar antipattern can be useful in practice - for example when advising a designer about a detected error that needs fixing. In case of quality grading, comparing two tested elements requires creation of a common denominator (i.e. a quality reference) which we aim to achieve with a library of antipatterns. The quantifiable values of similarity to antipatterns, as indicated by the output of the network, should consider each type of the antipatterns taught to the network. This approach, by encompassing all of the types of design mistakes provides a value representative for all used antipatterns.

Attempting to find additional features of numerical models enabling structural analysis of construction data we tested a custom designed neural network based on Hamming distance, probabilistic neural networks, and Kohonen maps. Among these three designs, only Kohonen maps allow for processing of structural data - through similarity of coordinates resulting from normalized positions feature instances by their class and node structure. Still, Kohonen maps proved to be challenging in practical applications, as they can't differentiate between antipatterns and require maps to function as standalone classifiers defining similarity to one antipattern at a time.

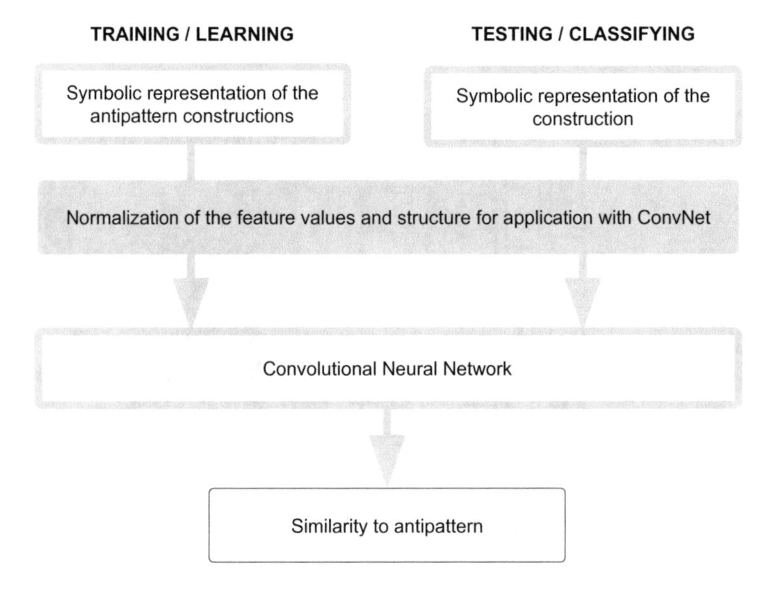

Fig. 6. General, visual representation of the process

When applying Convolutional Neural Network we mimicked the network designs used for text classification and took advantage of ConvNet's two main features: the location invariance and compositionality. The former one enables detection of patterns in any location (position) of the symbolic data representation, while the latter enables combining data patterns and non-direct data relations. Availability of such features in the numerical method used for data classification is a clear advantage compared to other models.

Having a Softmax function at the output, the distribution of results needs to be further processed. First, it is necessary to filter out values with confidence below a certain threshold. Having the Softmax activation distributed equally among all types of antipatterns means, there is no clear similarity to a specific antipattern class. Due to the fact, that Softmax output sums up to 1, it is expected for all of the values of the output matrix to be close to the inverse size of the output data array (M).

All k output values (V) exceeding such value ($\frac{1}{M}$), can be summed up and used to calculate the similarity to all classes of antipatterns used in the training process.

$$A = \sum_{n=1}^{k} V_n \tag{1}$$

Value A stands for similarity to trained dataset, and will always be less then 1. To represent the distance from the antipatterns and use it as a relative (to the library of antipatterns) we simply invert it.

$$Q_{ap} = \frac{1}{A} \tag{2}$$

The resulting quantitative value Q_{ap} can be used for defining a subjective (to the library of antipatterns and network configuration) quality grade. Such grade, beyond comparing designs against each other, provided a sufficient database of antipatterns can act as a rough (due to its non-deterministic nature) quality evaluation.

4 Conclusions

Calculation models, based on neural networks, have created an opportunity to widen the tool-set of mechanical component and assembly designers with automated algorithms preventing common design errors. We observe several data integration and processing challenges that still need to be addressed for the method of automated and intelligent quality assessment to be efficient. The ability to abstract over the structural data contained in antipatterns found in mechanical designs was observed partially in Kohonen maps, and Convolutional Neural Networks.

Compared with simple algebraic models, neural networks require additional configuration, data normalization to be performed adjusting the data and the model to fit each other. The benefits of applying neural networks become barely visible with ConvNet design, due to: local feature detectors (sliding window of convolutions) enabling structural analysis of the construction, and the layered design enabling varying scopes of convolutions on tested data. It is possible to further increase the efficiency of the design by introducing custom kernels and approach to data convolution.

Using Convolutional Neural Network as the classification algorithm in the proposed method of intelligent quality assessment, provides a novel (vs algebraic models) opportunity to detect patterns in the symbolic description of mechanical constructions. Support for detecting antipatterns can be observed in the ability to adapt the kernel and filters to work with the symbolic structure in the form of a list including basic information about the analyzed structure.

On the other hand, the pooling and fully connected layers reduce the ability to detect more complex structures. The limitations of the Convnet based designs (complexity of data normalization, required custom kernels, and filters), make the fitness of Convnet for quality assessment using proposed method promising but limited in terms of depth of analyzed structural complexity.

Algorithmic progress in machine learning, pattern detection and dimensionality reduction methods has enabled the ability for numerical models to build abstractions representing structural concepts of the mechanical construction's design. Such ability presents a unique opportunity to dramatically increase the spread of mechanical design knowledge and automated reduction of common design mistakes.

4.1 Future Research

Recently presented, capsule networks provide a unique design of capsules (i.e. neuron groups) to represent a specific type of an entity such as an object or object part [6]. Considering that the strengths of new neural network designs lie in the ability to detect structured patterns, proposed routing mechanisms will probably further increase the benefits of applying neural networks to detect antipatterns.

References

1. Farabet, C., Couprie, C., Najman, L., LeCun, Y.: Learning hierarchical features for scene labeling. IEEE Trans. Pattern Analy. Mach. Intell. **35**(8), 1915–1929 (2013)
2. Kacalak, W., Majewski, M., Tucholka, A.: A method of object-oriented symbolical description and evaluation of machine elements using antipatterns. J. Mach. Eng. **16**(4), 46–69 (2016)
3. Kacalak, W., Majewski, M., Tucholka, A.: Intelligent assessment of structure correctness using antipatterns. In: International Conference on Computational Science and Computational Intelligence, pp. 559–564. IEEE Xplore Digital Library. IEEE (2015)
4. Tucholka, A., Majewski, M., Kacalak, W.: Zorientowany obiektowo, symboliczny zapis cech, relacji i struktur konstrukcyjnych. Inzynieria Maszyn **20**(1), 112–120 (2015)
5. Kacalak, W., Majewski, M., Budniak, Z.: Intelligent automated design of machine components using antipatterns. In: Jackowski, K., Burduk, R., Walkowiak, K., Wozniak, M., Yin, H. (eds.) Intelligent Data Engineering and Automated Learning. Lecture Notes in Computer Science, vol. 9375, pp. 248–255. Springer, Cham (2015)
6. Sabour, S., Frost, N., Hinton, G.E.: Dynamic Routing Between Capsules. Computer Vision and Pattern Recognition, arXiv:1710.09829 (2017)

Spark-Based Classification Algorithms for Daily Living Activities

Dorin Moldovan(✉), Marcel Antal, Claudia Pop, Adrian Olosutean, Tudor Cioara, Ionut Anghel, and Ioan Salomie

Computer Science Department, Technical University of Cluj-Napoca, Cluj-Napoca, Romania
{dorin.moldovan,marcel.antal,claudia.pop,tudor.cioara,ionut.anghel, ioan.salomie}@cs.utcluj.ro

Abstract. Dementia is an incurable disease that affects a large part of the population of elders and more than 21% of the elders suffering from dementia are exposed to polypharmacy. Moreover, dementia is very correlated with diabetes and high blood pressure. The medication adherence becomes a big challenge that can be approached by analyzing the daily activities of the patients and taking preventive or corrective measures. The weakest link in the pharmacy chain tends to be the patients, especially the patients with cognitive impairments. In this paper we analyze the feasibility of four classification algorithms from the machine learning library of Apache Spark for the prediction of the daily behavior pattern of the patients that suffer from dementia. The algorithms are tested on two datasets from literature that contain data collected from sensors. The best results are obtained when the Random Forest classification algorithm is applied.

Keywords: Machine learning · Classification algorithms
Daily living activities · Sensors · Elderly

1 Introduction

Dementia is still an incurable disease and the management of this syndrome requires both pharmacological and non-pharmacological interventions. The treatment of dementia is a huge challenge because only a few drugs are approved for some forms of dementia. The medication prescription and the medication adherence tend to be a huge challenge in the case of dementia because it can be difficult for the medical professionals to assess the behavioral symptoms of the elderly patients. In addition, due to the cognitive decline, taking the appropriate medication at the right time becomes a challenge for the patients. In MedGUIDE [1] vision, the key to improve the medication process for people in the early stages of dementia is to continuously combine the automated monitoring of the daily living activities of the patients and the data provided by their informal network of caregivers.

R. Silhavy (Ed.): CSOC 2018, AISC 764, pp. 69–78, 2019.
https://doi.org/10.1007/978-3-319-91189-2_8

The project MedGUIDE is an innovative European project with partners from five countries and seven organizations that provides new approaches for supporting the seniors that have dementia with the medication adherence. The sensors and the reports provide information about the ADLs of the seniors. The ADLs that are monitored include nutrition, sleeping patterns, physical activity and movement habits. The Big Data analysis has as objective the minimization of the side effects of the medication by detecting the changes in the patients' routines. The most important challenges addressed by the project are the provision of insight in the actual needs of the elders with dementia, the provision of insight in the actual medication use, side effects and adherence and the provision of support for improving the care and the medication adherence.

The objective of this paper is to address one of the challenges of the MedGUIDE project, namely the analysis of the data that results from the monitoring of the daily living activities of the elders that have dementia in order to help them to adhere to the medication regime. In this context, the paper compares the feasibility of several machine learning classification algorithms for the detection of the ADLs of the elders that have dementia and of their behavioral pattern by analysing data collected through sensors. In order to test the proposed approaches, two datasets from literature were used: the ADLs Binary Sensors Dataset [2] and the DaLiAc Dataset [3].

The rest of the paper is structured as follows: Sect. 2 presents the background, Sect. 3 presents the research approach, Sect. 4 presents how the features are selected, Sect. 5 compares four Spark-based classification algorithms for the identification of the ADLs, Sect. 6 presents the experimental results and Sect. 7 presents the conclusions.

2 Background

A typical Ambient Assisted Living (AAL) system which is activity-based [4] usually consists of three stages. The first stage is represented by the acquisition of the raw-data, the second stage is represented by the processing of the sensor data in order to extract information about the context and the third stage is represented by the learning or the reasoning methods that have as objective the identification of the activities.

The activity recognition using binary sensors is the main research topic of [2]. The task of activity recognition is studied using two different schemes: the Artificial Neural Network (ANN) and the Support Vector Machines (SVM). The framework within which the two learning schemes are modeled is the Hidden Markov Model (HMM). The HMM can be used effectively for the recognition of the ADLs, but a higher complexity degree might be reached when the observable variables which are used in the model are defined using binary values. The proposed HMM is a hybrid one and it combines the discriminative characteristics of the machine learning techniques with the characteristics of the HMM. The training of the HMM is done using the Expectation-Maximization algorithm as in [5]. The results show that the recognition performance is increased when the hybrid models are used instead of the classical methods for activity recognition. The combination of the generative model and of the discriminative model is more accurate than considering either only the generative model or only the discriminative model. The results further demonstrate that the recognition of the ADLs can be achieved very easily by using a set of binary sensors. The SVM HMM hybrid approach is better in terms of performance than the other schemes that are evaluated. The recognition of the ADLs from sensor streams has been studied previously using different models such as the Bayesian Networks (BN) [6], the Evolving Classifiers (EC) [7] and the Conditional Random Field (CRF) [8].

3 Research Approach

The objective of this section is to present a high-level perspective of the research approach used for detecting the changes in the patients' routines in the MedGUIDE project. The first subsection presents the architecture of the prototype used in the experiments, the second subsection describes and compares the datasets used in the experiments and the third subsection enumerates the steps that are followed by the classification algorithms.

3.1 Architecture of the Prototype Used in Experiments

Figure 1 presents a high level view of the architecture of the prototype that is used in the experiments.

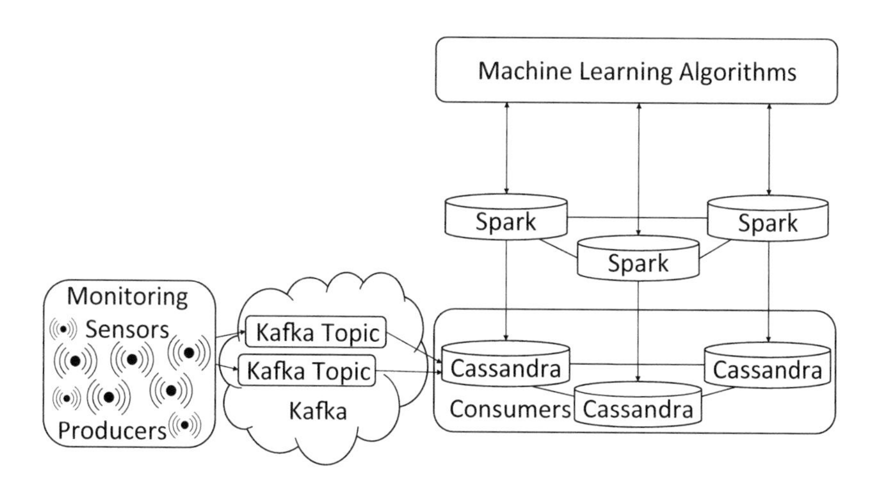

Fig. 1. High level view of the architecture of the prototype used in experiments

The data that comes from the monitoring sensors, in the form of messages, is inserted in Kafka topics. From the Kafka topics, the monitored data is inserted in Cassandra and finally the data is processed using Apache Spark.

3.2 Description of the Datasets Used in Experiments

The research approach proposed in this article uses two datasets that are available online and compared in Table 1:

Table 1. Datasets used in the experiments

	ADLs Binary Sensors Dataset [2]	DaLiAc Dataset [3]
Number of monitored subjects	2	19
Number of monitored activities	10	13
Number of features	2	25
Monitoring duration	35 days	90 min
Sensor types	Pressure, Magnetic, PIR, Flush, Electric	SHIMMER (Shimmer research, Dublin, Ireland)

The first dataset, the ADLs Binary Sensors Dataset, contains information about 2 subjects. The monitored activities include sleeping, breakfast, lunch and so on. The dataset has 2 features, the start time and the end time of the activities and one label, the name of the activity. The second dataset, the DaLiAc Dataset, contains information about 19 subjects. There are 13 activities that are monitored over a period of 90 min. The monitored data is generated by four sensors placed on the left ankle, right wrist, chest and right hip and each sensor has three gyroscope axes.

3.3 Classification Algorithms Steps for the Analysis of the Data

The machine learning steps which are followed in this research paper are presented in Fig. 2.

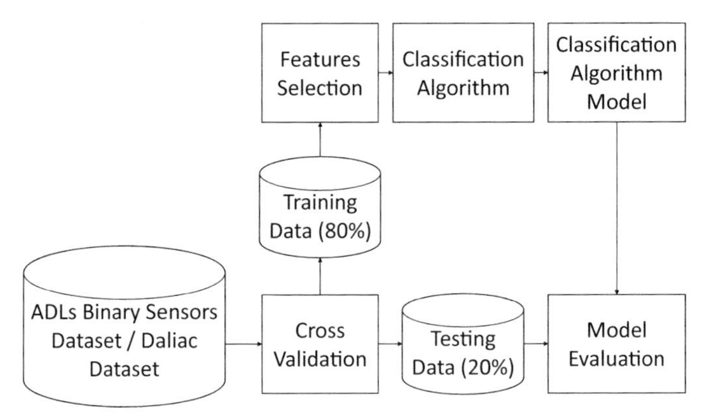

Fig. 2. Classification algorithms steps used in the experiments

The two datasets used in the experiments are the ADLs Binary Sensors Dataset and the DaLiAc Dataset. In the cross validation step the data is split in training data (80%) and testing data (20%). The training data is processed further and is used as input for the features selection step. In this step the features that are the most relevant for the prediction are selected. In the next step several classification algorithms are applied. The output of this step is a classification algorithm model which is further evaluated in the model evaluation phase using the testing data.

4 Features Selection

The first step of the proposed machine learning system is represented by the features selection step. The objective of this step is to select the features that are the most relevant for the prediction. A features selection algorithm takes as input a set of samples that are characterized by a number of features and ranks the features according to a metric. Then it selects a subset of those features that are the most relevant for the prediction of the labels. The features selection algorithm used in this paper is the Random Forest [9]. The Random Forest is an ensemble of decision trees and the results are more accurate when the number of decision trees is larger because it produces an overall estimate of the estimates generated by the decision trees. The Random Forest algorithm applied on the two datasets is the one included in the Caret package from R.

5 Classification Algorithms

The parameters that are tuned are presented in Table 2.

Table 2. Classification algorithms parameters description

Algorithm	Parameter
Logistic Regression	Maximum number of iterations
	Regularization parameter
	Elastic net parameter
Decision Tree	Maximum number of bins
	Subsampling rate
	MaxMemoryInMB
	Impurity
Random Forest	Number of trees
	Maximum depth
	Subsampling rate
	Feature subset strategy
Multilayer Perceptron Classifier	Neural network layers
	Block size
	Seed
	Maximum number of iterations

The objective of this step is to apply four classification algorithms on the two datasets that monitor the daily living activities: the ADLs Binary Sensors Dataset and the DaLiAc Dataset in order to determine the best algorithm in terms of F1 measure, accuracy, precision and prediction. The four algorithms that are applied on the two datasets are: Logistic Regression, Random Forest, Gradient Boosted Trees and Multilayer Perceptron Classifier and they use the implementations from the machine learning package of Apache Spark [10].

6 Experimental Results

This section presents the results obtained after performing the machine learning steps proposed in this paper on the two datasets: the ADLs Binary Sensors Dataset and the DaLiAc Dataset. In the case of the DaLiAc Dataset the number of samples was reduced in the experiments to 2611. The results are evaluated using the F1 measure, the precision, the recall and the accuracy which are defined next:

$$precision = \frac{TP}{TP + FP} \tag{1}$$

$$recall = \frac{TP}{TP + FN} \tag{2}$$

$$accuracy = \frac{TP + TN}{TP + TN + FP + FN} \tag{3}$$

$$F1 = 2 \times \frac{precision \times recall}{precision + recall} \tag{4}$$

These metrics are functions of the elements of the confusion matrix which is presented in Table 3.

Table 3. Confusion matrix

	p' (predicted)	n' (predicted)
p (actual)	True Positive (TP)	False Negative (FN)
n (actual)	False Positive (FP)	True Negative (TN)

6.1 Features Selection

Figure 3 presents the results obtained after applying the Random Forest algorithm on the ADLs Binary Sensors Dataset for the first person. The number of features represented in the graphic is 5 because 3 new features were derived from the 2 initial features: the start time and the end time. The 3 derived features are

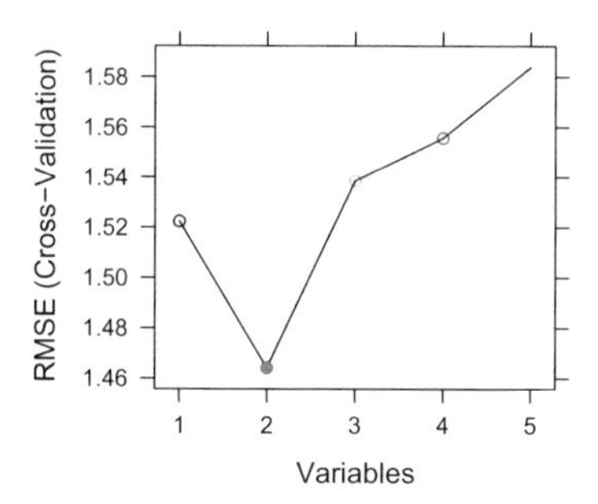

Fig. 3. Features selection for the ADLs Binary Sensors Dataset

the duration, the start hour and the end hour. The algorithm selects as relevant features the duration and the end time of the activities.

The Root Mean Squared Error (RMSE) is a metric for numerical predictions that has the formula:

$$RMSE = \sqrt{\frac{1}{n}\sum_{i=1}^{n}(y_i - \hat{y}_i)^2} \tag{5}$$

where n is the number of samples, y_i's are the labels and $\hat{y}_i$'s are the predictions. The algorithm selects the features with the smallest RMSE.

Figure 4 presents the results generated by the Random Forest algorithm on the DaLiAc Dataset for the first person. From the total number of 24 features, 21 are considered as relevant for the predictions.

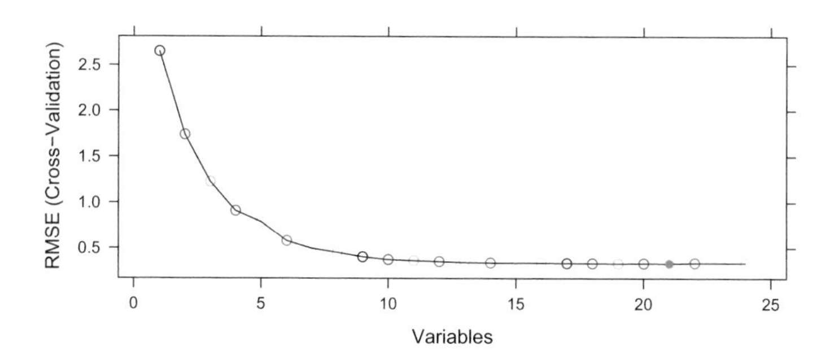

Fig. 4. Features selection for the DaLiAc Dataset

6.2 Classification Algorithms

Figure 5 shows the classification results for the first person from the ADLs Binary Sensors Dataset. Logistic Regression and Multilayer Perceptron Classifier give the worst results while Decision Tree and Random Forest give the best results.

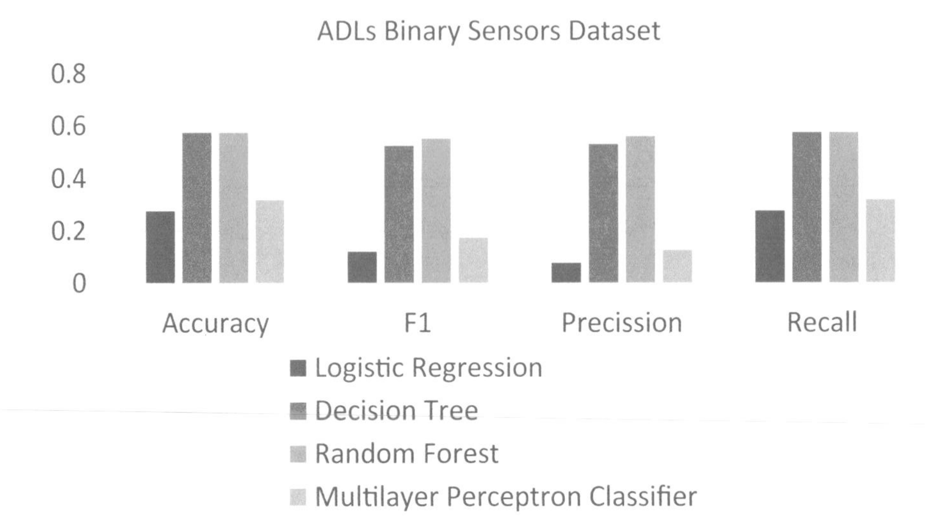

Fig. 5. Classification algorithms for the ADLs Binary Sensors Dataset

Figure 6 shows the classification results for the first person from the DaLiAc Dataset using the same four classification algorithms. In the case of the DaLiAc dataset, only a subset of 2612 samples from the initial number of 245576 samples was used in the experiments.

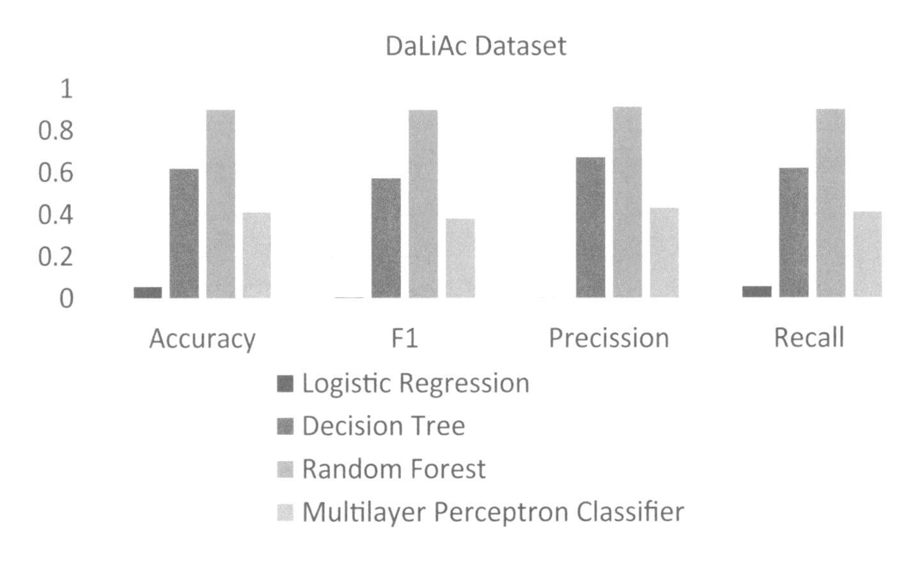

Fig. 6. Classification algorithms for the DaLiAc Dataset

The best values for the evaluation parameters in both cases were obtained when the Random Forest algorithm was used. In the case of the ADLs Binary Sensors Dataset the precision obtained by the Random Forest algorithm is 0.56 and in the case of the DaLiAc Dataset the precision obtained by the same algorithm is 0.91.

6.3 Analysis of the Experimental Results

In the case of the first dataset the best results are obtained when the Decision Tree and the Random Forest algorithms are applied. The similarities in the results between the Decision Tree and the Random Forest can be justified by the fact that both of them are based on decision trees. Also, Logistic Regression and Multilayer Perceptron Classifier give similar results as there are similarities between the Perceptron and the Logistic Regression.

In the case of the second dataset, the Random Forest algorithm behaves much better than the Decision Tree algorithm. This can be justified by the fact that a random forest is a collection of decision trees and thus its overall performance is better. The input data is also much different than the data from the first dataset because it is generated by sensors installed on the subjects and thus the activities are more correlated to specific body movements.

For our prototype the best classification algorithm from the four classification algorithms that were compared is the Random Forest algorithm because the data is monitored using sensors installed on different devices or on the bodies of the subjects and thus it is more similar to the second dataset that was used in the experiments.

7 Conclusions

This paper presented a machine learning approach for the prediction of the ADLs for elders developed in the context of the MedGUIDE AAL project. The approach focused on the batch processing and used two datasets from the literature in the experiments: the ADLs Binary Sensors Dataset and the DaLiAc Dataset. The features were selected using the Random Forest algorithm and the data was classified using Logistic Regression, Decision Tree, Random Forest and Multilayer Perceptron Classifier. The overall best results were obtained when the Random Forest algorithm was used for the classification of the data.

Acknowledgement. This work was supported by a grant of the Romanian National Authority for Scientific Research and Innovation, CCCDI UEFISCDI and of the AAL Programme with co-funding from the European Union's Horizon 2020 research and innovation programme project number AAL 44/2017 within PNCDI III [1].

References

1. MedGUIDE. http://www.aal-europe.eu/projects/medguide/
2. Ordonez, F.J., de Toledo, P., Sanchis, A.: Activity recognition using hybrid generative/discriminative models on home environments using binary sensors. Sensors **13**(5), 5460–5477 (2013). https://doi.org/10.3390/s130505460
3. Leutheuser, H., Schuldhaus, D., Eskofier, B.M.: Hierarchical, multi-sensor based classification of daily life activities: comparison with state-of-the-art algorithms using a benchmark dataset. PLoS ONE **8**(10), 1–11 (2013). https://doi.org/10.1371/journal.pone.0075196
4. Ni, Q., Hernando, A.B.G., de la Cruz, I.P.: The elderly's independent living in smart homes: a characterization of activities and sensing infrastructure survey to facilitate services development. Sensors **15**(5), 11312–11362 (2015). https://doi.org/10.3390/s150511312
5. Espana-Boquera, S., Castro-Bleda, M.J., Gorbe-Moya, J., Zamora-Martinez, F.: Improving offline handwritten text recognition with hybrid HMM/ANN models. IEEE Trans. Pattern Anal. Mach. Intell. **33**, 767–779 (2011). https://doi.org/10.1109/TPAMI.2010.141
6. Nazerfard, E., Cook, D.J.: CRAFFT: an activity prediction model based on Bayesian networks. J. Ambient Intell. Humanized Comput. **6**(2), 193–205 (2015). https://doi.org/10.1007/s12652-014-0219-x
7. Ordonez, J., Iglesias, J.A., de Toledo, P., Ledezma, A., Sanchis, A.: Online activity recognition using evolving classifiers. Expert Syst. Appl. **40**(4), 1248–1255 (2013). https://doi.org/10.1016/j.eswa.2012.08.066
8. Vail, D.L., Veloso, M.M., Lafferty, J.D.: Conditional random fields for activity recognition. In: AAMAS 2007 Proceedings of the 6th International Joint Conference on Autonomous Agents and Multiagent Systems, pp. 1–8 (2007). https://doi.org/10.1145/1329125.1329409
9. Lin, W., Wu, Z., Lin, L., Wen, A., Li. J.: An ensemble random forest algorithm for insurance big data analysis. IEEE Access (2017). https://doi.org/10.1109/ACCESS.2017.2738069
10. Meng, X., Bradley, J., Yavuz, B., Sparks, E., Venkataraman, S., Liu, D., Freeman, J., Tsai, D., Amde, M., Owen, S., Xin, D., Xin, R., Franklin, M.J., Zadeh, R., Zaharia, M., Talwalkar, A.: MLlib: machine learning in apache spark. J. Mach. Learn. Res. **17**(1), 1–7 (2016)

Fast Adaptive Image Binarization Using the Region Based Approach

Hubert Michalak(✉) and Krzysztof Okarma

Department of Signal Processing and Multimedia Engineering, Faculty of Elecrtical Engineering, West Pomeranian University of Technology, Szczecin, 26 Kwietnia 10, 71-126 Szczecin, Poland
{michalak.hubert,okarma}@zut.edu.pl

Abstract. Adaptive binarization of unevenly lightened images is one of the key issues in document image analysis and further text recognition. As the global thresholding leads to improper results making correct text recognition practically impossible, an efficient implementation of adaptive thresholding is necessary. The most popular global approach is the use of Otsu's binarization which can be improved using the fast block based method or calculated locally leading to AdOtsu method. Even faster adaptive thresholding based on local mean calculated for blocks is presented in the paper. Obtained results have been compared with some other adaptive thresholding algorithms, being typically the modifications of Niblack's method, for a set of images originating from DIBCO databases modified by addition of intensity gradients. Obtained results confirm the usefulness of the proposed fast approach for binarization of document images.

Keywords: Image binarization · Adaptive thresholding

1 Introduction

Image binarization is an important step of many image analysis algorithms, usually considering as a preprocessing operation, allowing to reduce significantly the amount of data present in the image. Since the applied binarization algorithm affects strongly the results of further shape recognition and object classification, it is essential to provide a proper algorithm of thresholding which should protect the shapes of individual objects visible on the image. It may be especially relevant for many applications typically utilizing binary image analysis such as Optical Character Recognition (OCR) [11], recognition of QR codes [18] as well as recognition of vehicles' register plate numbers [29].

Throughout the years many different thresholding methods have been proposed, typically used for relatively high quality grayscale images with bi-modal histograms. Probably the most popular global method is the classical Otsu's binarization [19] where the threshold is calculated by minimizing the intra-class variance and therefore maximizing the inter-class variance between two classes

R. Silhavy (Ed.): CSOC 2018, AISC 764, pp. 79–90, 2019.
https://doi.org/10.1007/978-3-319-91189-2_9

of pixels representing foreground and background. This method has also been further modified leading to multi-level thresholding and adaptive Otsu (AdOtsu) method applied for document image binarization [15]. Nevertheless, this adaptive approach requires the use of additional background estimation and the threshold is assigned to each pixel on the image domain so the computational complexity of the method increases noticeable, especially in the multi-scale version.

Some other well known local binarization methods has been developed by Niblack [16] and its improved version introduced by Sauvola [22]. Their further modification has been proposed by Gatos [5] who has applied further postprocessing operations allowing to remove noise from the resulting image. Another well-known approach to global thresholding, although based on image entropy, has been proposed by Kapur [6].

An interesting integrated approach to text extraction and recognition with binarization based on maximizing local contrast has been proposed by Wolf [28]. The proposed method is partially based on Niblack's approach which is generally based on the calculation of the local mean and variance of the image replaced by the local contrast in Wolf's method. Another adaptive binarization method of document images containing non-uniform illumination, undesirable shadows and random noise, based on the criterion of maximizing local contrast, has been proposed by Feng [4]. Nevertheless, this method uses the varying side of the sliding window depending on the normalized local standard deviation with additional median filter applied for impulse noise reduction and bilinear interpolation, increasing the overall computational complexity of the method. An interesting local binarization method based on dynamic windows has been proposed in the Bataineh's article [1]. Since the window size is considered as a critical factor in such local binarization methods, the Authors of the paper have proposed a method of its determination based on the global threshold and the standard deviation of the image pixels. However, the results obtained for handwritten document images have been much worse than for machine printed ones, obtained from well known DIBCO 2009 and 2010 datasets used also together with some newer datasets in this paper [20].

Considering Niblack's method as one of the simplest and fastest algorithms allowing to obtain reasonable results of binarization, the interest in some its modifications is still quite big. An overview of some of them has been provided by Khurshid [7] whereas the fast implementation of Niblack's method utilizing integral images has been proposed by Samorodova [21]. A comprehensive review of such modifications can be found in the most recent Saxena's paper [23].

Some other approaches for adaptive binarization have also been proposed recently, usually more specialized as e.g. Chou's method dedicated for images captured by cameras [3] based on the use of Otsu's thresholding for blocks of 3×3 pixels. As the Authors stated "*using Otsu threshold as a local threshold can yield poor results for regions containing background pixels only*" [3] and therefore Support Vector Machines have been applied for constructing decision functions. Another region based algorithm dedicated for barcodes has been proposed by Kulyuikin [8]. A simplified fast Balanced Histogram Thresholding (BHT) based on the use of the Monte Carlo method has been analyzed in the paper [9] whereas in the article [10] an iterative approach has been proposed utilizing the energy and entropy of the image estimated using random pixels drawn according to the Monte Carlo method. The combination of local and global Kapur and Otsu's methods has also been discussed in another paper [12].

In the paper [26] a robust method has been delivered based on adaptive image contrast with additional text stroke edge pixel detection followed by postprocessing operations related to the analysis of the pixels' connectivity. An interesting approach to binarization of non-uniform illuminated document images based on the Curvelet transform and Otsu's thresholding has also been proposed by Wen [27]. A more detailed review and comparison of many recently proposed algorithms can be found in survey papers [13,24].

2 Proposed Fast Adaptive Image Binarization

Considering the poor results obtained using the global thresholding for non-uniform illuminated document images a challenge of fast and reliable local binarization of such images is still up-to-date. As most of the adaptive methods is based on the analysis of the neighborhood of each pixel, typically using the sliding window approach, their computational complexity is relatively high. Therefore their application in lower performance systems may be strongly limited, especially for real-time applications not necessarily directly related to document images such as e.g. video based line following robots.

The solution proposed in this paper is the use of the region based approach based on the division of the image into relatively large blocks. The idea of the local thresholding is motivated by Niblack's method and its modifications, presented e.g. in the paper [7]. Since an attempt to combine the global and local methods has been made by Kulyukin [8], a quite similar idea of using the division into blocks (regions) has been used in our paper. This hybrid method gives the possibility to preserve the details (as in local methods) together with noise suppression (as in global thresholding). Nevertheless, the Kulyukin's algorithm has not been verified for non-uniform illuminated images at all.

One of the goals of our paper is to determine the most suitable size of the block for the proposed approach as the compromise between global thresholding and purely local methods leading to various threshold values for individual pixels being sensitive to noise contaminations. Since in many cases the local threshold obtained using the Niblack's formula causes the presence of noisy blocks in the resulting image, an appropriate lowering of the determined threshold has been assumed for improvement of final results, especially for the blocks with low variance representing the background.

3 Verification of the Method

The proposed method can be tuned using two parameters - size of the region (square blocks have been assumed in the paper) and the correction coefficient lowering the value of the threshold determined using Niblack's formula. However, since the best results have been obtained for its simplified version with the initial threshold calculated as the mean values of the pixels from the analyzed region, their variance have not been used lowering the required computational effort, similarly as in Bradley's method applied typically for each pixel [2]. To determine the appropriate combination of those parameters the verification of results have been made using the modified images from the available seven DIBCO databases [20] published since 2009.

Since the DIBCO datasets, developed as test images for Document Image Binarization Competitions organized during the International Conferences on Document Analysis and Recognition (ICDAR), contain quite uniform illuminated degraded color and grayscale document images together with Ground Truth (GT) binary images, for the purpose of our experiments, their modified versions have been prepared by the addition of grayscale gradients and conversion of all images into grayscale. Therefore their global binarization have caused bad results and adaptive methods are necessary.

Making use of the knowledge of GT images it is possible to calculate some widely known classification metrics, such as Precision, Recall, F-Measure etc., used typically for the comparison of binary images [25]. Another metric applied for comparison of the obtained result with GT images is the Distance Reciprocal Distortion (DRD) proposed in 2004 [14] by the analysis the surroundings of the changes pixels. For the n-th changed pixel it can be calculated locally using the formula:

$$DRD_n = \sum_{u=-2}^{2} \sum_{v=-2}^{2} |GT_n(u,v) - BW_n(u,v)| \times W(u,v) \tag{1}$$

where GT is the reference "ground-truth" and BW is the result of binarization. The normalized weight matrix W used in the formula (1) is defined as:

$$W = \begin{pmatrix} 0.0256 & 0.0324 & 0.0362 & 0.0324 & 0.0256 \\ 0.0324 & 0.0512 & 0.0724 & 0.0512 & 0.0324 \\ 0.0362 & 0.0724 & 0 & 0.0724 & 0.0362 \\ 0.0324 & 0.0512 & 0.0724 & 0.0512 & 0.0324 \\ 0.0256 & 0.0324 & 0.0362 & 0.0324 & 0.0256 \end{pmatrix} \tag{2}$$

Finally, the overall DRD value is computed as:

$$DRD = \frac{1}{K} \times \sum_{n=1}^{N} DRD_n \tag{3}$$

where N is the number of changed pixels in the image and the number of non-uniform blocks of size 8×8 pixels (with sum of values more than 0 and less than 64) in the GT image is denoted as K.

However, some widely known metrics often used for this purpose are Peak Signal-to-Noise Ratio (PSNR) and Misclassification Penalty Metric (MPM) introduced by Young [30]. More details about the typical performance evaluation methodology of binarization methods can be found in the paper [17].

4 Results of Experiments

The first conducted experiments have been made to determine the appropriate parameters (block size and the lowering coefficient) for the 86 images from 7 available DIBCO datasets. Three of them contain both handwritten and machine printed document images - DIBCO 2009 (5 each), DIBCO 2011 (8 each) and DIBCO 2013 (8 each) and four datasets contain only handwritten images - 10 images in each of DIBCO 2010, 2014, and 2016 datasets whereas DIBCO 2012 dataset contains 14 images. Considering various number of images, the weighted average metrics have been calculated for the overall merged DIBCO dataset (modified by the addition of non-uniform illumination). The metric used for the choice of the correct parameters of the proposed method the F-Measure metric has been chosen defined as:

$$\mathrm{FM} = 2 \cdot \frac{PR \cdot RC}{PR + RC} \cdot 100\% \,, \tag{4}$$

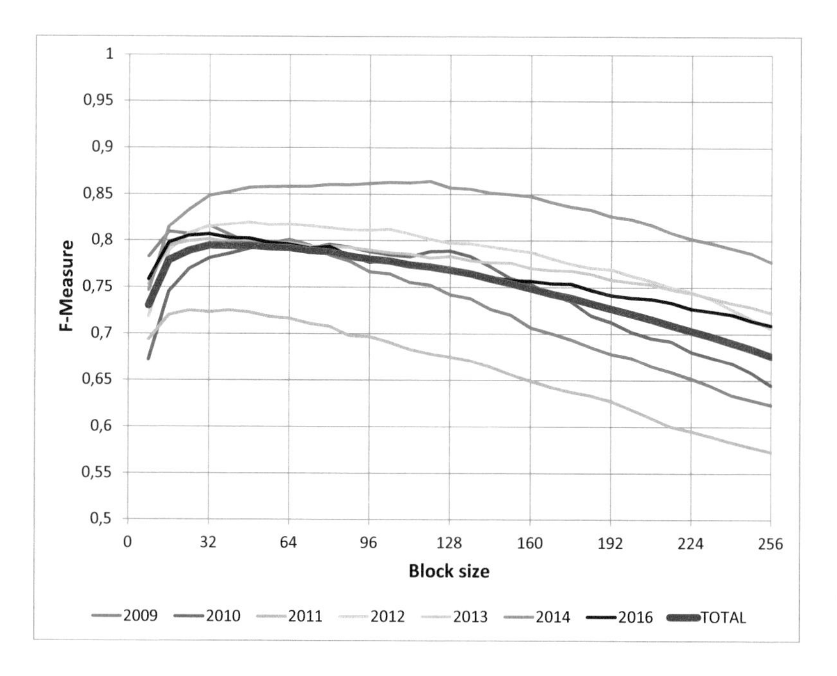

Fig. 1. F-Measure values for the images from various DIBCO datasets and various block size.

where Precision (PR) and Recall (RC) are calculated as the ratios of true positives to the sum of all positives and true positives to the sum of true positives and false negatives respectively.

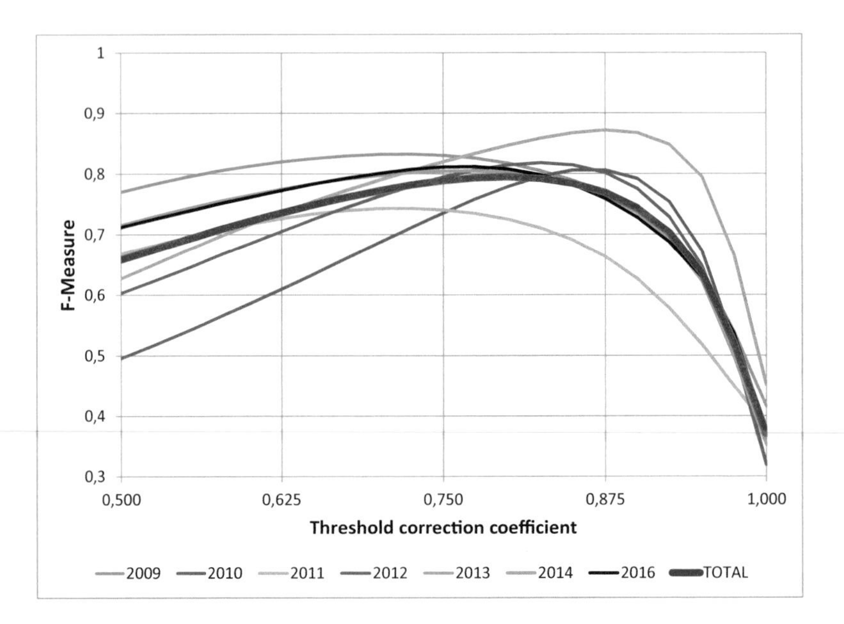

Fig. 2. F-Measure values for the images from various DIBCO datasets and various values of the lowering coefficient.

As shown in Figs. 1 and 2, the best results have been obtained for 32×32 pixels regions and the correction coefficient 0.8.

Further verification of the proposed approach has been made by comparison of the values of three metrics, namely F-Measure, DRD and PSNR, for all DIBCO databases with original images contaminated by the addition of grayscale gradients as previously. For comparison purposes the global Otsu's method and some of the most popular adaptive algorithms have been chosen introduced by Niblack [16], Sauvola [22], Wolf [28] and Bradley [2]. Additionally the modification of Bradley's method by using the Gaussian window has been used although its computational effort is much higher. Since the results achieved for the Otsu's method are much worse for clarity only the results obtained for adaptive algorithms are presented in Figs. 3, 4, 5 and 6.

Analyzing the results presented on these plots, it may be noticed that for some individual modified datasets the best results can be achieved using e.g. Sauvola method for DIBCO 2014 dataset measuring the results by means of F-Measure or DRD. Nevertheless, considering the PSNR values as well as all the metrics calculated for the whole merged dataset the advantages of the proposed approach are clearly visible.

The average values obtained using the proposed method are: F-Measure = 0.795, DRD = 11.106, MPM = 11.23 and PSNR = 15.473 dB.

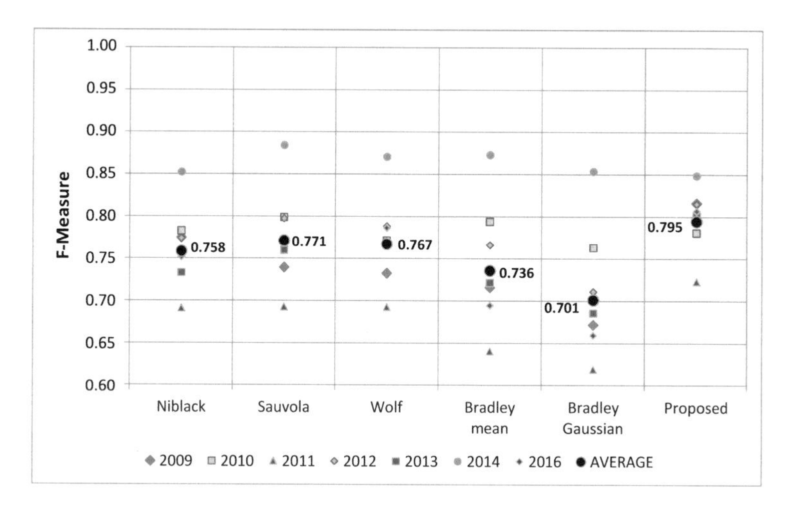

Fig. 3. F-Measure values for the images from various DIBCO datasets obtained using various adaptive binarization methods.

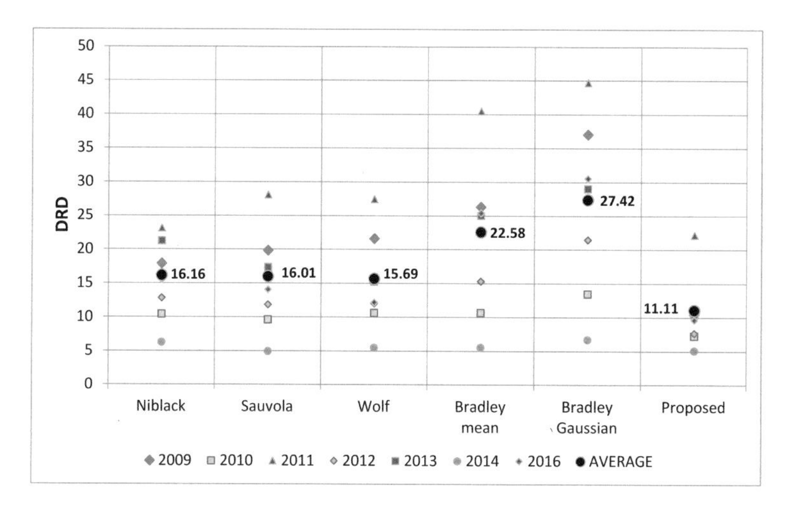

Fig. 4. DRD values for the images from various DIBCO datasets obtained using various adaptive binarization methods.

An additional analysis of execution time is shown in Fig. 7 (in logarithmic scale together with Otsu's method) where it can be easily observed that the proposed method is comparable with the fastest pixel based adaptive methods leading to better results. To compare the computational effort in a reliable way, all the methods have been implemented in the same way in Matlab R2017b environment. As can be noticed, the proposed method is almost the fastest one among adaptive algorithms. To illustrate the advantages of the proposed method some exemplary images have been chosen from DIBCO datasets and the results

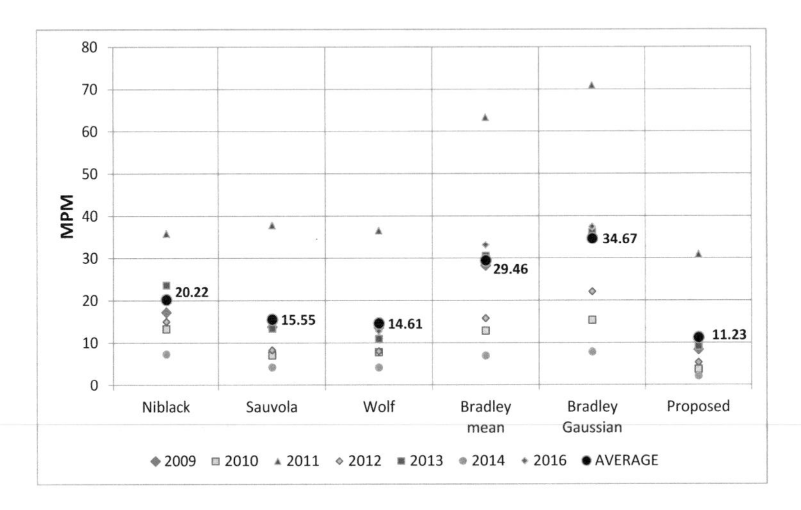

Fig. 5. MPM values for the images from various DIBCO datasets obtained using various adaptive binarization methods.

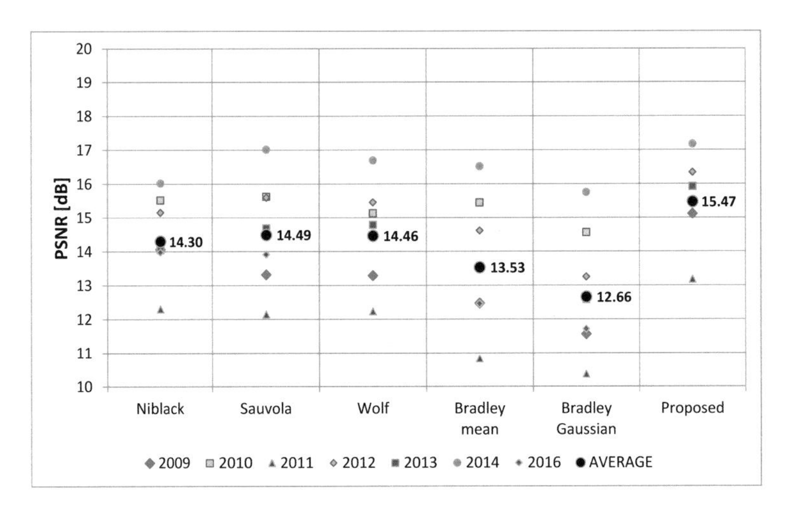

Fig. 6. PSNR values for the images from various DIBCO datasets obtained using various adaptive binarization methods.

obtained by means of various binarization algorithms have been presented in Figs. 8, 9, 10, 11 and 12 together with non-uniform illuminated input images and "ground-truth" ones.

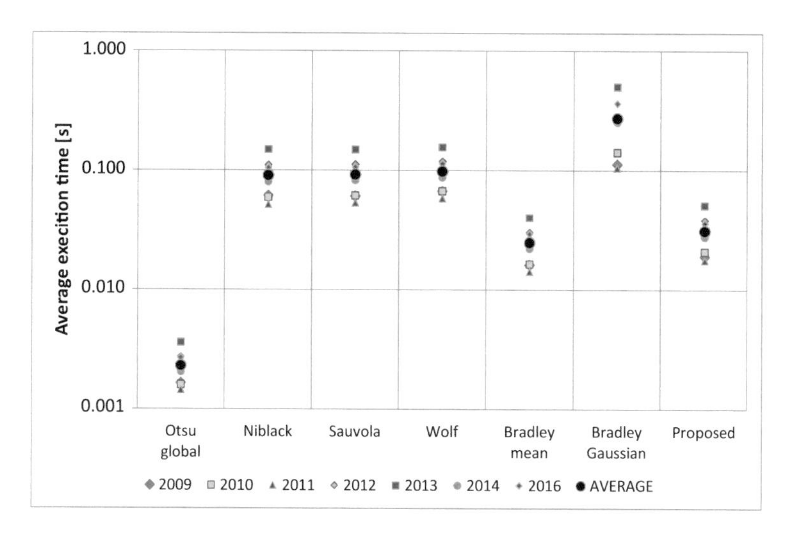

Fig. 7. Average execution time per image for the whole DIBCO datasets obtained using various binarization methods.

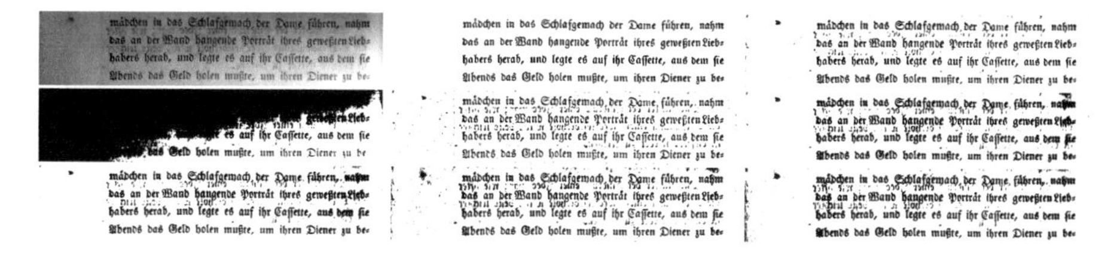

Fig. 8. Exemplary results for P01 image from DIBCO 2009 - from left to right: modified input image, GT and proposed method (top row), global Otsu, Niblack and Sauvola (middle row), Wolf, Bradley and Gaussian Bradley (bottom rom).

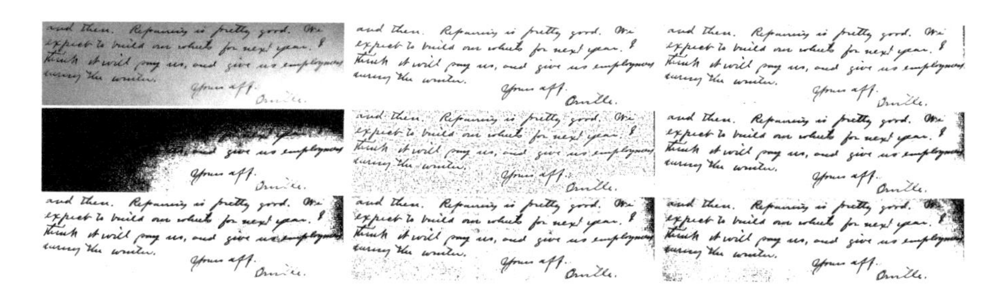

Fig. 9. Exemplary results for H07 image from DIBCO 2010 - from left to right: modified input image, GT and proposed method (top row), global Otsu, Niblack and Sauvola (middle row), Wolf, Bradley and Gaussian Bradley (bottom rom).

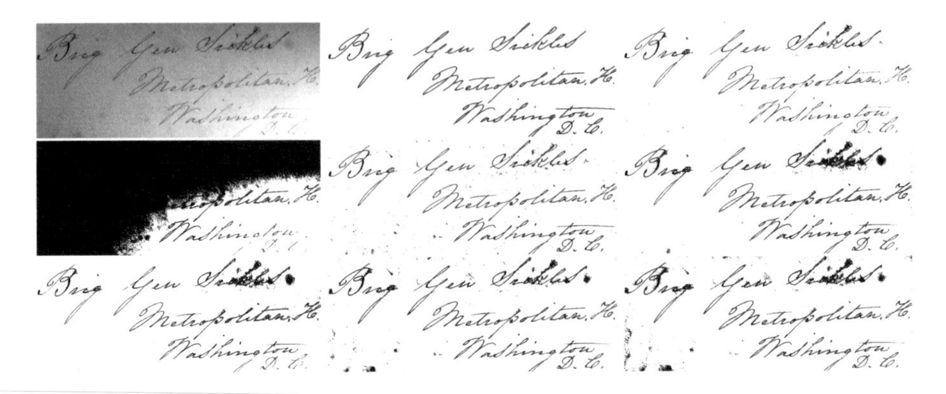

Fig. 10. Exemplary results for H06 image from DIBCO 2010 - from left to right: modified input image, GT and proposed method (top row), global Otsu, Niblack and Sauvola (middle row), Wolf, Bradley and Gaussian Bradley (bottom rom).

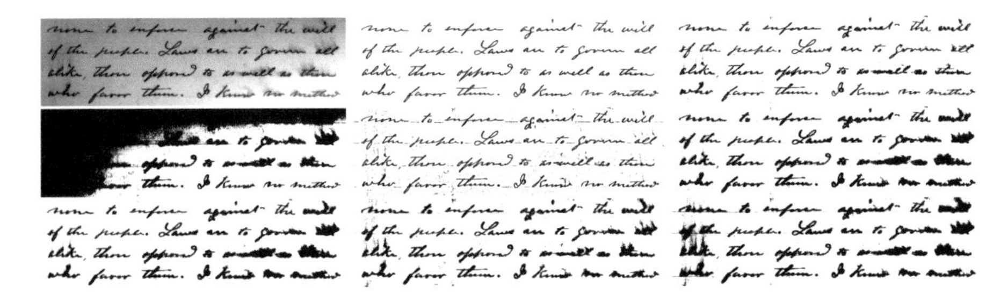

Fig. 11. Exemplary results for H08 image from DIBCO 2012 - from left to right: modified input image, GT and proposed method (top row), global Otsu, Niblack and Sauvola (middle row), Wolf, Bradley and Gaussian Bradley (bottom rom).

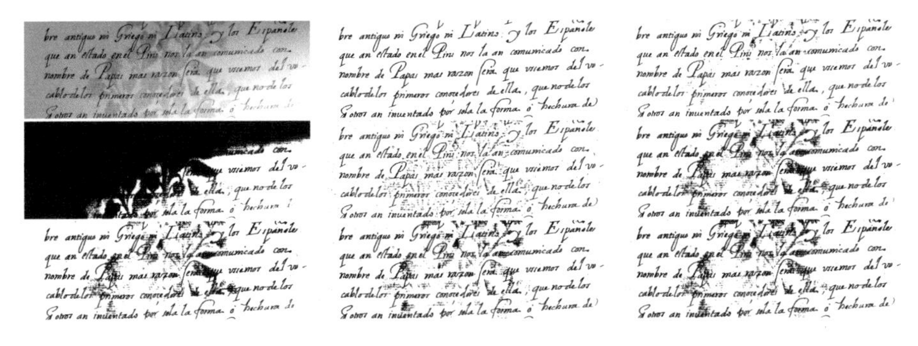

Fig. 12. Exemplary results for H09 image from DIBCO 2014 - from left to right: modified input image, GT and proposed method (top row), global Otsu, Niblack and Sauvola (middle row), Wolf, Bradley and Gaussian Bradley (bottom rom).

5 Concluding Remarks

The fast adaptive binarization method proposed in the paper combines its simplicity and high speed with the ability to preserve the details and noise suppression for non-uniform illuminated input grayscale images. Although the experiments have been conducted for the modified document images from DIBCO datasets, mainly due to the availability of the "ground-truth" binary images, the method is quite universal and may be applied also for some other types of non-uniform illuminated images e.g. captured by the cameras of autonomous robots e.g. vision based line-followers. Obtained results are encouraging for further modifications and out future research will be concentrated on its improvements e.g. for the OCR purposes.

References

1. Bataineh, B., Abdullah, S.N.H.S., Omar, K.: An adaptive local binarization method for document images based on a novel thresholding method and dynamic windows. Pattern Recogn. Lett. **32**(14), 1805–1813 (2011)
2. Bradley, D., Roth, G.: Adaptive thresholding using the integral image. J. Graph. Tools **12**(2), 13–21 (2007)
3. Chou, C.H., Lin, W.H., Chang, F.: A binarization method with learning-built rules for document images produced by cameras. Pattern Recogn. **43**(4), 1518–1530 (2010)
4. Feng, M.L., Tan, Y.P.: Adaptive binarization method for document image analysis. In: Proceedings of the 2004 IEEE International Conference on Multimedia and Expo (ICME), vol. 1, pp. 339–342, June 2004
5. Gatos, B., Pratikakis, I., Perantonis, S.: Adaptive degraded document image binarization. Pattern Recogn. **39**(3), 317–327 (2006)
6. Kapur, J., Sahoo, P., Wong, A.: A new method for gray-level picture thresholding using the entropy of the histogram. Comput. Vis. Graph. Image Process. **29**(3), 273–285 (1985)
7. Khurshid, K., Siddiqi, I., Faure, C., Vincent, N.: Comparison of Niblack inspired binarization methods for ancient documents. In: Document Recognition and Retrieval XVI, vol. 7247, pp. 7247–7247-9 (2009)
8. Kulyukin, V., Kutiyanawala, A., Zaman, T.: Eyes-free barcode detection on smartphones with Niblack's binarization and Support Vector Machines. In: Proceedings of the 16th International Conference on Image Processing, Computer Vision, and Pattern Recognition (IPCV 2012) at the World Congress in Computer Science, Computer Engineering, and Applied Computing WORLDCOMP, vol. 1, pp. 284–290. CSREA Press, July 2012
9. Lech, P., Okarma, K.: Fast histogram based image binarization using the Monte Carlo threshold estimation. In: Chmielewski, L.J., Kozera, R., Shin, B.S., Wojciechowski, K. (eds.) Computer Vision and Graphics. LNCS, vol. 8671, pp. 382–390. Springer, Switzerland (2014)
10. Lech, P., Okarma, K.: Optimization of the fast image binarization method based on the monte carlo approach. Elektronika Ir Elektrotechnika **20**(4), 63–66 (2014)
11. Lech, P., Okarma, K.: Prediction of the optical character recognition accuracy based on the combined assessment of image binarization results. Elektronika Ir Elektrotechnika **21**(6), 62–65 (2015)

12. Lech, P., Okarma, K., Wojnar, D.: Binarization of document images using the modified local-global Otsu and Kapur algorithms. Przegląd Elektrotechniczny **91**(1), 71–74 (2015)
13. Leedham, G., Yan, C., Takru, K., Tan, J.H.N., Mian, L.: Comparison of some thresholding algorithms for text/background segmentation in difficult document images. In: Proceedings of the 7th International Conference on Document Analysis and Recognition, ICDAR 2003, pp. 859–864, August 2003
14. Lu, H., Kot, A., Shi, Y.: Distance-reciprocal distortion measure for binary document images. IEEE Signal Process. Lett. **11**(2), 228–231 (2004)
15. Moghaddam, R.F., Cheriet, M.: AdOtsu: an adaptive and parameterless generalization of Otsu's method for document image binarization. Pattern Recogn. **45**(6), 2419–2431 (2012)
16. Niblack, W.: An Introduction to Digital Image Processing. Prentice Hall, Englewood Cliffs (1986)
17. Ntirogiannis, K., Gatos, B., Pratikakis, I.: Performance evaluation methodology for historical document image binarization. IEEE Trans. Image Process. **22**(2), 595–609 (2013)
18. Okarma, K., Lech, P.: Fast statistical image binarization of colour images for the recognition of the QR codes. Elektronika Ir Elektrotechnika **21**(3), 58–61 (2015)
19. Otsu, N.: A threshold selection method from gray-level histograms. IEEE Trans. Syst. Man Cybern. **9**(1), 62–66 (1979)
20. Pratikakis, I., Zagoris, K., Barlas, G., Gatos, B.: ICDAR 2017 Document Image Binarization COmpetition (DIBCO 2017) (2017). https://vc.ee.duth.gr/dibco2017/
21. Samorodova, O.A., Samorodov, A.V.: Fast implementation of the niblack binarization algorithm for microscope image segmentation. Pattern Recogn. Image Anal. **26**(3), 548–551 (2016)
22. Sauvola, J., Pietikäinen, M.: Adaptive document image binarization. Pattern Recogn. **33**(2), 225–236 (2000)
23. Saxena, L.P.: Niblack's binarization method and its modifications to real-time applications: a review. Artif. Intell. Rev., 1–33 (2017)
24. Shrivastava, A., Srivastava, D.K.: A review on pixel-based binarization of gray images. In: Satapathy, S., Bhatt, Y., Joshi, A., Mishra, D. (eds.) Proceedings of the International Congress on Information and Communication Technology. AISC, vol. 439, pp. 357–364. Springer, Singapore (2016)
25. Sokolova, M., Lapalme, G.: A systematic analysis of performance measures for classification tasks. Inf. Process. Manag. **45**(4), 427–437 (2009)
26. Su, B., Lu, S., Tan, C.L.: Robust document image binarization technique for degraded document images. IEEE Trans. Image Process. **22**(4), 1408–1417 (2013)
27. Wen, J., Li, S., Sun, J.: A new binarization method for non-uniform illuminated document images. Pattern Recogn. **46**(6), 1670–1690 (2013)
28. Wolf, C., Jolion, J.M.: Extraction and recognition of artificial text in multimedia documents. Formal Pattern Anal. Appl. **6**(4), 309–326 (2004)
29. Yoon, Y., Ban, K.D., Yoon, H., Lee, J., Kim, J.: Best combination of binarization methods for license plate character segmentation. ETRI J. **35**(3), 491–500 (2013)
30. Young, D.P., Ferryman, J.M.: PETS metrics: on-line performance evaluation service. In: Proceedings of 2005 IEEE International Workshop on Visual Surveillance and Performance Evaluation of Tracking and Surveillance, pp. 317–324 (2005)

Semantic Query Suggestion Based on Optimized Random Forests

Aytuğ Onan(✉)

Department of Software Engineering, Faculty of Technology, Manisa Celal Bayar University, 45400 Manisa, Turkey
aytug.onan@cbu.edu.tr

Abstract. Query suggestion is an integral part of Web search engines. Data-driven approaches to query suggestion aim to identify more relevant queries to users based on term frequencies and hence cannot fully reveal the underlying semantic intent of queries. Semantic query suggestion seeks to identify relevant queries by taking semantic concepts contained in user queries into account. In this paper, we propose a machine learning approach to semantic query suggestion based on Random Forests. The presented scheme employs an optimized Random Forest algorithm based on multi-objective simulated annealing and weighted voting. In this scheme, multi-objective simulated annealing is utilized to tune the parameters of Random Forests algorithm, i.e. the number of trees forming the ensemble and the number of features to split at each node. In addition, the weighted voting is utilized to combine the predictions of trees based on their predictive performance. The predictive performance of the proposed scheme is compared to conventional classification algorithms (such as Naïve Bayes, logistic regression, support vector machines, Random Forest) and ensemble learning methods (such as AdaBoost, Bagging and Random Subspace). The experimental results on semantic query suggestion prove the superiority of the proposed scheme.

Keywords: Query suggestion · Random Forests · Ensemble learning

1 Introduction

Search plays an essential role in identifying content, information, audios, videos, images, products and navigational links on the web [1]. Users can enter searches or queries into search engines to satisfy their information needs. Search engines are utilized as the starting point of entry to access information, Internet sites, services and many other sources of information on the web by many information seekers [2]. The queries are vital to the effective information retrieval from the web. Queries are short texts containing several words, which may be ambiguous [3]. Queries are often ill-defined representations of more complex information needs [4]. In addition, the same query can return a completely different result with the use of different search engines [5]. With the advances in Web search technologies, search engines provide information to users in a more efficient and effective way. However, it is still a challenging issue for users to express their intents in the form of queries [6].

R. Silhavy (Ed.): CSOC 2018, AISC 764, pp. 91–102, 2019.
https://doi.org/10.1007/978-3-319-91189-2_10

Query suggestion is a core task for enhancing the usability of search engines by providing alternative queries to the users, so that they can reach their information needs in a more effective way. Several approaches have been proposed in recent years for query suggestion. Data-driven approaches to query suggestion aim to identify more relevant queries to users based on the use of term frequencies, click-through or query session data and hence cannot fully reveal the underlying semantic intent of queries. Semantic query suggestion seeks to identify relevant queries by taking semantic concepts contained in user queries into account. The benefits of semantic query suggestion include acquiring contextual information to the user, suggesting related concepts/associated terms that could be utilized for search and providing valuable navigational recommendations [7].

In this paper, we propose a machine learning approach to semantic query suggestion. Random Forests are one of the promising algorithms on text and web mining with high predictive performance [8]. Hence, we utilized Random Forest algorithm as the base learner for machine learning based semantic query suggestion.

2 Related Work

This section briefly reviews the related work on query suggestion and Random Forest algorithms. One approach to query suggestion is based on clustering queries by using the clicked URLs. For instance, Beeferman and Berger [9] presented a query suggestion scheme, in which clustering is utilized to identify clusters of similar queries and similar URLs based on click-through data. In another study, clustering is utilized on the content of historical preferences of users, by which queries are identified and ranked according to a relevance criterion [10]. Cao et al. [5] proposed a two-staged query suggestion scheme. In this scheme, first, clustering is utilized to summarize queries into concepts based on a click-through bipartite graph. Then, sequence of queries written by a particular user is mapped into a sequence of concepts.

Another approach adopted in query suggestion is based on random walk and hitting time in a query. For instance, Mei et al. [11] presented a ranking based approach to query suggestion by using hitting time on a large scale bipartite. Similarly, Ma et al. [12] presented a two-staged query suggestion scheme which mines click through data in the forms of bipartite graphs. In another study, click and skip information of information seekers are optimized with the use of a random walk [13]. In another study, an optimized framework to query suggestion is presented based on click through and query session data [14].

The diversification mechanism is also employed in query suggestion to identify more relevant terms. For instance, Ma et al. [15] presented a Markov random walk and hitting time analysis based approach to suggest semantically relevant and diverse queries to information seekers. Similarly, Song et al. [16] presented a diversification based scheme for query suggestion based on random walks. In another study, Jiang et al. [17] presented a query suggestion scheme based on diversification and personalization. In this scheme, suggested queries were diversified to cover different aspects of input query.

Another approach to query suggestion is based on stemming terms. For instance, Kraft and Zien [18] presented a query refinement scheme based on anchor text for a large hypertext document collection. In another study, Jones et al. [19] presented a query substitution scheme to generate new queries by replacing original queries. Similarly, Wang and Zhai [20] presented a context-sensitive term substitution and term addition scheme. In another study, Jansen et al. [21] utilized a search log of records for identifying query-reformation patterns. More recently, Dang and Croft [22] utilized anchor text to substitute to a query log for query reformulation.

Data-driven query suggestion approaches cannot fully reveal the underlying semantic intent of queries. In this regard, a number of works on query suggestion focuses on semantic intentions. For instance, Meij et al. [7] presented a feature engineering and machine learning based approach to semantic query suggestion. In this scheme, term-based features were utilized in conjunction with history-based and concept-specific features. In another study, Anagnostopoulos, Razis, Mylonas and Anagnostopoulos [23] presented a semantic query suggestion scheme for Twitter entities. In another study, Kruschwitz et al. [24] presented an interactive query suggestion scheme by exploiting the interaction patterns collected from search engines of local websites. Kim et al. [25] introduced a phrasal-concept query suggestion approach for literature search. Recently, Momtazi and Lindenberg [26] presented a semantic approach to query suggestion based on latent Dirichlet allocation.

Random Forests are bootstrap aggregation based ensemble learning schemes [27]. Random Forests are effective techniques for classification, prediction, feature ranking and selection and outlier detection [28]. Bader-El-Den and Gaber [29] presented a genetic algorithm based scheme to optimize Random Forests. In this scheme, smaller forests are obtained by replacement form the initial random forests and genetic algorithm is utilized to evolve the initial population of trees based on fitness function. Elyan and Gaber [30] introduced a genetic algorithm based approach to optimize the parameters of Random Forests. In the presented scheme, class decomposition was employed to break down the labelled data into subclasses. Then, the number of trees forming classifier ensemble, number of features to split on at each node and a vector representing the number of clusters are optimized with the use of a genetic algorithm. Recently, Ronao and Cho [23] presented a Random Forest based classification scheme for identifying database access anomalies. In the presented scheme, principal component analysis was employed to obtain uncorrelated and relevant features. The individual trees of ensemble were combined via a weighted voting scheme.

3 The Framework

In this section, we present dataset, Random Forests algorithm, multi-objective optimization and multi-objective simulated annealing and proposed Random Forests based classification scheme.

3.1 Dataset

We utilized a publicly available dataset, i.e. queryDB [7]. The dataset is obtained from a set of 264,503 queries issued between 2008 and 2009 on sound and vision logs of users on the site. Four experts were asked to manually annotate 998 queries to DBpedia concepts. The experts were given a list of sessions and queries. For each session, experts were asked to identify the most relevant DBpedia concepts. In this way, queryDB dataset was constructed. The average query length of the dataset is 2.14 terms per query and each query contains 1.34 annotated concepts on average. In this scheme, each query is automatically linked to DBpedia concepts. In order to represent queries, different feature engineering schemes (namely, N-gram features, concept features, hybrid features and history features) are considered. N-gram features include number of terms in the query phrase, inverse document frequency of query, weighted information gain using top five retrieved concepts, number of times query appeared entirely query in the query log, number of times query appeared partially in the query log, ratio of number of times query appeared entirely query in the query log to number of times query appeared partially in the query log, flag indicating whether a sub-n-gram of query fully match with any concept label and flag indicating whether a sub-n-gram of query contained in any concept label [7]. Concept features include feature sets related to a DBpedia concept. Concept features include the number of concepts linking to a particular concept, the number of concepts linking from a particular concept, function of depth of concept in SKOS category hierarchy [31], number of associated categories and number of redirect pages linking to a particular concept [7]. Hybrid features include n-gram and concept based features. This set of features include the relative phrase frequency of query in a concept, position of nth occurrence of query in a concept, the distance between the last and first occurrences of query in a concept, TF-IDF measure value of query for a concept, the difference between expected and observed IDF, chi-square test of independence between query in a concept and in collection, flag indicating whether the query contain label of a particular concept, flag indicating whether a particular concept contain label of a particular query, flag indicating whether concept and query are equal, retrieval score of concept with regard to query and retrieval rank of concept with respect to query [7]. Finally, history features consider the earlier queries issues in the same session. Nine history features are considered, such as the number of occurrences of label of a concept appears as query in the history, the number of occurrences of label of a concept appears in any query in the history, the number of times concept is retrieved as a result of any query in the history.

3.2 Random Forests Algorithm

A random forest is a predictor consisting of a collection of tree classifiers $\{h(x, \Theta_k), k \geq 1\}$, where $\{\Theta_k\}$ are independent identically distributed random vectors and the most popular class at a particular input x is determined by voting of each tree in the ensemble. Random Forests algorithm is an ensemble classification scheme, which combines tree predictors such that each tree grows in randomly selected subspaces of data [27]. In this scheme, a random feature selection is employed to split each node. In Random Forests, each tree in the ensemble is grown based on a random parameter and the final prediction of ensemble is obtained by aggregation. The predictive performance has been enhanced by growing an ensemble of trees and aggregating the trees by voting for the most popular class. Random Forests achieves high diversity by employing bootstrap aggregation (i.e. simple random sampling with replacement) and node splitting from a subset of total feature set. Two critical parameters to the performance of Random Forests are the number of trees and the number of features examined at each node split in the tree. In the literature, the number of trees is set between 100 and 500, and the number of features is determined between $\sqrt{n}$ or $logn$, where n denotes the total number of features in the dataset [30]. Random Forests algorithm is based on the notion of Bagging and Random Subspace. However, it performs better than Bagging owing to the use of random features. Random Forests yield predictive performance comparable to AdaBoost algorithm and support vector machines [32]. Random Forests are fast and easy to implement and can yield high predictive performance without overfitting. Random Forests are faster than Bagging or Boosting. Random Forests are relatively robust to outliers and noise [27]. In addition, it is a robust classifier to deal with missing data and unbalanced datasets [33].

3.3 Multi-objective Simulated Annealing

Achieved multi-objective simulated annealing algorithm (AMOSA) is a generalized version of simulated annealing (SA) algorithm to deal with multi-objective optimization problems [34]. AMOSA algorithm utilizes the concept of an archive, where the non-dominated solutions seen so far are kept. Two limits are utilized to limit the size of the archive, namely a hard limit (HL) and a strict limit (SL). The non-dominated solutions are stored in the archive until the size of the archive reaches to SL. When the size of the archive increases to HL, the size is reduced by applying a single-linkage clustering scheme. The algorithm begins with the initialization of a number ($\gamma \times$ SL, $0 < \gamma < 1$) of solutions in the search space. Each solution is refined by using a simple hill-climbing technique and dominance relation. In AMOSA algorithm, clustering is employed to provide diversity of non-dominated solutions.

In AMOSA algorithm, the concept of amount of domination is utilized to compute the acceptance probability of a new solution. Given two solutions a and b, the amount of domination is defined as given by Eq. (1):

$$\Delta dom_{a,b} = \prod\nolimits_{i=1, f_i(a) \neq f_i(b)}^{M} \frac{|f_i(a) - f_i(b)|}{R_i} \tag{1}$$

where M denotes the number of objectives, R_i denotes the range of the ith objective and $f_i(a)$ and $f_i(b)$ are the ith objective values of two solutions. A randomly selected point from Archive, called *current-pt* is taken as the initial solution at temperature *temp* = *Tmax*. Based on the domination status between *current-pt* and *new-pt*, several cases may arise:

Case 1: *current-pt* dominates the *new-pt* and $k(k \geq 0)$ points from the archive dominate the new-*pt*. In this case, the *new-pt* is selected as the *current-pt* based on the probability as given by Eq. (2):

$$probability = \frac{1}{1 + \exp(\Delta dom_{avg} * temp)} \tag{2}$$

where Δdom_{avg} denotes the average amount of domination of the *new-pt* by the *current-pt* and k points from the archive.

Case 2: *current-pt* and *new-pt* are non-dominating each other. In this case, three situations may arise based on the domination relation of *new-pt* and points from the archive. If *new-pt* is dominated by $k(k \geq 0)$ points from the archive, the *new-pt* is selected as the *current-pt* based on the probability as given by Eq. 2. If *new-pt* is non-dominating with the other points from the archive, the *new-pt* is selected as the *current-pt.* Otherwise, if *new-pt* dominates $k(k \geq 0)$ points from the archive, the *new-pt* is selected as the *current-pt.*

Case 3: *new-pt* dominates *current-pt.* If *new-pt* dominates the *current-pt* but $k(k \geq 0)$ points in the archive dominate *new-pt, new-pt* is determined based on the minimum of the differences of domination (denoted by $\Delta dom_{\min}$) amounts between the *new-pt* and k points as given by Eq. (3):

$$probability = \frac{1}{1 + \exp(\Delta dom_{\min})} \tag{3}$$

Otherwise, the *new-pt* is selected as the *current-pt.*

In AMOSA algorithm, the process outlined above is repeated for a specified number of iterations for each temperature (*temp*). The value of temperature is reduced to ($\alpha \times temp$) by using the cooling rate of α, until the minimum temperature value is reached. Once the minimum temperature is obtained, the process is completed and the archive consists of non-dominated solutions to the problem. For multi-objective optimization problems, AMOSA can yield promising results. It has been empirically validated that AMOSA outperforms NSGA-II and some other well-known evolutionary multi-objective algorithms [34].

3.4 Improved Random Forests Algorithm

In Random Forests (RF) algorithm, the two main parameters critical to the predictive performance are the number of trees in the ensemble and the number of features to be examined at each node split. In the improved random forests, these two parameters are optimized with the use of archived multi-objective simulated annealing algorithm (AMOSA). In addition, the weighted voting is utilized to combine the predictions of trees based on their predictive performance.

In string representation and archive initialization phase of AMOSA algorithm, a real-valued vector V that represents the parameter set to be optimized is used to maximize the predictive performance of Random Forests algorithm. The real-valued vector is given by Eq. (4):

$$V = [mtry,\ ntrees] \tag{4}$$

where *mtry* denotes the number of features to be examined at each node split and *ntrees* denotes the number of trees in Random Forests ensemble. The benchmark values for these two parameters have been set based on empirical analysis [34]. In this scheme, the value of number of trees in Random Forests ensemble has been set between 100 and 1000. The size of maximum tree is confined to 100 trees and the size of minimum number of trees is taken as 100. Regarding the parameter value for *mtry,* Eq. (5) is utilized to find the range of possible values [31]:

$$\lceil 0.2 \times n \rceil \leq mtry \leq \lceil 0.8 \times n \rceil \tag{5}$$

where n is the total number of features in the dataset.

In the proposed Random Forests algorithm, a real-valued encoding scheme is employed and the values about *mtry* are randomly initialized to a real value based on Eq. 5 and the values about *ntrees* are taken between 100 and 1000. In the proposed scheme, the archive initialization has been done in this way. In evaluating the predictive performance of classifiers, there are several evaluation metrics, such as classification accuracy, precision, recall, error rate, F-measure and area under roc curve. Based on the empirical analysis with different evaluation metric combinations, F-measure and area under roc curve were utilized as the objective function.

In addition, a weighted voting scheme is utilized to combine the predictions of each tree in Random Forest ensemble. In the weighted voting scheme, each individual tree contributes to the overall predictive performance of the ensemble based on their individual predictive performance. In this manner, each tree contributes to the final prediction based on its local performance, which indicates how correctly a tree can classify its out-of-bag error cases [23, 25]. In the training phase, the weight value for each tree is computed as given by Eq. (6) [23]:

$$w_{t,r} = \frac{\sum_{OOBsamples} score_{t,r}(q)}{N_{OOBsample}} \tag{6}$$

$$score_{r,t}(q) = \begin{cases} 1, & \text{if tree } t \text{ outputs class } r \text{ for a} \\ 0, & \text{otherwise} \end{cases} \tag{7}$$

where q denotes the instance with class r in a set of out-of-bag error samples in a specified tree t. The weight values are normalized by *min-max* normalization and the final prediction of ensemble is determined based on Eq. (8) [23]:

$$r^{*}_{WRF} = \arg\max\ votes\ (q) \tag{8}$$

$$votes(q) = \sum_{t \in F} w_{t,f(norm)} \cdot p_r(q|t, leaf_t) \tag{9}$$

4 Empirical Analysis

In this section, we evaluate the predictive performance of the proposed Random Forests ensemble on semantic query suggestion datasets. This section briefly explains experimental procedure and the experimental results.

4.1 Experimental Procedure

In order to evaluate the predictive performance, classification accuracy is utilized as the evaluation measure. In the experimental analysis, 10-fold cross validation is employed. The experimental analysis are performed with the machine learning toolkit WEKA 3.9, which is an open-source platform implemented in Java with many machine learning algorithms [35]. For methods, the default parameters of WEKA are utilized. To examine the predictive performance of proposed scheme, genetic algorithms and simulated annealing are taken into consideration. The single-objective genetic algorithm based Random Forests (denoted by RFGA), multi-objective genetic algorithm based Random Forests (denoted by RFNSGAII), single-objective simulated annealing algorithm based Random Forests (denoted by RFSA) and archived multi-objective simulated annealing based Random Forests (denoted by RFAMOSA) are considered. For each case, the predictions of trees in the ensemble are combined by weighted and unweighted voting. For single-objective metaheuristics, F-measure is utilized as the objective function and for multi-objective metaheuristics, F-measure and area under roc curve are utilized as the objective functions. In Table 1, the parameter values for these schemes are presented.

4.2 Results and Discussions

The empirical analysis aims to examine the predictive performance of different feature sets and their combinations in semantic query suggestion. In Table 2, classification accuracies obtained by four different feature sets (i.e., N-gram features, concept features, hybrid features and history features) and their combinations are presented. The highest predictive performance in terms of classification accuracy is obtained by

Table 1. Parameter values for the metaheuristic algorithms.

Algorithms	Parameters
RFGA	Probability of mutation = 0.9, Probability of crossover = 0.5, Population size = 500, Number of generations = 50
RFSA	Maximum value of temperature(T_{max}) = 100, Minimum value of temperature (T_{min}) = 0.00001, Number of iterations = 100
RFNSGAII	Population size = 500, Probability of crossover = 0.7, Probability of mutation = 0.2, Number of generations = 50, Number of couples = 110, Pressure = 0.1
RFAMOSA	Maximum value of temperature(T_{max}) = 100, Minimum value of temperature (T_{min}) = 0.00001, Soft limit(SL) = 200, Hard limit (HL) = 100, cooling rate (α) = 0.8, Number of iterations = 100

Table 2. Classification accuracies obtained by different feature sets on the learning methods.

Feature Set	NB	LR	SVM	RF
Full set	82.81	87.4	80.17	93.5
N-gram	78.17	87.47	85.69	92.65
Concept	**87.48**	**87.99**	**87.9**	**94.17**
History	85.71	87.16	87.01	87.87
Hybrid	87.39	87.55	84.92	90.34
N-gram + Concept	83.94	87.47	73.84	94.13
N-gram + History	77.99	87.47	85.69	92.4
N-gram + Hybrid	79.05	87.79	87.67	92.65
Concept + History	84.19	87.25	83.46	93.82
Concept + Hybrid	85.72	87.5	69.89	92.89
History + Hybrid	85.53	87.54	79.59	90.15
N-gram + Concept + History	83.1	87.47	79.97	93.95
N-gram + Concept + Hybrid	83.88	87.79	80.13	93.64
N-gram + History + Hybrid	78.46	87.79	87.65	92.47
Concept + History + Hybrid	84.43	87.49	71.19	92.28

concept features. The second highest predictive performance on Naïve Bayes is obtained by hybrid features. The second highest predictive performance on logistic regression (87.79) is obtained by several combined feature sets, namely the feature set which combines N-gram and hybrid features, the feature set which combines N-gram, concept and hybrid features, the feature set which combines N-gram, history and hybrid features. The second highest predictive performance on support vector machines is obtained by the feature set which combines N-gram and hybrid features. The second highest predictive performance on Random Forests is obtained by the combined feature set, which integrates N-gram and concept features. Regarding the predictive performance of supervised learning algorithms on semantic query suggestion, Random Forests outperform other supervised classifiers. The second highest performance is generally achieved by logistic regression algorithm. Among all the configurations listed in Table 2, the highest predictive performance (94.17%) is obtained by concept based features when Random Forests algorithm is employed as the base learner.

Table 3 presents the empirical results (in terms of classification accuracy) for conventional classification algorithms, ensemble learning methods and Random Forests based schemes. As it can be seen from Table 3, the predictive performance of classifiers can be enhanced with the use of ensemble learners. Random Subspace method generally outperforms other ensemble learning schemes. Among all the configurations listed in Table 3, the highest (the best) predictive performance (98.98%) is obtained by concept based features when the proposed archived multi-objective simulated annealing based Random Forests is utilized as the classification algorithm. Random Forests based schemes outperform other classifiers and ensemble learners. In addition, the predictive performance of Random Forests enhance when the weighted voting scheme is employed to combine the predictions of each tree in the ensemble. As it can be observed from the results listed in Table 3, multi-objective metaheuristics based

Table 3. Classification accuracies obtained by classification algorithms and ensemble learners.

Algorithms	Full set	N-gram	Concept	History	Hybrid
NB	82.81	78.17	87.48	85.71	87.39
LR	87.40	87.47	87.99	87.16	87.55
SVM	80.17	85.69	87.90	87.01	84.92
RF (unweighted voting)	93.50	92.65	94.17	87.87	90.34
AdaBoost (NB)	82.88	83.00	87.55	85.78	87.46
AdaBoost (LR)	87.54	87.64	88.03	87.32	87.67
AdaBoost (SVM)	87.07	77.99	87.46	83.93	80.39
AdaBoost (RF)	93.62	92.82	95.25	87.80	90.37
Bagging (NB)	82.89	78.61	87.50	85.79	87.06
Bagging (LR)	87.47	87.55	87.87	87.23	87.63
Bagging (SVM)	87.72	87.47	87.88	87.07	85.40
Bagging (RF)	93.06	92.35	94.71	87.89	90.22
RS (NB)	85.73	87.01	87.71	86.25	87.66
RS (LR)	87.67	87.63	87.80	87.40	87.65
RS (SVM)	87.45	87.47	87.95	87.11	87.48
RS (RF)	93.83	92.73	95.26	88.89	91.24
RFGA (unweighted voting)	93.88	93.88	95.87	93.19	93.02
RFSA (unweighted voting)	94.51	94.49	95.94	94.36	94.29
RFNSGAII (unweighted voting)	95.41	95.25	96.18	94.98	94.96
RFAMOSA (unweighted voting)	96.45	96.44	96.60	96.37	96.35
RF (weighted voting)	95.29	95.22	95.22	95.21	95.21
RFGA (weighted voting)	96.04	95.99	96.89	95.89	95.85
RFSA (weighted voting)	97.00	96.94	96.91	96.82	96.54
RFNSGAII (weighted voting)	97.84	97.83	97.62	97.61	97.59
RFAMOSA (weighted voting)	98.97	98.97	98.98	98.78	98.52

Random Forests ensembles (i.e., RFNSGAII and RFAMOSA) yield better results compared to the single-objective Random Forests ensembles (i.e., RFGA and RFSA).

5 Conclusion

We presented a machine learning approach to semantic query suggestion based on optimized Random Forests. We explored the predictive performance of N-gram features, concept features, hybrid features and history features and their combinations in semantic query suggestion. Based on the empirical analysis, the highest predictive performance is obtained by concept features. In addition, the experimental results indicate that the combined feature sets cannot substantially improve the predictive performance of classifiers. Regarding the predictive performance of conventional classification algorithms and ensemble learning methods, Random Forests algorithm generally outperforms other schemes. In this regard, the predictive performance of Random Forests algorithm is further enhanced with a multi-objective evolutionary algorithm. In this scheme, the two main parameters of Random Forests algorithm, i.e. the number of trees forming the ensemble and the number of features to split at each node, are tuned by an archived multi-objective simulated annealing algorithm. The experimental results indicated that the weighted voting can enhance the predictive performance of Random Forests algorithm. The experimental results on semantic query suggestion indicate the superiority of the proposed scheme over the conventional classification algorithms and ensemble learning methods.

References

1. Parikh, N., Singh, G., Sundaresan, N.: Handbook of Statistics. Elsevier, New York (2013)
2. Jansen, B.J., Booth, D.L., Spink, A.: Determining the informational, navigational and transactional intent of web queries. Inf. Process. Manage. **44**(3), 1251–1266 (2008)
3. Wen, J.R., Nie, J.Y., Zhang, H.J.: Clustering user queries of a search engine. In: Proceedings of the 10th International Conference on World Wide Web, pp. 162–168. ACM, New York (2001)
4. Jansen, B.J., Spink, A., Bateman, J., Saracevic, T.: Real life information retrieval: a study of user queries on the web. ACM SIGIR Forum **32**(1), 5–17 (1998)
5. Cao, H., Jiang, D., Pei, J., He, Q., Liao, Z., Chen, E., Li, H.: Context-aware query suggestion by mining click-through and session data. In: Proceedings of the 14th International Conference on Knowledge Discovery and Data Mining, pp. 875–883. ACM, New York (2008)
6. Kato, M.P., Sakai, T., Tanaka, K.: When do people use query suggestion? A query suggestion log analysis. Inf. Retrieval **16**(6), 725–746 (2013)
7. Meij, E., Bron, M., Hollink, L., Huurnink, B., de Rijke, M.: Learning semantic query suggestions. In: Bernstein, A., Karger, David R., Heath, T., Feigenbaum, L., Maynard, D., Motta, E., Thirunarayan, K. (eds.) ISWC 2009. LNCS, vol. 5823, pp. 424–440. Springer, Heidelberg (2009)
8. Onan, A.: Classifier and feature set ensembles for web page classification. J. Inf. Sci. **42**(2), 150–165 (2016)
9. Beeferman, D., Berger, A.: Agglomerative clustering of a search engine query log. In: Proceedings of the sixth ACM Conference on Knowledge Discovery and Data Mining, pp. 407–416. ACM, New York (2000)
10. Baeza-Yates, R., Hurtado, C., Mendoza, M.: Query recommendation using query logs in search engines. In: Lindner, W., Mesiti, M., Türker, C., Tzitzikas, Y., Vakali, A.I. (eds.) EDBT 2004. LNCS, vol. 3268, pp. 588–596. Springer, Heidelberg (2004)
11. Mei, Q., Zhou, D., Church, K.: Query suggestion using hitting time. In: Proceedings of the 17th ACM Conference on Information and Knowledge Management, pp. 469–478. ACM, New York (2008)
12. Ma, H., Yang, H., King, I., Lyu, M.: Learning latent semantic relations from clickthrough data for query suggestion. In: Proceedings of the 17th ACM Conference on Information and Knowledge Management, pp. 709–718. ACM, New York (2008)
13. Song, Y., He, L.W.: Optimal rare query suggestion with implicit user feedback. In: Proceedings of the 19th International Conference on World Wide Web, pp. 901–910. ACM, New York (2010)
14. Anagnostopoulos, A., Becchetti, L., Castillo, C., Gionis, A.: An optimization framework for query recommendation. In: Proceedings of the Third ACM International Conference on Web Search and Data Mining, pp. 161–170. ACM, New York (2010)

15. Ma, H., Lyu, M.R., King, I.: Diversifying query suggestion results. In: Proceedings of AAAI (2010)
16. Song, Y., Zhou, D., He, L.W.: Post-ranking query suggestion by diversifying search results. In: Proceedings of the 34th International Conference on Research and Development in Information Retrieval, pp. 815–824. ACM, New York (2011)
17. Jiang, D., Leung, K.W.T., Yang, L., Ng, W.: Query suggestion with diversification and personalization. Knowl.-Based Syst. **89**, 553–568 (2015)
18. Kraft, R., Zien, J.: Mining anchor text for query refinement. In: Proceedings of the 13th International Conference on World Wide Web, pp. 666–674. ACM, New York (2004)
19. Jones, R., Rey, B., Madani, O., Greiner, W.: Generating query substitutions. In: Proceedings of the 15th International Conference on World Wide Web, pp. 387–396. ACM, New York (2006)
20. Wang, X., Zhai, C.: Mining term association patterns from search logs for effective query reformulation. In: Proceedings of the 17th Conference on Information and Knowledge Management, pp. 479–488. ACM, New York (2008)
21. Jansen, B.J., Booth, D.L., Spink, A.: Patterns of query reformulation during web searching. J. Am. Soc. Inform. Sci. Technol. **60**(7), 1358–1371 (2009)
22. Dang, V., Croft, B.W.: Query reformulation using anchor text. In: Proceedings of the Third International Conference on Web Search and Data Mining, pp. 41–50. ACM, New York (2010)
23. Ronao, C.A., Cho, S.B.: Anomalous query access detection in RBAC-administered databases with random forest and PCA. Inf. Sci. **369**, 238–250 (2016)
24. Kruschwitz, U., Lungley, D., Albakour, M.D., Song, D.: Deriving query suggestion for site search. J. Am. Soc. Inform. Sci. Technol. **64**(10), 1975–1994 (2013)
25. Kim, Y., Seo, J., Croft, W.B., Smith, D.A.: Automatic suggestion of phrasal-concept queries for literature search. Inf. Process. Manage. **50**(4), 568–583 (2014)
26. Momtazi, S., Lindenberg, F.: Generating query suggestions by exploiting latent semantics in query logs. J. Inf. Sci. **46**(2), 437–448 (2016)
27. Breiman, L.: Random forests. Mach. Learn. **45**(1), 5–32 (2001)
28. Verikas, A., Gelzinis, A., Bacauskinene, M.: Mining data with random forests: a survey and results of new test. Pattern Recogn. **44**(2), 330–349 (2011)
29. Bader-El-Den, M., Gaber, M.: GARF: towards self-optimised random forests. In: Huang, T., Zeng, Z., Li, C., Leung, C.S. (eds.) ICONIP 2012. LNCS, vol. 7664, pp. 506–515. Springer, Heidelberg (2012)
30. Elyan, E., Gaber, M.M.: A genetic algorithm approach to optimising random forests applied to class engineered data. Inf. Sci. **384**, 220–234 (2017)
31. Milne, D., Witten, I.H.: Learning to link with Wikipedia. In: Proceedings of the 17th Conference on Information and Knowledge Management, pp. 509–518, ACM, New York (2008)
32. Breiman, L.: Bagging predictors. Mach. Learn. **24**, 123–140 (1996)
33. Nisbet, R., Miner, G., Elder, J.: Handbook of Statistical Analysis and Data Mining Applications. Academic Press, New York (2009)
34. Caramia, M., Dell'Olmo, P.: Multi-objective management in freight logistics. Springer, London (2008)
35. Hall, M., Frank, E., Holmes, G., Pfahringer, B., Reutemann, P., Witten, I.H.: WEKA data mining software: an update. ACM SIGKDD Explor. Newsl. **11**(1), 10–18 (2009)

Financial Knowledge Instantiation from Semi-structured, Heterogeneous Data Sources

Francisco García-Sánchez[1(✉)], José Antonio García-Díaz[1], Juan Miguel Gómez-Berbís[2], and Rafael Valencia-García[1]

[1] Dpto. Informática y Sistemas, Facultad de Informática, Universidad de Murcia, Murcia, Spain
{frgarcia, joseantonio.garcia8, valencia}@um.es
[2] Departamento de Informática, Universidad Carlos III de Madrid, Leganés, Spain
juanmiguel.gomez@uc3m.es

Abstract. Decision making in the financial domain is a very challenging endeavor. The risk associated to this process can be diminished by gathering as much accurate and pertinent information as possible. However, most relevant data currently lies over the Internet in heterogeneous sources. Semantic Web technologies have proven to be a useful means to integrate knowledge from disparate sources. In this work, a framework to semi-automatically populate ontologies from data in semi-structured documents is proposed. The validation results in the financial domain are very promising.

Keywords: Ontology population · Knowledge-based decision support system Financial domain

1 Introduction

Decision making can be thought of as a composition of a certain amount of information and a portion of risk [1]. The risk taken when making a decision is tightly associated with uncertainty, that is, the lack of information (actually, with the lack of pertinent information to support the decision). The more the pertinent, timely and accurate information, the less the risk in decision making and the less the information, the higher the risk. Consequently, information as accurate and complete as possible is necessary to minimize the decision making process risk [2]. This involves the need for (1) identifying – among the huge amount of data – which data are relevant for decision making, (2) finding out the implicit relations among data, and (3) data processing to determine future actions.

In the financial sector the situation becomes even more critical due to the intrinsic complexity of the analytical tasks within this field [3]. Financial analysis and, more concretely, stock market price prediction is regarded as one of the most challenging tasks of financial time series prediction. The difficulty of forecasting arises from the inherent non-linearity and non-stationarity of the stock market and financial time series

R. Silhavy (Ed.): CSOC 2018, AISC 764, pp. 103–110, 2019.
https://doi.org/10.1007/978-3-319-91189-2_11

[4]. In the last few years different data mining technologies such as neuronal networks or support vector machines have been applied to solve this problem, but satisfactory results have not been achieved [5]. The financial domain is becoming a knowledge intensive domain, where a huge number of businesses and companies hinge on, with a tremendous economic impact in our society. The information required to make decisions in this field might come from disparate, heterogeneous data sources [6]. Thus, there is a need for more accurate and powerful strategies for storing data and knowledge in this domain to support decision making.

In view of the aforementioned facts, integrating financial information, as well as performing a faster and more accurate analysis across these disparate financial information sources, remains a fundamental challenge. The technologies associated to the Semantic Web [7] and Linked Data [8] have proved effective for the automatic treatment of information in different contexts [9, 10]. The underlying ontological models, which are based on logical formalisms, enable computer systems to somehow interpret the information that is being managed [11]. They also allow to carry out advanced reasoning and inferencing processes. The scientific community within these research fields have developed tools that make use of semantic technologies to both (i) integrate data from heterogeneous data sources [12, 13], and (ii) implement complex financial analyses [3, 14] with different degrees of success. In this paper we propose a framework to semi-automatically instantiate ontologies from semi-structured documents. The system supports the integration of information coming from heterogeneous Web resources in different formats (JSON, PDF, RSS, etc.). A proof-of-concept implementation of the framework has been validated in the financial domain.

The rest of the paper is organized as follows. In Sect. 2 background information on financial ontologies and ontology population approaches is provided. The framework proposed in this work to populate ontologies from semi-structured, heterogeneous data sources is described in Sect. 3. In Sect. 4 the validation of the framework in the financial domain is presented. Finally, conclusions and future work are put forward in Sect. 5.

2 Related Work

Today, most websites do not provide semantic information. Without such semantic information and given the ever-increasing size of the Web, the identification and automatic processing of relevant information is becoming increasingly difficult. In recent years, a number of approaches with the purpose of structuring in the form of ontologies non-structured and semi-structured data sources have appeared. In this section, the most representative ontologies in the financial domain will be listed and different ontology population approaches will be discussed.

2.1 Ontologies in the Financial Domain

In the last few years, several financial ontologies have been developed [15–18]. The eXtensible Business Reporting Language (XBRL, https://www.xbrl.org/) schema is an open international standard based on XML to define and exchange financial and business information. The Financial Regulation XBRL ontology (FigRegOnt) [15], is a one-to-one OWL/RDF representation of XBRL. FigRegOnt extends and aligns other ontologies in the legal and financial domains such as the Financial Industry Business Ontology (FIBO). FIBO [16] defines in OWL financial industry terms, definitions and synonyms. In [17] the authors describe a decision support system for financial analysis based on an ontology representing the elements required in the analysis processes. On the other hand, the authors in [18] propose an opinion mining tool for financial news that makes use of a domain ontology providing a controlled vocabulary of the concepts and relations in the financial news domain.

We have developed a financial ontology focused on the stock market domain and based on the above referred ontologies. The ontology has been defined in OWL and covers four main financial concepts, namely, '*Financial Market*', '*Financial Intermediary*', '*Asset*', and '*Legislation*'. The ontology has a total of 247 classes, 86 '*subClassOf*' properties, 34 datatype properties, 38 object properties and 87 restrictions.

2.2 Ontology Population

Ontology instantiation has to do with the extraction and classification of instances of the concepts and relations that have been defined in the ontology. Instantiating ontologies with new knowledge is a relevant step towards the provision of valuable ontology-based knowledge services. However, performing such task manually is time-consuming and error-prone. As a result, research has shifted attention to automating this process, introducing ontology population, which refers to a set of methodologies for automatically identifying and adding new instances of concepts from an external source into an ontology [19]. Ontology population does not affect the concept hierarchies and the relations in the ontology, leaving the structure of the ontology unmodified. What is affected are the individuals (a.k.a. concept instances) and the relationships between individuals in the domain.

Most ontology population tools focus on extracting knowledge from natural language text, offering significant advantages over traditional export formats [20, 21]. However, other sources of information more structured than free text are very often neglected [22]. In [20] the authors show a technique for extracting instances of a reduced number of ontological relations from agricultural texts by taking into account a set of axioms based on textual patterns. The authors in [21] propose a methodology for extracting ontological terms and relations from texts in the biomedical field based on linguistic and statistical methods. On the other hand, PROPhet, an instance extraction engine dealing with Linked Data sources, is described in [22]. PROPhet takes advantages of the already structured and semantically rich sets of Linked Data. In this paper, we present a semi-automatic method for ontology population from semi-structured texts. Our method deals with the large amount of Web-available free text documents, but it focuses on the semi-structured parts of such documents.

3 Knowledge Gathering in the Financial Domain

In this section, the framework proposed in this work is described in detail. It aims at populating an ontology with all the relevant knowledge gathered from semi-structured texts in a supervised way. Next, the architecture of the framework is presented and its main components are explained.

3.1 Proposed Framework

The architecture of the proposed system is shown in Fig. 1. It is composed of two main components, namely, the supervised document parser and the JSON2RDF module. The input of the system consists of a collection of Web-available information resources. As output, the system produces ontology instances, which are stored in the repository. In a nutshell, the system works as follows. The semi-structured content (i.e., tables) of data sources available on the Internet are parsed to extract the information that can be gathered from the text. Users are shown the elements identified by the parser and choose those that should be stored in the knowledge base. Then, the tables are transformed into an internal format in JSON. The produced JSON file is finally processed by the JSON2RDF module generating the correspondences between the data in the semi-structured texts and the concepts in the ontology. At last, the new discovered ontology instances are stored in the knowledge base.

Fig. 1. Proposed framework functional architecture

3.2 Supervised Document Parser

The main aim of this approach is to make the ontology population algorithm independent of the data source. To this end, the supervised document parser transforms the input data into a common representation format in JSON (a parser must be developed for each input format). The first step is to gather the information available in the processed documents (users must indicate the URLs of the sites that they want the system to analyze). Then, the parser looks for the tables that are contained within the source documents and end-users are shown the full list of identified tables. Users must choose the tables to be further processed and those tables are transformed into a JSON-based internal representation format. This shared data structure is described by a JSON Schema (http://json-schema.org/) and includes two main elements, namely, '*row*' and '*classGroup*'. Rows represent the rows in the input table and are composed of tuples defined by attribute-value pairs. '*classGroup*' elements contain a set of classes of the domain ontology.

The way each table is mapped into a JSON file complying with the JSON Schema is as follows. First, users must point out the classes of the domain ontologies that are represented in the table. Then, a '*row*' element is created for each row in the table. These rows are populated with the pairs attribute-value which correspond to the column name (attribute) and the content of the appropriate row/column cell (value).

3.3 From JSON to RDF: JSON2RDF Module

The JSON2RDF module populates the ontology by creating the ontology instances associated with the information previously gathered from the tables. First, a matching is produced to create the instances and decide the ontology classes to which they belong. Second, relations are established between the previously created instances. Finally, there is a consistency checking phase in which the system can identify contradictions.

In the matching phase an affinity function (see Eq. 1) is used to identify the group of classes closest to each row. The affinity of a row ('*R*') with a group of classes ('*G*') is the sum of the affinities of the row with each of the '*n*' classes that belong to such group ('*Affinity'*'), which is derived from the similarity between the attributes of each tuple in the row with the labels (existing annotations in the ontology) of the class ('*Class(i)*') and its attributes.

$$Affinity(R, G) = \sum_{i=1}^{n} Affinity'(R, Class(i)) \tag{1}$$

Once the closest group of classes ('*G*') has been identified, the instances within the referred row ('*R*') are created. An instance is created for each ontology class within '*G*'. The datatype properties of the instances are set by comparing the labels in them with the attribute-value pairs in '*R*'. To avoid redundancy, all classes in our ontology include a datatype property '*name*' that provides a unique identifier to each instance. If two instances of the same class have the same identifier, the system can conclude that they both refer to the same concept.

The relations (i.e., object properties) between instances are set after all the instances have been created. For this, the system first looks for object properties between the classes in '*G*' and establishes the corresponding relations between the previously created instances. Then, the system examines the object properties in all the other classes from different class groups creating the necessary relationships.

4 Validation

For evaluation purposes, we built a prototype of the framework and the parsers required to gather knowledge from different sites. It is important to note that Web sites often use the same format of tables to show data. Therefore, if the system works for a number of tables, it is likely that it can also support other, non-initially considered table formats. In the experiment, HTML tables from three different Web sites have been selected, namely, the Madrid Stock Exchange, the London Stock Exchange and Börse Frankfurt.

The aim of this experiment is to evaluate the precision and recall of the proposed framework. The precision score (see Eq. 2) is the result of dividing the number of instances that the system has correctly identified and introduced in the ontology by the total amount of instances obtained. The recall score (see Eq. 3) is the result of dividing the number of instances that the system has correctly identified and introduced in the ontology by the total amount of instances that are actually available in the processed table. Finally, the F1-measure (see Eq. 4) is the harmonic mean between precision and recall.

$$precision = (correct\ instances\ populated)/(total\ instances\ populated) \quad (2)$$

$$recall = (correct\ instances\ populated)/(total\ instances\ in\ table) \quad (3)$$

$$F1 = 2 * (precision * recall)/(precision + recall) \quad (4)$$

Table 1 shows the precision, recall and F1-measure scores obtained in this experiment for each Web site. It is possible to observe that the best results have been obtained for the Madrid Stock exchange site, where all the instances populated in the financial ontology are correct. The average (‘*AVG*’) precision, recall and F1-measure is equal or close to *0,9*, which represents a very satisfactory outcome as compared to other state-of-the-art ontology population tools. In summary, the validation experiment shows promising results.

Table 1. Results of the experiment

	Madrid Stock Exchange	London Stock Exchange	Börse Frankfurt	AVG
Precision	1,00	0,80	0,82	0,87
Recall	0,95	0,90	0,85	0,90
F1-measure	0,97	0,85	0,83	0,89

5 Conclusions and Future Work

To reduce the risk associated with uncertainty in decision making, it is necessary to have as much accurate, pertinent and timely information as possible. Even more so in the highly complex financial analysis field. To this end, in this paper we propose a framework capable of gathering knowledge from Web-available information in any format. Our framework focuses on the semi-structured elements of the documents, thus achieving remarkable precision results as compared with other ontology population techniques dealing with free text.

As future work, we plan to reduce user participation by automatically suggesting candidate classes associated with the tables in the documents. Also, more sophisticated approaches for redundancy checking and object properties creation will be examined. On the other hand, developing an ontology with the large set of annotations required by our approach can be costly; alternatively, the use of a thesaurus in addition to the semantic annotations will be studied. Finally, our next step is to investigate ontology-based financial analysis techniques that can take advantage of the gathered knowledge to provide proper insights to end users.

Acknowledgements. This work has been supported by the Spanish National Research Agency (AEI) and the European Regional Development Fund (FEDER/ERDF) through project KBS4FIA (TIN2016-76323-R).

References

1. Kochenderfer, M.J.: Decision Making Under Uncertainty: Theory and Application. Lincoln Laboratory Series. The MIT Press, Cambridge (2015)
2. Mercier-Laurent, E.: Decision making in dynamic world — facing new crisis and risks. In: 4th International Conference on Control, Decision and Information Technologies (CoDIT), Barcelona, Spain, pp. 433–438. IEEE (2017)
3. García-Sánchez, F., Paredes-Valverde, M.A., Valencia-García, R., Alcaraz-Mármol, G., Almela, A., Almela, Á.: KBS4FIA: leveraging advanced knowledge-based systems for financial information analysis. Proces. del Leng. Nat. **59**, 145–148 (2017)
4. Kazem, A., Sharifi, E., Hussain, F.K., Saberi, M., Hussain, O.K.: Support vector regression with chaos-based firefly algorithm for stock market price forecasting. Appl. Soft Comput. **13**, 947–958 (2013). https://doi.org/10.1016/J.ASOC.2012.09.024
5. Rodríguez-González, A., García-Crespo, Á., Colomo-Palacios, R., Guldrís Iglesias, F., Gómez-Berbís, J.M.: CAST: using neural networks to improve trading systems based on technical analysis by means of the RSI financial indicator. Expert Syst. Appl. **38**, 11489–11500 (2011). https://doi.org/10.1016/j.eswa.2011.03.023
6. Hernes, M., Bytniewski, A.: Integration of collective knowledge in financial decision support system. In: Nguyen, N.T., Trawiński, B., Fujita, H., Hong, T.-P. (eds.) Proceedings of Intelligent Information and Database Systems: 8th Asian Conference, ACIIDS 2016, Part I, Da Nang, Vietnam, 14–16 March 2016, pp. 470–479. Springer, Heidelberg (2016)
7. Shadbolt, N., Berners-Lee, T., Hall, W.: The semantic web revisited. IEEE Intell. Syst. **21**, 96–101 (2006). https://doi.org/10.1109/MIS.2006.62
8. Bizer, C., Heath, T., Berners-Lee, T.: Linked data - the story so far. Int. J. Semant. Web Inf. Syst. **5**, 1–22 (2009). https://doi.org/10.4018/jswis.2009081901
9. Lopez-Lorca, A.A., Beydoun, G., Valencia-Garcia, R., Martinez-Bejar, R.: Supporting agent oriented requirement analysis with ontologies. Int. J. Hum. Comput. Stud. **87**, 20–37 (2016). https://doi.org/10.1016/j.ijhcs.2015.10.007
10. Lagos-Ortiz, K., Medina-Moreira, J., Paredes-Valverde, M.A., Espinoza-Morán, W., Valencia-García, R.: An ontology-based decision support system for the diagnosis of plant diseases. J. Inf. Technol. Res. **10**, 42–55 (2017). https://doi.org/10.4018/JITR.2017100103

11. Studer, R., Benjamins, R., Fensel, D.: Knowledge engineering: principles and methods. Data Knowl. Eng. **25**, 161–197 (1998). https://doi.org/10.1016/S0169-023X(97)00056-6
12. García-Sánchez, F., Fernández-Breis, J.T., Valencia-García, R., Gómez, J.M., Martínez-Béjar, R.: Combining semantic web technologies with multi-agent systems for integrated access to biological resources. J. Biomed. Inform. **41**, 848–859 (2008). https://doi.org/10.1016/j.jbi.2008.05.007
13. Santipantakis, G., Kotis, K., Vouros, G.A.: OBDAIR: ontology-based distributed framework for accessing, integrating and reasoning with data in disparate data sources. Expert Syst. Appl. **90**, 464–483 (2017). https://doi.org/10.1016/j.eswa.2017.08.031
14. Rodríguez-González, A., Colomo-Palacios, R., Guldris-Iglesias, F., Gómez-Berbís, J.M., García-Crespo, A.: FAST: fundamental analysis support for financial statements. using semantics for trading recommendations. Inf. Syst. Front. **14**, 999–1017 (2012). https://doi.org/10.1007/s10796-011-9321-1
15. XBRL Ontology - Financial Regulation Ontology. http://finregont.com/xbrl/
16. EDM Council: Financial Industry Business Ontology^{TM}. https://spec.edmcouncil.org/fibo/
17. Korczak, J., Dudycz, H., Nita, B., Oleksyk, P.: Towards process-oriented ontology for financial analysis. In: Proceedings of the Federated Conference on Computer Science and Information Systems, pp. 981–987 (2017)
18. Salas-Zárate, M.P., Valencia-García, R., Ruiz-Martínez, A., Colomo-Palacios, R.: Feature-based opinion mining in financial news: an ontology-driven approach. J. Inf. Sci. **43**, 458–479 (2017). https://doi.org/10.1177/0165551516645528
19. Buitelaar, P., Buitelaar, P., Cimiano, P.: Ontology Learning and Population: Bridging the Gap Between Text and Knowledge. IOS Press, Amsterdam (2008)
20. Kaushik, N., Chatterjee, N.: Automatic relationship extraction from agricultural text for ontology construction. Inf. Process. Agric. (2017). https://doi.org/10.1016/j.inpa.2017.11.003
21. Medina-Moreira, J., Lagos-Ortiz, K., Luna-Aveiga, H., Apolinario-Arzube, O., Salas-Zárate, M.P., Valencia-García, R.: Knowledge acquisition through ontologies from medical natural language texts. J. Inf. Technol. Res. **10**, 56–69 (2017). https://doi.org/10.4018/jitr.2017100104
22. Mitzias, P., Riga, M., Kontopoulos, E., Stavropoulos, T.G., Andreadis, S., Meditskos, G., Kompatsiaris, I.: User-driven ontology population from linked data sources. In: Ngonga Ngomo, A.-C., Křemen, P. (eds.) Knowledge Engineering and Semantic Web, KESW 2016. Communications in Computer and Information Science, pp. 31–41. Springer, Cham (2016)

Hierarchical Fuzzy Deep Leaning Networks for Predicting Human Behavior in Strategic Setups

Arindam Chaudhuri[1(✉)] and Soumya K. Ghosh[2]

[1] Samsung R&D Institute Delhi Noida, Noida 201304, India
arindam_chau@yahoo.co.in, arindamphdthesis@gmail.com
[2] Department of Computer Science Engineering, Indian Institute of Technology Kharagpur, Kharagpur 701302, India
skg@iitkgp.ac.in

Abstract. The prediction of human behavior in strategic setups is important problem in many business situations. The uncertainty in human behavior complicates the problem to greater extent. It is generally assumed that players are rational in nature. But this assumption is far from actual scenarios. To address this, we propose hierarchical fuzzy deep learning network that automatically models cognitively without any expert knowledge. The architecture allows hierarchical network to generalize across different input and output dimensions by using matrix units rather than scalar units. The network's performance is significantly better than previous models which depend on expert constructed features. The experiments are performed using datasets prepared from RPS game played over specified network and responder behavior from CT experiments. The proposed deep learning network has superior prediction performance as compared to others. The experimental results demonstrate efficiency of proposed approach.

Keywords: Behavior prediction · Strategic setups · Game theory
Deep learning · Fuzzy sets

1 Introduction

The present business environment involves strategic interactions represented through multi-agent systems. These multi-systems are designed and analysed mathematically through game theory [1]. Some significant examples include search engines, spectrum auctions and security systems. In these applications choices are optimized considering assumptions about preferences, beliefs and players' capabilities [2]. The game theoretic approach assumes rational players which maximizes expected utility and optimizes human decision making. This asks for development of accurate model to predict human behavior in strategic settings incorporating biases and limitations considering play observations and cognitive psychology insights. The successful predictive models in this setup combines iterative reasoning and noisy best response notations. It uses hand-crafted features to model non-strategic players' behavior. Deep learning has played a significant role in increasing predictive accuracy by fitting highly flexible

R. Silhavy (Ed.): CSOC 2018, AISC 764, pp. 111–121, 2019.
https://doi.org/10.1007/978-3-319-91189-2_12

models learning novel representations. The success in using deep learning models lies in selecting good design towards encoding basic domain input knowledge with respect to its topological structure to learn better features. Some significant applications of deep learning in multi-agent settings include Clark and Storkey [3] and Silver et al. [4]. However modeling choice constrains more general architectures to solution space subset that contains optimal solutions.

This research work addresses solutions in human behavioral game theoretic framework. It identifies prior assumptions extending deep learning to predict behavior in strategic setups encoded as two player and normal-form games. A key property required from such model is invariance to game size. The current deep models typically assume either fixed-dimensional input or arbitrary-length fixed-dimensional inputs sequence. The outputs in both cases are fixed-dimensional output. It is prior belief that permutes rows and columns in input does not change output beyond corresponding permutation. Here focus is on constructing an architecture that generalizes across general-sum, normal form games. Invariance is enforced to network's input size. Fully convolutional networks [5] achieve invariance to image size in similar by manner replacing all fully connected layers with convolutions. However this architecture is derived independently using game theoretic invariances. This paper is organized as follows. In Sect. 2 computational method of HFDLN is highlighted. This is followed by experiments and results in Sect. 3. Finally in Sect. 4 conclusions are given.

2 Computational Method

In this section framework of proposed HFDLN model is presented. The schematic representation of the prediction system is given in Fig. 1.

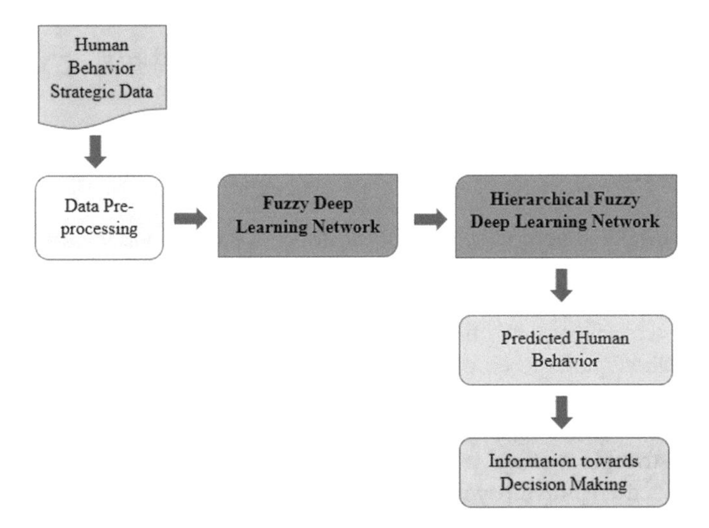

Fig. 1. Prediction framework through hierarchical fuzzy deep learning network

2.1 Problem Definition

The research problem entails in predicting behaviour of humans in strategic setups. In order to achieve this, we propose deep learning based predictor viz HFDLN to analyse human behavior data patterns at RPS game played over network [6] and responder behavior from CT experiments [7]. The prediction task on various strategic aspects of human behavior provide information for several decision-making activities.

2.2 Datasets

The experimental datasets are adopted and prepared from two different sources viz RPS game played over specified network [6] and CT experiments responded behavior [7]. RPS game is symmetric, two-player zero-sum game. The dataset was prepared through threads of 30 one-shot games. Every 6 s shot delay was considered. In case of non-reaction default gesture was considered. A thread consisted of 30×6 s = 3 min duration. This game has one mixed strategy equilibrium with identical probability distribution between three gestures. One hundred computer science undergraduates were recruited. They were in average 21 years old and 70 of them were male. They played thread twice against another test person. Between two threads they played other games. This way 3000 one-shot games or 6000 single human decisions were gathered. Each person got $ 3 for winning and $ 1 for draw. The persons who played against each other sat in two separate rooms. One of the players used cyber-glove and other one mouse as input for gestures. The information collected contained own last and actual choice, opponents last choice, timer and already gained money. CT experiments dataset was modified to contain 3796 single human decisions of 100 participating subjects. Positive decision of responder modifies monetary payoff of both players while negative respond does not change anything. Payoff varied between $10.50 and $–10.55 for responder. In 500 cases responder modification was zero. The responder's equilibrium accepts only proposals which increase his payoff regardless of proposer's payoff.

2.3 Fuzzy Deep Learning Networks

In this section FDLN shown in Fig. 2 is evaluated in terms of fuzzy sets [8] and deep learning networks. The inherent uncertainty in strategic setups is addressed using normal form game utility matrices represented through trapezoidal fuzzy numbers [8]:

$$trapezoid(x; a, b, c, d) = \begin{cases} 0 & x \leq a \\ \frac{x-a}{b-a} & a \leq x \leq b \\ 1 & b \leq x \leq c \; x \in \mathbb{R} \\ \frac{d-x}{d-c} & c \leq x \leq d \\ 0 & d \leq x \end{cases} \tag{1}$$

The game utility matrices are represented as $\widetilde{\boldsymbol{UM}}^{(row)}$ and $\widetilde{\boldsymbol{UM}}^{(col)}$. The weights associated with each row of $\widetilde{\boldsymbol{UM}}^{(row)}$ and $\widetilde{\boldsymbol{UM}}^{(col)}$ must be similar. The corresponding assumption about column player implies that weights associated with each column of $\widetilde{\boldsymbol{UM}}^{(row)}$ and $\widetilde{\boldsymbol{UM}}^{(col)}$ must be similar. Both assumptions can be satisfied by applying

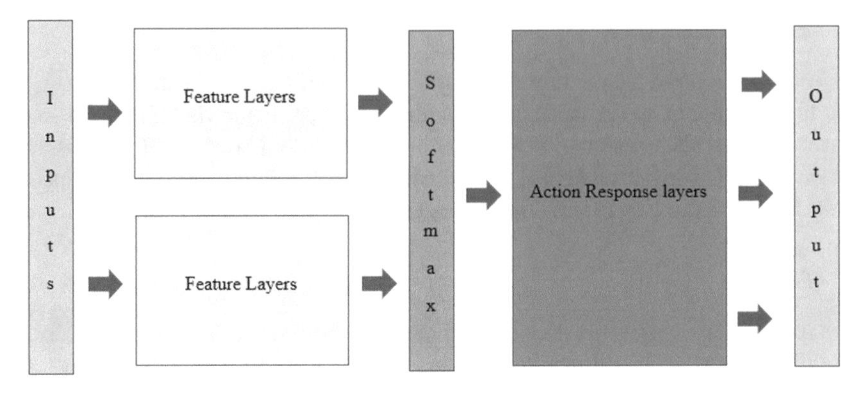

Fig. 2. Schematic representation of FDLN architecture

single scalar weight to each utility matrices such that $w_r\widetilde{UM}^{(row)} + w_c\widetilde{UM}^{(col)}$. This is generalized as standard feed-forward network fitting more complex functions. The utility values are flattened into single vector of length $mn + nm$. Then function is learned that maps to m-simplex output through multiple hidden layers. Here games are restricted to fixed size input. Feed-forward networks are generalized poorly as network over fitted training data which is handled using invariance through data augmentation. The assumption that players are indifferent to order in which actions are presented implies invariance to permutations of payoff matrix. Incorporating this assumption by randomly permuting rows or columns of payoff matrix at every training epoch improves generalization performance. However network is still limited to games of size it was trained on. Invariance is enforced in model architecture rather than through data augmentation. Then pooling units are added by incorporating iterative response ideas inspired through behavioral game theory models. This results in model that gives flexible representation to all models covering huge novel models space well identified automatically. The model is also invariant to input payoff matrix size, differentiable end to end and trainable using standard gradient-based optimization.

The model comprises of feature layers and action response layers. The feature layers take row and column player's normalized utility matrices $\widetilde{UM}^{(row)}$ and $\widetilde{UM}^{(col)}$ $\in \tilde{\mathbb{R}}^{m \times n}$ (with $\tilde{\mathbb{R}}$ as fuzzy real number set) as input where row player has m actions and column player has n actions. The feature layers consist of multiple levels of hidden matrix units $\widetilde{\mathbf{HM}}_{i,j}^{(row)} \in \tilde{\mathbb{R}}^{m \times n}$ each of which calculates weighted sum of units below and applies non-linear activation function. Each layer of hidden units is followed by pooling units which output aggregated versions of hidden matrices to be used by following layer. After multiple layers matrices are aggregated to vectors and normalized to distribution over actions $\tilde{\mathbf{a}}_i^{(row)} \in \tilde{\Delta}^m$ softmax units. These distributions are features because they encode higher level representations of input matrices that are combined towards output distribution. Since iterative strategic reasoning is important phenomenon in human decision making, this reasoning is incorporated in this model. This is achieved by computing features for column player in same manner by applying feature layers to transpose of input matrices which produces $\tilde{\mathbf{a}}_i^{(col)} \in \tilde{\Delta}^{\boldsymbol{n}}$. Each action response layer for given player then takes opposite player's preceding action response layers as input and uses them to construct distributions over respective players' outputs. The final output $\tilde{\mathbf{v}} \in \tilde{\Delta}^{\boldsymbol{m}}$ is weighted sum of all action response layers' outputs. A hidden matrix unit taking all preceding hidden matrix units as input are calculated as:

$$\widetilde{\mathrm{HM}}_{v,i} = \Phi\left(\sum\nolimits_j \tilde{\mathrm{w}}_{v,i,j}\widetilde{\mathrm{HM}}_{l-1,i} + \tilde{b}_{\boldsymbol{v},\boldsymbol{i}}\right) \qquad \widetilde{\mathrm{HM}}_{v,i} \in \tilde{\mathbb{R}}^{m \times n} \tag{2}$$

Here $\widetilde{\mathbf{HM}}_{\boldsymbol{v},i}$ is $\boldsymbol{i}^{th}$ hidden unit matrix for layer $\boldsymbol{v}$, $\tilde{\mathbf{w}}_{\boldsymbol{v},i,j}$ is $\boldsymbol{j}^{th}$ scalar weight, $\tilde{\boldsymbol{b}}_{\boldsymbol{v},i}$ is scalar bias variable and $\boldsymbol{\Phi}$ is non-linear activation function applied element-wise. Analogous to traditional feed-forward neural network output of each hidden unit is nonlinear transformation of weighted sum of preceding layer's hidden units. This architecture differs by considering matrix at each hidden unit instead of scalar. Here each hidden unit maps tensor of outputs from previous layer into matrix output. The number of parameters are considerably reduced by a factor $\boldsymbol{mn}$. It reduces the degree to which the network overfits and it makes model invariant to input matrix sizes. The number of parameters in each layer depends on number of hidden units in preceding layer but not on input and output matrix sizes. This leads generalization towards input sizes that do not appear in training data. A limitation of weight used in hidden matrix units is that it forces independence between elements of matrices preventing network from learning functions that compare values of related elements. Each element of matrices corresponds to an outcome in normal form game. A game theoretic notion compares playoffs set associated with each players' actions that results to outcome. This corresponds to row and column of each matrix associated with particular element. This serves as motivation for pooling units allowing information sharing by outputting aggregated versions of their input matrix which are used by later layers to learn and compare values of particular cell in matrix and its row or column wise aggregates.

$$\begin{aligned}\widetilde{\mathbf{HM}} &\rightarrow \left\{\widetilde{\mathbf{HM}}_{col}, \widetilde{\mathbf{HM}}_{row}\right\} \\ &= \left\{\begin{pmatrix} \max_i \widetilde{\mathrm{hm}}_{i,1} & \max_i \widetilde{\mathrm{hm}}_{i,2} & \ldots\ldots \\ \vdots & \vdots & \ldots\ldots \\ \max_i \widetilde{\mathrm{hm}}_{i,1} & \max_i \widetilde{\mathrm{hm}}_{i,2} & \ldots\ldots \end{pmatrix}, \begin{pmatrix} \max_j \widetilde{\mathrm{hm}}_{1,j} & \max_j \widetilde{\mathrm{hm}}_{1,j} & \ldots\ldots \\ \vdots & \vdots & \ldots\ldots \\ \max_j \widetilde{\mathrm{hm}}_{m,j} & \max_j \widetilde{\mathrm{hm}}_{m,j} & \ldots\ldots \end{pmatrix}\right\}\end{aligned} \tag{3}$$

A pooling unit takes matrix as input and outputs two matrices constructed from row and column preserving pooling operations respectively. A pooling operation could be any continuous function that maps from $\tilde{\mathbb{R}}^{\boldsymbol{n}} \rightarrow \tilde{\mathbb{R}}$. The *max* function is used because it is necessary to represent known behavioural functions and offers best empirical performance of functions. The equation above shows pooling layer with *max* functions for some arbitrary matrix $\widetilde{\mathbf{HM}}$. The first of the two outputs $\widetilde{\mathbf{HM}}_{\boldsymbol{col}}$ is column-preserving such that it selects maximum value in each column of $\widetilde{\mathbf{HM}}$ and then stacks resulting $\boldsymbol{n}$-dimensional vector $\boldsymbol{m}$ times where dimensionality of $\widetilde{\mathbf{HM}}$ and $\widetilde{\mathbf{HM}}_{\boldsymbol{col}}$ are identical. Similarly row-preserving output constructs vector of *max* elements in each column and stacks resulting m-dimensional vector n times such that $\widetilde{\mathbf{HM}}_{row}$ and $\widetilde{\mathbf{HM}}$ have similar dimensionality. The vectors that result from pooling operation in this fashion are stacked such that hidden units from next layer in network may have $\widetilde{\mathbf{HM}}$, $\widetilde{\mathbf{HM}}_{col}$ and $\widetilde{\mathbf{HM}}_{row}$ as input. This allows later hidden units to learn function where each output element is function of both corresponding element from matrices below as well as their row and column preserving maximums. The model predicts distribution over row player's actions. To achieve this mapping is required from hidden matrices in final layer $\widetilde{\mathbf{HM}}_{V,i} \in \widetilde{\mathbb{R}}^{m \times n}$ of network onto point on m-simplex $\tilde{\Delta}^m$. This mapping is applied through row-preserving sum to each final layer hidden matrices $\widetilde{\mathbf{HM}}_{V,i}$. The summation is performed uniformly over columns of matrix as described earlier. Then softmax function is applied to convert each resulting vectors $\mathbf{h}_i$ into normalized distributions. This produces k features $\mathbf{f}_i$ each of which is distribution over row player's m actions:

$$\begin{aligned}\tilde{\mathbf{a}}_i &= \text{softmax}\left(\widetilde{\mathbf{hm}}^{(i)}\right) \text{where } \widetilde{\mathbf{hm}}_j^{(i)} \\ &= \sum\nolimits_{k=1}^{n} \widetilde{\text{hm}}_{j,k}^{(i)} \forall j \in \{1, \ldots\ldots, m\},\ \widetilde{\text{hm}}_{j,k}^{(i)} \in \widetilde{\mathbf{HM}}^{(i)} i \in \{1, \ldots\ldots, k\} \end{aligned} \tag{4}$$

The output of features ar_0 are produced using weighted sum of individual features $\text{ar}_0 = \sum_{i=1}^{k} \tilde{\text{w}}_i \tilde{\mathbf{a}}_i$ where $\tilde{\text{w}}_i$ are optimized through simplex constraints $\tilde{\text{w}}_i \geq 0$ and $\sum_i \tilde{\text{w}}_i \neq 1$. Here each $\tilde{\mathbf{a}}_i$ is a distribution and weights $\tilde{\text{w}}_i$ are points on simplex with output of feature layers is mixture of distributions. The feature layers stated meets the objective of mapping from input payoff matrices towards distribution over row player's actions. This architecture does not explicitly represent iterative strategic reasoning. This ingredient is incorporated using action response layers. The first player responds to second's beliefs. The second responds back to this response by first player. This continues to some finite depth. The proportion of players in population who iterate at each depth is parameter of model. The architecture thus learns not to perform iterative reasoning. Considering output of feature layers as $\mathbf{ar}_0^{(row)} = \sum_{i=1}^{k} \tilde{\text{w}}_{0i}^{(row)} \tilde{\mathbf{a}}_i^{(row)}$ an index (row) is included to refer to output of row player's action response layer $\text{ar}_0^{(row)} \in \tilde{\Delta}^m$. By applying feature layers to transposed version of input matrices, the model outputs corresponding $\text{ar}_0^{(col)} \in \tilde{\Delta}^n$ for column player which expresses row player's beliefs about which actions column player will choose. Each action response layer composes its output by calculating expected value of an internal representation of utility with respect to its belief distribution over opposition actions. For internal utility representation weighted sum of final layer of hidden layers $\sum_i \tilde{\text{w}}_i \widetilde{\mathbf{HM}}_{V,i}$ because each $\widetilde{\mathbf{HM}}_{V,i}$ is already some non-linear transformation of original payoff matrix. This allows model to express utility as transformation of original payoffs. With matrix that results from this sum, expected utility is computed with respect to vector of beliefs about opposition's choice of actions $\mathbf{ar}_j^{(col)}$ considering dot product of weighted sum and beliefs. When this process of responding to beliefs about one's opposition more than once is iterated, higher-level players will respond to beliefs $\mathbf{ar}_i \forall i$ less than their level and outputs weighted combination of these responses using some weights $\tilde{\text{s}}_{v,i}$. The vth action response layer for row player (row) is:

$$\mathbf{ar}_v^{(col)} = \text{softmax}\left(\lambda_v\left(\sum\nolimits_{j=0}^{v-1}\tilde{s}_{v,j}^{(row)}\left(\sum\nolimits_{i=1}^{k}\tilde{w}_{v,i}^{(row)}\widetilde{\mathbf{HM}}_{V,i}^{(row)}\right).\mathbf{ar}_j^{(col)}\right)\right), \mathbf{ar}_v^{(col)} \in \tilde{\Delta}^m, \\ v \in \{1, \ldots\ldots, K\} \tag{5}$$

Here l indexes action response layer, λ_v is scalar sharpness parameter that sharpens resulting distribution, $\tilde{w}_{v,i}^{(row)}$ and $\tilde{s}_{v,j}^{(row)}$ are fuzzy scalar weights, $\widetilde{\mathbf{HM}}_{V,i}$ are row player's k hidden units from final hidden layer V, $\mathbf{ar}_j^{(col)}$ is output of column player's j^{th} action response layer and K is total number of action response layers. The parameters $\tilde{w}_{v,i}^{(row)}$ and $\tilde{s}_{v,j}^{(row)}$ are constrained to simplex and λ_l is used to sharpen output distribution so that sharpness of the distribution is optimized and relative weighting of its terms independently. The column player's action response layer $\mathbf{ar}_l^{(col)}$ is build using column player's internal utility representation $\widetilde{\mathbf{HM}}_{V,i}^{(col)}$ responding to row player's action response layers $\mathbf{ar}_l^{(row)}$. These layers are not used in final output directly but are relied upon by subsequent action response layers of row player. The model's final output is weighted sum of outputs of action response layers. This output needs to be valid distribution over actions. Because each action response layers also outputs distribution over actions, this can be achieved by constraining these weights to simplex. This ensures that output is mixture of distributions.

2.4 Hierarchical Fuzzy Deep Learning Networks for Predicting Human Behavior in Strategic Setups

The hierarchical version of FDLN viz HFDLN is proposed here. The computational benefits serve the major motivation. HFDLN is different from FDLN in terms of efficient classification accuracy based on similarities and running time when data volume grows. The model architecture is shown in Fig. 3. Some of the important descriptive statistics involving HFDLN are highlighted in Table 1. These statistics forms the deep learning networks' training parameters. The figures in Table 1 represent the optimal values of network parameters. The working of HFDLN is discussed below.

Table 1. Some important descriptive statistics of HFDLN

Network parameters	Parameter description
Number of network layers	8
Input units	8
Feature layers	8
Number of hidden layers	6
Number of hidden units per layer	14
Softmax units	8
Action response layers	4
Output unit	1

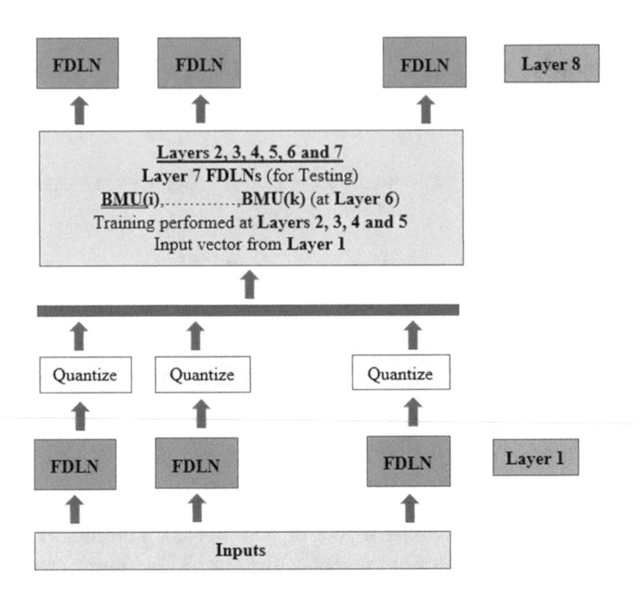

Fig. 3. Hierarchical fuzzy deep learning networks architecture

HFDLN architecture correlates data behavior across multiple features of relevance. This facilitates distribution of computational overheads in predictor construction. At 1st and 2nd layers relatively small FDLNs are used (8×8). The number of FDLNs are increased at layers 3, 4, 5, 6 and 7. FDLN in last layer are constructed over subset of examples for which neuron in 7th layer is Best Matching Unit (BMU). FDLN at last layer can be larger (40×40) than used in 1st to 7th layers. It improves resolution and discriminatory capacity of FDLN with less training overhead. Building HFDLN requires several data normalization operations. This provides for initial temporal pre-processing and inter-layer quantization between 1st to 7th layers. The pre-processing provides suitable representation for data and supports time based representation. The 1st DLN layer treats each feature independently with each data instance mapped to sequential values. In case of temporal representation standard FDLN has no capacity to recall histories of patterns directly. A shift register of length l is employed in which tap is taken at predetermined repeating interval k such that $l\%k = 0$ where $\%$ is modulus operator. The 1st level FDLNs only receive values from shift register. Thus, as each new connection is encountered (at left), content of each shift register location is transferred one location (to right) with previous item in l^{th} location being lost. In case of $\boldsymbol{n}$-feature architecture it is necessary to quantize number of neurons between 1st to 7th level FDLNs. The purpose of 2nd to 7th level FDLN is to provide an integrated view of input feature specific FDLNs developed in 1st layer. There is potential for each neuron in 2nd to 7th layer FDLN to have an input dimension defined by total neuron count across all 1st layer FDLNs. This is brute force solution that does not scale computationally. Given topological ordering provided by FDLN, neighboring neurons respond to similar stimuli. The topology of each 1st layer DLN is quantized in terms of fixed number of neurons using potential function classification algorithm. This reduces number of inputs in 2nd to 7th layers of FDLN. The neurons in 6th layer acts as BMU for examples with same class label thus maximizing detection rate and minimizing false positives. However, there is no guarantee for this. In order to resolve this 6th layer DLN neurons that act as BMU for examples from more than one class are used to partition data. The 7th layer FDLNs are trained on subsets of original training data. This enables size of 7th layer FDLNs to increase which improves class specificity while presenting reasonable computational cost. Once training is complete 6th layer BMUs acts to identify which examples are forwarded to corresponding 7th layer FDLNs on test dataset. A decision rule is required to determine under what conditions classification performance of BMU at 6th layer FDLN is judged sufficiently poor for association with 7th layer FDLN. There are several aspects that require attention such as minimum acceptable misclassification rate of 6th layer BMU relative to number of examples labeled at 6th layer BMU and number of examples 6th layer BMU represent. The basic implication is that there must be optimal number of connections associated with 4th layer BMU for training of corresponding 7th layer FDLN and misclassification rate over examples associated with 6th layer BMU exceeds threshold. HFDLN is characterized in terms of success probability in recovering true hierarchy H^* and runtime complexity. Some restrictions are placed to similarity function S [2] such that similarities scale with hierarchy upto some random noise: (a) For each $y_i \in Cs_j \in Cs^*$ and $j' \neq j$: $\min_{y_p \in Cs_j} \mathbb{Exp}[S(y_i, y_p)] - \max_{y_p \in Cs'_j} \mathbb{Exp}[S(y_i, y_p)] \geq \gamma > 0$. Here expectations are taken with respect to noise on S. (b) S2 For each $y_i \in Ct_j$, a set of V_j words of size v_j drawn uniformly from Cs_j satisfies: $\mathbb{Prob}\left(\min_{y_p \in Cs_j} \mathbb{Exp}[S(y_i, y_p)] - \sum_{y_p \in V_j} \frac{S(y_i, y_p)}{v_j} > \epsilon\right) \leq 2e^{\left\{\frac{-2v_j\epsilon^2}{\sigma^2}\right\}}$. Here $\sigma^2 \geq 0$ parameterizes noise on similarity function S. From viewpoint of feature learning stacked FDLNs extracts temporal features of sequences in weblog data. Various trade-offs are done towards improving representation ability and avoiding data over fitting. It is easy to overfit the network with limited data training sequences. This algorithm can be fine-tuned with heuristics. This model's functional form has interesting connections with existing deep architectures. Firstly invariance-preserving hidden layers can be encoded as MLP convolution layers [2] with two channel 1×1 input $x_{i,j}$ corresponding to two players' respective payoffs when actions i and j are played. Secondly pooling units are superficially similar to pooling units used in convolutional networks. Here pooling is used for sharing information between cells in matrices that are processed through network by taking maximums across entire rows or columns. The architecture highlights quantal cognitive hierarchy and quantal level-k models. As their sharpness tends to infinity their best-response equivalents are cognitive hierarchy and level-k [2]. The architecture extends universality by considering vagueness in real-life strategic scenarios.

3 Experiments and Results

In this section experimental results are presented for human behaviour prediction. The dataset combines observations from used dataset combining observations from 12 human subject experimental studies by behavioral economists where subjects are paid to select actions in normal-form games. The payment depended on subject's actions and actions of unseen opposition choosing an action simultaneously. The prima face is in model's ability to predict distribution over row player's action. The models are fitted to maximize training data likelihood $\mathbb{Prob}(\mathcal{Dataset}|\Theta)$ where Θ denotes model parameters and $\mathcal{Dataset}$ is experimental dataset. The evaluation is done in terms of negative log-likelihood on test set. All models here are optimized using [2] with initial learning rate of 0.0004, $\beta_1 = 0.96$, $\beta_2 = 0.999$ and $\epsilon = 10^{-7}$. The models are regularized with drop probability = 0.4 and $L_1 = 0.04$. They are trained until there is no training set improvement upto 30000 epochs and parameters from iteration with best training set performance is obtained. The architecture imposes simplex constraints on mixture weight parameters. The simplex constraints fall within simple constraints class that is efficiently optimized using [2] which modifies standard SGD by projecting

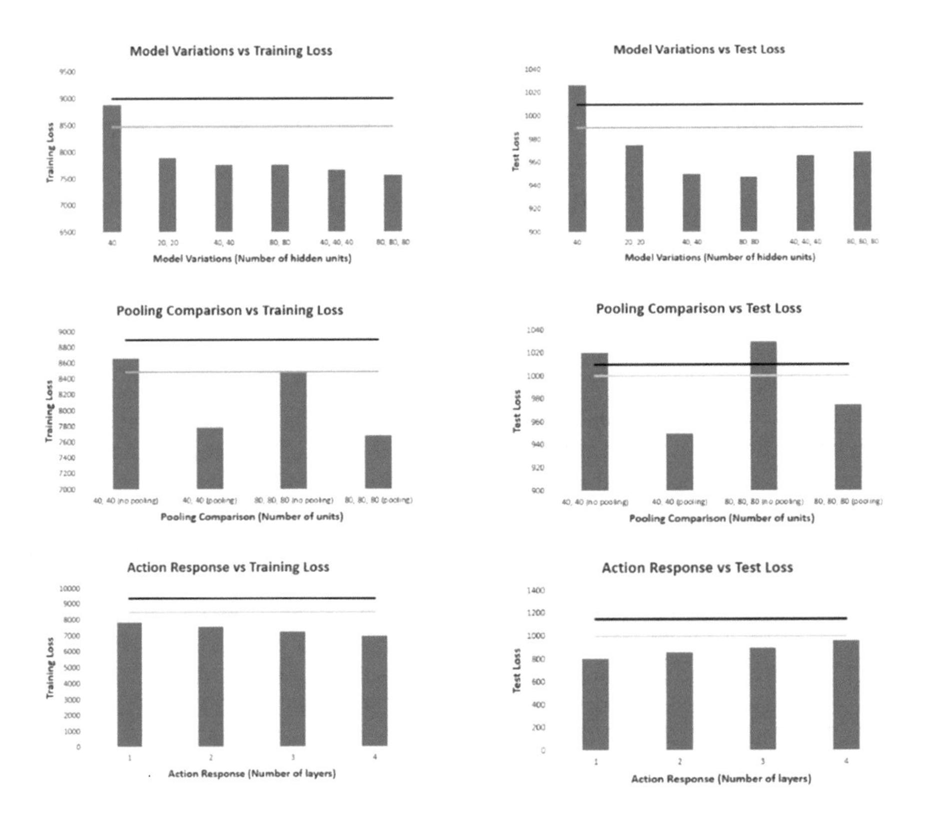

Fig. 4. The negative log likelihood performance (the error bars represent 95% confidence intervals across 10 rounds of 10-fold cross-validation)

Table 2. The network performance with respect to training/test loss

Network performance (wrt training/test loss)	Optimal values
Model variation vs training loss	80
Model variation vs test loss	80
Pooling comparison vs training loss	80
Pooling comparison vs test loss	40
Action response vs training loss	4
Action response vs test loss	1

relevant parameters onto constraint set after each gradient update. The Fig. 4 presents several interesting results which are further highlighted in Table 2.

4 Conclusion

An accurate model of boundedly rational behavior is required in order to design systems that efficiently interact with human players. We present hierarchical architecture for learning models that significantly improves upon state-of-the-art performance. The full architecture includes action response layers to explicitly incorporate iterative reasoning modeled by level-k-style models. This model achieves new performance benchmark. This indicates model's performance towards mapping from payoffs to distributions over actions that is different from previous models. The experimental results demonstrate superiority of proposed method.

References

1. Shoham, Y., Leyton Brown, K.: Multiagent Systems: Algorithmic, Game-Theoretic and Logical Foundations. Cambridge University Press, Cambridge (2008)
2. Chaudhuri, A.: Studies in applications of soft computing to some optimization problems. Ph. D. Thesis, Netaji Subhas Open University, Kolkata, India (2010)
3. Clark, C., Storkey, A.J.: Training deep convolutional neural networks to play go. In: Proceedings of 32nd International Conference on Machine Learning, vol. 37, pp. 1766–1774 (2015)
4. Silver, D., et al.: Mastering the game of go with deep neural networks and tree search. Nature **529**, 484–489 (2016)
5. Long, J., Shelhamer, E., Darrell, T.: Fully convolutional networks for semantic segmentation. IEEE Trans. Pattern Anal. Mach. Intell. **39**(4), 640–651 (2017)
6. Tagiew, R.: Hypotheses about typical general human strategic behavior in a concrete case. In: Serra, R., Cucchiara, R. (eds.) AI*IA 2009. LNCS (LNAI), vol. 5883, pp. 476–485. Springer, Heidelberg (2009)
7. Gal, Y., Pfeffer, A.: Modeling reciprocal behavior in human bilateral negotiation. In: Proceedings of the 22nd National Conference on Artificial Intelligence, vol. 1, pp. 815–820 (2007)
8. Chaudhuri, A.: Solving rectangular fuzzy games through. Open Comput. Sci. **7**(1), 46–50 (2017)

Fuzzy-Expert System for Customer Behavior Prediction

Monika Frankeová[1], Radim Farana[2](✉), Ivo Formánek[3], and Bogdan Walek[1]

[1] University of Ostrava, 30. dubna 22, 701 03 Ostrava, Czech Republic
frankeovam@seznam.cz, bogdan.walek@osu.cz
[2] Mendel University in Brno, Zemědělská 1, 613 00 Brno, Czech Republic
radim.farana@mendelu.cz
[3] University of Entrepreneurship and Law, Michálkovická 1810/181, 710 00 Ostrava, Czech Republic
ivo.formanek@vspp.cz

Abstract. The paper deals with the modelling of customer's behavior in the shop of the retail chain. The paper shows that the fuzzy-expert system is a good tool for describing the behavior of a system, where the customer's behavior is influenced by weather conditions and by events in the surroundings of the shop. The article also offers a procedure that allows dividing the system into logical units and reducing the number of necessary rules. The paper also details how the individual parts of the system have been verified. On specific real-time data the paper also presents the detection of incorrect (stereotypical) steps done by experts in compiling the knowledge base. The procedures that have been used have enabled effective identification and elimination of the errors. The advantage of our procedure was also that the IF-THEN rules that have been used were easily readable and understandable. At the end of the research work the expert system has been tested by means of available historical sales forecast data to optimize inventory, reduce storage costs, and reduce the risk of depreciation due to exceeding maximum warranty period. Achieved results have proved that fuzzy-expert systems are suitable also for the modelling of customer's behavior and can provide us good results.

Keywords: Behavior · Prediction · Customer · Fuzzy-expert system

1 Introduction to the Problem

The core task of the expert system is to hand on the knowledge or experience of an expert to another person who does not have such expert knowledge, but needs him or her to fulfil a task in good quality. In the ordinary life, a less experienced person consults a problem with a more experienced colleague. The expert system can mediate such advice without the expert being available. Expert experience is appropriately formalized and maintained in order to be used when querying the system by the user. Knowledge preservation is useful not only for decision support, but also for planning,

R. Silhavy (Ed.): CSOC 2018, AISC 764, pp. 122–131, 2019.
https://doi.org/10.1007/978-3-319-91189-2_13

diagnosing or predicting future developments. The first expert systems emerged in the early 1970s [1].

Knowledge or experience can best be passed through a natural language, as it has the best expressivity. However, natural language is problematic for computer processing. It contains a lot of constraints arising from context or vague terms. The practice has been proven that such binding can best be addressed using IF-THEN rules together with fuzzy sets and fuzzy logic. Fundamentals of the theory of fuzzy sets were laid by Lotfi Asker Zadeh in the 1960s [2]. The basic idea of fuzzy theory is the "degree of belonging" of an element to a set of elements. Thus, in a simplified way, apart from deciding whether an element belongs to or does not belong to a set, it also introduces a partial affiliation of the element into a set.

In this paper, we focus on using a fuzzy-expert system to predict customer behavior in retail. This model is used to determine the amount of inventory of individual products. Contribution follows the results of the research work of the research workplace, see e.g. [3].

One of the main business goals in general is to maximize product sales. It is obviously the core task of each organization that is based on prosperity, competitiveness, profitability and other attributes that determine the overall success of the organization. In order to achieve this goal, it must be ensured that the product sold is in sufficient quantity (and quality of course but it is no longer related to this work). Thus, a simple solution is available – always to have more products than can be sold in a given time period. But that is not a good solution. Storage of such goods is demanding for premises where proper conditions are necessary and costly. In inventory, the company can have also considerable amount of financial funds. Therefore, one of the goals of supply chain optimization is to minimize the inventory. Also, goods with short warranty period, typically food, cannot be ordered in this way because the customer always prefers fresh goods to the old ones. There is a risk of expiration of the warranty period and this can lead to loss of profit. The ideal thing is to have the goods as much as we are able to sell them at a certain time.

An experienced salesperson can handle this problem easily. The longer he or she works in a particular location, the more accurate he can predict and manage the amount of orders. He does it mostly intuitively. However, something can happen that the conditions are changed due to the price change of goods, traffic restrictions, bad weather conditions or change of the location, a new colleague from another department etc. In this case the old experience is not valid any more.

2 Current Inventory Planning Methods

There are several methods that support inventory strategic management. The two the most important ones are the ABC method and the JIT method. Both of them are based on statistical analysis over the reference period (usually over a year).

The ABC method is based on the consideration that there is no need to devote the same attention to each type of inventory. The method therefore divides inventories into three groups and work with each group in a different way.

Inventories are broken down by their share of total annual turnover. The method is also sometimes modified by the so-called XYZ analysis, which evaluates inventories in terms of smoothness and predictability of their consumption or sales. Group X means inventory with a regular and predictable consumption or sale. Group Y represents inventory with uneven but still predictable consumption or sales (e.g. seasonal fluctuations). Group Z means inventory with uneven and unpredictable consumption or sales. The inventory can then be determined as follows in Table 1.

Table 1. Product categories by ABC–XYZ method.

	X	Y	Z
A	Low	Low	Middle
B	Low	Middle	High
C	Middle	Middle	High

The Just In Time (JIT) method belongs to the LEAN group of methods and its aim is to deliver the goods exactly at the agreed time to a precisely agreed place. JIT minimizes inventory, does not have fixed delivery dates or fixed quantities. It is characterized by very frequent deliveries. JIT does not have any set of rules or procedures, but rather a philosophical approach that was created by Toyota around 1926. Although JIT is most often used in production, many of its principles and ideas can be applied to business. The JIT expert system would be a suitable support tool for determining the amount of orders. In retail practice, two approaches are generally applied:

- First approach: we order routinely as many items as sold for the past period since last delivery.
- Second approach: we first define the maximum number of pieces. When ordering, we will find out how many pieces are missing to the maximum and order them.

Both approaches have their shortcomings. For example, the first approach fails to respond to the differences in consumption at different times and does not respond flexibly enough to the increase in demand. Yet it is more flexible than the second approach, which in turn is not able to respond to the drop in demand.

3 Analysis of System Behavior, Determination of Main Parameters, Identification of Outputs

For the design of the expert system (EC), based on previous research, it was determined that the quantity to be tracked would be the quantity of goods sold. The EC will analyze the results of sales for previous periods and, on the basis of these, will estimate future sales figures, assessing the possible factors that may affect the amount of sales.

Sales of particular types of goods primarily affect:

- advertising, promotion,
- price,
- significant days, holidays,
- seasonality,
- change in store availability,
- organizing important events near the shop (e.g. fairs, concerts, etc.),
- weather.

Each of these influences has its own specific features and impacts on expert system knowledge base. The influences are well known and therefore we will not deal with them in detail. Just only a note.

The most significant measured quantity in our case will be the difference between estimated sales and actual sales. It can be assumed that this difference will always be greater at the beginning of the prediction process than at the end. Only after the system has been debugged and after the necessary adjustments can be expected that the difference may be small. At the same time, it is understandable that permanent zero differences can never be achieved. The reason is that the sale of goods affects various random factors and influences. Definition of main linguistic variables:

- A: units (very small, small, medium, high, very high)
- B: price (very small, small, medium, high, very high)
- C: holiday (no, one day, two days, Easter, Christmas)
- D: event (yes, no)
- E: availability (yes, no)
- F: weather (very bad, bad, normal, nice, very nice)
- G: difference (very small, small, medium, high, very high)

Linguistic variables will be specified by the names of parameters and by the names of fuzzy sets. The names of the fuzzy sets will be in the brackets.

To determine the number of rules for a complete system description, we need to quantify the number of fuzzy sets for individual parameters.

$$N = n_A.n_B.n_C.n_D.n_E.n_F.n_G = 5.5.5.2.2.5.5 = 12\,500 \qquad (1)$$

If we use shallow (subjective) reasoning [4]; the number of rules can be reduced by using a deep (objective) reasoning with one weight set. The evaluating process will therefore be carried out in two steps. In the first step, a second group of variables (i.e. C, D, E, F, G), whose significance for the overall result should be lower, and it will be evaluated. In the second step, the first group of variables (A, B, H) will be evaluated. To evaluate the first set of variables, a new linguistic variable called weight will be introduced, which will be the output from the first step of the evaluation process and the input into the second step.

Linguistic variable weight and its fuzzy set:
H: weight (very small, small, medium, high, very high)
The number of rules for a full system description will be significantly reduced:

$$N = n_A.n_B.n_H = 5.5.5 = 125,$$
$$n_H = n_C.n_D.n_E.n_F.n_G = 5.2.2.5.5 = 500. \quad (2)$$

So the knowledge base will not contain more than 625 rules. As part of system verification, it is possible to modify the structure of individual sets. Figure 1 shows the definition of fuzzy sets for the variable weight. On the vertical axis, there is a function of degree of membership, on the horizontal axis then the range of values that this linguistic variable can acquire.

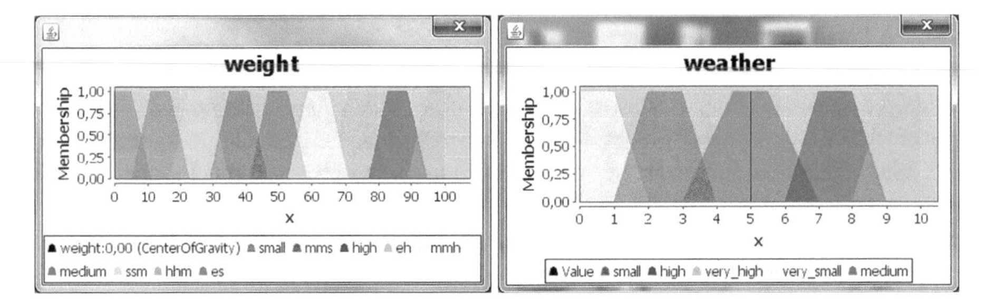

Fig. 1. Definition of fuzzy sets for variable *weight* and variable *weather*.

The data base for verification of the created prediction system was based on data from 1 June 2012 to 8 September 2014 from a specific supermarket that does not wish to be identified. This is therefore a long enough period to obtain a relevant description of the system's behavior. The monitored data are:

- the number of pieces sold,
- goods price,
- date of sale.

The following was added to the database:

- barcode,
- day of the week,
- month,
- prediction,
- the difference between the prediction and reality,
- fuzzy rule numbers enabled to analyze system behavior.

In addition, important events organized around the shop were identified where only the annual pilgrimage, organized during the second Sunday of September, was practically taken into account here. Other important events were not mentioned by shop staff.

For a fuzzy set of weather conditions, the weather was found in [5]. The scale 0–10 was used to evaluate the weather, where 0 is very bad, 10 is very beautiful. This scale has been covered by fuzzy sets, see Fig. 1. Historical records for the area under review had to be evaluated on a daily basis by temperature, sunshine, precipitation, wind strength and season.

For the most important variable units and price, a variable context method was used, which is used at the workplace, especially in the LFLC [6] tool. The range of actual values, minimum and maximum, and the corresponding context were always set for the tracking period.

4 Knowledge Base, Inference Mechanism

The knowledge base is compiled using IF-THEN rules. Each rule consists of an antecedent and a sequencer. The antecedent forms the conjunction of the associated input language variables with their valuation, followed by the evaluation of the output variable.

The knowledge base consists of two parts, the weight parameter will be evaluated first, it will be the output of the first set of rules and the entry for the second set of rules.

Example rule for the first group:

```
IF price IS very_small AND weather IS very_bad AND holiday IS no AND event IS yes AND availability IS yes THEN weight IS very_small;
```

Example rule for the second group:

```
IF units IS very_small AND difference IS high AND weight IS medium THEN tip IS very_small;
```

The rules were based on the experience of an expert working in a supermarket, whose job is ordering the goods. Based on his experience, an initial set of rules was developed. These rules were then verified and possibly modified during the test development phase of the system.

One of the most important parts of the whole system is the inferential mechanism. That is why the choice of the inference mechanism was given the greatest attention. In fuzzy systems there are two basic options. These are the logical deduction [4] and the fuzzy approximation [7]. The fuzzy approximation is suitable for fuzzy control because it is more about characterizing the course of function in a local area. Thus, fuzzy sets are defined to best describe the course of a function. Although this method would have been used by the solver in the area of industrial systems management and could be used for the exemplary solution, its principle is intended to achieve other values of EC behavior. Therefore, the principle of logical deduction was used for the EC. Various methods were tested for defuzzyfication in both calculation steps. Based on the test results, the Center of Gravity - COG method was chosen.

5 Expert System Testing

To verify the suitability of possible procedures and methods, prototyping and subsequent evaluation of the results of the procedures used were chosen.

Methods of selecting data from the source database to determine the contexts of key variables have been verified. Three variants were compared:

- **m1**: set consisting of all the occurring values.
  Disadvantage: it was not possible to predict the maximum/minimum that has not yet occurred.
- **m2**: as in the previous point and increased by 10%.
  Predictions of minimum/maximum were allowed, but in the course of time only the growth of sets was made, not the estimation of scales through the reduction of sets.
- **m3**: set consisting of a selection of values and increased by 20%.
  The choice of values allowed for the flexibility of the sets. Because the set contained much less elements, the magnification increased to 20%.

Methods for evaluating language variables for multiple source data have also been verified:

- **o1**: valuation is the average of the values from the selection.
- **o2**: the median of choice is the rating.
- **o3**: evaluation is the last occurring value of the selection.

The graphs in Fig. 2 show a comparison of two methods. Both are applied to the same type of goods in the same period:

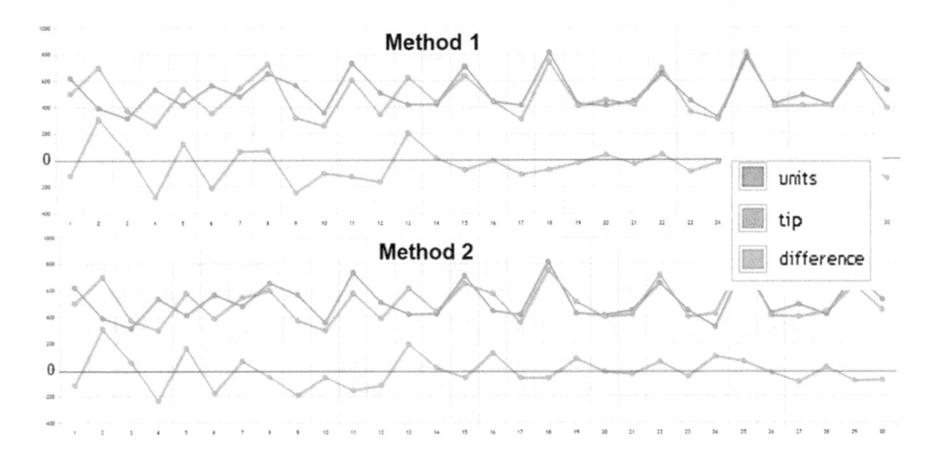

Fig. 2. Results of testing two evaluation methods for the first monitored month.

- **method 1**: m2, o2, trapezoid set shape.
- **method 2**: m2, o3, trapezoidal shape.

First day values were manually entered based on a qualified estimate, second and other days were already performed by an expert system. The graphs show that the estimate for the first week was evaluated in both cases in a similar way. The next weeks the estimates (tip) were refined. It can be seen on the difference value, which is the difference between the estimate and the units.

The method of calculating the standard deviation of the differences according to the following formula was used to compare the results:

$$s = \sqrt{\frac{1}{n-1}\sum_{i=1}^{n}(x_i - \bar{x})^2} \quad (3)$$

The results obtained by all three methods are presented in Table 2. In the evaluation of other months and other methods, **method 3** with parameters obtained by means of **m3**, **o3** was chosen because it gave the best results.

Table 2. Test results of three evaluation methods for the first monitored month.

Date	Method 1	Method 2	Method 3
2. VI	308	308	308
3. VI	57	**6**	57
...			
29. VI	36	31	**25**
30. VI	–64	**–51**	–81
Sum of differences	–476	–225	**43**
Number of nearest estimates	9	9	**13**
Standard deviation	123,54	128,98	**118,80**
Sum of absolute values	**2728**	2857	2737

The bases of the rules have been tested and edited during sales estimates over a portion of the test data base. The extreme differences between estimate and reality occurred in some days. Such cases were searched, then the active rules were analyzed, the rules were modified after the analysis. In this case, estimates for the days of 11/6, 18/6 and 25/6/2012 were analyzed.

Example of Modified Rule 355:

original:

IF price IS very_high AND weather IS medium AND holiday IS no AND event IS no AND availability IS yes THEN weight IS **ssm**

new:

IF price IS very_high AND weather IS medium AND holiday IS no AND event IS no AND availability IS yes THEN weight IS **medium**

The testing revealed inaccuracies in predictions, especially before holidays, when sales significantly exceeded normal values. However, this does not always apply. If the holiday falls on the weekend, the amount of sales will change only a little. If the holiday is on Monday, sales will be increased on Friday, or three days in advance. Based on the test results, part of the holiday rules were adjusted so that the highest values would be selected.

To verify rules for different weather, the number of undervalued estimates was compared to the number of overvalued estimates in the reporting period. The analysis showed that the ratio of undervalued and overvalued estimates is approximately the same for each kind of weather. For this reason, there was no need to modify the rules.

To verify the rules for major events, the results in the period of the annual pilgrimage were monitored. After evaluating the results, a total of 19 rules were modified. Repeated prediction was then performed. There has been a significant improvement in results. The deterioration of the results occurred in only 7 cases out of 36, with two differences being minimal.

After verification of all parts of the expert system, its results were compared with the commonly used inventory prediction methods.

- **Method 1**: To ensure a minimum amount of inventory, the same quantity of goods as was sold in the period under review was ordered. Of course, the existing inventory status was also taken into account and expected order has been reduced by inventory status.
- **Method 2**: A maximum limit has been set and the goods have been replenished to this maximum.
- **Method 3**: Prediction done by using the expert system created.

In the evaluation of the achieved results, two parameters were evaluated:

- Balance of unsold goods, surplus. Ideally, the balance should be close to zero.
- An unsupplied demand. This could be evaluated using historical data. In real operation, a special system for tracking unsatisfied requirements would have to be created.

Earlier research has shown that prediction systems can only be used effectively for fast-moving goods. For this reason, all three methods were tested for short-term goods, purchased in larger quantities, typically on pastries (Table 3).

Table 3. Activated rules in challenging days.

	11. 6. 2012	18. 6. 2012	25. 6. 2012
Rules No1	19	387	355
Rules No2	258, 259	174, 175, 178, 180	111, 112

Table 4 indicates that **method 2** shows the least undervalued orders but at the expense of higher inventory. When choosing the first two ways of ordering goods, especially for fresh food, there is a high risk of depreciation leading to losses. Method 1 shows a much higher number of undervalued orders than **method 2**. This would negatively affect revenue that would be lower due to shortage of goods.

Table 4. Prediction results of goods sales: bread roll.

	Method 1	Method 2	Method 3
Sum of positive surpluses	71138	71750	**32064**
Number of underestimations	160	**33**	68

If we ordered goods according to the prediction of the expert system, we would be able to optimize both factors. Compared with **method 1**, we would reduce the number of undervalued orders by 58%; surpluses would be decreased by 55%.

Compared to Method 2, the surpluses would also fall by 55%. The number of undervalued orders would be increased, but would be still lower than in case of **method 1**. In case of other types of goods, very similar results were achieved.

6 Conclusions

By applying an expert system for customer behavior modeling and then using it to predict commodity orders, there has been a significant improvement in inventory and savings amount. The system was designed in such a way to be easily expandable for other commodities, the fuzzy prediction rules would be easily modifiable and the remote access would be enabled via a client [8]. Thanks to the use of real data, a testing system for individual parts of the expert system was also validated. This testing system has revealed the wrong (stereotypical) user approaches and helped to remove them.

Acknowledgment. This work was supported during the completion of a Student Grant with student participation, supported by the Czech Ministry of Education, Youth and Sports.

References

1. Cingolani, P., Alcalá-Fdez, J.: JFuzzyLogic: a robust and flexible fuzzy-logic inference system language implementation. In: WCCI 2012 IEEE World Congress on Computational Intelligence, 2012, pp. 1090–1097. IEEE, Brisbane (2012)
2. Zadeh, L.A.: Fuzzy sets. Inf. Control **8**, 338–353 (1965)
3. Walek, B., Farana, R.: Proposal of an expert system for predicting warehouse stock. In: 4th Computer Science On-line Conference, 2015, CSOC 2015, pp. 85–91. UTB ve Zlíně, Zlín (2015)
4. Dvořák, J.: Expert Systems (in Czech). VUT v Brně, Brno (2004)
5. ČHMÚ Homepage. http://portal.chmi.cz/historicka-data. Accessed 12 Nov 2016
6. Novák, V.: Genuine linguistic fuzzy logic control: powerful and successful control method. In: Hüllermeier, E., Kruse, R., Hoffmann, F. (eds.) Computational Intelligence for Knowledge-Based Systems Design, 2010, pp. 634–644. Springer, Heidelberg (2010)
7. Novák, V., Knybel, J.: Fuzzy Modelling (in Czech). University of Ostrava, Ostrava (2005)
8. Frankeová, M.: Expert system with system behavior prediction (in Czech). Diploma Thesis, head prof. Ing. Radim Farana, CSc. (2016)

A Binary Grasshopper Algorithm Applied to the Knapsack Problem

Hernan Pinto, Alvaro Peña(✉), Matías Valenzuela, and Andrés Fernández

School of Engineering, Pontificia Universidad Católica de Valparaíso, Valparaíso, Chile
{hernan.pinto,alvaro.pena,matias.valenzuela,andres.fernandez}@pucv.cl

Abstract. In engineering and science, there are many combinatorial optimization problems. A lot of these problems are NP-hard and can hardly be addressed by full techniques. Therefore, designing binary algorithms based on swarm intelligence continuous metaheuristics is an area of interest in operational research. In this paper we use a general binarization mechanism based on the percentile concept. We apply the percentile concept to grasshopper algorithm to solve multidimensional knapsack problem (MKP). Experiments are designed to demonstrate the utility of the percentile concept in binarization. Additionally we verify the efficiency of our algorithm through benchmark instances, showing that binary grasshopper algorithm (BGOA) obtains adequate results when it is evaluated against another state of the art algorithm.

Keywords: Combinatorial optimization · KnapSack
Metaheuristics · Percentile

1 Introduction

The combinatorial problems have great relevance at industrial level, we find them in various areas such as Civil Engineering, Bio Informatics [1], Operational Research [2,3], resource allocation [4], scheduling problems [5], routing problems among others. On the other hand, in recent years, algorithms inspired by nature phenomena to solve optimization problems have been generated. As examples of these algorithms we have Cuckoo Search, Black Hole, Bat Algorithm, and grasshopper Algorithms [6] among others. Many of these algorithms work naturally in continuous spaces and therefore must be adapted to solve combinatorial problems. In the process of adaptation, the mechanisms of exploration and exploitation of the algorithm can be altered, having consequences in the efficiency of the algorithm.

Several binarization techniques have been developed to address this situation. In a literature search, the main binarization methods used correspond to transfer functions, angle modulation and quantum approach, for more detail refer to [7]. In this article we present a new binarization method that uses the percentile

R. Silhavy (Ed.): CSOC 2018, AISC 764, pp. 132–143, 2019.
https://doi.org/10.1007/978-3-319-91189-2_14

concept to group the solutions and then perform the binarization process. To verify the efficiency of our method, we used the grasshopper (BGOA) algorithm.

Grasshoppers are insects and under certain conditions can be considered a pest, the main characteristic of these insects is related to their movement. When it is in a larval state, its movement is slow in contrast with the large and abrupt movements of the adults.

The mathematical model to simulate the movement of grasshoppers are presented in Eq. 1

$$X_i^d = c\left(\sum_{j=1, j\neq i}^{N} c\frac{ub_d - lb_d}{2} s(|x_j^d - x_i^d|)\frac{x_j - x_i}{d_{ij}}\right) + \hat{T}_d \tag{1}$$

The first c reduces the movements of grasshoppers around the target. In other words, this parameter balances exploration and exploitation of the entire swarm around the target. The second c decreases the attraction zone, comfort zone, and repulsion zone between grasshoppers. The coefficient c is calculated as: $cmax - l\frac{cmax-cmin}{L}$, where L = 500 is the maximum number of iteration. The cmax, and cmin coefficients are 1 and 0.00001 respectively. The component $s(r) = fe^{\frac{-r}{l}} - e^{-r}$ indicates if a grasshopper should be repelled from (explore) or attracted to (exploitation) the target. For this case $f = 0.5$ and $l = 1.5$. Finally $\hat{T}_d$ represents the best solution at this iterations.

To check our binary Grasshopper algorithm (BGOA), we use the well-known knapsack problem. Experiments were developed using a random operator to validate the contribution of the percentile technique in the binarization process of the grasshopper algorithm. In addition, local search operator is used to strengthen the results. Moreover, the binary artificial algae (BAAA) and K-means transition ranking (KMTR) algorithms were used to compare our results. BAAA was developed in [8] and uses transfer functions to perform the binarization process. KMTR was developed in [9] and uses a K-means algorithm to perform the binarization. The results show that the percentile technique obtains results superior to those obtained by the random operator and that our BGOA algorithm shows competitive results against the BAAA and KMTR algorithms.

2 KnapSack Problem

The multidimensional knapsack problem (MKP), models resource allocation situations. The goal is to find the subset of objects that produce the greatest benefit by satisfying a set of constraints. MKP is one of the most studied NP-hard class combinatorial problems, however it remains as a challenge due to the difficulties encountered in solving medium and large size instances. A search for the last few years in the literature, we find that MKP has been solved for example in [10] using a quantum binarization technique, in [9] applied the k-means clustering technique to perform binarization, in [11] used a differential algorithm with transfer functions, and in [12] a modification of the PSO equations was used. MKP can be set as:

$$\text{maximize} \sum_{j=1}^{n} p_j x_j \tag{2}$$

$$\text{subjected to} \sum_{j=1}^{n} c_{ij} x_j \leq b_i \ , \ i \in \{1, ..., m\}. \tag{3}$$

With $x_j \in \{0, 1\}$, $j \in \{1, ..., n\}$. p_j corresponds to the profit of element j. c_{ij} represents a cost associated with dimension i and element j. The constraints in each dimension i are represented by b_i. The solution can be modelled using a binary representation, in this representation a 0 means the element is not included in the knapsack.

3 Binary Grasshopper Algorithm

The application of statistics and data mining, can be found in many areas such as transports, smart cities, agriculture and computational intelligence [9,13–16]. In this section, we explore the application of the percentile concept to the binarization of continuous metaheuristics.

The binary grasshopper algorithm(BGOA) uses 4 operators to achieve its goal. An initialization operator, A local search operator, the binary operator using the percentile technique, and a repair operator for the case of solutions that do not meet any of constraints. The general scheme of the algorithm is shown in Fig. 1. In the first step, BGOA initializes the solutions, this is detailed in Sect. 3.1. Once we have all the solutions initialized, we check the compliance of the stopping criterion of the algorithm associated to the maximum number of iterations. The grasshopper algorithm is then run in conjunction with the percentile operator, this is described in Sect. 3.2. Subsequently once the binarization was performed, the results are compared with the best solution obtained so far. In the case of finding a superior solution, it replaces the previous solution and is submitted to a local search operator. In the following subsection we will explain in detail the initialization method, the binary operator and the repair operator.

3.1 Initialization and Element Weighting

As BGOA is a swarm algorithm, to begin the exploration and exploitation of the search space, the list of solutions must be initialized. For the generation of each solution, an element is randomly chosen first. The next step is to check if other elements can be incorporated, for this we must evaluate the constraints of our problem. To select the new element, we generate a list of possible elements that comply with the constraints. Each element is calculated a weight and the best weight element is selected. This procedure is repeated until no additional element can be incorporated. In the Fig. 2, the initialization algorithm is displayed.

To calculate the weight of each element, several methods have been developed. In [17] a pseudo-utility in the surrogate duality approach was proposed.

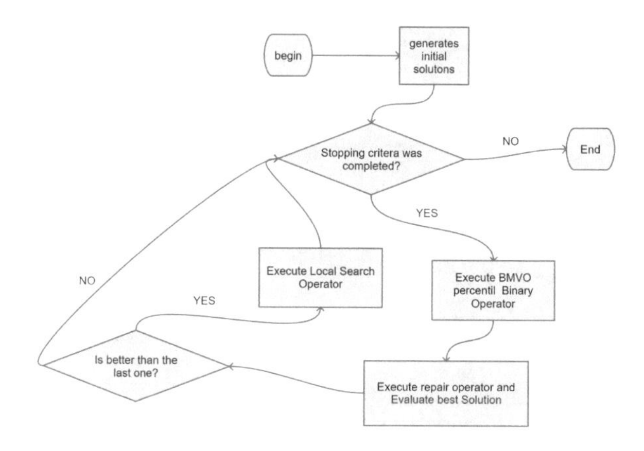

Fig. 1. Flowchart of the binary grasshopper algorithm.

The way to calculate it is shown in Eq. 4. In this equation the variable w_j corresponds to the surrogate multiplier whose value is between 0 and 1. This multiplier can be interpreted as a shadow prices of the j-th constraint.

$$\delta_i = \frac{p_i}{\sum_{j=1}^{m} w_j c_{ij}} \tag{4}$$

A more intuitive measure focused on the average resource occupancy was proposed by [18]. It is shown in Eq. 5.

$$\delta_i = \frac{\sum_{j=1}^{m} \frac{c_{ij}}{m b_j}}{p_i} \tag{5}$$

In this article we used a variation of this last measure focused on the average occupation and proposed in [9]. This variation considers the elements that exist in knapsacks to calculate the average occupancy. In each iteration depending on the selected items in the solution the measure is calculated again. The expression of this new measure is shown in Eq. 6.

$$\delta_i = \frac{\sum_{j=1}^{m} \frac{c_{ij}}{m(b_j - \sum_{i \in S} c_{ij})}}{p_i} \tag{6}$$

3.2 Percentile Binary Operator

Due to the iterative nature of swarm intelligence algorithms and considering that GOA works in continuous space. The velocity and position of the solutions are updated in $\mathbb{R}^n$. A general way of writing the update is shown in Eq. 7. In this equation x_{t+1} represents the position of the particle x at time $t+1$. To obtain the position, we consider the function Δ, which is specific to each algorithm. For example in Black Hole $\Delta(x) = \text{rand} \times (x_{bh}(t) - x(t))$, in Cuckoo Search

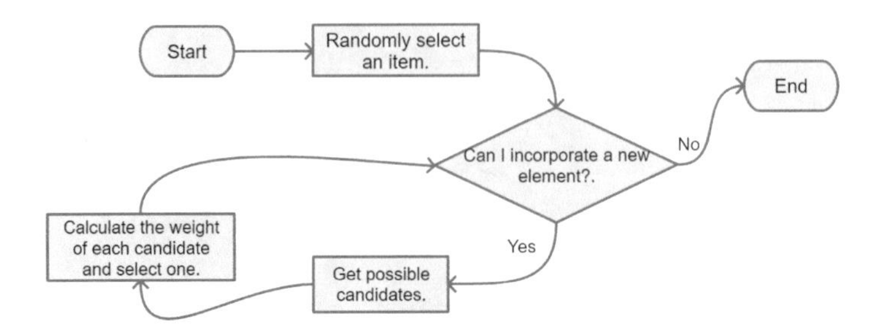

Fig. 2. Flowchart of generation of a new solution.

$\Delta(x) = \alpha \oplus Levy(\lambda)(x)$, and in PSO, Firefly, and Bat algorithms Δ can be written in simplified form as $\Delta(x) = v(x)$.

$$x_{t+1} = x_t + \Delta_{t+1}(x(t)) \tag{7}$$

For the case of GOA, it is considered a binary percentile operator to perform the passage of the continuous space to the binary space. Given the particle x, let us consider the magnitude of the displacement $Delta^i(x)$ in the i-th component and we group these magnitudes for all solutions in order to obtain the values for the percentiles {20, 40, 60, 80, 100}. At each percentile, we will assign a transition probability where the values are shown in Eq. 8. Using these transition probabilities together with the Eq. 9 binarization of the solutions is performed. The algorithm is detailed in Algorithm 1.

$$P_{tr}(x^i) = \begin{cases} 0.1, & \text{if } x^i \in \text{group } \{0,1\} \\ 0.5, & \text{if } x^i \in \text{group } \{2,3,4\} \end{cases} \tag{8}$$

$$x^i(t+1) := \begin{cases} \hat{x}^i(t), & \text{if } rand < P_{tg}(x^i) \\ x^i(t), & \text{otherwise} \end{cases} \tag{9}$$

Algorithm 1. Percentile binary operator

1: **Function** percentilebinary(vList, pList)
2: **Input** vList, pList
3: **Output** pGroupValue
4: pValue = getPercentileValue(vList, pList)
5: **for each** value in vList **do**
6: pGroupValue = getPercentileGroupValue(pValue,vList)
7: **end for**
8: **return** pGroupValue

3.3 Repair Operator

Local search and binary percentile operators can generate infeasible solutions. There are different mechanisms to address these infeasible solutions. In this article, the repair of the solutions was considered. To perform the repair, the measure described in the Eq. 6 was used. If the solution requires repair, the element with the maximum measure is chosen and it is removed from the solution, this process is iterated until a valid solution is obtained. The solution is then improved by exploring the possibility of incorporating new elements. This stage of improvement is iterated until there are no elements that can be incorporated without violating the constraints. The pseudocode is shown in Algorithm 2

Algorithm 2. Repair Algorithm

1: **Function** Repair(S_{in})
2: **Input** Input solution S_{in}
3: **Output** The Repair solution S_{out}
4: $S \leftarrow S_{in}$
5: **while** needRepair(S) == True **do**
6: $s_{max} \leftarrow$ getMaxWeight(S)
7: $S \leftarrow$ removeElement(S, s_{max})
8: **end while**
9: state $\leftarrow$ False
10: **while** state == False **do**
11: $s_{min} \leftarrow$ getMinWeight(S)
12: **if** $s_{min} == \emptyset$ **then**
13: state $\leftarrow$ True
14: **else**
15: $S \leftarrow$ addElement(S, s_{min})
16: **end if**
17: **end while**
18: $S_{out} \leftarrow S$
19: **return** S_{out}

4 Results

4.1 Insight of BGOA Algorithm

This section aims to find out the contribution of the percentile binary operator to the performance of the algorithm. To carry out the comparison problems cb.5.250 from the OR-library were selected. Violin charts and the Wilcoxon non-parametric signed-rank test were used to perform the statistical analyzes. In the charts, the X axis identifies the studied instances and the Y axis the % -Gap described in the Eq. 10. The Wilcoxon test is run to determine if the results obtained by BGOA have significant difference with respect to other algorithms. The parameter settings and browser ranges are shown in Table 1.

$$\% - Gap = 100 \frac{BestKnown - SolutionValue}{BestKnown} \tag{10}$$

Table 1. Setting of parameters for Binary grasshopper Algorithm.

Parameters	Description	Value	Range
N	Number of solutions	30	[20, 25, 30]
G	Number of percentiles	5	[4,5,6]
Iteration Number	Maximum iterations	1000	[1000]
cmax	Maximum for c coefficient	1	1
cmin	Minimum for c coefficient	0.00001	0.00001
f	f parameter for s function	0.5	0.5
l	l parameter for s function	1.5	1.5

Evaluation of Percentile Binary Operator. A random operator was designed to evaluate the contribution of the percentile binary operator. This random operator has the peculiarity of performing the transitions with a fixed probability of 0.5 without considering in which percentile the variable is located. Two configurations were studied: The first one where the local search operator is included and the second where the local operator is not considered. BGOA corresponds to our standard algorithm. *random.ls* is the random variant that includes the local search operator. *BGOA.wls* corresponds to the version with percentile binary operator without local search operator. Finally *random.wls* describes the random algorithm without local search operator.

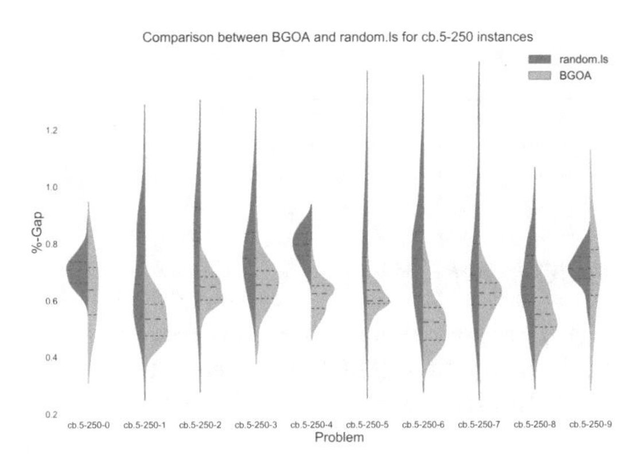

Fig. 3. Evaluation of percentile binary operator with Local Search operator

When we compared the Best Values between BGOA and *random.ls* which are shown in Table 2. BGOA outperforms to *random.ls*. However the Best Values between both algorithms are very close. In the Average comparison, BGOA outperforms *random.ls* almost in all problems. The comparison of distributions is shown in Fig. 3. We see the dispersion of the *random.ls* distributions are bigger than the dispersions of BGOA. In particular this can be appreciated in the problems 1, 2, 3, 5, 6 and 7. Therefore, the percentile binary operator together with local search operators, contribute to the precision of the results. Finally, the BGOA distributions are closer to zero than *random.ls* distributions, indicating that BGOA has consistently better results than *random.ls*. When we evaluate the behaviour of the algorithms through the Wilcoxon test, this indicates that there is a significant difference between the two algorithms.

Our next step is trying to separate the contribution of local search operator from the percentile binary operator. For this, we compared the algorithms *wls* and *random.wls*.

Table 2. Evaluation of percentile binary operator

Set	Best Known	Best *random.ls*	Best BGOA	Best *random.wls*	Best *wls*	Avg *random.ls*	Avg BGOA	Avg *random.wls*	Avg *wls*
cb.5.250-0	59312	59211	59225	59158	59175	59132.1	59152.4	59071.8	59132.1
cb.5.250-1	61472	61435	61435	61409	61409	61324.6	61387.1	61288.3	61370.7
cb.5.250-2	62130	62036	62074	61969	61969	61894.4	61978.5	61801.6	61924.6
cb.5.250-3	59463	59367	59446	59365	59349	59257.8	59322.1	59136.1	59269.3
cb.5.250-4	58951	58914	58951	58883	58930	58725.6	58824.6	58693.6	58758.5
cb.5.250-5	60077	60015	60056	59990	60015	59904.6	59952.3	59837.8	59945.9
cb.5.250-6	60414	60355	60355	60348	60349	60208.2	60332.1	60230.6	60311.1
cb.5.250-7	61472	61436	61436	61407	61407	61290.8	61331.9	61233.9	61342.3
cb.5.250-8	61885	61829	61885	61790	61782	61737.1	61787.7	61644.9	61738.1
cb.5.250-9	58959	58832	58866	58822	58787	58769.1	58782.8	58653.7	58770.4
Average	60413.5	60343	60372.9	60314.1	60317.2	60224.4	60285.15	60159.1	60256.3
p-value							1.14 e−05		2.43 e−06

When we check the Best Values shown in Table 2, we note that *wpe* performs better than *05.wpe* in all problems except 1, 6 and 7. However the results are quite close. In the case of the average indicator, *wpe* outperforms in all problems to *05.wpe*. The Wilcoxon test indicates that the difference is significant. This suggests that *wpe* is consistently better than *05.wpe*. In the violin chart shown n the Fig. 4 it is further observed that the dispersion of the solutions for the case of *05.wpe* is much larger than in the case of *wpe*. This indicates that the percentile binary operator plays an important role in the precision of the results.

4.2 BGOA Comparisons

In this section we evaluate the performance of our BGOA with the algorithm BAAA developed in [8]. BAAA uses transfer functions as a general mechanism of binarization. In particular BAAA used the $\tanh = \frac{e^{\tau|x|}-1}{e^{\tau|x|}+1}$ function to perform the transference. The parameter τ of the tanh function was set to a value 1.5. Additionally, a elite local search procedure was used by BAAA to improve solutions. As maximum number of iterations BAAA used 35000. The computer

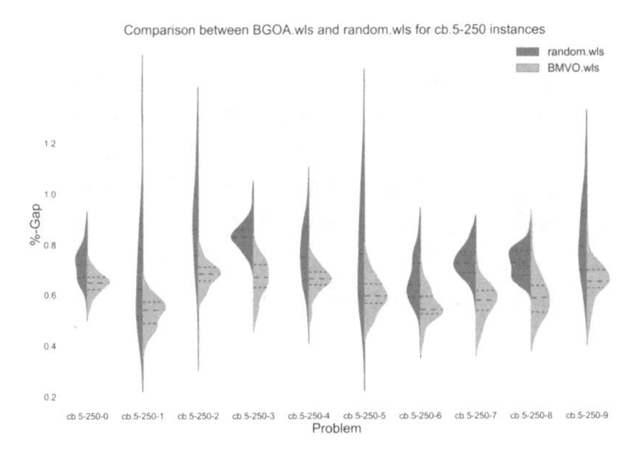

Fig. 4. Evaluation of percentile binary operator without Local Search operator

configuration used to run the BAAA algorithm was: PC Intel Core(TM) 2 dual CPU Q9300@2.5 GHz, 4 GB RAM and 64-bit Windows 7 operating system. In our BGOA algorithm, the configurations are the same used in the previous experiments. These are described in the Table 1. Additionally we made the comparison with KMTR-BH and KMTR-Cuckoo binarizations. KMTR uses the unsupervised K-means learning technique to perform the binarization process. In the article [9], the Black Hole and Cuckoo Search algorithms were binarized using KMTR.

The results are shown in Table 3. The comparison was performed for the set cb.5.500 of the OR-library. The results for BGOA were obtained from 30 executions for each problem. In black, the best results are marked for both indicators the Best Value and the Average. For the best value indicator BAAA was higher in 4, KMTR-BH in 11, KMTR-Cuckoo in 7 and BGOA in 11. We should note that the sum is greater than 30 because in some cases there was a tie between some of the algorithms. In the Average indicator BAAA was higher in 2 instances, KMTR-BH in 9, KMTR-Cuckoo in 4 and BGOA in 15. We should also note that the standard deviation in most problems was quite low, indicating that BGOA has good accuracy.

Additionally, to evaluate the robustness of BGOA, we experiment with the problems cb.10.500 and cb.30.500. These problems, correspond to the most difficult problems of the OR-library. A summary of the results are shown in Table 4. In this table we also incorporate the results for KMTR-BH and KMTR-Cuckoo algorithms.

Table 3. OR-Library benchmarks MKP cb.5.500

Instance	Best	BAAA	Avg	KMTR-BH	Avg	KMTR-Cuckoo	Avg	BGOA	Avg	Time (s)	std
	Known	Best		Best		Best		Best			
0	120148	120066	120013.7	**120096**	120029.9	120082	**120036.8**	120082	120011.3	449	42.3
1	117879	117702	117560.5	**117730**	**117617.5**	117656	117570.6	117656	117609.8	503	41.2
2	121131	120951	120782.9	**121039**	**120937.9**	120923	120855.1	120923	120877.3	521	44.7
3	120804	120572	120340.6	120683	**120522.8**	120683	120455.7	**120683**	120504.1	505	51.2
4	122319	122231	122101.8	**122280**	**122165.2**	122212	122136.4	122212	122122.8	512	48.3
5	122024	121957	121741.8	121982	**121868.7**	121946	121824.6	**121982**	121845.7	529	57.9
6	119127	**119070**	118913.4	119068	**118950.0**	118956	118895.5	118956	118917.3	509	44.2
7	120568	120472	120331.2	120463	**120336.6**	120392	120320.4	**120463**	120315.8	498	57.3
8	121586	121052	120683.6	**121377**	121161.9	121201	121126.3	121295	**121202.3**	545	51.9
9	120717	120499	120296.3	**120524**	120362.9	120467	120335.5	120467	**120391.7**	501	51.3
10	218428	218185	217984.7	218296	218163.7	218291	**218208.9**	**218296**	218187.5	478	47.6
11	221202	220852	220527.5	220951	220813.9	220969	**220862.3**	**220951**	220835.7	492	56.1
12	217542	217258	217056.7	217349	217254.3	217356	217293.0	**217356**	**217298.9**	468	48.3
13	223560	223510	223450.9	**223518**	223455.2	223516	223455.6	223516	**223459.7**	495	41.8
14	218966	218811	218634.3	218848	218771.5	**218884**	218794.0	218848	**218816.2**	541	49.5
15	220530	220429	**220375.9**	**220441**	220342.2	220433	220352.7	220410	220360.2	504	44.9
16	219989	219785	219619.3	219858	219717.9	**219943**	219732.8	219858	**219768.1**	499	62.3
17	218215	**218032**	217813.2	218010	217890.1	218094	217928.7	218010	**217952.6**	478	59.3
18	216976	**216940**	**216862.0**	216866	216798.8	216873	216829.8	216866	216820.9	429	37.7
19	219719	219602	219435.1	219631	219520.0	**219693**	219558.9	219631	**219582.1**	495	33.7
20	295828	295652	295505.0	**295717**	295628.4	295688	295608.8	295688	**295643.6**	401	36.5
21	308086	307783	307577.5	307924	307860.6	**308065**	**307914.8**	307924	307881.3	443	32.9
22	299796	299727	299664.1	299796	**299717.8**	299684	299660.9	**299796**	299702.7	402	41.4
23	306480	306469	306385.0	**306480**	**306445.2**	306415	306397.3	306415	306374.8	401	37.8
24	300342	300240	300136.7	**300245**	300202.5	300207	300184.4	**300245**	**300223.9**	408	37.2
25	302571	**302492**	302376.0	302481	302442.3	302474	302435.6	302481	**302482.6**	395	29.7
26	301339	301272	301158.0	301284	301238.3	**301284**	301239.7	**301284**	**301252.6**	401	26.8
27	306454	306290	306138.4	306325	306264.2	**306331**	306276.4	**306331**	**306311.9**	386	29.3
28	302828	302769	302690.1	302749	302721.4	**302781**	302716.9	302771	**302729.1**	367	26.3
29	299910	299757	299702.3	299774	299722.7	**299828**	299766.0	**299828**	**299774.8**	378	28.5
Average	214168.8	214014.2	213862.0	214059.5	213964.1	214044.2	213959.1	214040.8	213975.2	464.4	43.3

Table 4. Summary of OR-Library benchmarks MKP cb.10.500 and cb.30.500.

Problem set	KMTR-BH	std	KMTR-Cuckoo	std	BGOA	std
	Average %-Gap		Average %-Gap		Average %-Gap	
cb.10.500.25	0.34	0.08	0.37	0.1	0.41	0.16
cb.10.500.50	0.22	0.03	0.20	0.03	0.22	0.11
cb.10.500.75	0.11	0.03	0.09	0.03	0.11	0.06
cb.30.500.25	0.62	0.12	0.59	0.10	0.65	0.12
cb.30.500.50	0.27	0.05	0.26	0.05	0.23	0.07
cb.30.500.75	0.17	0.03	0.16	0.03	0.19	0.04
Average	0.28	0.05	0.28	0.05	0.30	0.09

5 Conclusions

In this article, the percentile technique is used to perform the binarization of the GOA algorithm. It should be noted that the percentile technique can be applied in the binarization of any continuous swarm-intelligence algorithm. We used the knapsack problem to evaluate our binary algorithm. The contribution of the percentile binary operator was studied, observing that this operator contributes to the precision and quality of the solutions obtained. Additionally, we develop the comparison with the BAAA and KMTR algorithms, in this comparison, BGOA was able to outperform the other algorithms in the best value indicator in 11 problems and in the average indicator in 15.

As a future line of research, it is interesting to compare the performance of the percentile operator with the K-means operator used in KMTR and with the transfer functions used in BAAA, all applied in the binarization of the GOA algorithm. Also, it is interesting to use the percentile operator to binarize other algorithms in addition to solving other NP-hard problems. From the theoretical point of view, it would also be interesting to understand how the exploration and exploitation of space are altered when we introduce the percentile operator. Another interesting research line is to explore adaptive techniques to automate the selection of parameters.

References

1. Barman, S., Kwon, Y.-K.: A novel mutual information-based Boolean network inference method from time-series gene expression data. PLoS ONE **12**(2), e0171097 (2017)
2. Crawford, B., Soto, R., Monfroy, E., Astorga, G., García, J., Cortes, E.: A meta-optimization approach for covering problems in facility location. In: Workshop on Engineering Applications, pp. 565–578. Springer (2017)
3. García, J., Crawford, B., Soto, R., Astorga, G.: A percentile transition ranking algorithm applied to binarization of continuous swarm intelligence metaheuristics. In: International Conference on Soft Computing and Data Mining, pp. 3–13. Springer (2018)
4. García, J., Crawford, B., Soto, R., Astorga, G.: A percentile transition ranking algorithm applied to knapsack problem. In: Proceedings of the Computational Methods in Systems and Software, pp. 126–138. Springer (2017)
5. García, J., Crawford, B., Soto, R., García, P.: A multi dynamic binary Black Hole algorithm applied to set covering problem. In: International Conference on Harmony Search Algorithm, pp. 42–51. Springer (2017)
6. Saremi, S., Mirjalili, S., Lewis, A.: Grasshopper optimisation algorithm: theory and application. Adv. Eng. Softw. **105**, 30–47 (2017)
7. Crawford, B., Soto, R., Astorga, G., García, J., Castro, C., Paredes, F.: Putting continuous metaheuristics to work in binary search spaces. Complexity **2017** (2017)
8. Zhang, X., Wu, C., Li, J., Wang, X., Yang, Z., Lee, J.-M., Jung, K.-H.: Binary artificial algae algorithm for multidimensional knapsack problems. Appl. Soft Comput. **43**, 583–595 (2016)

9. García, J., Crawford, B., Soto, R., Carlos, C., Paredes, F.: A k-means binarization framework applied to multidimensional knapsack problem. Appl. Intell. **48**, 1–24 (2017)
10. Haddar, B., Khemakhem, M., Hanafi, S., Wilbaut, C.: A hybrid quantum particle swarm optimization for the multidimensional knapsack problem. Eng. Appl. Artif. Intell. **55**, 1–13 (2016)
11. Liu, J., Wu, C., Cao, J., Wang, X., Teo, K.L.: A binary differential search algorithm for the 0–1 multidimensional knapsack problem. Appl. Math. Model. **40**, 9788–9805 (2016)
12. Bansal, J.C., Deep, K.: A modified binary particle swarm optimization for knapsack problems. Appl. Math. Comput. **218**(22), 11042–11061 (2012)
13. García, J., Pope, C., Altimiras, F.: A distributed k-means segmentation algorithm applied to Lobesia botrana recognition. Complexity **2017** (2017)
14. Graells-Garrido, E., García, J.: Visual exploration of urban dynamics using mobile data. In: International Conference on Ubiquitous Computing and Ambient Intelligence, pp. 480–491. Springer (2015)
15. Peredo, O.F., García, J.A., Stuven, R., Ortiz, J.M.: Urban dynamic estimation using mobile phone logs and locally varying anisotropy. In: Geostatistics Valencia 2016, pp. 949–964. Springer (2017)
16. Graells-Garrido, E., Peredo, O., García, J.: Sensing urban patterns with antenna mappings: the case of Santiago, Chile. Sensors **16**(7), 1098 (2016)
17. Pirkul, H.: A heuristic solution procedure for the multiconstraint zero? One knapsack problem. Naval Res. Logistics **34**(2), 161–172 (1987)
18. Kong, X., Gao, L., Ouyang, H., Li, S.: Solving large-scale multidimensional knapsack problems with a new binary harmony search algorithm. Comput. Oper. Res. **63**, 7–22 (2015)

Artificial Neural Networks Implementing Maximum Likelihood Estimator for Passive Radars

Timofey Shevgunov[1,2(✉)] and Evgeniy Efimov[1]

[1] Moscow Aviation Institute (National Research University), Moscow, Russia
tshevgunov@gmail.com, shevgunov@gmail.com,
omegatype@gmail.com
[2] National Research University Higher School of Economics, Moscow, Russia

Abstract. This paper introduces the maximum likelihood estimator (MLE) based on artificial neural network (ANN) for a fast computation of the bearing that indicates the direction to the source of the electromagnetic wave received by a passive radar system equipped with an array antenna. Authors propose the cascade scheme for ANN training phase where the network is fed with the pair-wise delays of received stationary or cyclostationary signals and the output of the network goes to the input of the target function being maximized together with the same data. The designed ANN topology has the modified output layer consisting of the custom neuron that implements argument function of a complex number rather than linear or sigmoid-like ones used in the conventional multilayer perceptron topologies. The simulation carried out for the ring array antenna shows that a single estimation obtained via ANN MLE takes 12 times less computational time comparing to the MLE implemented via the numerical optimization technique. The degradation of accuracy measured as the increase of mean-squared error does not exceed 10% of the potential value for the particular signal-to-noise ratio (SNR) and that difference has no tendency to decrease for higher SNR. The estimation error appeared to be independent from the true value in the wide range of bearings.

Keywords: Artificial neural network · Maximum likelihood estimation
Time difference of arrival · TDOA · Cyclostationary signals

1 Introduction

The recent advance in electronics has created a trend in the development of passive radar systems performing real-time or near-real-time digital signal processing towards the use of customizable and on-the-fly reconfigurable computing platform where the complicated algorithms could be run. The well-known algorithms for parameter estimation are based on the maximum likelihood (ML) approach promising to potentially achieve the theoretical minimum of the variance.

Despite the fact that the problem of ML estimation looks quite obvious being written analytically, its solution, except for some special cases, cannot be found in a closed form as a function of a measured sample. The underlying problem is a set of

R. Silhavy (Ed.): CSOC 2018, AISC 764, pp. 144–153, 2019.
https://doi.org/10.1007/978-3-319-91189-2_15

nonlinear equations that require numerical optimization methods to be solved. The implementation of some numerical methods could be challenging for real-life, not experimental, passive radars. The enhanced numerical optimization technique requires a lot of computational power to perform all specific calculations engaged in them while the simpler one usually suffers from the absence of convergence if the real world sample data are used.

A possible way to design the estimator that is only a little less accurate than ML estimator consists in its approximation in the functional sense by a feed-forward ANN. Although the chosen topology of the ANN, e.g. the multilayer perceptron, is conventional, the training procedures for an ANN used as a substitution for the ML estimator are not introduced in the comprehensive sources on ANN, e.g. in [1, 2]. The first attempt to design ML estimators containing ANN seems to be made in [3], but the training algorithm proposed there was suitable only for solving binary classification problem. The further investigation of this approach was made in [4] with some analytical suggestion about the network structure and its training. Later in [5] the more thorough analytical foundation was proposed leading to the synthesis of the ANN fitting ML estimators of distribution law parameters for random variables under normal, lognormal and Rayleigh distributions.

This paper deals with the design and training problem of the ANN-based algorithm suitable for implementation as part of software for a passive radar system equipped with an active electronically scanning array antenna. The algorithm is to perform the estimation of the direction of arrival of electromagnetic wave emitted by the source of communication signal usually modeled as realization of a stationary or cyclostationary random process. The authors use previously developed [6] ANN synthesis procedure based on universal adaptive elements which is simple and clear yet powerful implementation of the basic rules of diagrammatic methods described in [7]. This is a necessary step for construction the custom neurons computing the variable of an angular nature. Neuron of that type is placed in the output layer and it makes the network more relevant to the physical model of the direction-finding problem than any of standard neurons such as linear or sigmoid. The proposed approach to design of ANN approximating ML estimator is presented in Sect. 2. The training process for ANN puts it in the cascade connection to maximize the target function rather than in parallel where the minimization of mean-square error is performed. The example of the ANN synthesized and trained according to the proposed approaches for the case of passive radar system with seven-element ring antenna array is presented in Sect. 3. The performance of the presented estimator as a substitution for the ML estimator is evaluated by means of numerical simulation in the wide range of bearings. The paper ends with Discussion and Conclusion where a brief comparison of ANN and numerical solution are given.

2 Artificial Neural Network Implementing Maximum Likelihood Estimation

The estimation of the parameter vector $\boldsymbol{\theta}$ is assumed to be obtained via an observed realization of the random vector $\boldsymbol{\xi}$. Its probability density $p_{\xi}(\mathbf{x};\boldsymbol{\theta})$ is determined according to an a priori known probabilistic model that links vector of parameters $\boldsymbol{\theta}$ and the observed vector ξ. The maximum likelihood (ML) approach is used to obtain estimation $\boldsymbol{\theta}_{ML}(\mathbf{x})$ so that maximum of $p_{\xi}(\mathbf{x};\boldsymbol{\theta})$ over $\boldsymbol{\theta}$ is achieved for the given $\mathbf{x}$ as far as the realization of random vector ξ is known:

$$\boldsymbol{\theta}_{ML}(\mathbf{x}) = \arg\max_{\boldsymbol{\theta}} p_{\xi}(\mathbf{x};\boldsymbol{\theta}) \tag{1}$$

In practice, the procedure of maximum likelihood estimation is carried out by finding the maximum of the target function $T(\mathbf{x};\boldsymbol{\theta})$ that can be either the probability density function $p_{\xi}(\mathbf{x};\boldsymbol{\theta})$ itself or its monotone transformation such as a logarithm. It may also be possible to go even further and factorize the likelihood function and then exclude parts independent from the estimated parameter vector [12].

In rare but important cases such as normal and uniform distribution, ML estimator can be expressed in a closed form as a function depending on the observed realization $\mathbf{x}$. However, in most cases of practical interest, the maximization of the target function over $\boldsymbol{\theta}$ will yield the set of nonlinear equations and the closed-form solution cannot be found and numerical methods are employed.

Typically, a series of observations $\{\mathbf{x}_n\}$ of the random vector ξ is available and a ML estimation has to be found for each observation. Employing iterative numerical optimization techniques for each observation noticeably increases the required computational power and requires high performance. Another point specific to an iterative numerical solution is that the actual number of iterations is not known in advance because it depends on many factors such as the true value of $\boldsymbol{\theta}$, particular values taken in $\mathbf{x}$, the initial point and parameters operating the particular algorithm chosen as the implementation of the particular numerical method. Additionally, the application of the numerical method will also require monitoring for the conditions of convergence and limitations of running time.

Looking for a substitution for the numerical techniques for ML estimation we consider a feed-forward artificial neural network (ANN) characterized by the set of its synaptic weights $\mathbf{w}$. In that plot, the observation vector is $\mathbf{x}$ presented to the network as its input data vector and the estimation $\boldsymbol{\theta}_{NN}$ will be found as the output data vector. The maximization procedure of the target function $T(\mathbf{x};\boldsymbol{\theta})$ can be substituted by the direct calculation performing by ANN if the estimation $\boldsymbol{\theta}_{NN}$ approximates ML estimations $\boldsymbol{\theta}_{ML}$ well enough.

In case the quadratic form is used for calculation of the error $\varepsilon(\mathbf{x})$ between two estimators under some $\mathbf{x}$ can be expressed as:

$$\varepsilon(\mathbf{x}) = (\boldsymbol{\theta}_{ML}(\mathbf{x}) - \boldsymbol{\theta}_{NN}(\mathbf{x}))^H \mathbf{Q} (\boldsymbol{\theta}_{ML}(\mathbf{x}) - \boldsymbol{\theta}_{NN}(\mathbf{x})) \tag{2}$$

where $\boldsymbol{\theta}_{NN}$ and $\boldsymbol{\theta}_{ML}$ are column vectors, «H» stands for Hermitian transposition and $\mathbf{Q}$ is a positive-definite matrix. If the unit matrix is used to substitute the latter, $\varepsilon(\mathbf{x})$ becomes the Euclidian distance between vectors $\boldsymbol{\theta}_{NN}$ and $\boldsymbol{\theta}_{ML}$; in case of an arbitrary positive-definite matrix, the Mahalanobis distance takes place. In more practical cases, the diagonal matrix $\mathbf{Q}$ with all positive elements could account for the difference in range or importance of elements in the estimator vector. This easily implemented case is both useful and unavoidable when the elements of the estimated vector have different units. The reason for the diagonal form is clear if the components of vector $\boldsymbol{\theta}$ are uncorrelated, but it could be the second best choice if no prior information about the joint distribution of components $\boldsymbol{\theta}$ is available.

The scheme in Fig. 1 shows the concept of the parallel connection where numerical ML estimator and ANN-based estimator are placed according to (2).

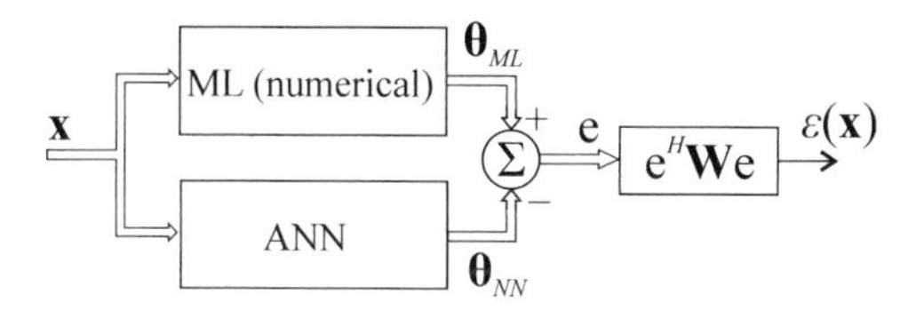

Fig. 1. Parallel work of the numerical ML and ANN estimators

Neural network training task consists in finding the vector of parameters $\mathbf{w}$ that minimizes the total error J over the training set Λ, i.e. the sum of the errors of each training pattern:

$$J = \sum_{n \in \Lambda} \varepsilon(\mathbf{x}_n) \tag{3}$$

The disadvantage of above-mentioned approach is that the ML estimations must be applied to each pattern in the training set. It is also obvious that if the training set is generated using a particular model incorporating a priori known parameter vector $\boldsymbol{\theta}$, a better scheme could be drawn if one substitutes the estimated vector $\boldsymbol{\theta}_{ML}$ by the known vector $\boldsymbol{\theta}$.

Here we introduce an alternative scheme with cascade connection in order to synthesize ANN approximating ML estimations; the proposed connection used for ANN training process is shown in Fig. 2.

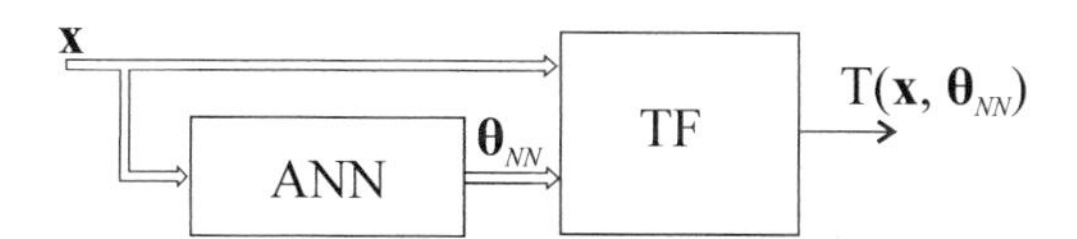

Fig. 2. The cascade scheme for the training phase of ANN

The vector $\mathbf{x}$ is introduced to the ANN as the input data vector and it also goes to the block of target function (TF) calculation. The other input of this block is the output of the ANN. The ANN training task in case of the connection described above is then defined to be the task of obtaining a vector of ANN internal parameters (vector of the synaptic weights $\mathbf{w}$) such that the each training pattern of the training set yields estimation $\boldsymbol{\theta}_{NN}$ that maximizes target function $T(\mathbf{x}, \boldsymbol{\theta})$.

If the target function is logarithmic transformation of the likelihood function $p_{\xi}(\mathbf{x};\boldsymbol{\theta})$ we choose the total performance of the network over the training set Λ as a sum:

$$G = \sum_{n \in \Lambda} T(\mathbf{x}_n, \boldsymbol{\theta}_{NN}(\mathbf{x}_n)). \tag{4}$$

This choice allows one to carry out the training procedure of ANN with well-known error backpropagation approach in its generic form. We should note that backpropagation learning requires neither the true value of vector $\boldsymbol{\theta}$ nor its ML estimation $\boldsymbol{\theta}_{ML}$. The only necessary value is the target function gradient $\nabla_{\boldsymbol{\theta}} T = \text{grad}\{T(\mathbf{x}, \boldsymbol{\theta})\}$ over vector of parameters $\boldsymbol{\theta}$; the latter can usually be expressed in closed form provided the target function possesses one or it can be calculated using finite differences scheme which engages values of function T itself in some complicated cases.

The gradient of the target function can be expressed as a column-vector function of the synaptic weights $\mathbf{w}$:

$$\nabla_{\mathbf{w}} T(\mathbf{x}, \boldsymbol{\theta}_{NN}(\mathbf{x})) = \left\{ \frac{\partial T(\mathbf{x}, \boldsymbol{\theta}_{NN}(\mathbf{x}))}{\partial w_k} \right\}^T, \tag{5}$$

where its components are calculated according to the chain rule of differentiation:

$$\frac{\partial T(\mathbf{x}, \boldsymbol{\theta}_{NN}(\mathbf{x}))}{\partial w_k} = \sum_{n=1}^{N} \left\{ \frac{\partial T(\mathbf{x}, \boldsymbol{\theta}_{NN}(\mathbf{x}))}{\partial \theta_m} \cdot \frac{\partial \theta_m}{\partial w_k} \right\}, \tag{6}$$

where $\partial T(\mathbf{x}, \boldsymbol{\theta}_{\mathbf{NN}})/\partial \theta_m$ are partial derivatives of the target function with respect to the m-th component of the vector $\boldsymbol{\theta}_{NN}$, $\partial \theta_m / \partial w_k$ are partial derivatives of the m-th output of the ANN with respect to the k-th element of the synaptic weight vector $\mathbf{w}$. The latter could be split up into some expression using the chain rule or by means of the diagrammatical approach describing the back propagation of the error in the ANN graph.

Since each of the terms in (4) has its upper bound, the maximum of G could be achieved when the maximum of each component takes place. For that reason, the alteration in the synaptic weights vector $\mathbf{w}$ during the training procedure should be made in the direction of the growth of the target function T that is in the direction of its gradient (5). If the batch training with the first order training methods is used, the general rule of weight vector adaptation will be defined as:

$$\mathbf{w}^{q+1} = \mathbf{w}^q + \mathbf{L}(q) \sum_{n \in \Lambda} \nabla_{\mathbf{w}} T(\mathbf{x}_n, \boldsymbol{\theta}_{NN}(\mathbf{x}_n)), \tag{7}$$

where $\mathbf{L}(q)$ is a square matrix incorporating the learning rate coefficients for each weight in q-th epoch. In the simplest case of the steepest decent the matrix $\mathbf{L}(q)$ is set to be a diagonal matrix with equal coefficients on the main diagonal remaining unchanged throughout the training process.

The increase of the learning speed could be achieved using procedures engaging second-order and mixed-order training methods [13]. If the most promising Levenberg-Marquardt [14] method is chosen the rule of changing **w** becomes:

$$\mathbf{w}^{q+1} = \mathbf{w}^{q} + \left\{ \sum_{n \in \Lambda} \left[\nabla_{\mathbf{w}} T(\mathbf{x}_n, \boldsymbol{\theta}) \cdot \nabla_{\mathbf{w}}^{T} T(\mathbf{x}_n, \boldsymbol{\theta}) \right] + \mu(q)\mathbf{I} \right\}^{-1} \nabla_{\mathbf{w}} T(\mathbf{x}_n, \boldsymbol{\theta}), \quad (8)$$

where **I** is the identity matrix of size N, $\mu(q)$ is the coefficient of the steepest descent component contributed to the inverse Hessian approximation.

3 Simulation Results

The direction-of-arrival approach consists in obtaining estimations of two bearing angles: azimuth α and elevation β that are the angles between the direction of arrival of the front of the electromagnetic wave and the reference direction of the system. The model under investigation includes the radio signal source and a narrow-band passive radar system [8] implementing the direction-of-arrival estimator [9]. The presented results are limited to the two-dimensional case when only the bearing is estimated. In order to synthesize the estimator able to approximate ML estimations the authors propose the use of the feed-forward ANN build using the adaptive element approach described in [6]. The topology of the network is shown in Fig. 3.

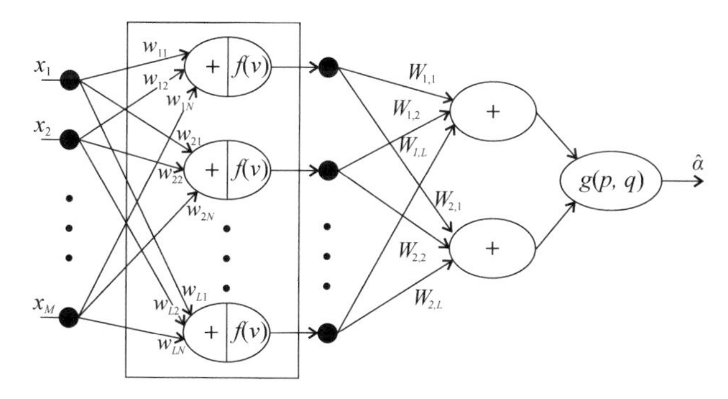

Fig. 3. Topology of the proposed ANN. W and w are the weights being adapted within output and hidden layers accordingly

The input data is a time-difference vector named **X** which element X_{mn} is the delay between signals in two antenna elements m and n which is simply evaluated via the mean phase difference of their spectra, $S_m(f)$ and $S_n(f)$ respectively, within narrow spectral band (F_{min}, F_{max}):

$$X_{mn} = \frac{1}{\pi(F_{\max} + F_{\min})} \arg\left(\frac{1}{F_{\max} - F_{\min}} \int_{F_{\min}}^{F_{\max}} S_m(f) S_n^*(f) df\right). \quad (9)$$

Since the signal is considered to be a realization of random communication signal it exhibits cyclostationary properties [10] with cyclic frequencies $\alpha = k/T$ (k is integer) and $\alpha = 2F_c$ where T is a synchronous period and F_c is a carrier frequency. This allows to enhance the accuracy of the delay estimation taking into account all signal-specific α belonging to the set Λ. The spectral correlation function $S_m^{\alpha}(f)$ is evaluated [11] for each α and the final value is an average over all elements in the set:

$$X_{mn} = \frac{1}{|\Lambda|} \sum_{\alpha \in \Lambda} \left\{ \frac{1}{2\pi\alpha} \arg\left(\frac{1}{F_{\max}(\alpha) - F_{\max}(\alpha)} \int_{F_{\max}(\alpha)}^{F_{\max}(\alpha)} S_m^{\alpha}(f) S_n^{\alpha *}(f) df \right) \right\}. \quad (10)$$

The output of the network is the estimation of the bearing $\hat{\alpha}$. The neurons in the hidden layer of the network use sigmoid activation functions $f(v) = [1 + \exp(-v)]^{-1}$.

Activation function g of the output layer neuron differs from the convenient networks since it is an argument function of a complex number by means of which the angle of the vector defined in Cartesian coordinates (p, q): $g(p,q) = \arg(p + jq)$. This form of the activation function was introduced in [15] and it allows assembling the output neuron shown in Fig. 3.

A numerical simulation was carried out for the case of ring antenna system consisting of seven sensors: two concentering rings with three sensors per ring and an additional sensor at the center of system. In order to train the ANN, a series of 271 learning patterns was calculated for the bearing taking values in the range between –45° to 225° with a degree step. Each learning pattern contains a vector of time-difference-of-arrival values calculating using the known geometry of the antenna system.

The proposed neural network topology was chosen in the accordance with the physical model of the problem. Thus, the number of output neurons is set to one and the number of pairs built up of the antenna sensors determines the number of net inputs. The capacity of the network, i.e. number of neurons in the hidden layer, varies from 2 up to 30. A set of ANNs possessing the topology shown in Fig. 3 was trained with various number of neurons in their hidden layers using second-order Levenberg-Marquardt training algorithm and the weight adaptation rule (8). Upon the completion of the training process, the residual errors over all learning patterns were calculated for each network with particular capacity. Finally, we ended up with the ANN holding 12 neurons in its hidden layer since the networks with larger layers have provided an insignificant increase of accuracy.

For the practical purposes of implementation of the TDOA algorithms, the errors of the proposed estimator were compared to the direct approach based on the numerical solution of ML optimization problem. The dependency of mean squared error (MSE) of the obtained estimations on signal-to-noise ratio is shown in Fig. 4. Each point of the graph is obtained by averaging the errors of ten thousand simulation runs. The plot also

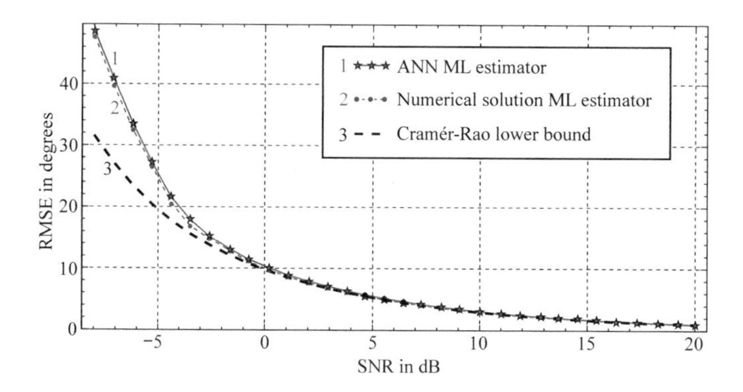

Fig. 4. The dependency of accuracy on SNR

includes Cramér-Rao lower bound for the reference, the corresponding analytical formula for ring antenna is derived, e.g., in [16].

The brief analysis of the dependencies presented by curves in Fig. 4 shows that as SNR increases both ML-estimators asymptotically approach Cramér-Rao lower bound (CRLB) and thus the minimum possible MSE for the model. The ANN estimator shows less accuracy of the estimation in comparison with the direct numerical solution. Generally speaking, MSE increases by a factor no greater than 1.1 in the SNR range between –8 and 20 dB and there is no tendency for this accuracy degradation to decrease for higher SNR.

The results of numerical simulation also indicate the absence of the dependency between accuracy of the ANN estimator and the true value of bearing. The polar plots in Fig. 5 show the accuracy against the true value for two SNRs along the 270° segment.

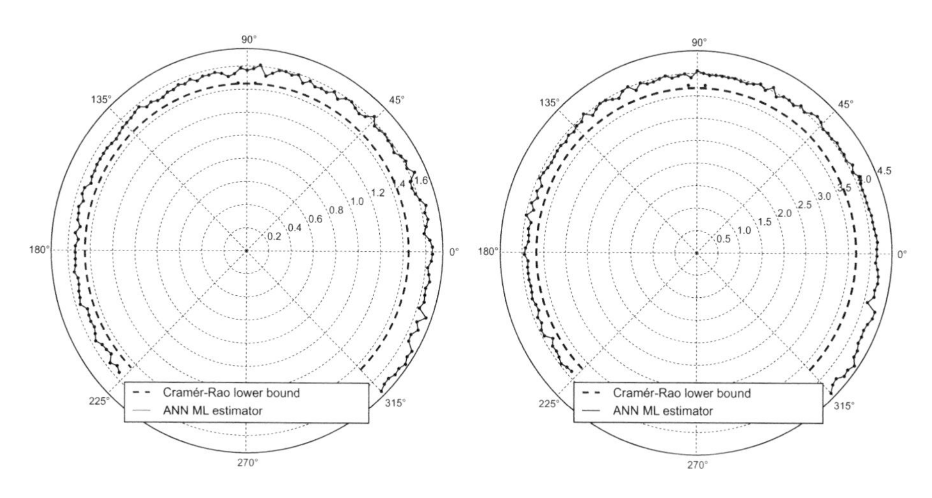

Fig. 5. The dependency of estimator accuracy on the true value for SNR: 8 dB (left) and 10 dB (right).

Table 1. Comparison of ANN and numerical MLE.

Parameter		ANN ML estimator	ML estimator
Number of computational operations		Determined by the topology of the network and the number of neurons in the hidden layer	Determined by the likelihood function and the number of iteration in the numerical algorithm
Dependency of the estimation time on the initial point of the approximation		No	Yes
Time to obtain a single estimation		2.74 ms	33.4 ms
Accuracy loss relative to CRLB	–4 dB	30%	26%
	0 dB	13%	4%
	10 dB	9%	2%

4 Conclusion

As far as the computational power required for the implementation of ANN ML estimator and the ML estimator based on the numerical solution are concerned, Table 1 gives key differences; we use a personal computer equipped with Intel Core i5 2.0 GHz processor and 4 GB RAM. We should notice that the training step for the ANN ML estimator has to be made only once before the regular use.

The application of the ANN based approach allows to synthesize ANN ML estimator for the passive radar system performing TDOA methods. The proposed solution requires at least 10 times less computational power for the single estimation in comparison with ML estimator base on the numerical solution of the underlying ML optimization problem. The results of the numerical simulation run for seven-element circular array antenna indicate that the application of ANN allows reducing the time required for a single estimation in 12 times. The simulation also shows that ANN ML estimator provides no significant dependency of the estimation accuracy on the true values in the wide range of bearing values. The degradation of accuracy could be estimated by means of the error increased by ANN comparing to the direct numerical approach to the optimization problem which is no more than 10% for the practical SNR range of –8 dB to 20 dB.

The practical signal processing scenario engaging ANNs suggests the most demanding for computational power stages of ANN synthesis such as finding the sufficient size of hidden layer and its further training should be run on a powerful workstation or even data processing cluster. After the best option is found, synaptic weights matrix is obtained. It allows replicating already trained ANN and using it as an algorithm with finite time execution time in the software of the radars with same geometry of antenna system wherever they are to be placed.

Acknowledgement. The work was supported by state assignment of the Ministry of Education and Science of the Russian Federation (project 8.8502.2017/BP).

References

1. Hassoun, M.: Fundamentals of Artificial Neural Networks. MIT series. A Bradford Book. MIT Press, Cambridge (2003)
2. Haykin, S.: Neural Networks and Learning Machines, 3rd edn. Pearson, New York (2008)
3. Baum, E.: Supervised learning of probability distributions by neural networks. American Institute of Physics, pp. 52–61 (1988)
4. Setiono, R.: A neural network construction algorithm which maximizes the likelihood function. Connection Sci. **7**(2), 147–166 (1995)
5. Cervellera, C., Maccio, D., Muselli, M.: Deterministic learning for maximum-likelihood estimation through neural networks. IEEE Trans. Neural Netw. **19**(8), 1456–1467 (2008)
6. Efimov, E.N., Shevgunov, T.Y.: Feedforward artificial neural networks constructed with the use of adaptive elements. J. Radio Electron. **2012**(8), 1–16 (2012)
7. Wan, E.A., Beaufays, F.: Diagrammatic methods for deriving and relating temporal neural network algorithms. In: Adaptive Processing of Sequences and Data Structures. LNCS, vol. 1387. Springer, Heidelberg (1998)
8. Dubrovin, A.V., Sosulin, Y.: One-stage estimation of the position of a radio source by a passive system consisting of narrow-base subsystems. J. Commun. Electron. **49**(2), 139–153 (2004)
9. Zekavat, R., Buehrer, R.M.: Handbook of Position Location: Theory, Practice and Advances. Wiley-IEEE Press, Hoboken (2011)
10. Gardner, W.A.: Cyclostationarity in Communications and Signal Processing. IEEE Press, New York (1994)
11. Brown, W.A., Loomis, H.H.: Digital implementations of spectral correlation analyzers. IEEE Trans. Sig. Process. **41**(2), 703–720 (1993)
12. Kay, S.M.: Fundamentals of Statistical Signal Processing: Estimation Theory. Prentice Hall, Upper Saddle River (1993)
13. Battiti, R.: First- and second-order methods for learning: between Steepest Descent and Newton's Method. Neural Comput. **4**(2), 141–166 (1992)
14. Marquardt, D.W.: An algorithm for least-squares estimation of non-linear parameters. J. Soc. Ind. Appl. Math. **11**, 431–441 (1963)
15. Efimov, E., Shevgunov, T., Filimonova, D.: Angle of arrival estimator based on artificial neural networks. In: Proceedings of 17th International Radar Symposium (IRS), pp. 1–3 (2016)
16. Dubrovin, A.V.: Potential direction-finding accuracy of systems with antenna arrays configured as a set of an arbitrary number of rings. J. Commun. Technol. Electron. **51**(3), 252–254 (2006)

Using Query Expansion for Cross-Lingual Mathematical Terminology Extraction

Velislava Stoykova[1(✉)] and Ranka Stankovic[2]

[1] Institute for Bulgarian Language, Bulgarian Academy of Sciences, 52, Shipchensky proh. str., bl. 17, 1113 Sofia, Bulgaria
vstoykova@yahoo.com

[2] Faculty of Mining and Geology, University of Belgrade, 7, Dusina, 11000 Belgrade, Serbia
ranka@rgf.bg.ac.rs

Abstract. The paper presents approach to knowledge discovery by using query expansion to search for cross-lingual mathematical terminology extraction. It employs information retrieval and statistically-based techniques to extract and process keyword collocations in large comparable cross-lingual web electronic text corpora in the domain of mathematics in Bulgarian and in Serbian language. It, also, offers examples and survey of used techniques for semantic search and clustering by comparing keyword collocations to build a cross-lingual thesauri. The results of semantic keyword search for the two web electronic text corpora using Sketch Engine software are presented and analyzed with respect to the types of keyword collocations processing and to multilingual application of the approach.

1 Introduction

The search-based knowledge discovery techniques widely use statistically-based approaches to extract semantic conceptual relations. The techniques adopted are based on search, retrieval and clustering of statistically similar words and recent improvements of that approaches use more complex and elaborated statistics for keyword search. At the same time, the structure of data sources, also, reflects the search results and taken together with the use of different types of keyword search allow extraction of more sophisticated semantic relations. For example, the approach to integrate keyword search, semantics and information retrieval gives promising results [1] for discovering more complex semantic relations.

Further, we are going to present and analyze applications of such techniques by comparing and discussing several results of semantic keyword search in cross-lingual comparable electronic text corpora using statistical functions of the Sketch Engine search software.

2 The Sketch Engine (SE)

The SE software [4] allows approaches to extract semantic properties of words most of which are with multilingual application. Concordance generation and

R. Silhavy (Ed.): CSOC 2018, AISC 764, pp. 154–164, 2019.
https://doi.org/10.1007/978-3-319-91189-2_16

keyword search are widely used techniques to extract terms of a particular domain by using statistically-based techniques. Also, semantic relations can be extracted by estimation and comparison of related keyword contexts through generation and processing of collocations for that keyword.

Generally, collocations are defined as words which are most probably to be found with a related keyword. They assign the semantic relations between the keyword and its particular collocated word which might be of similarity or of a distance. The statistical approaches used by SE to search for collocated words are based on defining the probability of their co-occurrences by using various techniques for scoring. We use techniques of $T - score$, $MI - score$ and $MI^3 - score$ for corpora processing and searching. For all, the following terms are used: N - corpus size, f_A - number of occurrences of keyword in the whole corpus (the size of concordance), f_B - number of occurrences of collocated keyword in the whole corpus, f_{AB} - number of occurrences of collocate in the concordance (number of co-occurrences). The related formulas for defining $T - score$, $MI - score$ and $MI^3 - score$ are as follows:

$$\text{MI-Score} \log_2 \frac{f_{AB}N}{f_A f_B}$$

$$\text{T-Score} \frac{f_{AB} - \frac{f_A f_B}{N}}{\sqrt{f_{AB}}}$$

$$\text{MI}^3\text{-Score} \log_2 \frac{f_{AB}^3 N}{f_A f_B}$$

The $T - score$, $MI - score$ and $MI^3 - score$ are applicable for processing multilingual comparable corpora as well.

3 Traditional and Computational Approaches to Terminology

Terminology presents knowledge of a certain domain by organizing related concepts and connecting them through semantic relations forming conceptual hierarchy. Traditional terminology representations are classifications (classes, types, hierarchies, etc.) which are basic knowledge representation content schemata of encyclopedias, terminological dictionaries, encyclopedic dictionaries, reference sources, etc.

However, language affect term representation which is language-dependent and unification was introduced by using: Latin language, terms from Old Greek have been explained by giving their etymology, stylistically neutral lexica, etc. Recent AI-based approaches are used not only for knowledge representation but also for knowledge extraction (knowledge discovery, text mining, etc.) by applying combined search and retrieval techniques.

The general tasks to be conducted are connected to evaluate: (i) which are the concepts and their related terms (language representations: single or multi-words) of a certain domain, (ii) which are the types of their semantic relations:

unification (synonymy, hierarchy, part-of, etc.) and opposition (antonymy). However, the synonymy has a quite controversial semantics. From the logical point of view, it means semantic equivalence as for definitions and formulas. From the lexical semantics point of view, it means partly semantic equivalence because some keyword's synonyms are closer semantic equivalents than others. From terminological point of view (for knowledge discovery) it means to discover hierarchical conceptual relations of certain domain and to build a thesauri, i.e. generating semantic hierarchy by extracting keywords semantic conceptual relations.

Generally, the IR approaches interpret that problem by using statistical similarity and relating it to semantic similarity by estimating the semantic closeness [3] or using only Wikipedia semantic representations [8]. However, that general methodology gives as a result the mixture of above mentioned relations (including some grammatical relations as well) and a more elaborate and complex search techniques can be used to disambiguate the results.

The most commonly used approach is a collocations extraction (mainly for multi-word terms or hyponyms extraction). However, within that approach there are some differences in semantic disambiguation techniques used. Thus, in [9] the collocations extraction uses the idea of fixed syntactic order and adopts technique of syntactic parsing which is with a multilingual scope of application. The approach used in [6] adopts only statistically-based techniques for parallel collocation extraction. Also, approaches using combined techniques of lexical function and log likelihood have been applied for extraction of specialized collocations [7].

More sophisticated approaches use the idea to search and compare collocations of related keyword by evaluating their relational similarity. Additionally, the approach is defined as extracting semantically related pairs of words to discover hidden semantic relations [5] by using relational semantic properties as a main technique.

Finally, the idea of distributional representation [2] and keyword frequency distributional search gives reliable results in knowledge discovery for more complex semantic representations [14].

Further, we shall present and analyze results for concordance generation, keyword frequency distributional search, semantic clustering by using collocations comparison search, and keyword sketch differences evaluation using the SE software options. We shall compare related results with respect to both texts sources – comparable electronic text corpora in the domain of mathematics in Bulgarian and in Serbian language – and to the semantic types of received results.

4 The Bulgarian Mathematical Wikipedia Corpus (MathWikiBG) and the Serbian Mathematical Wikipedia Corpus (MathWikiSR)

The Bulgarian Mathematical Wikipedia Corpus (MathWikiBG) and the Serbian Mathematical Wikipedia Corpus (MathWikiSR) are the first Bulgarian–Serbian/Serbian–Bulgarian comparable electronic text corpora. They were designed and used for running cross-lingual keyword search experiments within the framework of COST Action IC1302 "Semantic keyword-based search on structured data sources (KEYSTONE)" using pipeline method of compilation and keyword web-browsing of mathematical texts from Wikipedia in Bulgarian and in Serbian language.

The preliminary work which reflect the creation and structure of both corpora contain research on mono- and multilingual terminology extraction [12,13], including that for Serbian language [10,11]. Both corpora were created using specific tag sets for Bulgarian and for Serbian language, which allow performing language-specific and cross-lingual keyword search. The language-specific tag encodings allow, also, keyword-based search for both semantic and grammatical relations and are used for semantic disambiguation.

The MathWikiBG corpus includes mathematical texts in Bulgarian language at about 90 000 words distributed randomly across several broad domain areas of mathematics. Alternatively, the MathWikiSR corpus includes mathematical texts in Serbian language at about 103 000 words distributed randomly across several broad domain areas of mathematics. Both corpora were uploaded into SE allowing the use of its incorporated options for storing, sampling, searching and filtering corpora texts according to different criteria.

5 Cross-Lingual Keyword Search Results

For our research, we ran several experiments to discover some basic semantic conceptual relations. As a keyword search machine, we have used the SE standard approach, and particulary its options to generate keyword concordances, collocations and their distribution, grammatical and semantic relations for disambiguating the results.

We present the general approach by demonstrating related search results for the keyword *mathematics* for which we first generate comparable cross-lingual concordances. The concordances present all occurrences of the keyword within its related quantitative contexts. Figure 1 presents the occurrences of keyword *mathematics* within MathWikiBG and MathWikiSR corpora, respectively. The generated results show that keyword hits 262 occurrences in MathWikiBG corpus and 230 occurrences in MathWikiSR corpus including the inflected keyword forms.

However, the results do not reveal information about keyword frequency distribution which is important structural criterion (that uses mutual information for disambiguation). It can be evaluated by using keyword collocations comparison, so to identify more complex semantic relations.

The SE has options to evaluate different types of keyword distribution, which are created on the base of comparing keyword's common collocations. That type of keyword search is based on the idea that common collocations share common semantic relations and identifying them can reveal hidden keyword's semantic relations [14].

Thus, Fig. 2 shows results from comparable cross-lingual frequency distribution search of keyword *mathematics* over MathWikiBG and MathWikiSR corpora, respectively. The generated results present similar semantics of keyword *mathematics* (BG – математика, SR– matematika). They, also, include common collocations of that keyword for both languages like *domain* (BG – дисциплина, SR – oblast), *function* (BG – функция, SR – funkcija), etc. which

Fig. 1. The concordances of keyword *mathematics* from MathWikiBG and MathWikiSR corpora, respectively.

relate semantically. Obviously, the keyword frequency distribution search gives reliable results which reveal hidden semantic relations of keyword *mathematics* and can be used to build terminological conceptual hierarchy. Moreover, the technique is with multilingual application.

The distributional thesauri keyword search generates keyword semantic clusters using distributional similarity, i.e. by grouping words, that appear in similar contexts of related keyword. It uses the idea of estimating statistical similarity of keyword contexts to extract synonymic semantic relations.

The results of comparable cross-lingual distributional thesauri search for keyword *mathematics* performed over MathWikiBG and MathWikiSR corpora are given at Fig. 3 and present the generated semantic clusters of keyword *mathematics* for both languages. Compared to the results from frequency distribution keyword search, they reveal more sophisticated semantic relations enlarging the number of words semantically related to keyword for both languages.

Thus, for example, the keyword *mathematics* relates not only to *domain* (BG – дисциплина, SR – oblast) but through these already generated semantic relations, it relates also for MathWikiBG to $sub - domain$ (BG – клон by con-

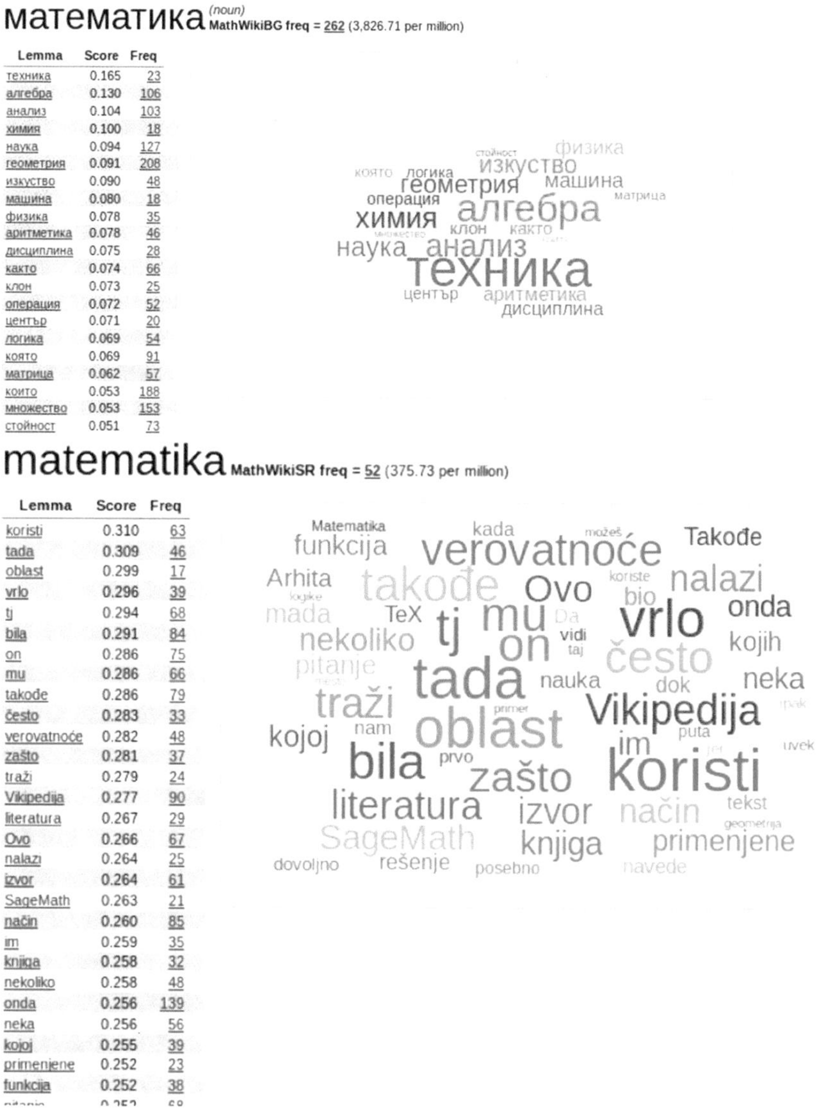

математика *(noun)* MathWikiBG freq = 262 (3,826.71 per million)

Lemma	Score	Freq
техника	0.165	23
алгебра	0.130	106
анализ	0.104	103
химия	0.100	18
наука	0.094	127
геометрия	0.091	208
изкуство	0.090	48
машина	0.080	18
физика	0.078	35
аритметика	0.078	46
дисциплина	0.075	28
както	0.074	66
клон	0.073	25
операция	0.072	52
център	0.071	20
логика	0.069	54
който	0.069	91
матрица	0.062	57
които	0.053	188
множество	0.053	153
стойност	0.051	73

matematika MathWikiSR freq = 52 (375.73 per million)

Lemma	Score	Freq
koristi	0.310	63
tada	0.309	46
oblast	0.299	17
vrlo	0.296	39
tj	0.294	68
bila	0.291	84
on	0.286	75
mu	0.286	66
takođe	0.286	79
često	0.283	33
verovatnoće	0.282	48
zašto	0.281	37
traži	0.279	24
Vikipedija	0.277	90
literatura	0.267	29
Ovo	0.266	67
nalazi	0.264	25
izvor	0.264	61
SageMath	0.263	21
način	0.260	85
im	0.259	35
knjiga	0.258	32
nekoliko	0.258	48
onda	0.256	139
neka	0.256	56
kojoj	0.255	39
primenjene	0.252	23
funkcija	0.252	38

Fig. 2. The frequency distribution of keyword *mathematics* and its semantically related words from MathWikiBG and MathWikiSR corpora, respectively.

nection дисциплина) and for MathWikiSR to *science* and *logics* (SR – nauka, logika by connection oblast).

The distributional thesauri keyword search reveals more complex hidden semantic relations of keyword through the use of hidden connections of that keyword to its semantically related words. It can, also, be used to enlarge and elaborate the terminological conceptual hierarchical thesauri representation.

математика *(noun)* **MathWikiBG freq = 262** (3,826.71 per million)

Lemma	Score	Freq	Cluster
техника	0.165	23	машина [0.08, 18] операция [0.072, 52]
алгебра	0.130	106	изкуство [0.09, 48]
анализ	0.104	103	логика [0.069, 54]
химия	0.100	18	
наука	0.094	127	
геометрия	0.091	208	
физика	0.078	35	
аритметика	0.078	46	както [0.074, 66]
дисциплина	0.075	28	клон [0.073, 25]
център	0.071	20	
която	0.069	91	които [0.053, 188]
матрица	0.062	57	
множество	0.053	153	
стойност	0.051	73	

matematika **MathWikiSR freq = 52** (375.73 per million)

Lemma	Score	Freq	Cluster
koristi	0.310	63	tada [0.309, 46] vrlo [0.296, 39] tj [0.294, 68] mu [0.286, 66] takođe [0.286, 79] on [0.286, 75] često [0.283, 33] zašto [0.281, 37] traži [0.279, 24] literatura [0.267, 29] nalazi [0.264, 25] izvor [0.264, 61] neka [0.256, 56] referenca [0.249, 47] TeX [0.244, 21] rešenje [0.242, 27] navede [0.24, 24] vidi [0.238, 42] dovoljno [0.238, 28] prvo [0.236, 38] posebno [0.235, 33] puta [0.234, 36] ipak [0.231, 30] uvek [0.229, 42] koriste [0.228, 45] možeš [0.226, 42] primer [0.226, 70] mesto [0.222, 23]
oblast	0.299	17	knjiga [0.258, 32] nekoliko [0.258, 48] primenjene [0.252, 23] funkcija [0.252, 38] nauka [0.245, 35] tekst [0.239, 51] logike [0.224, 26]
bila	0.291	84	Vikipedija [0.277, 90] način [0.26, 85] onda [0.256, 139] bio [0.25, 151] kada [0.24, 119] jer [0.233, 188] taj [0.229, 85] geometrija [0.225, 37]
verovatnoće	0.282	48	Ovo [0.266, 67] SageMath [0.263, 21] im [0.259, 35] kojoj [0.255, 39] mada [0.252, 24] pitanje [0.252, 68] kojih [0.251, 43] Arhita [0.251, 23] Takođe [0.25, 50] dok [0.244, 56] Da [0.242, 105] nam [0.24, 47]
Matematika	0.226	20	

Fig. 3. The semantic clusters of keyword *mathematics* generated from MathWikiBG and MathWikiSR corpora, respectively.

The main idea is to apply words distributional similarity by using bag-of-words technique to compare and extract those words which have similar contexts and are semantically related to the keyword. That keyword search technique can be used to discover hidden and very sophisticated semantic relations for knowledge discovery tasks.

In our approach, it is possible to evaluate all semantically related words to keyword *mathematics* by extracting not only semantic but also grammatical relations between the keyword and its related words. The main point is that, we have expanded keyword search with language-specific tagging (allowing processing different part-of-speech, inflection, etc.) for both Bulgarian and Serbian language (as previously described in Sect. 4).

Thus, we use keyword sketch differences option to generate all grammatical and semantic relations of keyword *mathematics* over MathWikiBG and MathWikiSR corpora, respectively. Moreover, it is possible to generate that type of relations for pairs of semantically related words.

Figure 4 shows generated results for the pair words *mathematics/logics* (BG – математика/логика, SR – matematika/logika) from MathWikiBG and MathWikiSR corpora, respectively. The results show that the generated semantically related words to that pair of words reveal various types of semantic relations like: *and/or*, $obj - of$, *pp*–, $modifies$, etc.

The results from MathWikiBG show that according to the relation *and/or* (synonymy) the word pair *mathematics/logics* (BG – математика/логика relates) the same way also to $informatics$, *physics*, *science*, etc. (BG – информатика, физика, наука). Also, according to the relation $modifies$ it relates to *applied*, *pure*, *discrete*, *computational*, etc. (BG – приложен, чист, дискретен, изчислителен) generating multi-word terms *applied mathematics*, *pure mathematics*, *discrete mathematics*, *computational mathematics*, etc. (BG – приложна математика, чиста математика, дискретна математика, изчислителна математика).

Similar results were obtained from MathWikiSR where extracted words which semantically relate to pair words *mathematics/logics* (SR – matematika/logika) are *discrete*, $financial$, *applied*, etc. (SR – diskretna, financijska, primenjena) generating multi-word terms like *discrete mathematics*, $financial$ *mathematics*, *applied mathematics*, etc. (SR – diskretna matematika, financijska matematika, primenjena matematika), etc.

The search results obtained for both languages show that generating keyword sketch differences (expanded with language-specific tagging) gives more complex and semantically rich results. It, practically, extracts all words which are semantically related to pair words. That technique is suitable for multi-word terms cross-lingual extraction generating predominantly hyponyms of related keyword (*mathematics*).

Thus, the performed keyword search experiments display results of several types which were obtained form MathWikiBG and MathWikiSR corpora. The above presented examples use the general approach of distributional keyword search expanded with language-specific tagging. The different types of search like comparable cross-lingual distribution frequency keyword search, comparable cross-lingual distributional thesauri keyword search, and generation of cross-

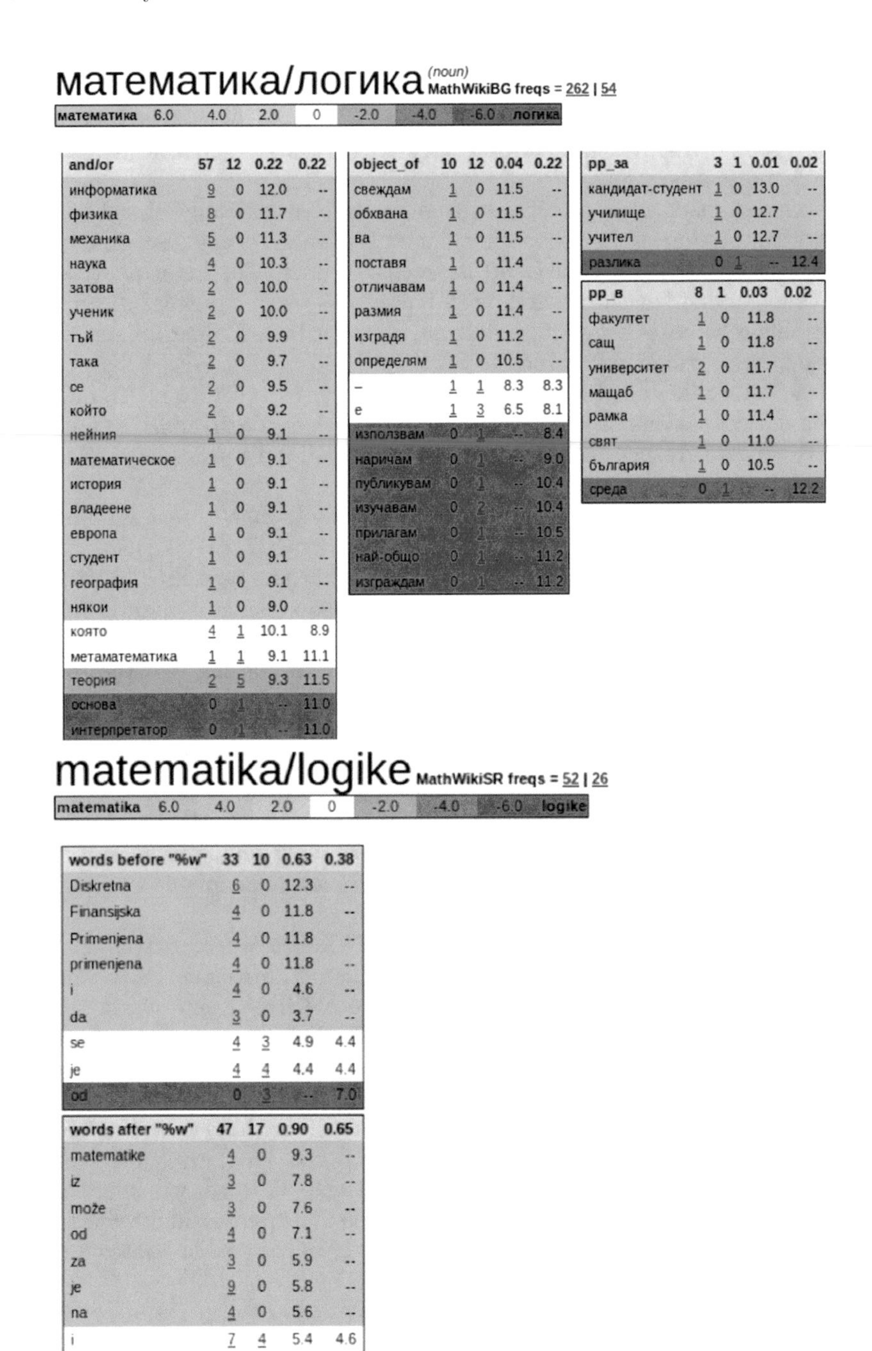

математика/логика (noun) MathWikiBG freqs = 262 | 54

математика	6.0	4.0	2.0	0	-2.0	-4.0	-6.0	логика

and/or	57	12	0.22	0.22
информатика	9	0	12.0	--
физика	8	0	11.7	--
механика	5	0	11.3	--
наука	4	0	10.3	--
затова	2	0	10.0	--
ученик	2	0	10.0	--
тъй	2	0	9.9	--
така	2	0	9.7	--
се	2	0	9.5	--
който	2	0	9.2	--
нейния	1	0	9.1	--
математическое	1	0	9.1	--
история	1	0	9.1	--
владеене	1	0	9.1	--
европа	1	0	9.1	--
студент	1	0	9.1	--
география	1	0	9.1	--
някои	1	0	9.0	--
която	4	1	10.1	8.9
метаматематика	1	1	9.1	11.1
теория	2	5	9.3	11.5
основа	0	1	--	11.0
интерпретатор	0	1	--	11.0

object_of	10	12	0.04	0.22
свеждам	1	0	11.5	--
обхвана	1	0	11.5	--
ва	1	0	11.5	--
поставя	1	0	11.4	--
отличавам	1	0	11.4	--
размия	1	0	11.4	--
изградя	1	0	11.2	--
определям	1	0	10.5	--
–	1	1	8.3	8.3
е	1	3	6.5	8.1
използвам	0	1	--	8.4
наричам	0	1	--	9.0
публикувам	0	1	--	10.4
изучавам	0	2	--	10.4
прилагам	0	1	--	10.5
най-общо	0	1	--	11.2
изграждам	0	1	--	11.2

pp_за	3	1	0.01	0.02
кандидат-студент	1	0	13.0	--
училище	1	0	12.7	--
учител	1	0	12.7	--
разлика	0	1	--	12.4

pp_в	8	1	0.03	0.02
факултет	1	0	11.8	--
сащ	1	0	11.8	--
университет	2	0	11.7	--
мащаб	1	0	11.7	--
рамка	1	0	11.4	--
свят	1	0	11.0	--
българия	1	0	10.5	--
среда	0	1	--	12.2

matematika/logike MathWikiSR freqs = 52 | 26

matematika	6.0	4.0	2.0	0	-2.0	-4.0	-6.0	logike

words before "%w"	33	10	0.63	0.38
Diskretna	6	0	12.3	--
Finansijska	4	0	11.8	--
Primenjena	4	0	11.8	--
primenjena	4	0	11.8	--
i	4	0	4.6	--
da	3	0	3.7	--
se	4	3	4.9	4.4
je	4	4	4.4	4.4
od	0	3	--	7.0

words after "%w"	47	17	0.90	0.65
matematike	4	0	9.3	--
iz	3	0	7.8	--
može	3	0	7.6	--
od	4	0	7.1	--
za	3	0	5.9	--
je	9	0	5.8	--
na	4	0	5.6	--
i	7	4	5.4	4.6
se	7	4	5.8	5.0

Fig. 4. The words sketch differences of pair words *mathematics/logics* generated from MathWikiBG and MathWikiSR corpora, respectively.

lingual keyword sketch differences can deal with more complex semantic structures and can be successfully used as a reliable technique for multilingual knowledge discovery.

6 Conclusion

The research presented includes results of keyword search for cross-lingual mathematical terminology extraction from the first comparable web-generated corpora in Bulgarian and in Serbian language (MathWikiBG and MathWikiSR). The approach presented uses distributional keyword search in order to deal with complex semantic representations. Also, the approach relates grammar and semantic representations by improving distributional keyword search with language-specific tagging and uses it for semantic disambiguation. The encodings and search experiments use the SE software to process comparable electronic text corpora.

The analyzed cross-lingual keyword search results show that using different types of distributional keyword search is a reliable technique to extract complex keyword semantic and grammatical relations and to discover new knowledge in return.

References

1. Azzopardi, J., et al.: Back to the sketch-board: integrating keyword search, semantics, and information retrieval. In: Cali, A., Gorgan, D., Ugarte, M. (eds.) Semantic Keyword-Based Search on Structured Data Sources, KEYSTONE 2016. LNCS, vol. 10151, pp. 49–61. Springer (2017)
2. Baroni, M., Lenci, A.: Distributional memory: a general framework for corpus-based semantics. Comput. Linguist. **36**(4), 673–721 (2010)
3. Bova, V., et al.: The combined method of semantic similarity estimation of problem oriented knowledge on the basis of evolutionary procedures. In: Artificial Intelligence Trends in Intelligent Systems - Proceedings of the 6th Computer Science On-line Conference 2017 (CSOC 2017), vol. 1. Advances in Intelligent Systems and Computing (AISC), vol. 573, pp. 74–83. Springer (2017)
4. Killgarriff, A., et al.: The sketch engine: ten years on. Lexicography **1**, 17–36 (2014)
5. Levy, O., Goldberg, Y.: Linguistic regularities in sparse and explicit word representations. In: Proceedings of the Eighteenth Conference on Computational Natural Language Learning, pp. 171–180, Baltimore (2014)
6. Novitskiy, V.: Automatic retrieval of parallel collocations for translation purposes. In: Kuznetsov, S.O., Mandal, D.P., Kundu, M.K., Pal, S.K. (eds.) Pattern Recognition and Machine Intelligence, PReMI 2011. LNCS, vol. 6744. Springer, pp. 261–267 (2011)
7. Orliac, B.: Extracting specialized collocations using lexical functions. In: Granger, S., Mennier, F. (eds.) Phreseology: An Interdisciplinary Perspective, pp. 377–390 (2008)
8. Nguyen, D., et al.: WikiTranslate: query translation for cross-lingual information retrieval using only Wikipedia. In: Peters, C., et al. (eds.) Evaluating Systems for Multilingual and Multimodal Information Access, CLEF 2008. LNCS, vol. 5706, pp. 58–65. Springer (2009)

9. Seretan, V., Wehrli, E.: Accurate collocation extraction using a multilingual parser. In: Proceedings of the 21st International Conference on Computational Linguistics and the 44th Annual Meeting of the Association for Computational Linguistics, pp. 953–960 (2006)
10. Stankovic, R., et al.: Developing termbases for expert terminology under the TBX standards. In: Pavlovic-Lazetic, G., Krstev, C., Obradovic, I., Vitas, D. (eds.) Natural Language Processing for Serbian – Resources and Applications, pp. 12–26, University of Belgrade, Faculty of Mathematics (2014)
11. Stankovic, R., et al.: Keyword-based search on bilingual digital libraries. In: Cali, A., Gorgan, D., Ugarte, M. (eds.) Semantic Keyword-Based Search on Structured Data Sources, KEYSTONE 2016. LNCS, vol. 10151, pp. 112–123. Springer (2016)
12. Stoykova, V., Mitkova, M.: Conceptual semantic relationships for terms of Precalculus study. WSEAS Trans. Adv. Eng. Educ. **8**(1), 13–22 (2011)
13. Stoykova, V.: Using statistical search to discover semantic relations of political Lexica - evidences from Bulgarian-Slovak EUROPARL 7 corpus. In: Kotsireas, I., Rump, S., Yap, C. (eds.), Mathematical Aspects of Computer and Information Sciences. LNCS, vol. 9582, pp. 335–339. Springer (2016)
14. Stoykova, V.: Discovering distributional Thesauri semantic relations. In: Proceedings of the 3rd International Workshop on Knowledge Discovery on the Web, CEUR-WS, Cagliari, Italy (2017). http://ceur-ws.org/Vol-1959/paper-11.pdf

Text Summarization Techniques for Meta Description Generation in Process of Search Engine Optimization

Goran Matošević(✉)

Faculty of economics and tourism "Dr. Mijo Mirković",
P. Preradovića 1/1, 52100 Pula, Croatia
gmatosev@unipu.hr

Abstract. Search engine optimization involves various techniques which web site creators and marketers can apply on web pages with goal of ranking higher in popular search engines. One of important ranking factors is "meta description" - a short textual description of the website which is put inside web page header accessible to web spiders. In this paper we investigate if existing text summarization techniques can be used to artificially build "meta description" for websites which are missing it. Also, we propose a simple query based algorithm for generation of "meta description" content based on some summarization techniques. The experimental results and expert evaluation show that proposed algorithm for text summarization can successfully be used in this context. Results of this research can be used to build recommender system for improvement of search engine optimization of a webpage.

Keywords: Search engine optimization · Text summarization
Website summarization · Meta description · SEO · On-page SEO

1 Introduction

The process of Search engine optimization (SEO) involves "on-page" and "off-page" techniques that are applied to a webpage with an aim to rank as high as possible in search engine results pages (SERP). "On-page" represents work on the page content, while "off-page" is all about links and trying to build quality back links to our website. Today's search engines use more than 200 factors in their ranking algorithms to present the user most reliable search results for given query. Queries that users type in search engine interface represent keywords for which the results are shown. Therefore, if we want to rank higher, it's important to use selected keywords in appropriate places inside webpage content and in backlinks. One of important "on-page" factor is "meta description": a short textual description of web page content with keywords included. This text is displayed in SERP under the website link and after the web page title it's the second thing that users sees when deciding to click the link and visit the site. Unfortunately many websites don't have (or have poor) "meta description" text. The research question that we are asking in this paper is: How can we use the text in webpage content to generate "meta description" which will comfort SEO practice?

R. Silhavy (Ed.): CSOC 2018, AISC 764, pp. 165–173, 2019.
https://doi.org/10.1007/978-3-319-91189-2_17

Obviously, the text summarization algorithms can be helpful, but are they good enough from SEO point of view?

Text summarization is a process of generating shorter text which well represents document topic [1]. There are two techniques: abstractive and extractive text summarization. The goal of both is to create a document summary. Extractive techniques work by extracting important sentences in order to create a summary, while abstractive summarization can create summaries that contain sentences and words not included in original document – but semantically and topically related. Therefore, abstractive summarization is much harder then extractive - it must address various AI problems and natural language processing (NLP). Extractive summarization is more straightforward: find most frequent words and choose sentences which use them to include in summary. In information retrieval a query-based summarization is used when the keywords for detecting important sentences are not extracted from the document, but from given query. This approach is interesting from SEO view and will be used in this research.

In this research we are using extractive summarization techniques to create summary text that is included in "meta description" tag for webpages without defined "meta description".

2 Previous Research

In [2] authors are exploring web page summarization based on keyword frequency and web page structure. They are weighting sentences based on its position in HTML document (headings, subtitles, beginning of paragraph, etc.) and also containing indication phrases like "to sum up", "in a word", "this paper presents" etc. From SEO aspect, the main problem of this approach is that it relies on keywords extracted from text (web page content) which can differ from target keywords in SEO process. In addition, the indication phrases that will detect good sentence candidates for "meta description" are different from SEO point of view – they should include some kind of "call to action" like "visit us", "click to view", etc. We believe that combination of such phrases and target keywords will detect better sentences for "meta description" than what this and similar summarization algorithms are suggesting. We will show this in following chapters.

Using context of the webpage for summarization tasks is explored in [3]. Their work is based on [4] which states that webpages backlinks and its anchor text can be used to determine web page topic. While this is true, the small length of anchor text cannot be used for summarization, instead the wider context of the backlink must be taken in consideration. Authors of [3] are proposing algorithms for summarization by context and conclude that this method is promising but needs more refinement.

In [5] authors are experimenting with text summarization in field of contextual advertising. They implemented a few enhancements to extractive summarization that improves performance in matching ads to the text of web page.

The usage of "meta description" tag in web pages was analyzed in [6] in 2004. Their findings pointed that only 30% of webpages were using "meta description" tag. They are also exploring the differences of "meta description" between domains (directory categories), length in words and the directories themselves (Google and Yahoo).

In [7] the process of generating better page titles by using anchor text is explored. Along with algorithm proposed in this research, this can be a good step towards automatic, semi-automatic or recommendation-based system for SEO experts or any other interested in this topic.

3 Text Summarization

The text summarization became an area of research long before web was invented. Summarization algorithms can be categorized in multiple ways: depending on type of extraction can be extractive or abstractive, based on number of documents can be one-document or multi-document summarization, they can be graph based, semantic based or word frequency based, etc.

3.1 Luhn and Edmundson Algorithms

One of first algorithms was from Luhn in 1958 for generating summaries of scientific texts [8]. His work relied on scoring sentences based on word frequency - high scored sentences are used to generate summary. Edmundson [9] used a similar approach for calculating sentence weight: using location of key phrases (title of document, beginning of document, cue dictionary). Both Luhn and Edmundson approaches were extractive methods of text summarization, which in most cases delivers better results than abstractive methods which struggle with a lot of semantic and natural language generation problems [10]. All extractive summarization techniques work the same way: they determine important sentences in document(s) and use it to construct a summary, which should express main key topics of document. There are two different approaches in determining the score of a sentence: topic approach and indicator approach. In topic approach sentence is scored based on how well it represents main topics of a document. In indicator approach, score is calculated based on multiple evidence (indicators), like sentence lengths, position in document, containing key phrases and similar. Topics can be represented by topic words which best represents the topic of document. Sentences that contain topic words in high frequency are candidates for inclusion in summary – this is the basic idea behind Luhn's topic approach.

3.2 TextRank and LexRank Algorithms

TextRank [11] and LexRank [10] were developed at the same time by different authors. Both work in a similar way by making a graph representation of sentences in document. In this graph, sentences are nodes while edges are similarity relationship between sentences. Similarity in LexRank is calculated based on word frequencies using modified TF-IDF formula, and in TextRank a similar formula is used based on words in common between two sentences. This enables construction of a similarity matrix. Sentences that have more connections with high similarity scores are treated as more important therefore good candidates for summary. By applying a score threshold a number of sentences are selected to construct a summary. TextRank and LexRank are developed on top of the well-known Google PageRank algorithm which is used to rank webpages on search engines by constructing graph to find important webpages.

In literature is often used the algorithm described in [12] as base line method, and KL-Sum method described in [13].

3.3 Latent Semantic Analysis

Latent semantic analysis (LSA) is another topic representation approach introduced in [14, 15]. LSA lies on assumption that words with similar meaning appear together in similar parts of text. In natural language multiple words can mean the same thing, and same words can mean different things based on context and hence in language all sorts of ambiguities can exists, when doing summarization we really want to compare topics and concepts which are behind words. LSA method solves this by mapping words and documents in concept space by constructing term with sentences matrix and applying singular value decomposition (SVD) to reduce the size of matrix without any significant loss of information.

4 Search Engine Optimization

Search Engine Optimization (SEO) is an important tool for internet marketers and website creators for boosting their webpage, social media page or blog on search engine results pages (SERP). Ranking higher on selected keywords leads to more clicks, meaning more visitors that can convert to buyers (for e-comer web pages) [16]. There is a lot of SEO tips & tricks, recommendations and guides published on the web, but the most important ones are those published by search engine companies, like Google and Bing guides [17, 18]. To understand how search engines rank webpages, we must understand factors that influence rankings. Ranking factors are subject of research of many papers in domain of SEO. In [19] authors are trying reverse engineering to get the ranking function of Google. Their findings reveal that PageRank [20] is most relevant ranking factor, followed by page title, page snippet (contains meta description) and URL. Several research have been done in finding ranking factors in specific domains, like in [21] where authors point to important characteristics of article pages – stating that title and abstracts should be optimized for good results. Authors of [22] are investigating SEO usage in media webpages in Greece by performing a survey. They found that most common SEO factors used are images optimization, meta description and URL optimization.

Meta description – which is the objective of this study – is identified as important ranking factor in literature [23, 24]. "Meta description" is a HTML tag inserted in web page header with a short description of a webpage topic. It's a "meta" type of tag – which means its content is not visible to web page visitors, but to search engines which displays it in their SERP as a snippet. Here's an HTML example of "meta description" and "page title" tags:

<head>
<title>7th Computer Science On-line Conference 2018</title>
<meta name="description" content="The 7th Computer Science On-line Conference 2018 brings together researchers and practitioners (young and experienced) interested in strengthening the scientific foundations in computer science, informatics, and software engineering. Indexed in Web of Science Core Collection - Conference Proceedings Citation Index - Science and in SCOPUS since 2014." />
</head>

Optimal length of meta description text should be around 150 characters, and should include some type of "call to action" – since this short description is first text (after page title) which users see when browsing SERP pages and deciding which result to click. Web pages that do not have "meta description" defined are risking low positions in search engines results pages.

5 Methods

To create a dataset for testing summarization algorithms we randomly extracted 100 URLs from DMOZ directory[1], along with category name, with the rule that only webpages in minimum third category depth and pages without "meta description" are taken into account. This is important because in our research keywords are needed to perform summarization. Keywords are important in SEO – they are identified and selected by SEO expert and are first step in SEO process. They could be extracted from text, but if text is poorly written, we can end with wrong or misleading keywords. Better approach is to manually define keywords for each website, and this is done by SEO experts after discussing with website owners about their goals. In our research, we set keywords from DMOZ category. We believe that websites in particular category should contain keywords from category name – if they belong to that category then is logically to conclude that the topics of websites in category should match the category title. To avoid too broad keywords we have set the rule of minimum 3-level category depth which enable us to for each URL have minimum 3 target keywords.

We parsed data from dataset and generated summaries using popular summarization algorithms (described in previous chapters). Also, we developed and used a special query based summarization algorithm (presented in Listing 1).

[1] http://dmoztools.net/.

```
Listing 1: Proposed query based summarization algorithm:
Input: K=list of target keywords, T=text to summarize, N=number of sentences to use,
S=word similarity threshold
Output: SUM=summary
T=remove stop words from T
K=STEM(K)
For each word in T
        tag=POS_TAG(word)
        if tag=noun then
                word=STEM(word)
                word_s=WordNet_Synset(word)
                for each keyword in K
                        sim=similarity(word or word_s, keyword)
                        if sim>S then replace word with keyword in T
For each sentence in T
        if sentence contains keyword from K then score(sentence)+=1
Return SUM=top N sentences ordered by score (sentence) DESC
```

Proposed algorithm presented in Listing 1 is simple – it counts keyword frequency in sentences to score them and selects top N sentences in summary. However, there is one feature of above algorithm that empowers it: it uses lexical database WordNet[2] to find synonyms to identify potential replacements with keywords so that sentence score is more reliable in capturing sentence topic. POS tagging used in our algorithm refers to "part of speech" tagging process which identifies word category (noun, verb, adverb, adjective etc.). Stemming is a process of reducing word to "base" or "root" form (so that is easier to compare it). We use popular Porter Stemmer for this task. Word similarity is calculated based on distance in hyper-hyponyms tree. In our experiment we used values $N = 5$ and $S = 0.8$ as inputs in proposed algorithm.

6 Results

For each website a summary using algorithm from Listing 1 is generated and compared with summaries created with popular summarization algorithms. The results are evaluated according to experts scoring. Three independent SEO expert (from three different companies in field of internet marketing) have evaluated generated summaries. They were asked to select most appropriate summary for "meta description" tag out of six summaries offered for each website according to selected target keywords. In Table 1 we present one record (as an example) of generated summaries.

From Table 1 we can see that some summaries have issues with sentence length – selecting too short sentences can lead to poor summaries like ones in Basic algorithm. On the other hand, if algorithm selects sentences that are too long we can get too long

[2] https://wordnet.princeton.edu/.

Table 1. Example of one record from results (generated summaries)

Keywords	Recreation, pets, dogs
Webpage title	Dog owner's guide: welcome to dog owner's guide
Proposed algorithm	You have a dog, want a dog, or just plain. Like dogs, you've come to the right place for all kinds of information. About living with and loving dogs. Along with dozens of articles about dogs, we also have a bookstore
LexRank	If an article fits in more than one topic it is listed. See our list of topics; choose of all topics and articles. Books for help choosing the right dog
LSA	Chock full of books for further reading and videos for family enjoyment and a mall for doggie purchases, than 300 pages of features, breed profiles, training tips, health information, how to teach that puppy good manners, how to choose a veterinarian, a boarding. Kennel, a groomer, or a trainer, what to expect at a dog show, and much, and be sure to check these special pages: gift books and movies for the whole family
Luhn	You'll find articles to help you choose a breed, a breeder and a puppy; the right dog, manners and training, dog owner's. See our list of topics; choose. See our list of related books, amazon's lists of best-selling dog books, or browse for anything in the entire amazon catalog
KL	Chock full of books for further reading and videos for family enjoyment and a mall for doggie purchases, than 300 pages of features, breed profiles, training tips, health information. Looking for information on one subject? Be sure to check these special pages: gift books and movies for the whole family. Books for new owners
TextRank	You have a dog, want a dog, or just plain. If an article fits in more than one topic it is listed. See a list of all articles listed by title. See our list of related books, amazon's lists of best-selling dog books, or browse for anything in the entire amazon catalog
Basic	Dog. See a list. All topics and articles. Books for new owners

summaries (e.g. LSA), which is also not good from SEO point of view. Adding sentence length restrictions or scoring to algorithms could improve the results. This can be a topic for further research.

The results of expert evaluations are presented in Table 2. Experts selected our proposed algorithm in 95 cases as most appropriate algorithm for generating "meta

Table 2. Expert selections of most appropriate summarization algorithm

Algorithm	Expert 1	Expert 2	Expert 3	Total
Proposed algorithm	24	39	32	95
LexRank	21	34	13	68
LSA	19	12	15	46
Luhn	11	10	9	30
KL	5	11	9	25
TextRank	10	7	5	22
None	5	7	8	20
Basic	4	4	7	15

description" text, following by LexRank (68 times) and LSA (46). TextRank and Basic algorithm had worst performance, while 20 times experts selected "none" of presented summaries.

Low "none" frequency tell us that in most cases one of existing summarization algorithms can help in the generation of meta description tag.

7 Conclusion

According to research in [16, 19, 22, 24] "meta description" is important "on-page" SEO ranking factor. It represents a short textual description of webpage topic that is visible in search engine results pages. This highlights the importance of having "meta description" tag in webpage HTML. However, many webpages are missing "meta description" tag because of poor SEO optimization or lack of knowledge about this factor. In this paper we investigated text summarization techniques that can be used to artificially build "meta description" and proposed our algorithm for this task. Experimental results based on SEO expert scores showed that text summarization can be successfully used in SEO process. These results can be used towards building an automatic or semi-automatic software agents for SEO optimization. Future research can include algorithms for meta description generation of webpages that don't have any text. Page context and methods from natural language generation can be used for this task.

References

1. Radev, D.R., Hovy, E., McKeown, K.: Introduction to the special issue on summarization. Comput. Linguist. **28**(4), 399–408 (2002)
2. Zheng, S., Yu, J.: Automatic summarization of web page based on statistics and structure. In: Tan, H. (ed.) Knowledge Discovery and Data Mining, vol. 135, pp. 643–649. Springer, Heidelberg (2012)
3. Delort, J.-Y., Bouchon-Meunier, B., Rifqi, M.: Enhanced web document summarization using hyperlinks. In: Proceedings of the Fourteenth ACM Conference on Hypertext and Hypermedia. ACM (2003)
4. Davison, B.D.: Topical locality in the web. In: Proceedings of the 23rd Annual International ACM SIGIR Conference on Research and Development in Information Retrieval. ACM (2000)
5. Armano, G., Giuliani, A., Vargiu, E.: Experimenting text summarization techniques for contextual advertising. In: IIR (2011)
6. Craven, T.C.: Variations in use of meta tag descriptions by Web pages in different subject areas. Libr. Inf. Sci. Res. **26**(4), 448–462 (2004)
7. Matosevic, G.: Using anchor text to improve web page title in process of search engine optimization. In: Central European Conference on Information and Intelligent Systems. Faculty of Organization and Informatics Varazdin (2015)
8. Luhn, H.P.: The automatic creation of literature abstracts. IBM J. Res. Dev. **2**(2), 159–165 (1958)
9. Edmundson, H.P.: New methods in automatic extracting. J. ACM (JACM) **16**(2), 264–285 (1969)

10. Erkan, G., Radev, D.R.: Lexrank: graph-based lexical centrality as salience in text summarization. J. Artif. Intell. Res. **22**, 457–479 (2004)
11. Mihalcea, R., Tarau, P.: Textrank: bringing order into text. In: Proceedings of the 2004 Conference on Empirical Methods in Natural Language Processing (2004)
12. Vanderwende, L., Suzuki, H., Brockett, C., Nenkova, A.: Beyond SumBasic: task-focused summarization with sentence simplification and lexical expansion. Inf. Process. Manag. **43** (6), 1606–1618 (2007)
13. Haghighi, A., Vanderwende, L.: Exploring content models for multi-document summarization. In: Proceedings of Human Language Technologies: The 2009 Annual Conference of the North American Chapter of the Association for Computational Linguistics, pp. 362–370. Association for Computational Linguistics, May 2009
14. Deerwester, S., et al.: Indexing by latent semantic analysis. J. Am. Soc. Inf. Sci. **41**(6), 391 (1990)
15. Gong, Y., Liu, X.: Generic text summarization using relevance measure and latent semantic analysis. In: Proceedings of the 24th Annual International ACM SIGIR Conference on Research and Development in Information Retrieval. ACM (2001)
16. Van Looy, A.: Search engine optimization. In: Social Media Management, pp. 113–132. Springer International Publishing, Cham (2016)
17. Search engine optimization starter guide. https://www.google.com/webmasters/docs/search-engine-optimization-starter-guide.pdf
18. Bing Webmaster Guidelines. https://www.bing.com/webmaster/help/webmaster-guidelines-30fba23a
19. Luh, C.-J., Yang, S.-A., Dean Huang, T.-L.: Estimating Google's search engine ranking function from a search engine optimization perspective. Online Inf. Rev. **40**(2), 239–255 (2016)
20. Page, L., et al.: The PageRank citation ranking: bringing order to the web. Stanford InfoLab (1999)
21. Marks, T., Le, A.: Increasing article findability online: the four Cs of search engine optimization. Law Libr. J. **109**, 83 (2017)
22. Giomelakis, D., Veglis, A.: Investigating search engine optimization factors in media websites: the case of Greece. Digit. J. **4**(3), 379–400 (2016)
23. Ledford, J.L.: Search Engine Optimization Bible, vol. 584. Wiley, Hoboken (2015)
24. Matošević, G.: Measuring the utilization of on-page search engine optimization in selected domain. J. Inf. Organ. Sci. **39**(2), 199–207 (2015)

Integration of Models of Adaptive Behavior of Ant and Bee Colony

Boris K. Lebedev, Oleg B. Lebedev(✉), Elena M. Lebedeva, and Andrey I. Kostyuk

Southern Federal University, Rostov-on-Don, Russia
lebedev.b.k@gmail.com, lebedev.ob@mail.ru,
lebedeva.el.m@mail.ru, aikostyuk@sfedu.ru

Abstract. The paradigm of the swarm algorithm is proposed on the basis of integration of models of adaptive behavior of ant and bee colony. Integration of models is reduced to the creation of a hybrid agent alternately performing the functions of adaptive behavior of ant and bee colony. The proposed class of hybrid algorithms can be used to solve a wide range of combinatorial problems Based on the hybrid paradigm, a partitioning algorithm has been developed. Also article give a comparison hybrid algorithms with other methods of solution problem. Compared with the existing algorithms, the improvement of results is achieved by 5–10%. The probability of obtaining the global optimum was 0.9.

Keywords: Swarm intelligence · Adaptive behavior · Ant and Bee Colony Hybrid algorithm · Optimization

1 Introduction

The architectures and principles of the functioning of biological control systems that ensure the ability of animals to adapt, adapt to the constantly changing conditions of the external environment are the subject of active research in leading scientific centers. With these studies, the prospects for the development of new information technologies of a breakthrough nature are connected. In this regard, the studies conducted within the framework of this work are relevant. Such methods are iterative, heuristic methods of random search. Among them, the methods of Swarm Intelligence [1, 2] are especially active, in which a set of comparatively simple agents constructs a strategy of their behavior without the presence of a global management. The idea of an ant algorithm is to simulate the behavior of ants associated with their ability to quickly find the shortest path from an anthill to a food source [3, 4]. With its movement, the ant marks the path with a pheromone, and this information is used by other ants to select the path. The concentration of pheromones determines the desire of the individual to choose one or the other way. However, with this approach, it is inevitable to hit the local optimum. This problem is solved by the evaporation of pheromones, which is a negative feedback.

The basis of the ant algorithm is the modeling of the movement of ants over the decision graph [5]. This approach is an effective way of finding rational solutions for optimization problems that allow graph interpretation. The path passed by the ant is

R. Silhavy (Ed.): CSOC 2018, AISC 764, pp. 174–185, 2019.
https://doi.org/10.1007/978-3-319-91189-2_18

displayed when the ant visits all the nodes of the graph. The process of finding solutions with the ant algorithm is iterative.

One of the newest multi-agent methods of intellectual optimization is the bee colony method [6–8]. The main mechanisms of behavior of bees are as follows. First, a number of bee-scouts fly out of the hive in a random direction, who search for sources where there is nectar. After some time, the bees return to the hive and in a special way inform the others - where and how much they have found nectar. After that, other bees are sent to the sources found, and, the more a nectar is supposed to be found on a certain source, the more bees fly in this direction. And scouts again fly away to look for other sources, after which the process repeats. In the work, in accordance with the mathematical models of the behavior of bees and ants proposed in [1–8], modifications and methodology of representing combinatorial problems in the form of swarm algorithms are proposed. To the merit of the bee colony method it is possible to include an in-depth study of the areas in which already found sources of nectar are found (achieved with the help of employed foragers). That is, there are solutions located in the search space near the considered solution.

The drawbacks of the bee colony method include: the fact that the search for new sources of nectar (new areas) by bees-scouts is carried out randomly, that is, they randomly choose a possible solution in the search space.

The advantages of the ant colony method include the guaranteed convergence to the optimal solution; and a higher speed of finding the optimal solution than traditional methods. The drawbacks are that the result of the work of the method depends quite strongly on the initial search parameters, which are selected experimentally, as well as the lack of detailed research of the search space. To strengthen the merits and mitigate the shortcomings of the methods considered, the paradigm of the swarm algorithm is proposed on the basis of integrating the models of adaptive behavior of the ant and bee colonies, which consists in creating a hybrid agent – bee-keeping, alternately performing the functions of ants and bees. Experimental studies have been carried out. confirmed the effectiveness of the proposed approach.

1.1 Representation of Problems in Algorithms Based on an Ant Colony

In general, the method of ant colony can be applied to any combinatorial problem, which can be reconciled with the following requirements [9–12].

Corresponding representation of the problem: the solution space must be represented as a graph with a set of vertices and edges between vertices; There must be a correspondence between the solution of the combinatorial problem and the route in the graph.

It is necessary to develop rules (methods):
initial placement of ants in the tops of the graph;
construction of feasible alternative solutions (route in the graph);
a rule that determines the probability of an ant moving from one vertex of the graph to the other;
the rule of updating pheromones on the edges (vertices) of the graph;
rule of evaporation of pheromones.

The search for the solution of the problem is carried out by the collective of ants $Z = \{z_k | k = 1, 2, \ldots, 1\}$. At each iteration of the ant algorithm, each ant z_k constructs its specific solution to the problem [9–12]. The solution is a route in the graph $G = (X, U)$, constructed in accordance with the conditions and constraints. In this case, the constructed route is represented as a legitimate expression D.

To uniformly distribute the ants and create equal starting conditions, the vertices of the graph $G = (X, U)$ are used as the initial vertices for the ants formed by the ants. Simulation of the behavior of the ants is related to the distribution of pheromone on the edges of graph G. At the initial stage, all the edges of graph G are laid on the same (small) amount of pheromone Q/v, where $v = |U|$. The parameter Q is given a priori.

The process of finding solutions is iterative. Each iteration l consists of three steps. At the first stage, each ant finds a solution – its own route D_k. At the second stage lays pheromone, at the third stage the pheromone is evaporated. The work uses the ant-cycle method of ant systems. In this case, the ferromone is deposited by the agents on the ribs after all they form the solutions.

The process of constructing the route D_k is step-by-step. At each step t, the agent applies a probability rule for selecting the next vertex to include its generated route $D_k(t)$. To do this, a set $x_i \in X_k(t)$ of vertices of the graph G is formed such that each of the vertices $x_i \in X_k(t)$ can be added to the generated route $D_k(t)$ in accordance with the conditions and constraints. Let $e_k(t)$ be the last vertex of the path $D_k(t)$. The agent looks through all the vertices $x_i \in X_k(t)$. For each vertex $x_i \in X_k(t)$, the parameter f_{ik} is the total level of pheromone on the edge of the graph G that connects x_i to the vertex $e_k(t)$.

The probability P_{ik} of the inclusion of the vertex $x_i \in X_k(t)$ in the generated path $D_k(t)$ is determined by the following relation

$$P_{ik} = f_{ik} / \sum_i f_{ik} \tag{1}$$

The agent with probability P_{ik} chooses one of the vertices that includes the route $D_k(t)$.

At the second stage of the iteration, each ant lays pheromone on the edges of the constructed route.

The amount of pheromone $\Delta\tau_k(l)$ deposited by the ant z_k on each edge of the constructed route D_k, is defined as follows:

$$\Delta\tau_k(l) = Q/F_k(l), \tag{2}$$

where l is the iteration number, Q_i – is the total amount of pheromone deposited by the ant on the edges of the path D_k, $F_k(l)$ is the target function for the solution obtained by the z_k at the iteration l. The smaller $F_k(l)$, the more pheromone is deposited on the edges of the constructed route and, consequently, the greater the probability of selecting these edges when constructing routes at the next iteration.

After each agent has formed a solution and postponed the pheromone, in the third stage there is a general evaporation of the pheromone on the edges of the graph G in accordance with the formula (3).

$$f_{ik} = f_{ik}(1 - \rho), \tag{3}$$

where ρ is the refresh rate.

After performing all the actions on the iteration, there is an agent with the best solution that is remembered. Then the next iteration is performed.

1.2 Representation of Problems in Algorithms Based on Bee Colony

The basis of the behavior of the bee swarm is self-organization, which ensures achievement of the common goals of the swarm on the basis of low-level interaction. The main idea of the paradigm of a bee colony is to use a two-level search strategy. At the first level, a number of promising areas (sources) are formed with the help of scouts, at the second level, studies of the neighborhoods of these regions (sources) are carried out with the help of bees of foragers. The goal of the bee colony is to find a source containing the maximum amount of nectar.

In the algorithms of the problems under consideration [13–15], each solution is represented as a point (position) in the search space. The amount of nectar found is the value of the objective function at this point. The solution represents a combination of unique components (vertices and edges of the solution search graph), chosen, as a rule, from a finite set of competing components. The values of the objective function F are determined by the combinations selected by the agents. The goal is to find the optimal combination of components.

In heuristic algorithms of roving intelligence, the process of finding solutions consists in sequentially moving agents in the search space. The process of finding solutions is iterative. At each iteration, agents move to new positions.

Development of the behavioral model of the self-organization of the colony of bees, consist in the development of methods and mechanisms:

- forming the search space,
- formation of the quantitative composition of a swarm of scout agents and a swarm of foraging agents,
- search for intelligence agents prospective positions,
- selection of basic positions among the prospective for their study neighborhoods,
- selection of foraging agents by basic positions,
- formation of neighborhoods of base positions,
- the choice of foraging agents positions in the vicinity of the base positions,
- the general structure of the optimization process.

The first task in developing an algorithm based on the paradigm of a bee colony is to form a search space.

The key operation of the bee algorithm is to study the prospective positions and their environs in search space. Let us dwell on the notion of a neighborhood. The meaning originally implied in the notion of a neighborhood is that the solutions lying in the neighborhood of some position have a high degree of similarity and, as a rule, are slightly different from each other. However, many researchers, in the author's opinion, make the mistake connected with the fact that in the neighborhood include solutions that differ in the proximity of their codes. In this case, the solutions, in which the codes are close, turned out to be very different when decoding. For this reason, it is often unjustified to use to determine the degree of closeness between solutions of the Hamming distance between decision codes.

The main parameters of the bee colony method are: the number of agents n_b, the maximum number of iterations L, the initial number of reconnaissance agents n_r, the limitation of the maximum number of reconnaissance agents, the threshold value of the size of the neighborhood λ.

At the beginning of the search process, all agents are located in the hive, i.e. outside the search space.

At the first iteration ($l = 1$), scout agents in the number n_r are randomly allocated in the search space. This operation consists in randomly generating a set of differing positions $R = \{r_s | s = 1, 2, \ldots, n_r\}$. For each position, the value of the objective function $F^*(l)$ is calculated. It is chosen n_σ of the basic (best) solutions $X^\sigma = \{X_s\}$, for which the value of the objective function is not less than the value of the objective function for any solution not chosen. A number of basic (best) $R^\sigma(1) = \{r_s^\sigma(l) | s = 1, 2, \ldots, n_\sigma\}$ are formed, corresponding to the set of basic (best) solutions of X^σ.

Three approaches are proposed for determining the number of foraging agents sent in the vicinity of each base position. At the first approach, forage agents are distributed at basic positions evenly. In the second approach, forage agents are distributed along the base positions in proportion to the value of the objective function of the position. With the third approach, a probabilistic choice is realized. The probability $P(r_s^\sigma)$ of the forage agent's choice of the base position $r_s^\sigma \in R^\sigma$ is proportional to the value of the objective function F_s^σ in this position and is defined as $P(r_s^\sigma) = F_s^\sigma / \sum s(F_s^\sigma)$.

In the first and second approaches, the number of solutions in the vicinity is calculated, with the third approach, it is determined randomly.

After selecting the basic position $r_s^\sigma(l) \in R^\sigma(l)$ by the forage agent b_z, the probabilistic choice of the position $r_z(l)$, located in the vicinity of the base position $r_s^\sigma(l)$ is realized. We denote the set of positions chosen by foraging agents in the vicinity of the position $r_s^\sigma(l)$ as $O_s(l)$. We call the set of positions $O_s(l) \cup r_s^\sigma(l)$ the domain $O_s^*(l) = O_s(l) \cup r_s^\sigma(l)$. In each domain $O_s^*(l)$ the best position $r_s^*(l)$ with the best estimate $F^*(l)$ is chosen. We call $F^*(l)$ the estimate of the domain $O_s^*(l)$. Among $F^*(l)$ the best solution with the estimate $F^*(l)$ found at this iteration together with a swarm of scouts and a swarm of foragers, is chosen. The best solution with the estimate $F^*(l)$ is preserved, and then the next iteration is performed.

2 Integration of Models of Adaptive Behavior of Ant and Bee Colonies

The main goal in the development of the new paradigm of the swarm algorithm ABC (Ant and Bee Colony) is to integrate the metaheuristics inherent in the bee and ant algorithms. The essence of integration is that in the process of performing the search procedure, alternating procedures of bee and ant algorithms are performed, and bees and ants exchange their functions, that is, in separate stages ants perform the functions of bees, and bees are functions of ants. In this case, the solution is alternately reflected either in the solution search graph or in the form of a position in the solution search space. In other words, the integration of models of adaptive behavior of ants and bee colonies is reduced to the creation of a hybrid agent - "bee-ant" alternately performing the functions of ants and bees.

The work of the search procedure begins with the construction, in accordance with the specificity of the solved problem, of the solution search graph (Fig. 1).

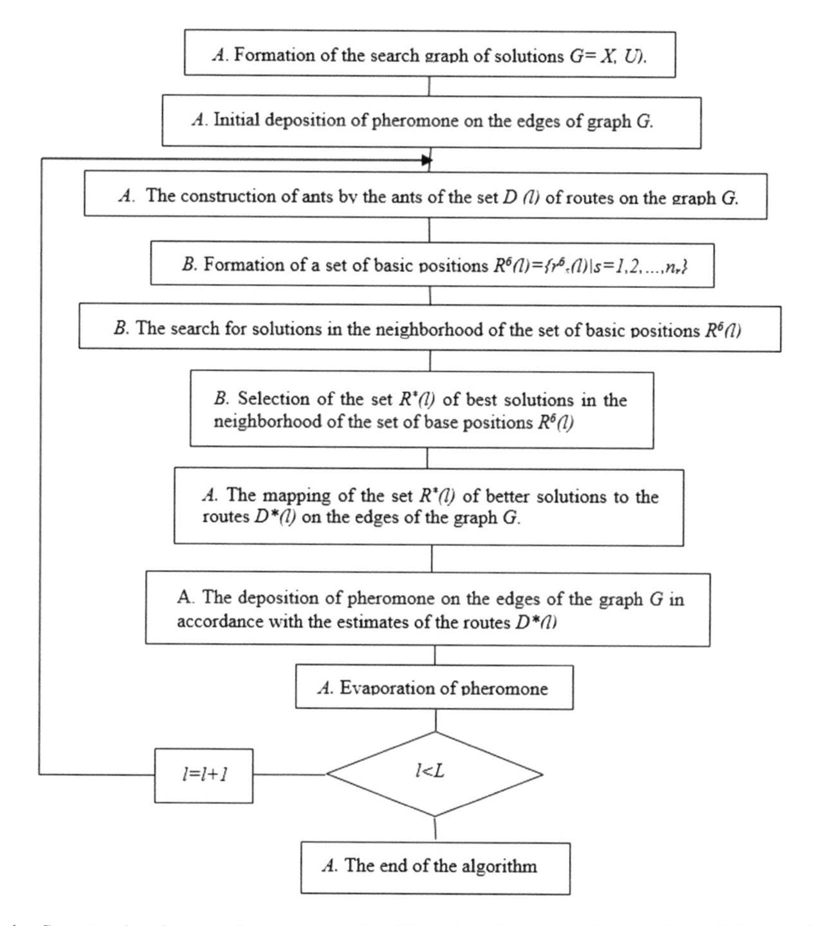

Fig. 1. Structural scheme of a swarm algorithm based on the integration of bee and ant algorithms

In Fig. 1, for convenience, the block related to the ant algorithm is marked with the letter A, and the block related to the bee algorithm is marked with the letter B.

At the initial stage, the same (small) amount of pheromone Q/v is deposited on all edges of graph G, where $v = |U|$. The parameter Q is given a priori. The process of finding solutions is iterative. Each iteration l includes 3 stages. At the first stage of each iteration the procedures of the ant algorithm are performed. Each agent z_k forms its own route $D_s(l)$ on the edges of the graph G, the solution corresponding to the route $D_s(l)$, and the solution estimate $F_s(l)$ are determined.

At the second stage of each iteration the procedures of the bee algorithm are performed. A lot of solutions corresponding to routes built by ants agents are considered as the set of base positions of $R^{\sigma}(l)$ found by intelligence agents at the iteration l. Each of the n_f foraging agents b_z sequentially selects the base position $r_s^{\sigma}(l) \in R^{\sigma}(l)$ with probability $P(r_s^{\sigma}) = F_s^{\sigma} / \sum s(F_s^{\sigma})$. Then it performs the probabilistic choice of the position $r_z(l)$ located in the vicinity of the base $r_s^{\sigma}(l)$ with the corresponding solution $X_z(l)$. Note that the neighborhood of the base position is formed taking into account the specifics of the problem being solved.

In Fig. 1, for convenience, the block related to the ant algorithm is marked with the letter *A*, and the block related to the bee algorithm is marked with the letter *B*.

At the initial stage, the same (small) amount of pheromone Q/v is deposited on all edges of graph G, where $v = |U|$. The parameter Q is given a priori. The process of finding solutions is iterative. Each iteration l includes 3 stages. At the first stage of each iteration the procedures of the ant algorithm are performed. Each agent z_k forms its own route Ds (l) on the edges of the graph G, the solution corresponding to the route $D_s(l)$, and the solution estimate $F_s(l)$ are determined.

At the second stage of each iteration the procedures of the bee algorithm are performed. A lot of solutions corresponding to routes built by ants agents are considered as the set of base positions of $R^{6}(l)$ found by intelligence agents at the iteration l. Each of the n_f foraging agents b_z sequentially selects the base position $r_S^{6}(l) \in R^{6}(l)$ with probability $P(r_s^{6}) = F_s^{6} / \sum_s (F_s^{6})$. Then it performs the probabilistic choice of the position $r_z(l)$ located in the vicinity of the base $r_s^{6}(l)$ with the corresponding solution $X_z(l)$. Note that the neighborhood of the base position is formed taking into account the specifics of the problem being solved.

Next, the estimate $F_z(l)$ of the solution $X_z(l)$ is calculated, and the position $r_z(l)$ is included in the set $O_s(l)$. After selecting agents foragers positions in the vicinity of the base positions for each base position $r_s^{\sigma}(l)$ the region $O_s^*(l) = O_s(l) \cup r_s^{\sigma}(l)$ is formed.

In each region $O_s^*(l)$, the best position $r_s^*(l)$ is chosen with the best solution $X_s^*(l)$, which is placed in the set of the best positions $R^*(l)$.

At the third stage of each iteration, the procedures of the ant algorithm are performed. The set $R^*(l)$ of the best solutions is mapped to the set $D^*(l)$ of routes on the graph G. On the edges of each route $D_s^*(l) \in D^*(l)$, the pheromone is deposited according to the route estimates. The final operation of the third stage is the evaporation of pheromone on the edges of graph G.

The best solution found after l iterations is fixed.

Thus, in the search process at each iteration, the agents alternately perform the functions of ants, bees of scouts and foraging bees.

Based on the ABC (Ant and Bee Colony) paradigm, a partitioning algorithm has been developed. Let the hypergraph $G = (X, U)$ be given, where X is the set of vertices, $|X| = n$, and U is the set of edges. It is necessary to divide the set X into two non-empty and disjoint subsets X_1 and X_2, $X_1 \cup X_2 = X$, $X_1 \cap X_2 = \varnothing$, $X_i \neq \varnothing$. The following nodes are imposed on the nodes (blocks, components) being formed: $|X_1| = n_1, |X_2| = n_2, n_1 + n_2 = n$. The optimization criterion is the number of links - F between X1 and X2. The purpose of optimization is to minimize the criterion F. To find the solution of the partition problem, use the complete graph of solutions $R(X, E)$, where E is the set of all edges of the complete graph connecting the set of vertices of X. Each agent a_k forms the set X_{1k}. Formation of the set X_{1k} is carried out sequentially (step by step). Simulation of the behavior of ants in the partition problem is associated with the distribution of pheromone on the edges of graph R. Moreover, the probability of including the vertex $x_j \in G$ in the set $X_{1k}(t)$ formed by the individual ant is proportional to the total amount of pheromone on the edges connecting the vertex x_j with $X_{1k}(t)$. The local goal of the ant a_k in the search for a solution is to form a set X_{1k} such that on the edges of the complete subgraph $R_{1k} \subset R$, constructed on the vertices of the node X_{1k}, the maximum amount of pheromone is postponed. This corresponds to maximizing the number of internal links in the node X_{1k} and minimizing the number of external links of the node X_{1k} with the remaining nodes.

The key operation of the bee algorithm is to investigate the neighborhoods of solutions in search space. Solutions lying in the vicinity of a certain position have a high degree of similarity and, as a rule, differ slightly from each other. Solutions lying in the neighborhood of the base position are formed by random pairwise permutations between the nodes X_{1k} and X_{2k}.

Experimental researches were conducted on IBM PC. The main goal was to compare the results of the hybrid algorithm with algorithms implemented on the basis of one of the methods - ant or bee. In general, the hybrid algorithm received solutions for 5% better.

3 Experimental Researches

When creating program, Visual Studio 2005 was used for 32-bit Windows® 9x family, and debugging, testing and experiments were conducted on IBM® compatible computer with AMD Athlon-64® 3000+ processor and 512 MB DIMM PC-3200 memory.

Experimental studies of program ABC consist of the following stages:

- Conducting a series of experiments to determine the best values of control operators;
- Conducting a series of experiments with the best values of the general parameters of the algorithms;
- Determination of the parameters of the algorithm in which the solution of the problem of VLSI partitioning is most optimal, both in the quality of the solution and in the time of its obtaining;
- Comparative analysis with other partitioning algorithms on standard test cases and schemes (benchmarks).

For the experiments, a procedure was used for the synthesis of control examples with the known optimum F_{opt}, in analogy with the known method of BEKU (Partitioning Examples with Tight Upper Bound of Optimal Solution) [16].

The experiments were carried out on ten generated graphs. Investigations were made of examples containing up to 1000 vertices. The weight of the vertices was assumed to be zero, and the weight of all the edges was equal to one. In this case, the graph are parted into two subgraphs with an equal number of vertices in each subgraph.

Based on the processing of experimental studies, the average dependence of the quality of solutions on the number of iterations (Fig. 2) and on the size of the population was constructed (Fig. 3). The quality estimate is the value F_{opt}/F, where F is the estimate of the solution obtained.

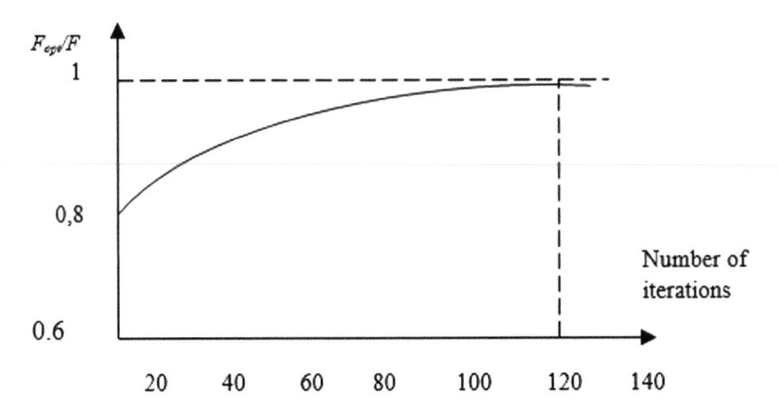

Fig. 2. Dependence of the quality of the decisions of the ABC algorithm on the number of iterations

Comparative analysis with other partitioning algorithms was performed using standard test cases and schemes (benchmarks) located on the sites: http://www.cad.polito.it/tools/9.html, http://www.cbl.ncsu.edu. These examples are standard industrial circuits (blocks, circuits, IS, LSI, VLSI).

As a result of the conducted researches it was established that the quality of the solutions of the hybrid algorithm is 10–15% better than the quality of the solutions obtained by ant and bee algorithms separately.

The number of iterations for which the algorithm found the best solution lies in the range 115–125. It can be seen from the graph that, on average, at the 120 iteration, the solution estimate is close to optimal.

The comparison was made by the parameter Cut - the number of chains that fell into the cut. Table 1 compares the results of testing the developed ABC algorithm with the standard multi-level algorithm hMetis (hM) [17], the FM algorithm, and the algorithm based on the Kernighan-Lin heuristic (KL) [18].

In Table 1, the cut column shows the number of connections in the section obtained by the ABC algorithm. The results of the other algorithms are described as a percentage of the partition found using the ABC. As can be seen from Table 1, the ABC significantly improved the results of FM.

It is noticeable that the results of the ABC improve compared to hMetis, which is a very strong algorithm. For example, in the ibmO4 scheme, the results of hMetis are 4.8% worse than the results of the ABC.

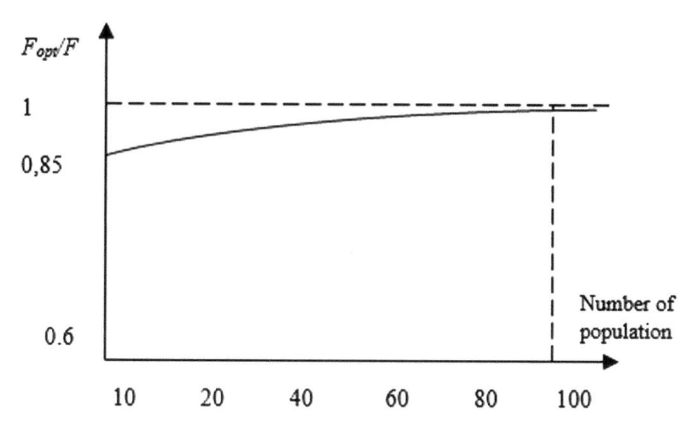

Fig. 3. Dependence of the quality of solutions of the ABC algorithm on the population size

Table 1.

Test	ABC Cut	FM %	hM %	KL %
ibmOl	180	6.1	0.0	0.0
ibmO2	262	1.5	0.0	0.0
ibmO3	931	23.1	2.6	2.0
ibmO4	516	16.5	4.8	1.1
ibmO5	1640	14.1	4.6	2.1
ibmO6	876	10.9	1.3	1.0
ibmO7	815	27.3	4.9	1.2
ibmO8	1112	14.2	1.7	1.6
ibmO9	620	47.1	0.6	0.0
ibmO10	1249	20.3	1.6	1.1
ibmO11	941	53.7	1.6	1.2
ibmO12	1835	22.8	4.8	2.3
ibmO13	823	43.0	1.9	0.8
ibmO14	1758	67.3	4.5	2.1
ibmO15	2587	99.8	4.7	3.2
ibmO16	1680	41.0	5.4	2.8
ibmO17	2126	43.1	5.9	3.5
ibmO18	1490	14.2	3.3	2.0

Analyzing the data in Table 1, we can conclude that the algorithm developed finds solutions that are not inferior in quality, and sometimes superior to their counterparts on average by 5–10%. The probability of obtaining the optimal solution was 0.9.

The overall estimate of the time complexity lies in the range $O(n^2) - O(n^3)$.

4 Conclusion

A new paradigm of the swarm algorithm was developed. The key moments of integration of models of adaptive behavior of bee and ant colonies are considered. The proposed class of hybrid algorithms can be used to solve a wide range of combinatorial problems. The metaheuristics inherent in each model, when integrated, strengthen the capabilities and effectiveness of algorithms. Integration of models extends the scope of the application of roving intelligence. Such an approach is an effective way to find rational solutions for optimization problems that allow graph interpretation.

The developed hybrid algorithm was successfully applied to solve complex optimization problems. A typical example of a solution to this problem is the problem of VLSI crystal planning, the allocation problem, the problem of constructing the shortest connecting networks, the problem of synthesizing mathematical expressions, learning and recognizing patterns. Experimental studies have shown that algorithms based on the proposed approach can give better results than using methods of bee and ant colonies separately.

Acknowledgements. This research is supported by grants of the Russian Foundation for Basic Research of the Russian Federation, the project № 18-07-00737.

References

1. Engelbrecht, A.P.: Fundamentals of Computational Swarm Intelligence. Wiley, Chichester (2005)
2. Dorigo, M., Stützle, T.: Ant Colony Optimization. MIT Press, Cambridge (2004)
3. Dorigo, M., Stützle, T.: Ant colony optimization: overview and recent advances. In: Gendreaum, M., Potvin, J.Y. (eds.) Handbook of Metaheuristics, vol. 146, pp. 227–263. Springer, Boston (2010)
4. Kureichik, V.M., Lebedev, B.K., Lebedev, O.B.: Search Adaptation: Theory and Practice. Fizmatlit, Moscow (2006)
5. Poli, R.: Swarm optimization. J. Artif. Evol. Appl., Article ID 685175, 10 pages (2008)
6. Lučić, P., Teodorović, D.: Computing with bees: attacking complex transportation engineering problems. Int. J. Artif. Intell. Tools **12**, 375–394 (2003)
7. Teodorović, D., Dell'Orco, M.: Bee colony optimization – a cooperative learning approach to complex transportation problems. In: Advanced OR and AI Methods in Transportation: Proceedings of the 16th Mini-EURO Conference and 10th Meeting of the EWGT, 13–16 September 2005, pp. 51–60. Publishing House of the Polish Operational and System Research, Poznan (2005)
8. Quijano, N., Passino, K.M.: Honey Bee Social Foraging Algorithms for Resource Allocation: Theory and Application. Publishing House of the Ohio State University, Columbus (2007)
9. Lebedev, O.B.: Tracing in the channel by the method of an ant colony. Izvestiya SFedU **2**, 46–52 (2009)
10. Lebedev, B.K., Lebedev, O.B.: Modeling of the adaptive behavior of an ant colony in the search for solutions interpreted by trees. Izvestia SFedU **7**, 27–35 (2012)

11. Lebedev, O.B.: Coverage by the method of the ant colony. In: Proceedings of the Twelfth National Conference on Artificial Intelligence with International Participation AI-2010, vol. 2, pp. 423–431. Fizmatlit, Moscow (2010)
12. Kureichik, V.M., Lebedev, B.K., Lebedev, O.B.: Hybrid partitioning algorithm based on natural decision-making mechanisms. In: Artificial Intelligence and Decision Making, pp. 3–15. Published by Institute for System Analysis of Russian Academy of Sciences, Moscow (2012)
13. Lebedev, V.B.: The method of a bee colony in combinatorial problems on graphs. In: Proceedings of the Thirteenth National Conference on Artificial Intelligence with International Participation AIS-2012, vol. 2, pp. 414–422. Fizmatlit, Moscow (2012)
14. Lebedev, V.B.: Tracing in the channel on the basis of the bee colony method. In: Proceedings of the Congress on Intelligent Systems and Information Technologies "AIS-IT 2011", vol. 2, pp. 7–14. Fizmatlit, Moscow (2011)
15. Lebedev, B.K., Lebedev, V.B.: Accommodation based on the bee colony method. Izvestia SFedU **12**, 12–18 (2010)
16. Cong, J., Romesis, M., Xie, M.: Optimality, scalability and stability study of partitioning and placement algorithms. In: Proceedings of the International Symposium on Physical Design. Monterey, CA, pp. 88–94, April 2003
17. Selvakkumaran, N., Karypis, G.: Multi-objective hypergraph partitioning algorithms for cut and maximum subdomain degree minimization. In: ICCAD (2003)
18. Karypis, G.: Multilevel hypergraph partitioning. In: Cong, J., Shinnerl, J. (eds.) Multilevel Optimization Methods for VLSI, Chap. 6. Kluwer Academic Publishers, Boston (2002)

Optimization of Multistage Tourist Route for Electric Vehicle

Joanna Karbowska-Chilinska(✉) and Kacper Chociej

Computer Science Department, Bialystok University of Technology,
Wiejska Street 45 A, 15-333 Bialystok, Poland
j.karbowska@pb.edu.pl

Abstract. This paper presents heuristics approach for the problem of generation an optimal multistage tourist route of electric vehicle (EV). For the given starting and a final point (being EV charging stations) the points of interests (POIs) are included which maximizing the tourist attractiveness. Furthermore the intermediate EV charging stations are selected to the route, in order to after specified number of kilometers a tourist could recharge the batteries and move on to the next stage of a route. Greedy algorithm strengthened the local search methods is proposed by us. Computational tests are conducted on realistic database including POIs and EV charging stations. Results and the execution time of the algorithm show that the presented solution could be a part of software module which generates the most interesting route taking into consideration driving range of EV battery.

Keywords: Optimization · Tourist trip design problem
Electric vehicle

1 Introduction

The optimization problem described in the paper belongs to the class of Tourist Trip Design Problem (TTDP) [2]. In this graph problem the vertices are represented by points of interests (POIs). The profits connected with vertices are assigned on the basis of POIs attractiveness. The main objective of TTDP is to select POIs which maximizes tourist satisfaction, taking into consideration the different limitations (e.g. the length of the route, the visiting time of each POI, the cost of hotels, entrance fees).

Single trip variant of TTDP can be modelled as Orienteering Problem (OP) [9].

In OP n vertices are given, each has a profit p_i. The starting and ending vertices are given and the distance from vertex i to j is known for all vertices. The limit of the route length is also given as T_{max}. The goal of OP is to find a single route, limited in the given constraint T_{max}, that visits a subset of vertices and maximizes the collected profits.

Extensions of OP are e.g. OP with Time Windows(OPTW) [5] or Time Dependent OP (TDOP) [7] may model more real constraints in a planning single

R. Silhavy (Ed.): CSOC 2018, AISC 764, pp. 186–196, 2019.
https://doi.org/10.1007/978-3-319-91189-2_19

route. A good example of use TDOP is a determining a tour visiting some POIs and using public transport to travel between these points. In this case the travel times between given points depend on bus/train timetable and it changes in time. Additionally adding time window to POIs allows model a route that take into account opening and closing hours of POIs.

Multitrips variants of TTDP can be modeling as Team Orienteering Problem (TOP) [10] or Orienteering Problem with Hotel Selection (OPwHS) [1]. TOP is extension of OP and the solution is a tour composed of d connected trips, each limited by T_{max}. Additionally the collected profit in a tour is maximized. In OPwHS a set of hotel, a set of vertices with profits, D-the number of a tour stages are given. The goal is to determine a tour that maximizes the total collected profit. The tour is composed of $D+1$ connected trips. Every trip starts and ends in one of hotels.

OP and problems related to it belong to NP hard class [3]. This implies that exact algorithms are very time consuming. In practice approximation heuristic methods like e.g. iterated local search [10], tabu search method, genetic algorithms [8,12], memetic algorithm [1], greedy method are used. Comparison of the methods can be found in the surveys [4,11].

The optimization of multistage tourist route of EV belongs to multiple variants of TTDP. The described problem is closely related to OPwHS but instead of hotels we have charging stations. Imagine a owner of electric vehicle (EV) who wants to visit the region with many attractions. He knows the starting point, the ending point of the tour and the number of brakes D (in order to charge battery). The goal is to determine a tour, containing $D+1$ stages (trips). The collected POIs could maximize a tour attractiveness. The EV charging station could be optimally selected that the number of kilometers between the selected charging stations and the selected POI do not exceed capacity of EV battery.

In the past work we presented genetic algorithm to the problem of maximization of attractiveness EV tourist route [6]. The approach described there includes only one stage trip ($D=0$) when POIs have time windows. In this article we give a solution the extended problem to the multistage tourist route, in the case that POIs have not the time windows.

In the present work we describe the problem as a graph optimization problem and give the greedy method strengthened by the local search operators such as: 2opt, replace, insert. Computational tests are conducted on realistic database POIs and EV charging stations of Silesian region in Poland.

The paper is organized as follows. Section 2 presents a formal definition of the mention problem. In Sect. 3 the algorithm is described in detail. The results of computational experiments conducted on realistic database are discussed in Sect. 4. Conclusions are drawn and further work is suggested in Sect. 5.

2 Problem Description

The problem described in the paper is a graph routing optimization problem. In an output a set A of n POIs and a set B of m EV charging stations are given.

POIs and charging stations are vertices of a graph. The distance between each pair of POIs and each POIs and EV charging stations are denoted by t_{ij} for $i, j \in \{1, .., n+m\}$. Each POI i has a profit $p_i \geq 0$ interpreted as attractiveness. Each EV charging station has profit equal to 0. The number of kilometers that the car can travel on a single battery charging is given and denoted by T_{max}. Moreover the number of brakes D (in order to charge batteries) is also given. The starting and the ending points, denoted by S and E respectively, are selected from the set B of EV charging stations. An ordered set of POIs between two EV charging stations is called by us as a trip. The all tour is composed of *D+1* trips. In output the algorithm finds *D+1* connected trips visiting POIs and the charging stations at most once and the sum of collected profit is maximized. In addition, the total distance of each trip does not exceed the constraint T_{max}.

3 Proposed Algorithm

For the presented problem the greedy approach strengthened by local search operators is used. Trips between next charge points are built in the greedy way. First S and E are loaded. Next in the greedy way the next POI and the nearest charging station are included to the trip. After this operator 2opt is used to shorten length of the trip. Inclusion of new POIs to the trip and the use of 2opt are repeated to the moment when length of the trip do not exceed T_{max}. In the same way *D+1* connected trips are built. In the last step the insert and replace operators are used to the each of *D+1* trips in order to improve an attractiveness of the tour.

The basic structure of the proposed greedy algorithm is given as follows. Let *attractions* mean list of all attractions, *charges* mean list of all charging stations, *MOD* mean parameter used to build sorted lists.

```
foreach(attraction in attractions)
{attractions_to_attractions[attraction] =
  attractions.SortDescending(MOD, attraction);
  attractions_to_charges[attraction]=
  attractions.Sort(attraction);
 }
foreach(chargePoint in charges)
 charges_to_attractions[chargePoint]=
 charges.SortDescending(MOD, chargePoint);
i = 0; actual = S;
Routes = {}; // empty list of trips
while (i <= D)
{ recently_charged = true; tripDistance = 0;
   route = {actual}; // a list of attractions representing single trip
   while (true)
    { if (recently_charged)
         { nextAttraction =
            BestUnvisitedAttraction(charges_to_attractions[actual]);
           recently_charged = false;
         }
         else
```

```
          nextAttraction =
          BestUnvisitedAttraction(attractions_to_attractions[actual]);
        if (i == D) lastInTrip = E;
          else
          lastInTrip =
           NearestUnvisitedChargePoint(attractions_to_charges[nextAttraction]);
           distanceA = distance(actual, nextAttraction);
           distanceB = distance(nextAttraction, lastInTrip);
        if (tripDistance + distanceA + distanceB <= Tmax)
           { route.Add(nextAttraction);
             tripDistance += distanceA;
             actual = nextAttraction; last = lastInTrip;
             route = 2opt(route);
           }
        else
            {actual = last;  route.Add(actual);
             break;
            }
    }
    Routes.Add(route);
    i++;  //  next trips are built
}
insert(); replace();
print(Routes.Total); // print total profit of the tour
```

The subsequent steps of the algorithm are described in detail in 3.1, 3.2, 3.3, 3.4 and 3.5 subsections.

3.1 Sorted Lists

The important step of the presented greedy approach is sorting list of attractions and charging stations. For every attraction the list of other attractions is built. Lists are sorted in descending way depending on the following ratio:

$$power(object.Profit, MOD)/distance(attraction, object) \tag{1}$$

where *attraction* - the current tourist attraction, the list of other attractions is created depending on it, *object* - any attraction from the sorted list, *MOD* - float value (it is a parameter), *distance()* - function computing distance between two points. Additionally for every attraction the sorted list of all charging stations (depending on distance between an attraction and a charging station) are created. The next step is to create for every charging station list of all attractions. The attractions is sorted in descending way depending on following value:

$$power(object.Profit, MOD)/distance(chargePoint, object) \tag{2}$$

where *chargePoint* - the point where car is being charged, the list of other attractions is created depending on it.

3.2 Trip Generation

The first trip starts and ends with charging stations S and E respectively. Every trip, except first, starts with the last point (charging station) of the previous trip. For an actual last point of the trip the function BestUnvisitedAttraction() searches the best no visited attraction. As parameter it gets list of sorted attractions for this point, generated in the previous step (see Subsect. 3.1). Next the algorithm finds nearest charging station which can be end point of actual building trip (if the last trip is built the point E is the last charging station). The function NearestUnvisitedChargePoint() finds charge station which is the nearest to last attraction in the trip. As parameter it gets list of sorted charging stations for this attraction. In the next step algorithm checks whether the length of the trip does not exceed limit T_{max}. From this reason the condition is checked:

$$tripDistance + distanceA + distanceB <= T_{max} \tag{3}$$

where *tripDistance* - actual length of trip, *distanceA* - distance between actual point and the found attraction, distanceB - distance between the found attraction and nearest charging station (the end of the trip). If (3) is satisfied the attraction is added to trip and the length of trip, the actual attraction and the charging station that can be end of trip are actualized. If (3) not satisfied the algorithm finishes building this trip and the saved charging point is the end of the trip and the beginning of the next trip.

3.3 2opt

Generally the standard 2opt operator [9] compares each pair of edges in a route (not lying next to each other) and next these pairs of edges are replaced by other edges not existing in the route (among all possibilities the option which the most shorten the route length is chosen).

In the presented approach after an attraction is added to the trip the algorithm checks if new edge does not cross with another edges in the trip. To detect intersection of two edges in a trip (v1, v2, ...,vi−1, vi, vi+1, vi+2,...,vj−2, vj−1, vj, vj+1, ..., vk) two pairs of two points are checked. Let it be: vi, vi+1, vj, vj+1 and sums of lengths of the following edges are calculated: *Old* = *distance* (i, i+1) +*distance* (j, j+1) and *New* = *distance* (i, j) + *distance* (i+1, j+1). If *Old* is bigger then *New* it means an order of all edges between points i+1 and j (including them) are changed to the following trip (v1, v2,..., vi−1, vi, vj, vj−1, ...,vi+2, vi+1, vj+1, vj+2,...,vk).

```
2opt(actualRoute) {
    j = actualRoute.Count - 2;
    for (i = actualRoute.Count - 3; i >= 0; i--)
    {Old = distance(actualRoute[i], actualRoute[i + 1]) +
       distance(actualRoute[j], actualRoute[j + 1]);
      New = distance(actualRoute[i], actualRoute[j]) +
```

```
      distance(actualRoute[i + 1], actualRoute[j + 1]);
    if (New < Old)
      { tripDistance -= Old; tripDistance += New;
          left = i + 1, right = j;
          while (left < right)
          {swap(actualRoute[left], actualRoute[right]);
            left++; right--;
          }
          break;
      }
  }return actualRoute;
```

3.4 Insert

When the tour composed of $D+1$ connected trips is completely built an operator insert tries to include an additional attractions to the each trip and not exceed limit T_{max}. For each subsequent points i and j a possibility to insert between them the point not included to the tour is checked. Among of all possibilities of insertions the possibility which maximizes the ratio: $profit/increase\_distance$ is chosen.

```
Insert()
    foreach(route in Routes)
      for (i = 1; i < route.Count; i++)
        {best_ratio = -1; best_attraction = null;
          foreach (not_visited in NotVisitedAttractions)
          {increase_distance = distance(route[i - 1], not_visited) +
           distance(not_visited, route[i]);
           new_dist = route.GetDistance() -
           distance(route[i - 1], route[i]) + increase_distance;
            if (not_visited.Profit /increase_distance >= best_ratio
               1&& new_dist <= Tmax)
                {best_ratio = not_visited.Profit / increase_distance;
                  best_attraction = not_visited;
                }
            } if (best_attraction != null)
                {route.InsertAttraction(best_attraction);
                 not_visited.RemoveAttraction(best_attraction);
                }
        }
```

3.5 Replace

In order to give a possibility to improve profit of the tour the attractions from trips are replaced by the attractions that are not visited yet. In the first step the algorithm finds among attractions in a trip (except charging points) the least valuable attraction (i.e. the minimum ratio of profit to the sum of length of edges, that are connected with this attraction). In the next step each not visited attraction is considered to put into the trip instead of the least valuable attraction. From the set of not visited attractions the most valuable attraction (i.e. the maximum ratio of profit to the sum of length of edges, that are connected with this attraction) is inserted instead of the least valuable attraction. The least valuable attraction is deleted from a trip and is marked as not visited. If there is no possibility to replace any not visited attraction the algorithm ends.

4 Experiments and Results

The presented algorithm was implemented in `C#` and run on an Intel(R) Pentium(R) CPU B960 @ 2.20 GHz. The computational experiments were carried out on realistic database 303 POIs and 21 EV charging stations Silesian region (Poland). Each POI is specified by coordinates, profit (random number from interval [1,10]).

The proposed method was tested in the following versions: greedy (GR), greedy with 2opt (GR+2opt), greedy with 2opt and insert (GR+2opt+in), greedy with 2opt, insert, replace (GR+2opt+in+re), greedy with insert (GR+in), greedy with insert and replace (GR+in+re).

Many tests were carried out to establish the parameter MOD in the ratio which was used in the sorting of lists attractions and the lists of battery charging stations in the greedy method. The parameter MOD = 1.2 represent the best trade-off between the best attractiveness of the tour and its length. In the tests different values of T_{max}(limits of kilometers possible to drive on single battery charge) and D (number of a battery charging during tour) are regarded.

The comparison of the results of all versions of the greedy algorithm for $D = 0$, 1 is presented in Tables 1, 2, 3 and 4. In tables the length is given in kilometers, execution time of the algorithm in milliseconds (ms). The sums of profits of the tours are denoted shortly by *pr*. The result in tables shows that using the local search operators improves greedy results. The using 2opt operator after the each insertion of attraction in some cases (e.g. in case $D = 0$, for $T_{max} = 120$, 170, 420 and $D = 1$ for $T_{max} = 120$, 150, 420) shorted the lengths of the tours and give better profits, about 2–16% ($D = 0$) and 1.5–8% ($D = 1$) in comparison to pure greedy results. It easy to note that the version GR+in gives better profits results than GR+2opt (about 1–16%($D = 0$), 1–7% ($D = 1$)). However GR+2opt+in give the better results in comparison to GR+2opt (about 5–27% and 2–11% in case $D = 0$ and $D = 1$ respectively). It is interesting that additionally GR+2opt+in completes the length of the trips to the specified limit T_{max} (see in Table 1 the column length3 and in Table 4 the column length12). Using the operator replace after GR+2opt+in do not change the profit of the tour. In the case $D=0$, the operator replace, when it is used after GR+in improves profit (about 1%–8%). The same profit/the length of the tour give the methods GR+2opt+in and GR+2opt+in+re. The execution time of GR+2opt+in+replace is higher only some milliseconds and in experiments for $D=0$, 1, 2, 3 do not excite 70ms. Figure 1 illustrate the example routes marked on map generated by GR + in and GR+2opt+in. In figures the points shaped as the plugs are charging stations and the other points are attractions.

Table 1. Comparison the results of GR and operators insert and replace ($D=0$)

T_{max}	GR			GR+in			%gap pr2 and pr1	GR+in+re			%gap pr3 and pr2
	length1	pr1	time1	length2	pr2	time2		length3	pr3	time3	
120	114.08	110.4	0.05	119.83	130.1	10.02	17.8	119.85	136.6	14.34	4.99
150	135.57	132.5	0.06	149.84	160.6	12.79	21.2	149.26	173.6	13.00	8.09
170	169.84	154.8	0.07	169.98	176.3	12.91	13.8	169.98	176.3	15.6	0.00
210	176.33	163.8	0.07	209.85	216.3	16.89	32.1	209.85	216.3	21.06	0.00
420	416.83	760.2	0.23	419.96	776.9	38.92	2.1	419.89	784.7	48.25	1.00

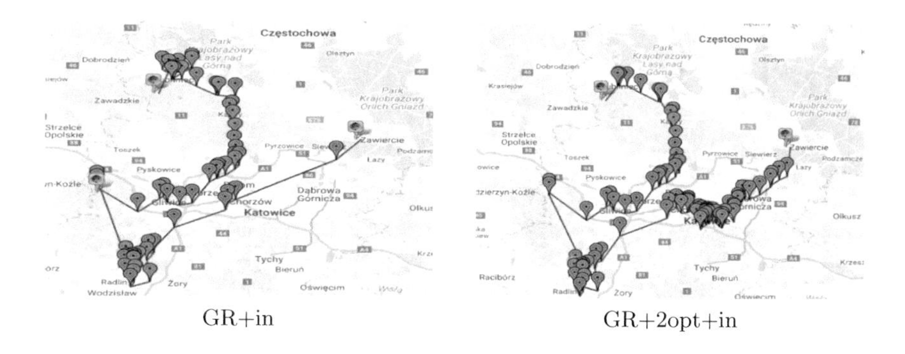

Fig. 1. The example routes generated by GR+in and GR+2opt+in ($D=1$, $T_{max}=150$)

Table 2. Comparison the results of GR and operators 2opt, insert and replace ($D = 0$)

T_{max}	GR			GR+2opt			%gap pr5	GR+2 opt+ in			%gap pr6	GR+2opt+ in+re			%gap pr7
	length1	pr1	time1	length5	pr5	time5	and pr1	length6	pr6	time6	and pr5	length7	pr7	time7	and pr6
120	114.08	110.4	0.05	113.25	110.4	0.22	0.00	119.73	132.8	10.53	20.28	119.73	132.8	13.66	0.00
150	135.57	132.5	0.06	146.33	142.1	0.33	7.25	149.21	180.9	13.00	27.30	149.21	180.9	16.99	0.00
170	169.84	154.8	0.07	167.75	159.4	0.49	2.97	168.81	198.2	14.82	24.34	168.81	198.2	17.51	0.00
210	176.33	163.8	0.07	206.53	190.2	0.81	16.2	209.85	246.7	17.52	29.70	209.85	246.7	21.79	0.00
420	416.83	760,2	0.23	395.47	774.6	15.3	1.89	419.91	818.3	51.48	5.64	419.91	818.3	63.84	0.00

Table 3. Comparison the results of GR and operators insert and replace ($D = 1$)

T_{max}	GR			GR+in			%gap pr9	GR+in+re			%gap pr 10
	length8	pr8	time8	length9	pr9	time9	and pr8	length10	pr10	time10	and pr9
120	226.42	401.5	0.14	239.1	430.5	27.95	7.22	231.9	430.5	27.95	0.00
150	297.52	597.0	0.19	299.52	610.8	34.43	2.31	299.52	610.8	42.72	0.00
170	333.23	419.6	0.16	339.81	460.1	30.41	9.65	339.81	460.1	37.53	0.00
210	401.72	721.9	0.24	419.765	757.0	38.24	4.86	419.76	757.0	48.25	0.00
420	811.27	1205.7	0.46	839,57	1236.8	31.23	2.57	839.57	1236.8	37.52	0.00

Table 4. Comparison the results of GR and operators 2opt, insert and replace ($D = 1$)

T_{max}	GR			GR+2 opt			%gap pr11	GR+2opt+in			%gap pr12	GR+2opt+in+re			%gap pr 13
	length8	pr8	time8	length11	pr11	time11	and pr8	length12	pr12	time12	and pr11	length13	pr13	time13	and pr12
120	226.42	401.5	0.14	219.54	401.5	2.20	0.00	239.71	447.2	37.48	11.38	239.71	447.2	39.35	0.00
150	297.52	597	0.19	283.37	605.1	4.69	1.35	299.63	642.0	48.13	6.09	299.63	642.0	49.63	0.00
170	333.23	419.6	0.16	334.5	429,3	2.51	2.31	339.85	467.4	39.50	8.94	339.85	467.4	41.97	0.00
210	401.72	721.9	0.24	415.05	784.4	8.44	8.65	419.12	798.9	55.67	1.84	419.12	798.9	57.38	0.00
420	811.27	1205.7	0.46	790.85	1205.7	24.99	0.00	839.75	1224.8	62.96	1.58	839.75	1224.8	62.47	0.00

5 Conclusions and Further Work

In the paper the problem of optimization of attractiveness of a tourist route of electric vehicle is discussed in the case when the number of charge batteries is fixed. The proposed algorithm uses greedy method during determining next attraction and charging station. Additionally after including new vertex to the trip, 2opt operator shorted length of a building trip. It gives a possibility to include most POIs by the insert or replace operators. The best attractiveness of the route and execution time gives the method GR+2opt+in. The execution time of the algorithm is relatively short to use it in practice e.g. as part of software which suggest a user the most attractiveness route and charging stations before a battery runs out.

Future work focus on the solving the problem where number of charging stations is not fixed. The main goal will be minimize a number of charging stations and maximize of attractiveness of a route.

Acknowledgements. The authors gratefully acknowledge support from the Polish Ministry of Science and Higher Education at the Bialystok University of Technology (grant S/WI/1/2014).

References

1. Divsalar, A., Vansteenwegen, P., Srensen, K., Cattrysse, D.: A memetic algorithm for the orienteering problem with hotel selection. Eur. J. Oper. Res. **237**(1), 29–49 (2014). https://doi.org/10.1016/j.ejor.2014.01.001
2. Gavalas, D., Konstantopoulos, C., Mastakas, K., Pantziou, G.: A survey on algorithmic approaches for solving tourist trip design problems. J. Heuristics **20**(3), 291–328 (2014). https://doi.org/10.1007/s10732-014-9242-5
3. Golden, B., Levy, L., Vohra, R.: The orienteering problem. Naval Res. Logistics. **34**, 307–318 (1987). https://doi.org/10.1002/1520-6750(198706)
4. Gunawan, A., Lau, H.C., Vansteenwegen, P.: Orienteering problem: a survey of recent variants, solution approaches and applications. Eur. J. Oper. Res. **255**(2), 315–332 (2016). https://doi.org/10.1016/j.ejor.2016.04.059
5. Karbowska-Chilinska, J., Zabielski, P.: Genetic algorithm with path relinking for the orienteering problem with time widows. Fundamenta Informaticae. **135**(4), 419–431 (2014). https://doi.org/10.3233/FI-2014-1132
6. Karbowska-Chilinska, J., Zabielski, P.: Maximization of attractiveness EV tourist routes. In: Computer Information Systems and Industrial Management, CISIM 2017, pp. 514–525 (2017). https://doi.org/10.1007/978-3-319-59105-6_44
7. Ostrowski, K.: Evolutionary algorithm for the time-dependent orienteering problem. In: Computer Information Systems and Industrial Management, CISIM 2017, pp. 50–62 (2017). https://doi.org/10.1007/978-3-319-59105-6_5
8. Ostrowski, K., Karbowska-Chilinska, J., Koszelew, J., Zabielski, P.: Evolution-inspired local improvement algorithm solving orienteering problem. Ann. Oper. Res. 1–25 (2017). https://doi.org/10.1007/s10479-016-2278-1
9. Tsiligirides, T.: Heuristic methods applied to orienteering. J. Oper. Res. Soc. **35**(9), 797–809 (1984). https://doi.org/10.1057/jors.1984.162

10. Vansteenwegen, P., Souffriau, W., Vanden Berghe, G., Van Oudheusden, D.: Iterated local search for the team orienteering problem with time windows. Comput. Oper. Res. **36**, 3281–3290 (2009). https://doi.org/10.1016/j.cor.2009.03.008
11. Vansteenwegen, P., Souffriau, W., Van Oudheusden, D.: The orienteering problem: a survey. Eur. J. Oper. Res. **209**(1), 1–10 (2011). https://doi.org/10.1016/j.ejor.2010.03.045
12. Zabielski, P., Karbowska-Chilinska, J., Koszelew, J., Ostrowski, K.: A genetic algorithm with grouping selection and searching operators for the orienteering problem. In: ACIIDS 2015, LNCS, vol. 9012, pp. 31–40 (2015). https://doi.org/10.1007/978-3-319-15705-4_4

Enhancing Stratified Graph Sampling Algorithms Based on Approximate Degree Distribution

Junpeng Zhu[1,2], Hui Li[1,2](✉), Mei Chen[1,2], Zhenyu Dai[1,2], and Ming Zhu[3]

[1] College of Computer Science and Technology, Guizhou University, Guiyang, People's Republic of China
jpzhu.gm@gmail.com, {cse.HuiLi,gychm,zydai}@gzu.edu.cn
[2] Guizhou Engineer Lab of ACMIS, Guizhou University, Guiyang, People's Republic of China
[3] National Astronomical Observatories, Chinese Academy of Sciences, Beijing, People's Republic of China
mz@nao.cas.cn

Abstract. Sampling technique has become one of the recent research focuses in the graph-related fields. Most of the existing graph sampling algorithms tend to sample the high degree or low degree nodes in the complex networks because of the characteristic of scale-free. Scale-free means that degrees of different nodes are subject to a power law distribution. So, there is a significant difference in the degrees between the overall sampling nodes. In this paper, we propose a concept of approximate degree distribution and devise a stratified strategy using it in the complex networks. We also develop two graph sampling algorithms combining the node selection method with the stratified strategy. The experimental results show that our sampling algorithms preserve several properties of different graphs and behave more accurately than other algorithms. Further, we prove the proposed algorithms are superior to the off-the-shelf algorithms in terms of the unbiasedness of the degrees and more efficient than state-of-the-art FFS and ES-i algorithms.

Keywords: Graph sampling · Sampling bias · Stratified sampling
Approximate degree distribution · Vector clustering

1 Introduction

Networks arise as a natural representation of data in various domains, such as social networks, biological and information networks. However, the real-world networks are massive, which makes traditional analytical approaches infeasible. The methods of data reduction [1] are indispensable. There are various methods have been proposed to cope with the challenge of data reduction, ranging from the principal components analysis to clustering analysis to sampling. Because sampling is an efficient and critical technique

R. Silhavy (Ed.): CSOC 2018, AISC 764, pp. 197–207, 2019.
https://doi.org/10.1007/978-3-319-91189-2_20

for solving massive data analysis bottlenecks, in this paper, we focus on researching graph sampling technique to accelerate the large network data analysis.

Conceptually, we could divide the graph sampling algorithms into three groups: node, edge, and topology-based sampling. In the complex networks, because of the characteristic of scale-free [2], the degrees of different nodes are subject to a power law distribution, the existing algorithms based on the node selection method often over-sample the low degree nodes [3–5, 7–9]. The number of edges corresponding to the high degree nodes is greater, resulting in objectivity the existing algorithms based on the edge selection method tend to draw the high degree nodes [3–5, 7–9]. Simultaneously, the algorithms based on the exploration techniques also have a biased problem to the high degree nodes [3, 5–10], which often results in the lower accuracy of sampling results.

There are several major challenges in the graph sampling. First, how to smooth the sampling bias between the high degree and low degree nodes in a sampling procedure? To address the issue, we proposed a concept of approximate degree distribution of nodes and devised a stratified strategy using the concept. Second, how to get the approximate degree distribution of nodes? Our experimental results showed that vector clustering algorithms [1] could tackle the problem. And we also found the approximate degree distribution results which are got by k-means are faster and more accurate. Subsequently, we also developed two stratified graph sampling algorithms combining the stratified strategy with the node selection method. Our major technical contributions are the following:

- We proposed the concept of approximate degree distribution (Sect. 4.2) in the complex networks, and we also devised a stratified strategy using the approximate degree distribution which is got by clustering algorithms. Further, we investigated the number of strata and presented the most superiority empirical value.
- We developed two new graph sampling algorithms combining our stratified strategy with the node selection method (NS). The experimental results showed that our proposed algorithms are more effective and achieved higher accuracy of the sampling results compared with others algorithms.

The rest of this paper is organized as follows. Section 2 presents the proposed algorithms in this paper. Mathematical analysis is in Sect. 3. Section 4 provides and analyzes the experimental results, and concluded remarks and future work are in Sect. 5.

2 Proposed Algorithms

Allow for the characteristic of scale-free in the complex networks, the difference between nodes is huge. So, we use the stratified sampling technique to smooth sampling error in this paper. It has been proved [5] the algorithms based on the topological structure are difficult to extend from static to streaming graphs. This is the main reason our proposed algorithms are based on the node selection sampling methods (NS). We briefly outline two algorithms and give a formal description in this section.

2.1 NS-d Algorithm

We propose Node Degree-Distribution Sampling (for brevity NS-d) algorithm based on the node selection method (NS [9]) and stratified strategy of the nodes. We give the sampling steps of the NS-d in Algorithm 1. The algorithm consists of three steps. First (lines 1–3), NS-d divides the nodes into three strata by k-means. We will explain the reason the number of clusters is 3 in Sect. 4.2. Second (lines 4–7), NS-d selects the nodes from the set of $\mathtt{N_{high\text{-}degree}}$, $\mathtt{N_{medium\text{-}degree}}$ and $\mathtt{N_{low\text{-}degree}}$ using NS by sampling fraction φ. Subsequently (line 7), NS-d gets a set of nodes V_s which joins the results of stratified sampling. Finally (line 8), using the graph induce to get a set of edges E_s.

ALGORITHM 1: NS-d (φ, N, S, D)

Input: Sample fraction φ, Node set N, Edge set S, Degree set D
Output: Sample Subgraph S = (V_s, E_s) // V_s □ N and E_s □ S

1. V_s = null set, E_s = null set
2. k-means = k-means (distanceFunction,3)
3. ($N_{high\text{-}degree}$, $N_{medium\text{-}degree}$, $N_{low\text{-}degree}$) = k-means.run(N, D)
4. $V_{high\text{-}degree}$ = NS(φ, $N_{high\text{-}degree}$) //NS: Node Sampling Algorithm
5. $V_{medium\text{-}degree}$ = NS(φ, $N_{medium\text{-}degree}$)
6. $V_{low\text{-}degree}$ = NS(φ, $N_{low\text{-}degree}$)
7. $V_s = V_{high\text{-}degree} \cup V_{medium\text{-}degree} \cup V_{low\text{-}degree}$
8. Get E_s(sourceNode, targetNode) and sourceNode $\in V_s$ and targetNode $\in V_s$

NS-d adds k-means step based on NS algorithm, and the time complexity of the k-means method is $\mathtt{O(kt|N|)}$, where k is the number of clusters, t is iterations, and there is $\mathtt{k} \ll |\mathtt{N}|$ and $\mathtt{t} \ll |\mathtt{N}|$ for the k-means procedure. Therefore, the time complexity of NS-d is $\mathtt{O(kt|N|)} + \mathtt{O(|N| + |E|)}$, where $|\mathtt{N}|$ is the number of nodes, $|\mathtt{E}|$ is the number of edges. The space complexity of NS-d is $\mathtt{O(|N| + |E|)}$.

2.2 NS-d+ Algorithm

NS-d+ is similar in spirit to the NS-d described in Sect. 2.1. We summarized the sampling steps of NS-d+ in Algorithm 2. Specifically (lines 4–7), NS-d+ sorts all nodes from the set of high degree nodes using counting sort [11] algorithm by descending and draws them into the sample by sample fraction φ.

ALGORITHM 2: NS-d+ (φ, N, S, D)

Input: Sample fraction φ, Node set N, Edge set S, Degree set D
Output: Sample Subgraph S = (V_s, E_s) // V_s □ N and E_s □ S

1. V_s = null set, E_s = null set
2. k-means = k-means (distanceFunction,3)
3. ($N_{high\text{-}degree}$, $N_{medium\text{-}degree}$, $N_{low\text{-}degree}$) = k-means.run(N, D)
4. CountingSort($N_{high\text{-}degree}$) ⟶ Queue // degrees $\in [0, d_{max}]$ and $d_{max} = O(n)$, so, using counting sort
5. While i < $|N_{high\text{-}degree}|$ *φ
6.   Queue.get()⟶ $V_{high\text{-}degree}$
7.   i++
8. $V_{medium\text{-}degree}$ = NS(φ, $N_{medium\text{-}degree}$) //NS: Node Sampling Algorithm
9. $V_{low\text{-}degree}$ = NS(φ, $N_{low\text{-}degree}$)
10. $V_s = V_{high\text{-}degree} \cup V_{medium\text{-}degree} \cup V_{low\text{-}degree}$
11. Get E_s(sourceNode, targetNode) and sourceNode $\in V_s$ and targetNode $\in V_s$

Because the NS-d+ adds counting sort [11] except for k-means step, and the counting sort is stable and its time complexity is $\theta(|\mathrm{N}|)$. The time complexity of NS-d+ is $\theta(|\mathrm{N}|)+\mathrm{O}(\mathrm{kt}|\mathrm{N}|)+\mathrm{O}(|\mathrm{N}|+|\mathrm{E}|)$. The space complexity of the NS-d+ is also $\mathrm{O}(|\mathrm{N}|+|\mathrm{E}|)$. Obviously, the time complexity of these two algorithms is linear time, and the space complexity of these compared with NS algorithm remains unchanged.

3 Mathematical Analysis

3.1 Analysis of the Validity of Algorithms

Let us suppose, the number of the nodes in the original graph is $|\mathrm{N}|$. These algorithms divide the population into i strata that are recorded as $\mathrm{N}_1, \mathrm{N}_2, \mathrm{N}_3 \ldots$, and N_i respectively. These strata satisfy the following conditions:

$$\begin{aligned} & N_p \cap N_q = \varnothing \,\text{and}\, p, q \in [1, i] \,\text{and}\, p \neq q \\ & N_1 \cup N_2 \cup \ldots \cup N_i = N \end{aligned} \tag{1}$$

According (1), these strata are no overlapping, together they compose all the population. So,

$$|N| = |N_1| + |N_2| + \ldots + |N_i| \tag{2}$$

Further,

$$|N| * \varphi = (|N_1| + |N_2| + \ldots + |N_i|) * \varphi \tag{3}$$

$$= |N_1| * \varphi + |N_2| * \varphi + \ldots + |N_i| * \varphi \tag{4}$$

Where $|N| * \varphi$ is the sampling results of the node selection method (NS). Equation (4) is the sampling results of our proposed algorithms. It is the fact to these results are equivalent, which proves the validity of our proposed algorithms.

3.2 Analysis of the Unbiasedness of Degrees

We explain the unbiasedness of proposed algorithms in this section. For the mean of the population, the estimate used in stratified sampling is:

$$\overline{y}_{st} = \frac{\sum_{k=1}^{i} |N_k| \overline{y}_k}{|N|} = \sum_{k=1}^{i} W_k \overline{y}_k \; and \; W_k = \frac{|N_k|}{|N|} \tag{5}$$

The $\overline{y}_{st}$ represents the estimate of the mean for the population. The $\overline{y}_k$ represents the estimated value of mean at each stratum.

Theorem 1. If in every stratum the sample estimate $\overline{y}_k$ is unbiased, then $\overline{y}_{st}$ is an unbiased estimate of the population mean $\overline{Y}$.

Proof. Our proposed algorithms divide nodes into three groups based on the approximate degree distribution of nodes. It results in a small difference of the inner stratum and the difference of the between different strata is large in the set of degree (it is proved in Sect. 4.2). So, the estimate $\overline{y}_k$ is unbiased for the set of the degree in the individual strata. So,

$$E(\overline{y}_{st}) = E\sum_{k=1}^{i} W_k\overline{y}_k = \sum_{k=1}^{i} W_k\overline{Y}_k \tag{6}$$

Where $\overline{Y_k}$ represents the mean at each stratum. Since the estimates are unbiased in the individual strata. So, the population mean $\overline{Y}$ may be written as:

$$\overline{Y} = \frac{\sum_{k=1}^{i}\sum_{l=1}^{i} y_{kl}}{|N|} = \frac{\sum_{k=1}^{i} |N_k|\overline{Y_k}}{|N|} = \sum_{k=1}^{i} W_k\overline{Y_k} \tag{7}$$

The key is the variance of $\overline{y}_{st}$ depends only on the variances of the individual stratum means $\overline{y}_k$.

4 Experiments

In this section, we explain the reasons our proposed algorithms use k-means and the number of clusters is 3 by experiments first. Next, we empirically compare the performance of our proposed algorithms with other methods, such as NS [9], ES [9], ES-i [5], and FFS [9]. We apply to the full network and sample subgraphs over a range of sampling fraction $\varphi = [5\%, 25\%]$, and the interval number is 5%. For each sampling fraction, we repeated at least 200 experiments to ensure the accuracy of the sampling results. The following section describes in details.

4.1 Datasets

We used three data sets from SNAP [12] in the experiments, namely the Facebook that consists of friends' lists, the Condmat that consists of collaboration network from the arXiv and the Amazon that network was collected by crawling the Amazon website. Table 1 summarizes the global statistics of these three real-world datasets.

Table 1. Characteristics of networks datasets

Graph	Nodes	Edges	Clustering coefficient	Diameter	Average density	Average degree
Facebook	4039	88234	0.6055	8	1.08×10^{-2}	46.6910
Condmat	23133	186936	0.6334	15	7.0×10^{-4}	16.1618
Amazon	334863	925872	0.2050	47	1.65×10^{-5}	5.5299

4.2 Why Do the Algorithms Use K-Means?

The essence of the concept of approximate degree distribution is to get a better cut for the degrees of different nodes. We consider that clustering technology is superior to other methods for getting the results of approximate degree distribution.

There are many vector clustering algorithms [1] in data mining, such as k-means, k-medoids, EM, DBSCAN. The time complexity of DBSCAN is higher than the k-means algorithm, and there are two parameters: one is Eps [1], the other is MinPts [1]. The time complexity of EM and k-medoids are also greater than the k-means algorithm. We consider the procedure is simple and efficient for getting approximate degree distribution results, and k-means satisfies these conditions. The distance function adopts correlation distance [13], which shows a better interpretation of clustered data [13]. Distance correlation is a measure of dependence between random vectors (x_i and x_j) and is defined as follows:

$$d_{ij} = 1 - \frac{(x_i - \overline{x_i})(x_j - \overline{x_j})'}{\sqrt{(x_i - \overline{x_i})(x_i - \overline{x_i})'}\sqrt{(x_j - \overline{x_j})(x_j - \overline{x_j})'}} \tag{8}$$

Silhouette coefficient [14] is a measure that can be used to study the separation distance between clusters. It is usually used to select the most superiority k when the number of clusters is unknown. The silhouette coefficient is defined as:

$$S(i) = \frac{b(i) - a(i)}{\max\{a(i), b(i)\}} \tag{9}$$

Where i represents any object in the datasets, $a(i)$ represents the average distance of i with all other data within the same cluster (cohesion), $b(i)$ represents the lowest average distance of i to all nodes in any other clusters (isolation). The average silhouette coefficient is defined as:

$$S_{avg} = \frac{\sum_{i=1}^{|N|} S(i)}{|N|} \tag{10}$$

The greater the average silhouette coefficient is, the more excellent the number of clusters k is. In this paper, the experimental results show that k is 2 or 3 outperforms [15, 16] other values in the Fig. 1(a). When k is 2 in the Fig. 1(b), it has not been better separated to the high degree nodes which are significant and small in Table 2. It is superior to others in the Fig. 1(b) for clustering results when k is 3, and the average silhouette coefficient is also close to the optimal value in the Fig. 1(a). Therefore, it is the most excellent empirical value to that k is 3.

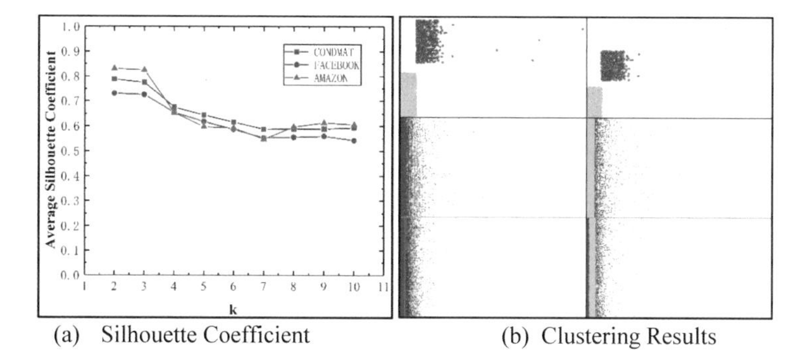

(a) Silhouette Coefficient (b) Clustering Results

Fig. 1. (a) Illustration the relation of k and average silhouette coefficient. (b) Illustration clustering results using k-means when k is 2 and 3 on these datasets in turn.

Table 2. The percentage of nodes of different clusters

Graph	Low degree	Medium degree	High degree
Facebook	91%	9%	0% (extremely tiny)
Condmat	81%	18%	1% (approximate)
Amazon	83%	16%	1% (approximate)

4.3 Accuracy Analysis of Topological Properties

In this section, we compare our proposed algorithms with state-of-the-art algorithms of three groups, which are NS, ES, ES-i, and FFS separately on the degree (distribution), clustering coefficient (distribution), density, and diameter.

We acquire that ES-i [5] outperforms ES by experiments in the Figs. 2, 3, 4 and 5. This is mainly due to the concept of graph induce [5], which shows that it is effective to the concept. It is well-known that high degree nodes are often the hubs nodes [3, 5], which serve as good navigators through the graph.

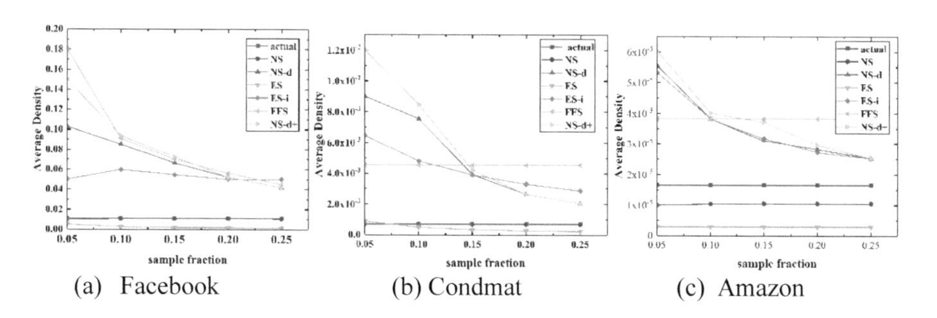

(a) Facebook (b) Condmat (c) Amazon

Fig. 2. Illustration of average density on the three real-world datasets.

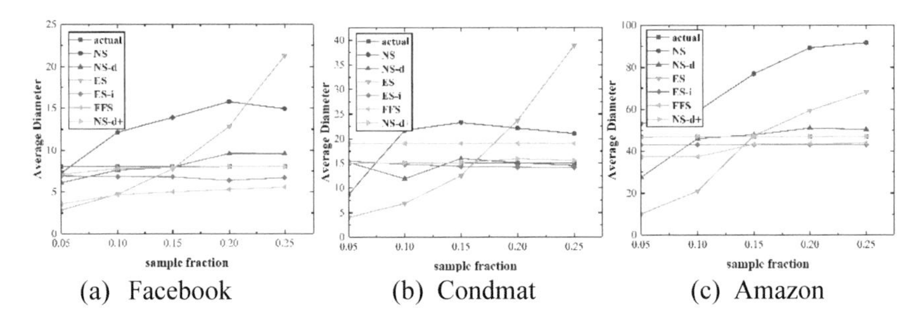

(a) Facebook (b) Condmat (c) Amazon

Fig. 3. Illustration of average diameter on the three real-world datasets.

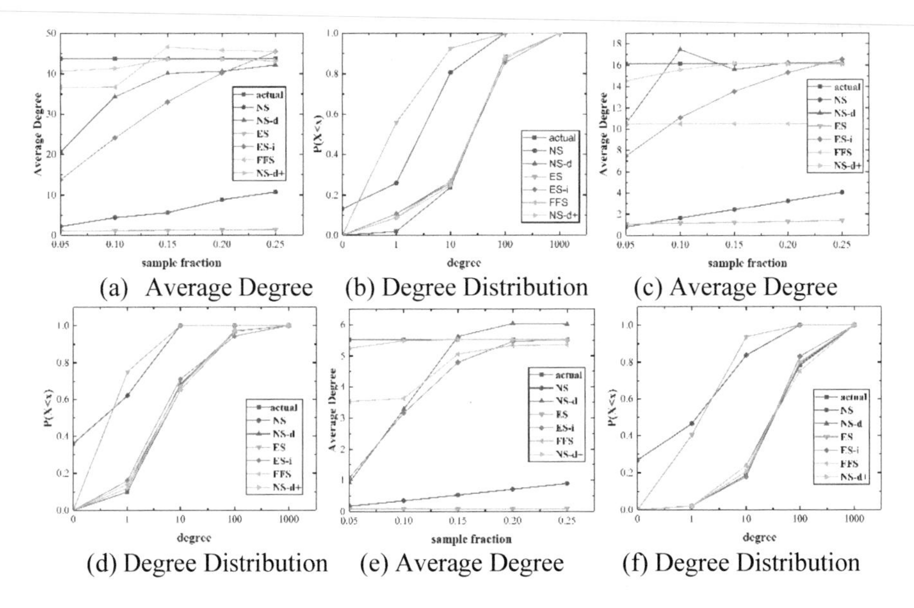

(a) Average Degree (b) Degree Distribution (c) Average Degree

(d) Degree Distribution (e) Average Degree (f) Degree Distribution

Fig. 4. Illustration of average degree, degree distribution on three real-world datasets in turn.

The density of NS-d and NS-d+ is greater than NS, and NS-d+ is superior to other methods in the Fig. 2. There are three primary reasons as follows: one is the stratified strategy; two is the concept of graph induce; three is high degree nodes. The density is proportional to the real number of edges according to its definition. So, our proposed stratified strategy enhances the connectivity of sample graphs. It is significant to balance the sampling proportion between high degree and low degree nodes.

The diameter of NS-d and NS-d+ is close to original graphs in the Fig. 3. The diameter is the maximum of the shortest path. In the graphs, many shortest paths usually pass through these high degree nodes. This is mainly due to that our stratified strategy achieves a balance between the low degree nodes and high degree nodes, which smooth the biasedness of NS that is likely to sample low degree nodes.

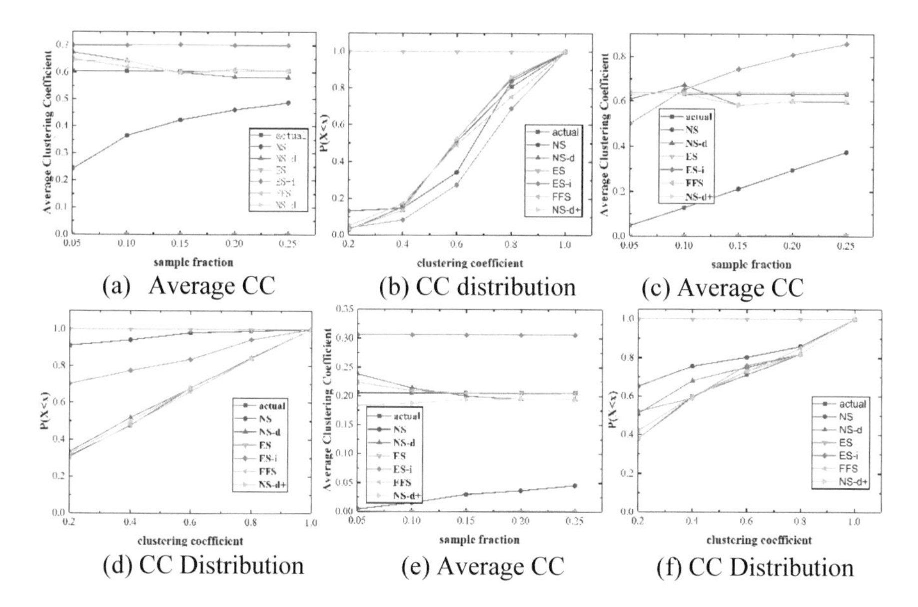

(a) Average CC (b) CC distribution (c) Average CC

(d) CC Distribution (e) Average CC (f) CC Distribution

Fig. 5. Illustration of average clustering coefficient and clustering coefficient distribution on the three real-world datasets in turn.

Next, we compare the average degree and degree distribution of 15% on three real-world datasets by experiments in the Fig. 4. The experimental results show NS-d and NS-d+ outperform the other algorithms. This is also mainly due to the stratified strategy which smooth the biasedness of degrees. And the results of NS-d+ are superior to NS-d. The primary reason is the greater the degrees of different nodes are, the more crucial the nodes are. NS-d+ sorts all nodes and draws them from higher to lower by φ, which results in that NS-d+ is more accuracy than NS-d.

Finally, Fig. 5 shows the average clustering coefficient and clustering coefficient distribution of 15% on the three datasets. We also acquire that NS-d and NS-d+ outperform the other methods, leading to these algorithms could capture the transitivity of the graph because of the stratified strategy of nodes.

4.4 Comparison of Runtime with State-of-the-Art Algorithms

We compare the NS-d and NS-d+ with FFS and ES-i in terms of the average runtime in the Fig. 6. Obviously, the average runtime of NS-d and NS-d+ are both much lower than FFS algorithm. The FFS is a BFS-based algorithm and requires many passes over the edges in a sampling procedure, which results in tangible the time complexity of FFS is polynomial time. ES-i requires two passes over the edges. The time of passing over edges is usually far greater than nodes because there is a huge difference between them on the majority real-world datasets. The NS-d adds k-means step, and NS-d+ algorithm adds k-means and sort steps. The time complexity of k-means is $O(kt|N|)$, and the time complexity of counting sort is $\theta(|N|)$. And NS-d and NS-d+ only require one pass over edges. So, the time complexity of our proposed algorithms is lower than FFS and ES-i.

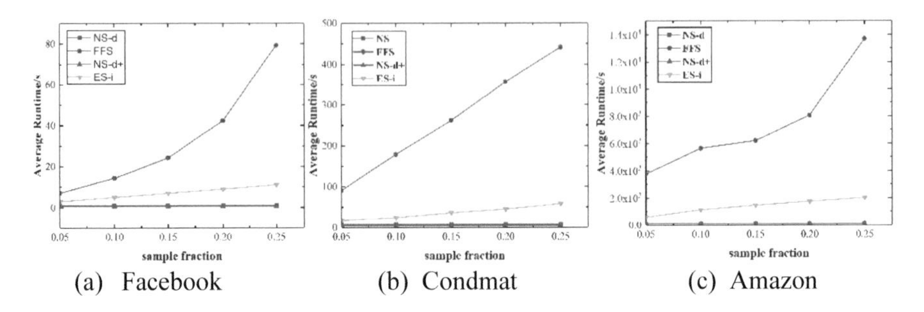

(a) Facebook (b) Condmat (c) Amazon

Fig. 6. Illustration average runtime on the three real-world datasets in turn.

5 Conclusion and Future Work

In this paper, we proposed the concept of approximate degree distribution and devised a stratified strategy using it. Further, we investigated the number of strata and presented the most superiority empirical value of k. Subsequently, we developed two new graph sampling algorithms combining the proposed stratified strategy with the node selection method (NS). The experimental results showed that these algorithms not only accurately preserve more topological properties than state-of-the-art algorithms, but also prove the effectiveness of the approximate degree distribution strategy. Meanwhile, our proposed algorithms outperformed the state-of-the-art FFS and ES-i in terms of average runtime.

Much work still remains. We have focused exclusively on the stratified sampling algorithms on graphs. More sampling strategies should be considered in the different strata. It also will be interesting to extend stratified graph sampling algorithms from static to streaming graphs.

Acknowledgements. This work was supported by the Fund by The National Natural Science Foundation of China (Grant No. 61462012, No. 61562010, No. U1531246), Guizhou University Graduate Innovation Fund (Grant No. 2017078), the Innovation Team of the Data Analysis and Cloud Service of Guizhou Province (Grant No. [2015]53), Science and Technology Project of the Department of Science and Technology in Guizhou Province (Grant No. LH [2016]7427).

References

1. Han, J.: Data Mining: Concepts and Techniques, 3rd edn. Morgan Kaufmann Publishers Inc., San Francisco (2005)
2. Clauset, A., Shalizi, C.R., Newman, M.E.J.: Power-law distributions in empirical data. Soc. Ind. Appl. Math. **51**(4), 661–703 (2009)
3. Yu, L.: Sampling and characterizing online social networks. Dissertation. The University of Bristol, England (2016)
4. Maiya, A.S., Berger-Wolf, T.Y.: Benefits of bias: towards better characterization of network sampling. In: ACM SIGKDD International Conference on Knowledge Discovery and Data Mining, pp. 105–113 (2011)

5. Ahmed, N.K., Neville, J., Kompella, R.: Network sampling: from static to streaming graphs. ACM Trans. Knowl. Discov. Data (TKDD) **8**(2), 1–56 (2014)
6. Stutzbach, D., et al.: Sampling techniques for large, dynamic graphs. In: Proceedings of 25th IEEE International Conference on Computer Communications, INFOCOM 2006. IEEE, pp. 1–6 (2006)
7. Gjoka, M., Kurant, M., Butts, C.T., Markopoulou, A.: Walking in Facebook: a case study of unbiased sampling of OSNs. In: INFOCOM, Proceedings, pp. 1–9. IEEE (2010)
8. Lee, C.H., Xu, X., Eun, D.Y.: Beyond random walk and metropolis-hastings samplers: why you should not backtrack for unbiased graph sampling. ACM SIGMETRICS Perform. Eval. Rev. **40**(1), 319–330 (2012)
9. Leskovec, J., Faloutsos, C.: Sampling from large graphs. In: Proceedings of the 12th ACM SIGKDD International Conference on Knowledge Discovery and Data Mining, pp. 631–636. ACM (2006)
10. Kurant, M., Gjoka, M., Butts, C.T., Markopoulou, A.: Walking on a graph with a magnifying glass: stratified sampling via weighted random walks. ACM SIGMETRICS Perform. Eval. Rev. **39**(1), 241–252 (2011)
11. Cormen, T.H., Leiserson, C.E., Rivest, R.L., et al.: Introduction to Algorithms, 3rd (edn.), 30 (00), 118–118 (2015)
12. SNAP homepage. http://snap.stanford.edu/data/index.html
13. Bora, D.J., Gupta, A.K.: Effect of different distance measures on the performance of k-means algorithm: an experimental study in Matlab. Computer Science (2014)
14. Kim, B., Kim, J.M., Yi, G.: Analysis of clustering evaluation considering features of item response data using data mining technique for setting cut-off scores. Symmetry **9**(5), 62 (2017)
15. Doran, D.: Triad-based role discovery for large social systems. In: Social Informatics, pp. 130–143 (2014)
16. de Heer, W.: Harmonic syntax and high-level statistics of the songs of three early Classical composers, EECS Department, University of California, Berkeley, 167 (2017)

MIC-KMeans: A Maximum Information Coefficient Based High-Dimensional Clustering Algorithm

Ruping Wang[1,2], Hui Li[1,2(✉)], Mei Chen[1,2], Zhenyu Dai[1,2], and Ming Zhu[3]

[1] College of Computer Science and Technology, Guizhou University, Guiyang, People's Republic of China
aygxywrp@gmail.com, {cse.HuiLi,gychm}@gzu.edu.cn
[2] Guizhou Engineering Lab of ACMIS, Guizhou University, Guiyang, People's Republic of China
[3] National Astronomical Observatories, Chinese Academy of Sciences, Beijing, People's Republic of China
mz@nao.cas.cn

Abstract. Clustering algorithms often use distance measure as the measure of similarity between point pairs. Such clustering algorithms are difficult to deal with the curse of dimensionality in high-dimension space. In order to address this issue which is common in clustering algorithms, we proposed to use MIC instead of distance measure in k-means clustering algorithm and implemented the novel MIC-kmeans algorithm for high-dimension clustering. MIC-kmeans can cluster the data with correlation to avoid the problem of distance failure in high-dimension space. The experimental results over the synthetic data and real datasets show that MIC-kmeans is superior to k-means clustering algorithm based on distance measure.

Keywords: Curse of dimensionality · Distance measure · MIC
k-means

1 Introduction

Clustering analysis is an important research topic in the field of data mining. As an independent analysis tool, it has been widely used in many fields, including biomedicine, pattern recognition, images processing and business intelligence, etc. [1]. Common clustering algorithms mostly analyze points and gather the data points into clusters according to some similarity measure. The most commonly used similarity measures of clustering algorithms are distance measures [2], which use the distance to indicate the degree of similarity between points. The smaller the distance is, the higher the similarity is. Commonly used distance measures are Euclidean distance, Manhattan distance, Cosine distance and so on. The goal of distance-based clustering algorithms is to maximize the distance between the clusters and minimize the distance within the cluster. However, in the high-dimension space, there is a challenge named the curse of dimensionality, it means almost all point-to-point distances are almost equal and almost

R. Silhavy (Ed.): CSOC 2018, AISC 764, pp. 208–218, 2019.
https://doi.org/10.1007/978-3-319-91189-2_21

all vectors are orthogonal [3, 4]. Therefore, using the distance measures as measure of similarity in the high-dimension space easily leads to inaccurate clustering results.

In order to solve the above problems which clustering algorithms based on distance measures cause in high-dimension space, we propose that using a correlation method replaces distance measures to further discover clusters which are defined by advanced correlation models. The traditional correlation methods mainly discover the correlation between variables with some specific function (such as linearity, exponential, periodic function, etc.). Towards the high-dimension data, the relationship of data often is complicated, such as nonlinear relationship [5]. Therefore, orthodox linear correlation analysis methods, such as Pearson Correlation Coefficient, are difficult to fully characterize the relationship within the data in high-dimension data.

Reshef et al. [6] proposed a method to detect novel associations of large-scale data sets, called the Maximum Information Coefficient (MIC). MIC is a method based on mutual information. MIC aims to find an optimal meshing, on which we can use it to obtain the maximized mutual information between two variables. It can measure the correlation between variables which satisfy any function relationship and is more robust than conventional method, meanwhile, with the same degree of noise [6].

Because MIC has the ability to effectively find and characterize complicated correlation in high-dimension data, we propose using MIC replaces distance measures of k-means clustering algorithm to solve the ubiquitous curse of dimensionality problem of clustering algorithm in high-dimension space.

The remainder of this paper is organized as follows. Section 2 briefly presents the related work. Section 3 describe the proposed MIC-kmeans algorithm in detail. We illustrate the experimental evaluation in Sect. 4 and finally draw the conclusion in Sect. 5.

2 Related Work

2.1 Maximal Information Coefficient

MIC is defined by the mutual information, measuring the correlation between two variables. Mutual information is derived from information entropy. Let the probability distribution of random variable X be $P(X = x_i) = p_i$. Then the entropy of X is

$$H(X) = -\sum\nolimits_{i=1}^{n} p_i \log p_i. \tag{1}$$

The larger entropy is, the greater the uncertainty of random variable is. Suppose that p_{ij} is the joint probability distribution of X and Y, that $p_{i\cdot}$, $p_{\cdot j}$ are the marginal distributions of X, Y, respectively. The conditional entropy of X, when Y is given, is $H(X|Y) = -\sum\nolimits_{i=1}^{n}\sum\nolimits_{j=1}^{m} p_{ij}\log\frac{p_{ij}}{p_{\cdot j}}$. Then, the mutual information of X and Y is

$$I(X, Y) = H(X) - H(X|Y). \tag{2}$$

The MIC is based on the idea that if a relationship exists between two variables X and Y, then a grid can be drawn on the scatterplot of the two variables that partitions the data to encapsulate that relationship. According to this partition, the mutual information of X and Y is the largest. The computing steps of MIC are as follows. Firstly, dataset D is divided to a grid G which is form of x columns and y rows. Hence each point corresponding to the random variable (X, Y) falls into a sub-grid. Secondly, according to the dividing of different locations, it can obtain multiple meshes. Each mesh is marked as g_{xy}, respectively, compute the maximum mutual information corresponding of each mesh. The formula is as follows

$$I(D,x,y) = \max_{g_{xy} \in G} I(D|g_{xy}) = \max_{g_{xy} \in G} \sum_{i \in x, j \in y} p_{ij} \log \frac{p_{ij}}{p_{i \cdot} p_{\cdot j}}. \tag{3}$$

Where $I(D|g_{xy})$ is the mutual information of each mesh g_{xy}, and the probability of $D|g_{xy}$ is approximated by the frequency of the point that falls into each mesh. p_{ij} is a joint probability function that is approximated by the proportion of samples that fall into the sub-grid. $p_{i \cdot}$, $p_{\cdot j}$ are the edge probability distribution functions that is approximated by the proportion of samples that fall into the intervals $(i, i+1)$ and $(j, j+1)$, respectively. Finally, for any grid of x column and y row, each maximum mutual information is standardized to obtain a characteristic matrix, as follows

$$M(D)_{x,y} = \frac{I(D,x,y)}{\log \min\{x,y\}}. \tag{4}$$

The maximum information coefficient (MIC) is defined as

$$MIC = \max_{xy < B(n)} \{M(D)_{x,y}\}. \tag{5}$$

Reshef et al. demonstrated that MIC described the correlation between variables in more detail than the classical method through a large number of experiments [6].

2.2 The K-Means Algorithm

The k-means clustering is an orthodox clustering algorithm based on partitioning strategy. It divides the dataset D of given n objects into specified k clusters. It aims to achieve the highest possible similarity in clusters and the lowest similarity among clusters. The procedure of k-means algorithm is as follows. First, k sample points are randomly chosen from the data set samples, representing k initial cluster centers respectively. Second, distances between the remaining sample points and k cluster centers are obtained and the aforementioned sample points will be assigned to the nearest cluster, respectively. Afterwards, k-means algorithm iteratively changes the intra-cluster variation. The samples assigned to each cluster in the last iteration are used to compute the mean of the samples in each cluster, respectively. The value of mean will be used as the new cluster center to redistribute all samples. Above process will iterated to achieve convergence until the sample of the cluster no longer changes or the iteration stop condition is satisfied.

2.3 Measurement of Clustering Quality

The measurement of clustering quality often divides into two categories [7]. One is the external measurement, which evaluates the quality of the clusters by comparing the clustering results with the class labels. The other is the internal measurement, which evaluates the quality of the clusters by computing the degree of separation between clusters. The external measurement adopted in this paper are homogeneity, completeness, V-measure, Adjusted Rand index (ARI) and Adjusted Mutual Information (AMI). And the internal measurement silhouette coefficient is adopted.

(1) **Homogeneity**. It represents that each cluster contains only members of a single class [1]. The less the class labels are, the higher the homogeneity index, the better the quality of clustering results. Homogeneity score is formally given by

$$h = 1 - \frac{H(C|K)}{H(C)}. \tag{6}$$

Where $H(C|K)$ is the conditional entropy of the classes given the cluster assignments, measuring the degree of uncertainty of the classes (C) for given a set of clusters (K). Where $H(C)$ is the entropy of the classes.

(2) **Completeness.** It is the symmetric definition of homogeneity [1], representing all members of a given class are assigned to the same cluster. Completeness score is formally given by

$$c = 1 - \frac{H(K|C)}{H(K)}. \tag{7}$$

Where $H(K|C)$ is the conditional entropy of clusters given class, $H(K)$ is the entropy of clusters.

(3) **V-measure**. It is the harmonic mean of homogeneity and completeness [8, 9], calculated according to the following formula

$$v = 2 \cdot \frac{h \cdot c}{h + c}. \tag{8}$$

(4) **ARI**. It is an improvement of the Rand index (RI) [9, 10]. Given the knowledge of the ground truth class assignments *labels_true* and our clustering algorithm assignments of the same samples *labels_pred*, the adjusted Rand index is a function that measures the similarity of the two assignments, ignoring permutations and with chance normalization. The range of RI is [0, 1], while the range of ARI is [−1, 1].

(5) **AMI**. It is a mutual information-based clustering results evaluation method, representing the degree of coincidence between clusters and real classes [11]. NMI and AMI are Mutual information-based evaluation methods. The range of NMI is [0, 1], while the range of AMI is [−1, 1].

(6) **Silhouette Coefficient.** If the ground truth labels are not known, evaluation must be performed using the model itself. The silhouette coefficient is an example of

such an evaluation, representing the degree of inter-cluster separation and the compactness within the cluster [1]. This method is suitable for cases where the class label is unknown. The silhouette coefficient s for a single sample is given as

$$s = \frac{b - a}{\max(a, b)}. \quad (9)$$

Where a represents the mean distance between a sample and all other points in the same class and b is the mean distance between a sample and all other points in the next nearest cluster. The silhouette coefficient for a set of samples is the mean of each sample's silhouette coefficient. The range of the silhouette coefficient is [−1, 1].

3 MIC-Kmeans Algorithm

Considering that MIC performs better under the condition of large amount of data [6, 12, 13], we devise a novel algorithm named MIC-kmeans which using MIC replaces distance measure for k-means clustering algorithm in high-dimension space, that is

$$d(x, y) = MIC(x, y), \quad (10)$$

Compared with the general clustering algorithms based on distance measures, MIC-kmeans considers the correlation between any sample point and cluster center. The objective function of the algorithm is as follows

$$E = \sum_{i=1}^{k} \sum_{p \in C_i} MIC(p, c_i), \quad (11)$$

Where $MIC(p, c_i)$ represents the MIC value of the sample and the representative of the cluster. MIC is obtained according to the formula in Sect. 2.1.

Pseudocode of MIC-kmeans algorithm is described as Algorithm 1. The detailed process of the MIC-kmeans is as follows. First, k sample points are randomly selected in the data set, representing k initial cluster centers respectively (line 1). For each of the remaining samples, computing the MIC values of each sample point to k cluster centers, respectively, and assign the sample points to the clusters which has the maximal MIC values with some cluster (line 3–4). Then, MIC-kmeans iteratively changes the value of the objective function E (line 5). The samples assigned to each cluster in the last iteration are used to compute the mean of each sample in each cluster. The means will be used as the new cluster centers to redistribute all samples (line 6). Iterate through the above process (lines 2–7) until the samples in the cluster no longer change or the iteration stop condition is satisfied.

Algorithm 1: MIC-kmeans

Require Input:

- ***k*:** the number of clusters,
- ***D*:** a data set containing ***n*** objects.

Output: a set of k clusters.

1: arbitrarily choose ***k*** objects from ***D*** as the initial cluster center
2: **repeat**
3: compute the ***MIC*** value of each object to ***k*** cluster centers
4: assign each object to clusters, which has the biggest ***MIC*** value with cluster center
5: change the value of the algorithm's objective function
6: update clusters' means, calculate the mean of the objects for each cluster
7: **until** stop

4 Experiments

4.1 Datasets

There are two categories of datasets in experiments. One is synthetic datasets, which are generated by machine learning framework scikit-learn data generator. The other is ground truth datasets, including UCI handwritten digital dataset and medical service data from community hospitals of Guizhou Province, China. Detailed parameters are shown in Table 1. Evaluation indexes of experiment results are given in Sect. 2.3.

Table 1. Experiment datasets

Data set	Sample no.	Feature no.	Label no.
Make-blobs-0.5	500	90	10
Make-blobs-1	1000	90	10
Make-blobs-2	2000	90	10
Make-blobs-4	4000	30	10
Make-blobs-4	4000	60	10
Make-blobs-4	4000	90	10
Make-blobs-4	4000	120	10
Make-blobs-4	4000	150	10
Make-blobs-8	8000	90	10
Make-blobs-16	16000	90	10
Make-blobs-32	32000	90	10
Handwritten digits	1797	64	10
Medical service	453	44	5

4.2 Experimental Results

Clustering Results on Synthetic Data Sets. In order to verify the validity of the MIC-kmeans algorithm, we compare the MIC-kmeans algorithm with the k-means algorithms which adopt Euclidean distance, Manhattan distance and cosine distance as similarity measure respectively. The dataset which is generated by scikit-learn data generator make_blobs has 4000 samples, 90 feature attributes and 10 cluster labels. The experimental results are shown in Fig. 1, in which Edist, Mdist, Cosine and MICDist denote four k-means algorithms based on Euclidean distance, Manhattan distance, cosine distance, MIC respectively. As it shown, MIC-kmeans algorithm has the higher accuracy than the other three algorithms on the clustering quality measurement of homogeneity, V-measure, ARI, AMI and silhouette. The completeness measure for the MIC-kmeans algorithm is only 0.024 lower than the Euclidean distance and higher than the algorithms based on the Manhattan distance and cosine distance.

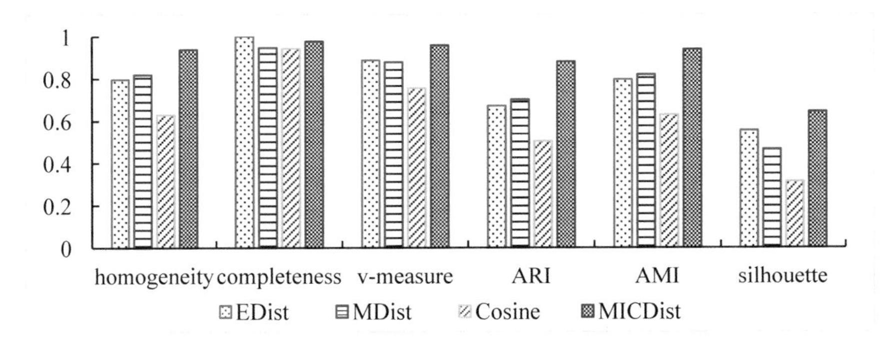

Fig. 1. Comparison of MIC-kmeans and distance-based k-means

In terms of scalability of MIC-kmeans for data size, we use 7 make_blobs datasets with different data sizes from 500 to 32000, all of which contain 90 feature attributes and 10 labels. Experimental results are shown in Fig. 2. As the amount of data increases, evaluation measurement of the MIC-kmenas algorithm show upward trend, which verifies that the MIC-kmeans algorithm is more suitable for large datasets. In Fig. 2, when the amount of data reaches 8000, homogeneity, completeness, ARI, AMI and V-measure are all close to the maximum value 1 and have no downward trend (homogeneity and AMI in the figure almost coincide). The silhouette coefficient is stable at 0.803. The reason for this result is that as the amount of data increases, the clustering tends to converge and the samples of clusters have no change.

Figure 3 illustrate the impact of dimensions of the data on the MIC-kmeans algorithm. The experiment use 5 make_blobs datasets, which size is 4000 each, dimension is 30, 60, 90, 120, 150 respectively. As the number of features increases, the clustering evaluation indexes show upward trend. When the number of features is 150, all the measurements except for silhouette coefficient are close to the maximum (the homogeneity and AMI in the figure almost coincide). Therefore, within a certain range,

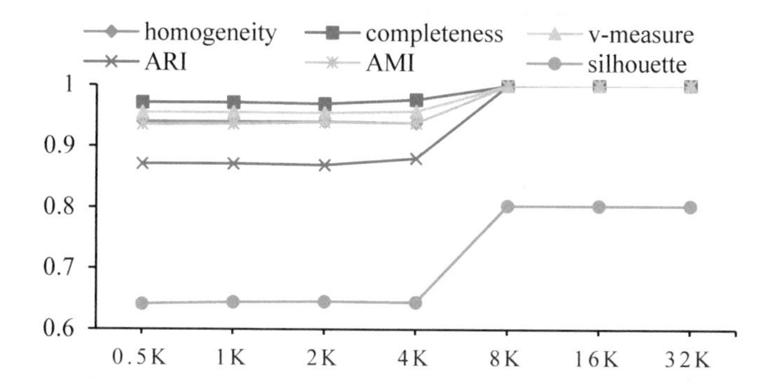

Fig. 2. The effect of data amount on MIC-kmeans

the clustering quality of MIC-kmeans algorithm become better with the increase of dimensionality. There are two reasons for this phenomenon. One is that the MIC calculation is to estimate the probability by the frequency. The more the dimension features are, the more accurate the probability estimation is and the MIC is easier to obtain the larger value. The other is that the more features are, the more complete the information of the data is and the better the clustering results are.

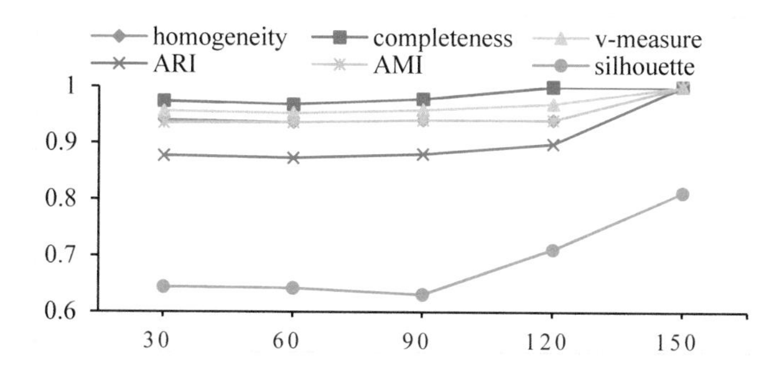

Fig. 3. The effect of dimensions on MIC-kmeans

Figure 4 shows the effect of algorithm parameter k on the MIC-kmeans algorithm. The experiment data has 10 labels, 4000 make_blobs samples and 90 dimension features. Experiment shows that the closer the k is to the number of cluster labels, the better the clustering results. When k is less than 10, all measurements show upward trend; when k is greater than 10, all indexes are generally declining. This indicates that the closer parameter k is to the number of labels, the better the performance of MIC-kmeans is.

Clustering Results on Real World Data Sets. Figure 5 shows the results of the MIC-kmeans algorithm on the handwritten digits dataset, which contains 1797 samples, 64 dimension features, 10 labels. Experiment shows the homogeneity and AMI of MIC-kmeans clustering results are higher than k-means based on Euclidean distance,

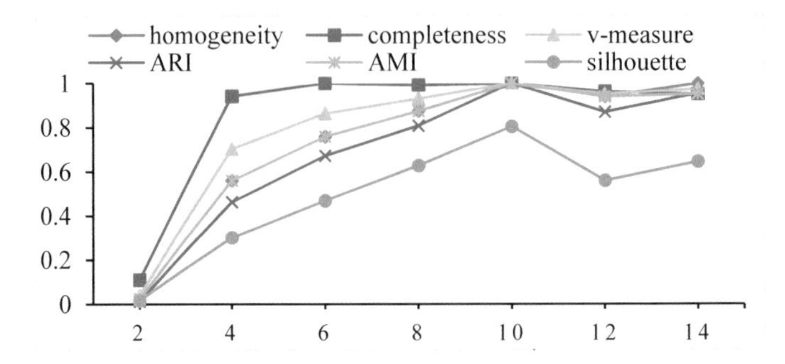

Fig. 4. The effect of algorithm parameter k on MIC-kmeans

Manhattan distance, and slightly lower than the cosine. For the completeness measurements, MIC-kmeans is slightly lower than the other three k-means. This occurs because MIC-kmeans algorithm assigns some clusters of the same class into different clusters. For the V-measure, MIC-kmeans is significant better than the k-means based on Euclidean distance. While it is slightly worse than the Manhattan distance and cosine distance, but the difference is relative small. On the ARI, MIC-kmeans is better than other three methods. MIC-kmeans does not perform well on the silhouette coefficients, since the calculation of the silhouette coefficients is based on the Euclidean distance.

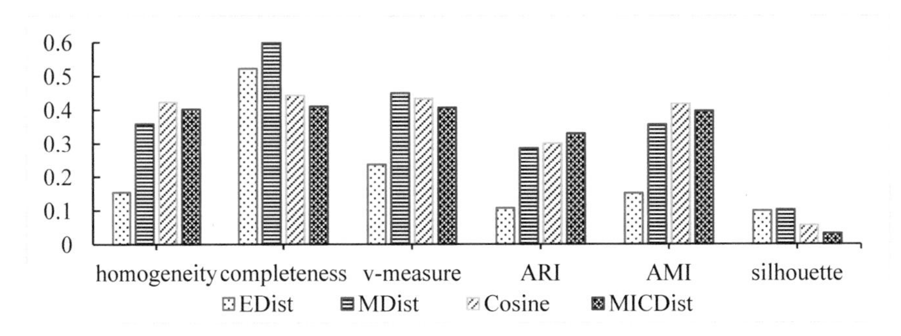

Fig. 5. The performance of MIC-kmeans on the handwritten digits dataset

Figure 6 shows the results of the MIC-kmeans algorithm on the medical service dataset. The experiment dataset includes 453 community hospitals. The features of medical service dataset are 44 common medical service statistics indexes, and the data contains 5 kinds of hospitals. It can be seen from the figure that the MIC-kmeans algorithm is better than the k-means based on three kinds of distances on the homogeneity, V-measure, ARI and AMI. The completeness index of clustering results by MIC-kmeans algorithm is lower than k-means based on Manhattan distance, and higher than other two kinds of method. Experiments indicate that the MIC-kmeans algorithm has better clustering quality for this real datasets.

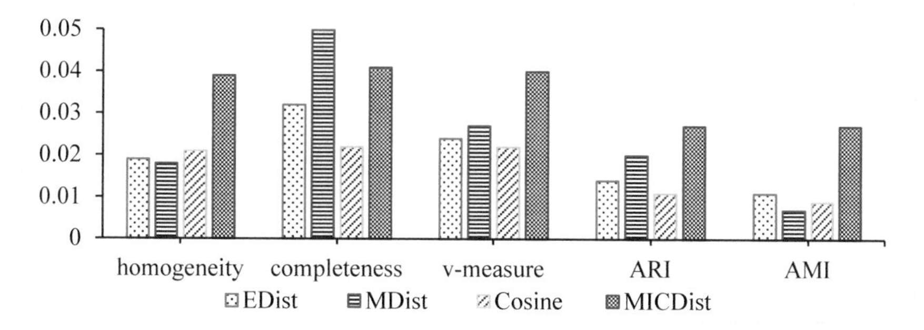

Fig. 6. The performance of MIC-kmeans on the medical service dataset

5 Conclusion

In this paper, we proposed a novel partitioning based clustering algorithm named MIC-kmeans, which takes the maximum information coefficient as similarity measure. MIC-kmeans algorithm aims to solve the common curse of dimensionality problem in high dimensional data clustering. During the evaluation, both the synthetic and real datasets are used to validate the effectiveness of the MIC-kmeans in both the data size and data dimensions aspects. The experiment results indicate that MIC-kmeans not only has the better generalization capability to high-dimension data and large datasets, but also has better clustering accuracy on the ground truth data. Since the computing of the MIC requires to divide the data into grids, which tends to resulting in costly processing [14, 15], we plan to introduce parallel paradigm of implementation in this sub-procedure to achieve better efficiency.

Acknowledgements. This research is supported by the National Natural Science Foundation of China (Grant No. 61462012, No. 61562010, No. U1531246), Guizhou University Graduate Innovation Fund Project (Grant No. 2017082), the Innovation Team of the Data Analysis and Cloud Service of Guizhou Province (Grant No. [2015]53), Science and Technology Project of the Department of Science and Technology in Guizhou Province (Grant No. LH [2016]7427).

References

1. Han, J., Kamber, M., Pei, J.: Data Mining: Concepts and Techniques, 3rd edn. Morgan Kaufmann Publishers Inc., San Francisco (2012)
2. Witten, I., Frank, E., Hall, M.: Data Mining: Practical Machine Learning Tools and Techniques, 2nd edn. Morgan Kaufmann Publishers Inc., San Francisco (2005)
3. Leskovec, J., Rajaraman, A., Jeffrey, D.U.: Mining of Massive Datasets, 2nd edn. Cambridge University Press, Cambridge (2014)
4. Steinbach, M., Ertöz, L., Kumar, V.: The challenges of clustering high dimensional data. In: New Directions in Statistical Physics, pp. 273–309 (2003)
5. Liang, J.Y., Feng, C.J., Song, P.: A survey on correlation analysis of big data. Chin. J. Comput. **1**, 1–18 (2016)

6. Reshef, D.N., Reshefet, Y.A., et al.: Detecting novel associations in large datasets. Science **334**(6062), 1518 (2011)
7. Zhihua, Z.: Machine Learning. Tsinghua University Press, Beijing (2016)
8. Rosenberg, A., Hirschberg, J.: V-Measure: a conditional entropy-based external cluster evaluation measure. In: Proceedings of the 2007 Joint Conference on Empirical Methods in Natural Language Processing and Computational Natural Language Learning, EMNLP-CoNLL 2007, 28–30 June 2007, Prague, Czech Republic DBLP, pp. 410–420 (2007)
9. Wagner, S., Wagner, D.: Comparing Clusterings - An Overview (2007)
10. Rand index. https://en.wikipedia.org/w/index.php?title=Rand_index&oldid=813287904. Accessed 15 Dec 2017
11. Vinh, N.X., Epps, J., Bailey, J.: Information theoretic measures for clusterings comparison: variants, properties, normalization and correction for chance. J. Mach. Learn. Res. **11**, 2837–2854 (2010)
12. Kinney, J.B., Atwal, G.S.: Equitability, mutual information, and the maximal information coefficient. Proc. Nat. Acad. Sci. U. S. Am. **111**(9), 3354 (2013)
13. Reshef, D., et al.: Equitability Analysis of the Maximal Information Coefficient, with Comparisons. abs/1301.6314 (2013)
14. Zhang, Y., et al.: A novel algorithm for the precise calculation of the maximal information coefficient. Sci. Rep. **4**(4), 6662 (2014)
15. Chen, Y., et al.: A new algorithm to optimize maximal information coefficient. PLoS ONE **11**(6), e0157567 (2016). PMC. Web. 18. Ed. Zhongxue Chen

DACC: A Data Exploration Method for High-Dimensional Data Sets

Qingnan Zhao[1,2], Hui Li[1,2(✉)], Mei Chen[1,2], Zhenyu Dai[1,2], and Ming Zhu[3]

[1] Department of Science and Technology, Guizhou University, Guiyang 550025, China
qingnan91@gmail.com, {cse.HuiLi,gychm}@gzu.edu.cn
[2] Guizhou Engineering Lab of ACMIS, Guizhou University, Guiyang 550025, China
[3] National Astronomical Observatories, Chinese Academy of Sciences, Beijing 100016, China

Abstract. Data exploration has been proved to be an efficient solution to learn interesting new insights from dataset in an intuitional way. Typically, discovering interesting patterns and objects over high-dimensional dataset is often very difficult due to its large search space. In this paper, we developed a data exploration method named Decision Analysis of Cross Clustering (DACC) based on subspace clustering. It characterize the data objects in the representation of decision trees over divided clustering subspace, which help users quickly understand the patterns of the data and then make interactive exploration easier. We conducted a series of experiments over the real-world datasets and the results showed that, DACC is superior to the representative data explorative approach in term of efficiency and accuracy, and it is applicable for interactive exploration analysis of high-dimensional data sets.

Keywords: Cross-clustering · High-dimensional dataset · Subspace clustering Interactive explorative analysis

1 Introduction

Exploratory data analysis is an important technique to discover patterns in large-scale datasets. This process usually involves multiple iterations of the data analysis and relies heavily on the intuition and patience of the analyst [1, 2]. Although there are many visual business intelligence tools with interactive discovery capabilities, such as Tableau and Qlik, they still rely heavily on users' intuitive knowledge of the data. In addition, because data analyst often do not know enough about the characteristics of the data to be analyzed, using of these tools still takes a lot of time of analyst to discover interesting insights by trial and errors [3].

In this paper, we proposed a data explorative algorithm named Decision Analysis of Cross Clustering (DACC) based on subspace clustering technology [4, 5]. It aims to solve the data exploration problem that users often have idea about what they want to find, but have no idea where to look. The proposed DACC approach presented a

R. Silhavy (Ed.): CSOC 2018, AISC 764, pp. 219–229, 2019.
https://doi.org/10.1007/978-3-319-91189-2_22

guiding manner to do exploration by providing users with analysis of recommendations, which characterize the data objects in a representation of decision trees. These features made that users quickly discover what they want easier, and saved lots of the time for costly high-dimensional data analysis and exploration.

The major procedure of DACC algorithm is divided into three parts: Firstly, we divided the data set into several subspaces based on vertical cluster and the obtained subspaces often contain several similar patterns which need to be further investigated to find interesting insights. Secondly, we clustering the data objects of each subspace into clusters and construct a decision tree model over each obtained cluster objects, and summarize the relationship between each attribute in the subspace. Finally, we ensemble all the decision tree models. It makes analyst to do data exploration easier due to the overall characteristics of datasets and attributes sets presented.

The rest of paper is organized as follows. In Sect. 2, we present a short description of related work. Section 3 mainly introduces the proposed DACC algorithm. And Sect. 4 presents our experimental evaluation. Section 5 is the conclusion.

2 Related Work

2.1 Variation of Information

Variation of information (VI) is a kind of distance measurement for information difference. It is mainly used to evaluate the distance between different partitions or different clusters of the same data set, it means the information difference between subsets. The design and implementation of VI are described in detail in [6]. VI is a measure of the distance between two variables. If the variables are similar, the measure is lower. If the variables are independent, the measure is higher. H(x) denotes entropy, H(x, y) denotes joint entropy, I(x, y) represents mutual information. The VI value is calculated as follows:

$$I(X, Y) = H(X) + H(Y) - H(X, Y) \tag{1}$$

$$VI(X, Y) = H(X) + H(Y) - 2I(X, Y) \tag{2}$$

2.2 Subspace Cluster

High-dimensional data clustering is a difficult problem and R. Agrawal firstly proposed the concept of subspace clustering [5] to address the issue. It is an extension of traditional clustering algorithms and its search is localized in the heavy relevant dimensions. During the search, subspace clustering often limit the scope of the evaluation criteria so as to consider different subspaces for each different cluster. According to the search strategy, subspace clustering approach can be divided into two types: Top-down algorithms and bottom-up algorithms. The former one often find an initial clustering in the full scope of the dimensions space and then evaluate the cluster of subspaces iteratively to obtain, e.g., PROCLUS algorithm, FINDIT algorithm and COSA algorithm. The latter one used to merge dense regions which found in low dimensional spaces into finally clusters, e.g., CLIQUE algorithm, ENCLUS algorithm and CLTREE algorithm [7].

2.3 PROCLUS

PROCLUS is a representative subspace clustering algorithm which randomly choose initial cluster centers and iteratively optimizes the clusters [7, 8]. For the current k center points, it is determined whether the dimension is related by calculating the distance between the current clustering center neighboring point and the first dimension center point. It iteratively improves the clustering results by using the new center point instead of the poorly performing central point. PROCLUS is a top-down subspace search clustering algorithm, and there is no overlap between the divided subspaces. The subspace clustering process is similar to the DACC. In addition, the related research [9] shows that PROCLUS has better efficient and accuracy among the subspace clustering algorithms.

2.4 DASC

Decision Analysis of Single Clustering (DASC) is an exploration methods we proposed in our previous work for low-dimensional data based on clustering technique. Figure 1. showed the major procedure of DASC. The frequently used clustering methods we used include K-means, K-mediods and CLARA [10]. Since the calculation of the mean in K-Means is vulnerable to outliers and the K-mediods algorithm [10] tends to has scalability issues in large data scenarios, we use the CLARA algorithm in this paper. CLARA obtains a plurality of sample data from the overall data by sampling, and applies the PAM algorithm to get the clusters on each sample dataset. Then we use CART algorithm to construct the classification decision tree for the user to analyze the data information and the clustering result. The CART decision tree is a full binary tree, which can clearly calculate the importance of the field, and is suitable for processing data sets with more eigenvalues. Besides, it can process continuous fields and category fields quickly. But DASC is only applicable to low-dimensional datasets. Therefore, we designed and implemented an algorithm named Decision Analysis of Cross Clustering (DACC) based on cross-clustering. We will describe the algorithm in detail in the next section.

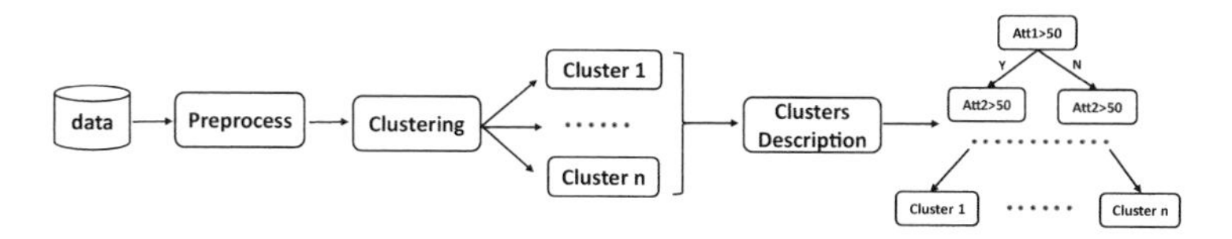

Fig. 1. Structure of DASC

3 DACC

3.1 Overview of Clustering

The procedure of DACC is as Fig. 2. It mainly includes three steps: vertical clustering, horizontal clustering and model ensemble. First, DACC cluster the data preprocessed vertically to obtain different subspaces which is isolated in each other and the dimensions in which usually have the theme (column group) relevance. When we focus on one of the themes, there only involve several attributes for analysis. Then we horizontally cluster the data items in each subspace and create the decision trees, the implementation of which is just the same as DASC. Finally, we ensemble the tree models of each subspace to characterize the dataset for further decision supporting. Because DACC illustrates the impact of each attribute on the overall data distribution and its correlation in an easily understanding tree-like view, it offers users a novel guided approach to make interactive exploration over high-dimensional data easier.

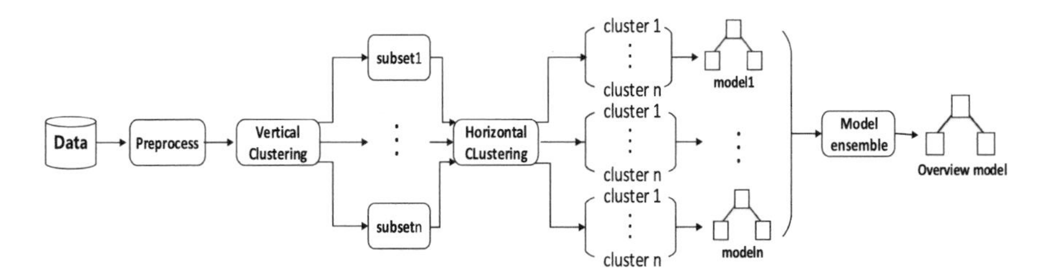

Fig. 2. Procedure of DACC

3.2 Cross Clustering

The Cross clustering of DACC includes two steps: vertical clustering and horizontal clustering.

Vertical Clustering. The data is divided into different subsets according to the correlation between the columns. There are two problems to be considered: One is the curse of dimensionality. All data items tend to be equidistant in the high-dimensional space. So the simple clustering will be no significance. The other one is the diversity of data. There may be completely unrelated attributes in the data users analyze. For example, irrelevant attributes such as skin color and working hours are often found in data analysis [11]. In DACC, vertical clustering is mainly based on clustering attributes. We cluster the similar attributes together, that is to say, the dataset is divided into multiple subspaces. The data in each subspace describe the same aspect. Then users could quickly get the main information in dataset.

Finding Groups Vertically with VI. In this paper, we form a set of groups by statistical analysis of the independence between the variables. If two variables are completely independent of each other, they are almost irrelevant in real life. On the contrary, they are similar and describe the same aspect. To divide the data by column, the DACC algorithm identifies the set of interdependent columns. We use VI to quantify the dependence between two variables. Because VI can only handle two variables, we introduced the concept of diameter in this paper. We define the diameter of a partition consisting of a set of attributes as the maximum VI for the distance between each pair of the attributes in the partition.

To identify the low diameter of each partition, we builds an undirected weighted graph by using VI. In this graph, the vertices represent the variables, and the edges represent the distances. We create a partition for each vertex and find the nearest partition to merge them. This process is repeated until all the partitions are merged, as shown in Fig. 3(a). We use Hierarchical Clustering [12] algorithm to implement the process in this paper. Then the decision tree constructed by the hierarchical clustering is dynamically cut [13]. We define the clipping height of the tree and divide the leaves into different clusters in which the variables are strongly correlated. Figure 3(b) depicts the process.

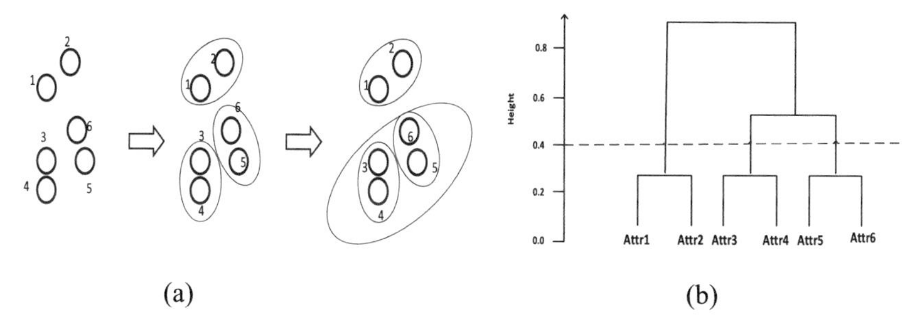

Fig. 3. Partitions merging and mechanism of dynamic cut

Horizontal Clustering. After vertically clustering, we cluster the data in each subspace horizontally, which is similar to DASC. Then we create the CART trees based on the clusters in each subspace. Each CART tree provides users the analysis of different themes, which could reduce the analysis dimensions and make the data exploration more efficient.

3.3 Model Ensemble

In DACC algorithm, we ensembles the subspace models based on the implementation of Random forest [14]. We suppose the input dataset has N samples and m dimensions. For instance, if m attributes are divided into *three* clusters, then *three* decision tree classifiers will be generated. After entering the sample to be classified, each tree classifier will classify it and there will be *three* decision trees presented to the user. When determining the classification result of each sample, all the classifiers will participate in the classification process. The final result will be judged according to the principle of majority compliance. Finally, the total decision tree of the complete dataset is constructed based on the classification results of each sample, as shown in Fig. 2.

The total decision tree provides a complete overview of the dataset information, showing the relationship between the attributes.

3.4 The Complexity of DACC

The implementation of DACC algorithm is shown in Table 1. We use R and C to represent the number of rows and columns for the dataset. Considering subspace partitioning needs to calculate the correlation between columns and dynamic clipping, the time complexity is $O(R * C^2)$. Then the time complexity of horizontal clustering is $O(N * R * C * (K + L)/\mathrm{N}) = O(R * C * (K + L))$. if there are N subspace. And the time complexity of model ensemble is $O(R * N)$. Therefore, the time complexity of the DACC algorithm is $O(R * C^2) + R * C * (K + L) + R * N)$.

Table 1. The Pseudo-Code of DACC

```
    Algorithm 2: DACC(D, K, L)
    Input: the data set D, the number of clusters K, the number of the leaves L
    Output: decision tree ensemble_tree
1.  preprocessed_data ← preprocess(D)
2.  VI_matrix ← VI(preprocessed_data)
3.  clusters_col ←clusterCol(VI_matrix)
4.  themes ← getThemes(clusters_col)
5.  for i in 1:length(themes) do
6.    clus_lab ← clusters(clusters_col[i], K)
7.    Tree ← createTree(clusters_col[i], themes[i], clus_lab, L)
8.    trees ←trees+tree
9.  end for
10. ensemble_tree← mergeTrees(trees)
11. Return ensemble_tree, trees
```

4 Performance Evaluation

4.1 Experimental Methods

In this paper, the performance evaluation of DACC is mainly focus on the efficiency and the accuracy. DACC is an improvement to the application of DASC, which is different in time consumption. And we compare DACC with PROCLUS to further verify the performance of DACC.

How to Evaluate the Time Consumption. We evaluate the consumed time in two aspects. Firstly, when the dimension of dataset increases linearly and regularly in the case of a certain samples, we evaluate the impact of the dimension on the algorithm's time consumption. Secondly, when the number of sample increases linearly and regularly in a certain dimensions, we evaluate the impact of samples on the algorithm's runtime.

How to Evaluate the Accuracy. The accuracy of DACC is mainly based on the prediction of hidden clusters. The higher the predicted result matches the hidden clusters, the more accurate the algorithm is. We suppose that *Res* represents the clusters and *O(C)* represents the object contained in a cluster *C*. For the training dataset, we use bit-vector to store the virtual binary to represent the objects belonging to the corresponding hidden clusters, $C_i \in$ Res. It means the object *O* in each cluster includes a bit-vector of length k = |Res|. If the value of the *i*th bit is 1, the *i*th object $o \in C_i$. And if the value of the ith bit is 0, it indicates that the *i*th objects do not belong to C_i.

4.2 Experimental Data and Settings

The experimental data used in the experiment mainly comes from OpenSubspace [15]. The official website provides the test dataset. The columns of the dataset are obviously grouped, and the columns of each group shows obviously strong correlation. Therefore, each dataset will contain several clustering subspaces. These test datasets has real clustering results. In this paper, the real clustering result is generated as a hidden clusters file, which is used to evaluate the accuracy of subspace clustering results. The hidden clusters file is mainly composed of three parts: virtual binary, the number of data items and the corresponding row number of each data item. The virtual binary is used to represent the current subspace. The number of data items represents the number of objects contained in a class of the subspace, and the row number represent the object in the class. The parameters of the experimental datasets are shown in Table 2. The experimental evaluation does not involve sampling operation.

Table 2. Parameters of the datasets

Describe	Dataset 1	Dataset 2
Dimensions	[10, 80]	20
Tuples	2000	[10000, 80000]
Clusters	5	5
Clusters of subspace	5	5
Dimensions of subspace	[2, 10]	[2, 10]

4.3 Efficiency Evaluation

How the Dimensions Impact Efficiency. The DASC simply clusters data and constructs a single decision tree. Although DASC consumes more and more time as the dimension increases, the time consumption of DACC is obviously much higher than that of DASC. AS shown in Fig. 4(a), it is mainly caused by the complexity of DACC. The analysis of the third quarter shows that the time consumption of DACC is mainly divided into three parts: subspace division, decision tree construction and model ensemble.

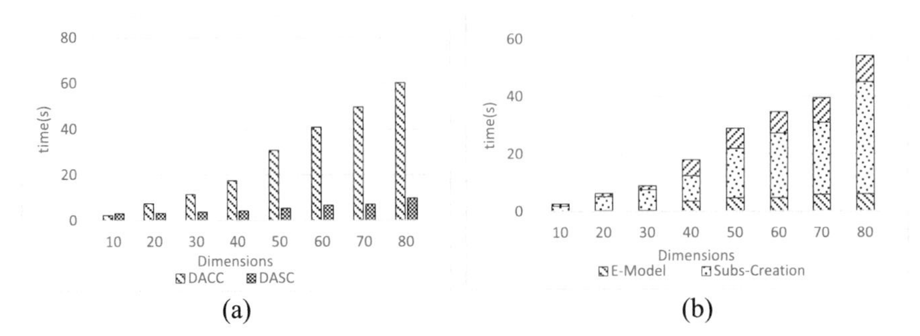

Fig. 4. The efficiency between DACC and DASC under different dimensions

As shown in Fig. 4(b), the analysis shows that the main runtime of DACC is spent on SubsCreation. The main task of this part is to calculate the distance measure of each column and to divide the dataset into several subspaces. And its' time complexity is high. Therefore, as the dimension increasing, the time spent on building DACC subspaces is significantly longer. The second time consumption is the construction of decision tree in each subspace. The more the dimensions are, the more the subspaces are divided, so the more decision trees need to be constructed. Finally, the decision tree model of each subspace is ensemble and the data overview model of the complete dataset is constructed. However, the experimental results show that the efficiency of the construction of the overview decision tree is far lower than the efficiency of the DASC.

How the Tuples Impact Efficiency. In this part, we evaluate the efficiency of DACC under different number of tuples in the case of certain dimensions. As shown in Fig. 5(a), the time consumption of the DASC and the DACC increases continuously with the number of tuples. However, the difference between the runtime of the two methods is not as long as Fig. 4. The main reason is that the influence of the number of tuples on the two methods is the same. The more samples is, the longer the runtime is. Moreover, it can be found from Fig. 5(b) that the time consumption of each part of DACC does not show the prominent change with the increase of the number of tuples. On the contrary, the time of each part increases uniformly.

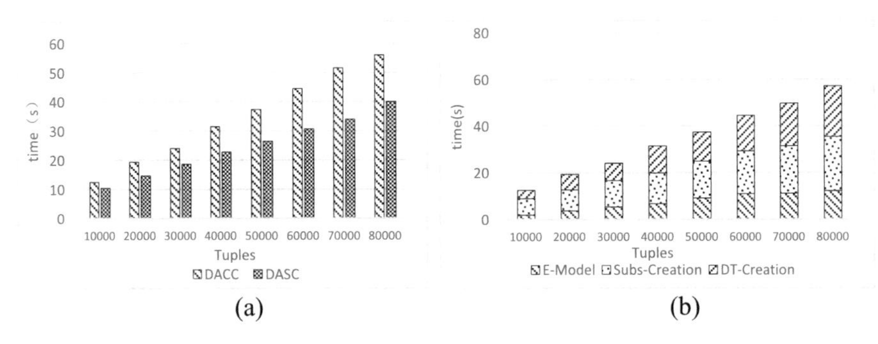

Fig. 5. The efficiency between DACC and DASC under different tuples

Conclusion of Efficiency. From above comparison of DASC and DACC, we know that although the implementation of DACC is far more complex than DASC, but the time consumption is not very high and it is acceptable, meanwhile, with a higher accuracy. In addition, the strategy of construct an overview decision tree in DACC is similar to DASC, but its overhead is lower than DASC. The main reason is that the construction of overview decision tree does not require clustering process. We only need to carry out statistics based on the results of the clustering of subspaces to obtain the class label of each data object. Therefore, it significantly saves the time overheads.

4.4 Accuracy Evaluation

The DACC algorithm is an improvement on the DASC algorithm. Figure 6(b) shows that the accuracy of DASC algorithm will gradually decrease as the dimension increasing. In the case of high-dimensional dataset, it is obvious that the accuracy of DACC is higher than DASC. And with the number of dimension increasing, the distance of the accuracy between DACC and DASC is more and more obvious.

PROCLUS algorithm has some randomness to the partition of subspace. However, the subspace division of DACC is divided by calculating the distance between each column, so the accuracy of DACC algorithm is slightly higher than PROCLUS as the experiments show. However, the accuracy of DACC comes at the expense of time, and the higher the number of dimension is, the longer the runtime is. Therefore, Fig. 6(a) shows that the time consumption of DACC is slightly higher than that of PROCLUS. But the difference in time consumption between the two algorithms is not obvious.

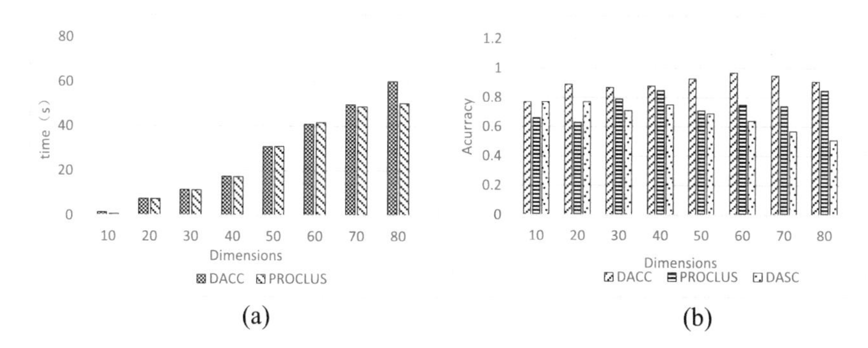

Fig. 6. The accuracy Analysis about DASC, DACC and PROCLUS

5 Conclusion and Future Work

In this paper, we proposed a data exploration method named DACC based on subspace clustering techniques. DACC aims at help explorers analyze the high-dimensional dataset efficiently and accurately. DACC presents the characteristics of data in the form of tree structures, which make analyst obtain data insights eaiser. The experiment results show that DACC outperformed PROCLUS and DASC in term of accuracy meanwhile with acceptable time consuming in high dimensional dataset scenarios. There still exist several performance issues of DACC which need further investigate. For instance, the partitioning of the subspace is costly in DACC, and the overhead of construct decision tree over subspaces is very expensive, we plan to implement these procedure in a parallel paradigm for further performance improvements.

Acknowledgment. This work was supported by the Fund by National Natural Science Foundation of China (Grant No. 61462012, No. 61562010, No. U1531246), Guizhou University Graduate Innovation Fund (Grant No. 2017081) and the Innovation Team of the Data Analysis and Cloud Service of Guizhou Province (Grant No. [2015]53), Science and Technology Project of the Department of Science and Technology in Guizhou Province (Grant No. LH [2016]7427).

References

1. Idreos, S., Papaemmanouil, O., Chaudhuri, S.: Overview of data exploration techniques, pp. 277–281 (2015)
2. Abouzied, A., Hellerstein, J., Silberschatz, A.: DataPlay: interactive tweaking and example-driven correction of graphical database queries. In: Proceedings of the 25th Annual ACM Symposium on User Interface Software and Technology, pp. 207–218. ACM (2012)
3. Dimitriadou, K., Papaemmanouil, O., Diao, Y.: Explore-by-example: an automatic query steering framework for interactive data exploration. In: Proceedings of the 2014 ACM SIGMOD International Conference on Management of Data, pp. 517–528. ACM (2014)
4. Kriegel, H.P., Kröger, P., Zimek, A.: Clustering high-dimensional data: a survey on subspace clustering, pattern-based clustering, and correlation clustering. ACM Trans. Knowl. Disc. Data (TKDD) **3**(1), 1 (2009)
5. Agrawal, R., Gehrke, J., Gunopulos, D., et al.: Automatic subspace clustering of high dimensional data for data mining applications. ACM (1998)
6. Meilă, M.: Comparing clusterings—an information based distance. J. Multivar. Anal. **98**(5), 873–895 (2007)
7. Parsons, L., Haque, E., Liu, H.: Subspace clustering for high dimensional data: a review. ACM SIGKDD Explor. Newsl. **6**(1), 90–105 (2004)
8. Aggarwal, C.C., Wolf, J.L., Yu, P.S., et al.: Fast algorithms for projected clustering. ACM SIGMoD Rec. **28**(2), 61–72 (1999)
9. Müller, E., Günnemann, S., Assent, I., et al.: Evaluating clustering in subspace projections of high dimensional data. Proc. VLDB Endow. **2**(1), 1270–1281 (2009)
10. Kaufman, L., Rousseeuw, P.J.: Finding Groups in Data: An Introduction to Cluster Analysis. Wiley, New York (1990)
11. Muller, E., Gunnemann, S., Farber, I., et al.: Discovering multiple clustering solutions: grouping objects in different views of the data. In: 2012 IEEE 28th International Conference on Data Engineering (ICDE), pp. 1207–1210. IEEE (2012)
12. Aghagolzadeh, M., Soltanian-Zadeh, H., Araabi, B., et al.: A hierarchical clustering based on mutual information maximization. In: IEEE International Conference on Image Processing, ICIP 2007, vol. 1, pp. I-277–I-280. IEEE (2007)
13. Langfelder, P., Zhang, B., Horvath, S.: Defining clusters from a hierarchical cluster tree: the dynamic tree cut package for R. Bioinformatics **24**(5), 719–720 (2007)

14. Ho, T.K.: Random decision forests. In: International Conference on Document Analysis and Recognition, p. 278. IEEE (2002)
15. Müller, E., Assent, I., Günnemann, S., et al.: OpenSubspace: an open source framework for evaluation and exploration of subspace clustering algorithms in weka. In: Open Source in Data Mining Workshop (OSDM 2009) in Conjunction with 13th Pacific-Asia Conference on Knowledge Discovery and Data Mining (PAKDD 2009), pp. 1–12 (2009)

Multi-targets Tracking of Multiple Instance Boosting Combining with Particle Filtering

Hongxia Chu[1,2(✉)], Kejun Wang[2], Yumin Han[1], Rongyi Zhang[1], and Xifeng Wang[1]

[1] College of Electrical and Information Engineering, Heilongjiang Institute of Technology, Harbin, Heilongjiang, China
chx0420@163.com

[2] College of Automation, Harbin Engineering University, Harbin, Heilongjiang, China

Abstract. In order to surmount the major difficulties in multi-target tracking, one was that the observation model and target distribution was highly non-linear and non-Gaussian, the other was varying number of targets bring about overlapping complex interactions and ambiguities. We proposed a kind of system that is able of learning, detecting and tracking the multi-targets. In the method we combine the advantages of two algorithms: mixture particle filters and Multiple Instance Boosting. The key design issues in particle filtering are the selecting of the proposal distribution and the handling the problem of objects leaving and entering the scene. We construct the proposal distribution using a compound model that incorporates information from the dynamic models of each object and the detection hypotheses generated by Multiple Instance Boosting. The learned Multiple Instance Boosting proposal distribution makes us to detect quickly object which is entering the scene, while the filtering process allows us to keep the tracking of the simple object. An automatic multiple targets tracking system is constructed, and it can learn and detect and track the interest object. Finally, the algorithm is tested on multiple pedestrian objects in video sequences. The experiment results show that the algorithm can effectively track the targets the number is changed.

Keywords: Multiple Instance Boosting Detection · Multi-targets tracking
Particle filtering

1 Introduction

Real time multi-targets tracking is an important process for some high level computer vision application, such as surveillance, behavior identity, event classification, automobile safety, virtual reality, etc. [1]. Although significant progress [2–5], automatic tracking of multiple targets remains largely unsolved problems in unconstrained and crowded environments. Noisy and imprecise measurements, long-term occlusions, complex dynamics and target interactions all contribute to their complexity [6–9]. In the process of multi-targets tracking, the target tracking will fail when the number of targets is changed. In addition, the interference between these targets and the background clutter will affect the accuracy of the tracking.

R. Silhavy (Ed.): CSOC 2018, AISC 764, pp. 230–240, 2019.
https://doi.org/10.1007/978-3-319-91189-2_23

Tracking-by-detection has proven to be the most successful strategies for solving the task of tracking multiple targets in unconstrained situations [10]. The most obvious drawback of this approach is that most of the available information in the image sequence is simply ignored due to weak threshold detection response and application of non maximal suppression. In [11] a multi-target tracker is proposed, which utilizes low-level image information and associates each (super) pixel to a specific target or classifies it as background. The results show the proposed algorithm gets the encouraging achievements in many standard benchmark sequences. The algorithm is significantly better than state-of-the-art tracking-by-detection methods in crowded scenes with long-time partial occlusions.

In the past few years, a kind of nonlinear particle filtering based on Bayesian estimation has many advantages in tracking the moving nonlinear multi-targets [12, 13]. The advantage of these methods is that they are simple and flexible and systematic to deal with nonlinear and non-Gaussianity. These methods improve the effectiveness of particle filtering technology application and expand its application field. As pointed out in [17], particle filters may perform weekly when the posterior is multi-modal because of the ambiguities and multiple targets. In order to solve this problem, Vermaak et al. introduce a mixture particle filter (MPF) [17]. In this paper, we extend the approach of Vermaak et al. [17, 18]. We use particularly a Multiple Instance boost algorithm (MIBoost) to learn models of the moving target [19, 20]. These detection models are used for leading the particle filter. The proposal distribution consists of a probabilistic mixture model that incorporates information from Multiple Instance boost and the dynamic models of the individual players. Target can be quickly detected and tracked in a dynamically changing background, despite the fact that the objects enter and leave the scene frequently. We call the resulting algorithm the Multiple Instance Boosted Particle Filter (MIBPF). We use MIBoost to replace adaBoost for three main reasons:

(1) MIBoost has no use for the knowing the location and scale of the objects, in order to generate positive examples in the classification algorithm.
(2) Object detection using MIBoost allows the detector to assign flexibly labels to the training set, thus reducing labels noise and improving performance.
(3) The MIBoost framework yields a classification with a much higher detection rates and fast computation times.

2 Hybrid Particle Filtering

2.1 Hybrid Bayesian Sequential Filtering

It is assumed that x_t is the state vector of the target. $y_{1:t} = (y_1, \cdots, y_t)$ is the observation vector until the time t. For the tracking problem, the distribution of the target is to realize the filtering distribution. In the Bayesian sequence estimation, the distribution is calculated using two step auto-regression [12]:

(1) Predicting

$$p(x_t|y_{1:t-1}) = \int p(x_t|x_{t-1})p(x_{t-1}|y_{1:t-1})dx_{t-1} \tag{1}$$

(2) Updating

$$p(x_t|y_{1:t}) = \frac{p(y_t|x_t)p(x_t|y_{1:t-1})}{\int p(y_t|x_t)p(x_t|y_{1:t-1})dx_t} \tag{2}$$

The dynamic model $p(x_t|x_{t-1})$ of the state transition is the auto-regression. According to the current observation $p(y_t|x_t)$, the likelihood model of any state is given. And it is initialized according to some initial distribution $p(x_0)$.

In order to track the multiple targets, we utilize a hybrid filtering method [17], which the filtering distribution is decomposed into a nonparametric mixed model with M component:

$$p(x_t|y_{1:t}) = \sum_{m=1}^{M} \pi_{m,t} p_m(x_t|y_{1:t}) \tag{3}$$

Weight is $\sum_{m=1}^{M} \pi_{m,t} = 1$. The non-parametric model is assumed to be a mixture of individual components. Using the filter distribution, $p(x_{t-1}|y_{1:t-1})$ is computed in the previous step. It is derived from a new predictive distribution.

$$\begin{aligned} p(x_t|y_{1:t-1}) &= \sum_{m=1}^{M} \pi_{m,t-1} \int p(x_t|x_{t-1})p_m(x_{t-1}|y_{1:t-1})dx_{t-1} \\ &= \sum_{m=1}^{M} \pi_{m,t-1} p_m(x_t|y_{1:t-1}) \end{aligned} \tag{4}$$

Therefore, the new predictive distribution is got directly from the prediction of each individual component. And then these distributions are combined on the basis of maintaining the prior weight.

The new filter distribution is obtained through adding the new predictive distribution into the formula (2).

$$\begin{aligned} p(x_t|y_{1:t}) &= \frac{\sum_{m=1}^{M} \pi_{m,t-1} p(y_t|x_t) p_m(x_t|y_{1:t-1})}{\sum_{n=1}^{M} \pi_{n,t-1} \int p_n(y_t|x_t)p_n(x_t|y_{1:t-1})dx_t} \\ &= \sum_{m=1}^{M} \left[\frac{\pi_{m,t-1} \int p_m(y_t|x_t)p_m(x_t|y_{1:t-1})dx_t}{\sum_{n=1}^{M} \pi_{n,t-1} \int p_n(y_t|x_t)p_n(x_t|y_{1:t-1})dx_t} \right] \times \left[\frac{p_m(y_t|x_t)p_m(x_t|y_{1:t-1})}{\int p_m(y_t|x_t)p_m(x_t|y_{1:t-1})dx_t} \right] \\ &= \sum_{m=1}^{M} \pi_{m,t} p_m(x_t|y_{1:t}) \end{aligned} \tag{5}$$

Second item of the second lows can be regarded as the new filter distribution of the m component in the above formula. That is: $p_m(x_t|y_{1:t}) = \frac{p(y_t|x_t)p_m(x_t|y_{1:t-1})}{\int p(y_t|s_t)p_m(ds_t|y_{1:t-1})}$.

The state vector x_t of first items in brackets is independent. Its new weight is:

$$\pi_{m,t} = \frac{\pi_{m,t-1} \int p_m(y_t|x_t)p_m(x_t|y_{1:t-1})dx_t}{\sum_{n=1}^{M} \pi_{n,t-1} \int p_n(y_t|x_t)p_n(x_t|y_{1:t-1})dx_t} = \frac{\pi_{m,t-1}p_m(y_t|y_{1:t-1})}{\sum_{n=1}^{M} \pi_{n,t-1}p_n(y_t|y_{1:t-1})} \tag{6}$$

Formula (3) shows that the new filtering distribution is the mixture of each individual component. As long as the mixed weights are updated according to the formula (6), the correct target distribution can be obtained. New weight is standardized likelihood function of the component weight. That is, there is M different likelihood distribution $\{p_m(y_{1:t}|x_t)\}_{m=1\ldots M}$. When one or more new targets is in the scene, the likelihood distribution can be obtained by initialized automatically observation model.

2.2 Mixture Particle Filtering

In [17], the particles are used to obtain the following approximation of the posterior distribution:

$$\bar{p}(x_t|y_{1:t}) = \sum_{m=1}^{M} \pi_{m,t} \sum_{i \in I_m} w_t^{(i)} \delta_{x_t^{(i)}}(x_t) \tag{7}$$

Where N is the number of particles. M is the number of mixture components that are the number of multiple targets. $\Pi_t = \{\pi_{m,t}\}_{m=1}^{M}$ is for the weight of the mixture component. $X_t = \{x_t^{(i)}\}_{i=1}^{N}$ is for the particle state, $W_t = \{w_t^{(i)}\}_{i=1}^{N}$ is for the particle weight. $\delta_a(\cdot)$ is Dirac measurement in the formula. I_m is a particle index set belonging to m mixture component. The sum of the weight of the mixture and the weight of each particle is equal to 1. That is $\sum_{m=1}^{M} \pi_{m,t} = 1$ and $\sum_{i \in I_m} w_t^{(i)} = 1, m = 1 \cdots M$.

The new sample should be produced by appropriate selection of the proposed distribution, and the selection of the proposed distribution is related to the old state and the new measurement. That is $x_t^{(i)} \sim q(x_t \middle| x_{t-1}^{(i)}, y_t), i \in I_m$. In order to retain the appropriate weighted sample set, the new particle weight:

$$w_t^{(i)} = \frac{\tilde{w}_t^{(i)}}{\sum_{j \in I_m} \tilde{w}_t^{(j)}}, \tilde{w}_t^{(i)} = \frac{w_{t-1}^{(i)} p(y_t \middle| x_t^{(i)}) p(x_t^{(i)} \middle| x_{t-1}^{(i)})}{q(x_t^{(i)} \middle| x_{t-1}^{(i)}, y_t)} \tag{8}$$

New sample set $\{x_t^{(i)}, w_t^{(i)}\}_{i \in I_m}$ is the likelihood distribution for $p_m(x_t|y_{1:t})$.

In order to obtain the new mixture weights, we have to compute the likelihood of the component $p_m(y_t|y_{1:t-1}), m = 1 \cdots M$. The likelihood distribution which is using particle Monte Carlo approximate to the first m component is expressed as [18]:

$$p_m(x_t|y_{1:t-1}) = \iint p(y_t|x_t)p(x_t|x_{t-1})p_m(x_{t-1}|y_{1:t-1})dx_t dx_t$$
$$\approx \sum_{i \in I_m} w_{t-1}^{(i)} \frac{p(y_t \middle| x_t^{(i)})p(x_t^{(i)}|x_{t-1}^{(i)})}{q(x_t^{(i)} \middle| x_{t-1}^{(i)}, y_t)} = \sum_{i \in I_m} \tilde{w}_t^{(i)} \tag{9}$$

The new approximation mixed weights was given through the substitution of results in Eq. (6).

$$\pi_{m,t} \approx \frac{\pi_{m,t-1}\tilde{w}_{m,t}}{\sum_{n=1}^{M} \pi_{n,t-1}\tilde{w}_{n,t}}, \tilde{w}_{m,t} = \sum_{i \in \mathrm{Im}} \tilde{w}_t^{(i)} \tag{10}$$

Like many particle filtering methods, the mixture particle filter is carried out from the prior density sampling, and the importance weights are obtained by the formula (10) to update particle.

In [17], the mixed expression is acquired and maintained using a simple K-means spatial clustering method. In the paper, we utilize boosting to supply a more effective solution to the problem.

3 MIL and Boosting

We will apply Noisy OR variants of AdaBoost to solve the MIL problem. The Any-Boost framework of Mason [21] is using for the derivation. The derivation is based on the appropriate cost function of the previous MIL, namely Noisy OR. The derivation of the Noisy OR is simpler and more intuitive.

3.1 Noisy-OR Boost

Boosting each example is recalled that is classified by a linear combination of weak classifiers. In MILBoost, examples are not separately marked. Instead, they exist in bags. So, an example is indexed with two variable: i, which indicates the bag, and j, which indicates the example within the bag. The score of the example is $y_{ij} = C(x_{ij})$ and $C(x_{ij}) = \sum_t \lambda_t c^t(x_{ij})$ a weighted sum of weak classifiers. The probability of a positive example is provided by the standard logistic function.

$$p_{ij} = \frac{1}{1 + \exp(-y_{ij})} \tag{11}$$

The probability of the positive bag is a "noisy OR". $p_i = 1 - \prod_{j \in i}(1 - p_{ij})$ [22, 23]. In this model the likelihood distributed to a set of training bags is:

$$L(C) = \prod_i p_i^{t_i}(1 - p_i)^{(1-t_i)} \tag{12}$$

Where $t_i \in \{0, 1\}$ is the label of bag i.

According to the AnyBoost method, the weight of each example is compared to the derivative of the cost function with respect to a change of score in the example. The derivative of the log likelihood is:

$$\frac{\partial \log L(C)}{\partial y_{ij}} = \omega_{ij} = \frac{t_i - p_i}{p_i} p_{ij} \tag{13}$$

Where, the weights are signed. The sign determines the example label. In each round of boosting, a classifier which maximizes $\sum_{ij} c(x_{ij})\omega_{ij}$ is searched. Where $c(x_{ij})$ is the score which is distributed to the example through the weak classifier for a binary classifier $c(x_{ij}) \in \{-1, +1\}$. The parameter λ_t is confirmed by utilizing a linear search to maximize $\log L(C + \lambda_t c_t)$.

3.2 Multiple Instance Boosting Detector

We adopt a Multiple Instance Boosting algorithm which combined Multiple Instance Learning (MIL) with the Viola-Jones method of object detection [24]. To do this, a new method called as MILBoost was established, which was adding MIL into the AnyBoost [24] framework. MIL boosting algorithm has a lot of advantages compared with adaBoost algorithm. Its advantage lies in not requiring exact target location to learn. Replacing an EM-like algorithm, we use MIL to build the system, which does not require iteration. Thus we hold a cascade of detectors for maximum speed.

Experiments were performed using on pedestrians in a digitized video sequence from broadcast television. We sample a collection of 3000 images from the video. We compared classical adaBoost with Noisy-OR variants of MIL boost. In the MIL algorithms there is one bag for each labeled body, including those positive windows which occlude that body. In addition, there is a negative bag for each image. After training, performance is evaluated on the body of a walking man (see Fig. 1). Estimating from the performance of ROC curve, MIL boost was better than Adaboost.

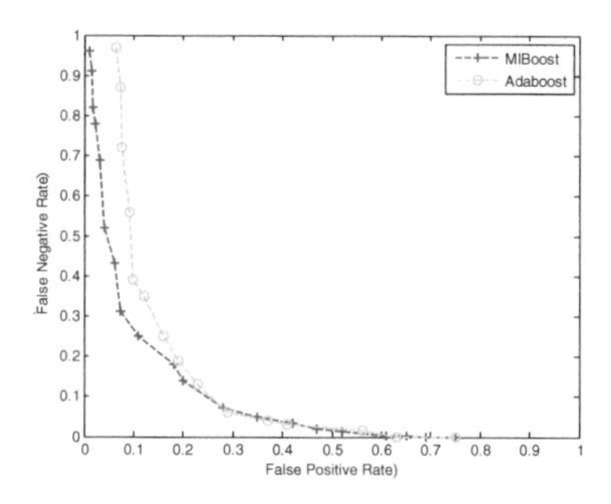

Fig. 1. ROC comparison between two boosting rules.

All the MIL test results are actually useful. A representative example of detection results are shown in Fig. 2. Results are shown for the noisy OR algorithm.

Fig. 2. Pedestrian detection result: This figure shows results of the MILBoost pedestrian detection. There are a set of the detection of errors or omission on pedestrian.

The results of MILBoost in our dataset are a more precise way to detect people than adaboost, but people are often confused, resulting in many false positives.

4 Multiple Instance Boosting Particle Filtering

The Multiple Instance Boosting particle filtering provides two major extensions of the MPF. First, it uses Multiple Instance Boosting for the constructing of the proposal distribution. This method improves the robustness of the algorithm. It is widely accepted that proposal distributions that incorporate the recent observations (in our case, through the Multiple Instance Boosting detecting) outperform original transition prior proposals [25, 26]. Second, Multiple Instance Boosted detecting provides a mechanism for obtaining and maintaining the mixture description. The method is more effective than the original K-means clustering plan in [17]. Particularly, it can detect efficiently objects entering and leaving the scene.

4.1 The Proposed Density Function Combined Multiple Instance Boosting Detection

The tracking quality can be greatly improved if the motion model of the target is combined with the Multiple Instance Boosting detection. In particular, the number of positive samples will be greatly reduced by using a reasonable plan. When one or more new targets enter in the scene, the likelihood distribution can be obtained by initializing automatically observation model. Therefore, Multiple Instance Boosted detecting is merged to the mixture particle filter as the framework of proposed density. The expression for the proposed density function is:

$$q^{*}_{MIBPF}(x_t|x_{1:t-1}, y_{1:t}) = \alpha q_{MIB}(x_t|x_{t-1}|y_t) + (1-\alpha)p(x_t|x_{t-1}) \tag{14}$$

Where: q_{MIB} is Gaussian distribution of the positive samples as the center of MIBoosting detection. Parameter α can be set dynamically and does not influence the convergence of particle filter. When $\alpha = 0$, algorithm is mixture particle filter. When the α is added, the Multiple Instance Boosted detecting function is stronger. We can adjust the value of α according to the tracking situation (including interaction, collision and occlusion).

4.2 The Process of Target Tracking of MIBoosting Detection Combining with Mixture Particle Filtering

(1) Initialization

N sample set $\{x_{m,0}^{(i)}, w_{m,0}^{(i)}\}_{i=1,\ldots,N,m=1,\ldots,M}$ is used to represent $P(X_0)$. Where $X_{m,0}^{(i)} \sim q(X_0), w_{m,0}^{(i)} = 1/N,\ M = 0$.

(2) MILBoost Detection

We detect targets using for MIBoosting detector by using Eq. (14). If there is M_{new} new target, it is from 1 to M_{new} for m. That is $M = M + M_{new}$, N particles $\{x_{m,t}^{(i)}\}_{i=1}^{N}$ is generated from the Gauss distribution with the MIBoosting detection as center.

(3) Sampling

(1) New particles $\{x_{m,t}^{(i)}\}_{i=1}^{N}$ is produced by the Eqs. (1)–(13).

(2) Calculating observation likelihood $\{x_{m,t}^{(i)}\}_{i=1}^{N}$.

(3) The importance of weight $\{w_{m,t}^{(i)}\}_{i=1}^{N}$ is updated by using Eqs. (1)–(8).

(4) Estimating

(1) $\{x_{m,t}^{(i)}\}_{i=1}^{N}$ is resampled according to the importance of weight $\{w_{m,t}^{(i)}\}_{i=1}^{N}$ and it results in an un-weighted sample $\{\tilde{x}_{m,t}^{(i)}\}_{i=1}^{N}$.

(2) $\bar{x}_{m,t} = mean(\{\tilde{x}_{m,t}^{(i)}\}_{i=1}^{N})$.

(5) End

When the trust degree of the MIBoosting detection is lower than a given threshold value, the first M_1 target is removed. When the two goals are overlap with each other, the first M_2 target is merged. There is $M = M - M_1 - M_2$ at this time.

5 Experimental Results and Analysis

5.1 State Model

The state sequence $\{x_t;\ t \in \mathbb{N}\}$, $x_t \in \mathbb{R}^{n_x}$ is a hidden Markov process with initial distribution $p(x_0)$ and transition distribution $p(x_t|x_{t-1})$, where n_x is the dimension of the state vector. In the tracking system, a standard autoregressive dynamic model is used for the transition model. The observations $\{y_t; t \in \mathbb{N}^*\}$, $y_t \in \mathbb{R}^{n_y}$ are conditionally independent in a given process $\{x_t;\ t \in \mathbb{N}\}$ with marginal distribution $p(y_t|x_t)$, where n_y is the dimension of the observation vector.

5.2 Observation Model

We use Hue-Saturation-Value color histograms. So, a feature vector x is consists of a set of HSV color histograms, each of which has $N = N_h N_s + N_v$ bins. A distribution $K(R) \triangleq \{k(n;R)\}_{n=1,\ldots,N}$ of the color histogram R is given around a window as follows:

$$K(n;R) = \eta \sum_{d \in R} \delta[b(d) - n] \tag{15}$$

Where d is any pixel position within R, and $b(d) \in \{1, \ldots, N\}$ as the bin index. δ is the delta function. If we describe $K* = \{k^*(n;x_0)\}_{n=1,\ldots,N}$ as the reference model and $K(x_t)$ as a candidate model, then we require to measure the similarity between $K*$ and $K(x_t)$. As in [18], we utilize the Bhattacharyya similarity coefficient to calculate a distance ξ on HSV histograms. The equation of the measure is given as (16):

$$\xi[K^*, K(x_t)] = \left[1 - \sum_{n=1}^{N} \sqrt{k^*(n;x_0)k(n;x_t)}\right]^{\frac{1}{2}} \tag{16}$$

When we get a distance ξ on the HSV color histograms, likelihood distribution is following as:

$$p(y_t|x_t) \propto e^{-\lambda\xi^2[K^*,K(x_t)]} \tag{17}$$

5.3 Experimental Setup

Color histogram is set as $N_h = N_s = N_v = 10$. That is HSV color histogram with dimension 110. The number of particles is 100 particles and $\alpha_{ada} = 0.6$.

5.4 Experimental Results and Analysis

First testing video sequence is OneLeaveShopReenter1cor sequence from [27]. The tracked people are multiple walking people. There is similar background interference in the process of the moving for the experimental sequence. Tracked target has occlusion and interaction each other. Object number also varies with time. Experiments are carried out using the BPF algorithm and MIBPF. From Fig. 3, it can be seen that the BPF algorithm is not enough to deal with the severe occlusion, so there is no accurate tracking of the target location. The selection of the target initialization also has certain dependence. Following tracking algorithm combined MIBoosting detecting with particle filter is used to re-test, taking the same key frame of the experimental. Tracking results is shown in Fig. 4. The tracking accuracy and processing speed can be improved effectively due to the addition of MIBoosting detection algorithm. In addition, using the online updating to update the target feature, feature template can timely response feature information in the current moment. The feature information of the target increases or decreases because of the changes in background conditions, feature template can provide abundant target feature information.

It makes more stable tracking. As a result, the ability of algorithm for similar background interference and occlusion is enhanced after the advantages of MIL boosting are combined with mixture particle filter.

Fig. 3. Results of multi-targets tracking based on BPF(73, 161, 268, 481)

Fig. 4. Results of multi-object tracking based on MIBPF(73, 161, 268, 481)

6 Conclusions

We have provided a method to merge the advantage of Multiple Instance Boosting for object detection into the mixture particle filtering for multi-target tracking. The combination is obtained by constituting the proposal distribution of the particle filtering from a mixture of the MIB detections between the dynamic model predicted from the previous time step and the current frame. The combination of the two methods results to lesser failures than any one on their own, along with solving both detecting and continuous track composition within the same framework. We have experimented using the boosted particle filter in the background of tracking pedestrian composition in actual video sequence. The results show that most humans are successfully detected and tracked, even if pedestrians enter into and out of the scene. Results can be thought further improved by employing a probabilistic model for target elimination, and it may also improve our MIBPF tracking.

Acknowledgments. This research was supported by the research foundation of young reserve talents project in scientific and technical bureau in Harbin (No. 2017RAQXJ134) and the project of the National Natural Science Foundation of China (No. 61573114).

References

1. Luo, W., Zhao, X., Kim, T.-K.: Multiple object tracking: a review. arXiv:1409.7618 [cs] (2014)
2. Wen, L., Li, W., Yan, J., Lei, Z., Yi, D., Li, S.: Multiple target tracking based on undirected hierarchical relation hypergraph. In: CVPR (2014)
3. Pirsiavash, H., Ramanan, D., Fowlkes, C.C.: Globally optimal greedy algorithms for tracking a variable number of objects. In: CVPR (2011)
4. Segal, A.V., Reid, I.: Latent data association: Bayesian model selection for multi-target tracking. In: ICCV (2013)

5. Tang, S., Andres, B., Andriluka, M., Schiele, B.: Subgraph decomposition for multi-target tracking. In: CVPR, pp. 5033–5041. IEEE Computer Society (2015)
6. Milan, A., Roth, S., Schindler, K.: Continuous energy minimization for multitarget tracking. IEEE Trans. Pattern Anal. Mach. Intell. **36**(1), 58–72 (2014)
7. Zhang, L., Li, Y., Nevatia, R.: Global data association for multi-object tracking using network flows. In: CVPR (2008)
8. Andriyenko, A., Schindler, K., Roth, S.: Discrete continuous optimization for multi-target tracking. In: CVPR (2012)
9. Bae, S.-H., Yoon, K.-J.: Robust online multi-object tracking based on tracklet confidence and online discriminative appearance learning. In: CVPR (2014)
10. Zamir, A.R., Dehghan, A., Shah, M.: GMCP-tracker: global multi-object tracking using generalized minimum clique graphs. In: ECCV (2012)
11. Milan, A., Leal-Taixe, L.: Joint tracking and segmentation of multiple targets. In: CVPR, pp. 5397–5406 (2015)
12. Arulampalam, S., Maskell, S., Gordon, N., Clapp, T.: A tutorial on particle filters for on-line non-linear/non-Gaussian Bayesian tracking. IEEE Trans. Sig. Process. **50**, 174–188 (2002)
13. Isard, M., MacCormick, J.: BraMBLe: a Bayesian multipleblob tracker. In: ICCV, vol. 2, pp. 4–41 (2001)
14. Vermaak, J., Perez, P.: Monte Carlo filtering for multi-target tracking and data association. IEEE Trans. Aerosp. Electron. Syst. **41**(1), 309–330 (2005)
15. Ooi, A., Vo, B.-N., Doucet, A.: Particle filtering for multi-target racking using jump Markov systems. In: Proceedings of the IEEE International Conference on Intelligent Sensors, Sensor Networks and Information Processing (ISSNIP), pp. 131–136 (2004)
16. Sidenbladh, H.: Multi-target particle filtering for the probability hypothesis density. In: Proceedings of the 6th International Conference on Information Fusion, Cairns, Australia, pp. 800–806 (2003)
17. Vermaak, J., Doucet, A., Pérez, P.: Maintaining multi-modality through mixture tracking. In: International Conference on Computer Vision (2003)
18. Kenji, O., Ali, T., De Freitas, N., Little, J.J., Lowe, D.G.: A boosted particle filter: multitarget detection and tracking. In: The European Conference on Computer Vision in ECCV (2004)
19. Viola, P., Platt, J.C., Zhang, C.: Multiple instance boosting for object detection. Adv. Neural. Inf. Process. Syst. **18**, 1419–1426 (2005)
20. Dollár, P., Babenko, B., Belongie, S.: Multiple component learning for object detection. In: ECCV, vol. 5303, pp. 211–224 (2008)
21. Mason, L., Baxter, J., Bartlett, P., Frean, M.: Boosting algorithms as gradient descent in function space (1999)
22. Heckerman, D.: A tractable inference algorithm for diagnosing multiple diseases. In: Proceedings of UAI, pp. 163–171 (1989)
23. Maron, O., Lozano-Perez, T.: A framework for multiple-instance learning. Proc. NIPS **10**, 570–576 (1998)
24. Viola, P., Jones, M.: Robust real-time object detection. Int. J. Comput. Vis. **57**(2), 137–154 (2002)
25. Rui, Y., Chen, Y.: Better proposal distributions: object tracking using unscented particle filter. In: IEEE Conference on Computer Vision and Pattern Recognition, pp. 786–793 (2001)
26. van der Merwe, R., Doucet, A., de Freitas, J.F.G., Wan, E.: The unscented particle filter. Adv. Neural. Inf. Process. Syst. **8**, 351–357 (2000)
27. http://homepages.inf.ed.ac.uk/rbf/CAVIAR

An Enhance Approach of Filtering to Select Adaptive IMFs of EEMD in Fiber Optic Sensor for Oxidized Carbon Steel

Nur Syakirah Mohd Jaafar[1(✉)], Izzatdin Abdul Aziz[1], Jafreezal Jaafar[1], Ahmad Kamil Mahmood[1], and Abdul Rehman Gilal[2]

[1] Centre for Research in Data Science, Universiti Teknologi PETRONAS, Seri Iskandar, Malaysia
{nur_16001470, izzatdin, jafreez}@utp.edu.my

[2] Department of Computer Science, Sukkur IBA University, Sukkur, Pakistan
a-rehman@iba-suk.edu

Abstract. Number of existing signal processing methods can be used for extracting useful information. However, receiving desired and eliminating undesired information is yet a significant problem of these methods. Empirical Mode Decomposition (EMD) algorithm shows promising results in comparison to other signal processing methods especially in terms of accuracy. For example, it shows an efficient relationship between signal energy and time frequency distribution. Though, EMD algorithm still has a noise contamination which may compromise the accuracy of the signal processing. It is due to the mode mixing phenomenon in the Intrinsic Mode Function's (IMF) which causes the undesirable signal with the mix of additional noise. Therefore, it has still a room for the improvements in the selective accuracy of the sensitive IMF after decomposition that can influence the correctness of feature extraction of the oxidized carbon steel. This study has used two datasets to compare the parameters analysis of the Ensemble Empirical Mode Decomposition (EEMD) algorithm for constructing the signal signature.

Keywords: Signal processing method · Empirical Mode Decomposition Ensemble Empirical Mode Decomposition · Intrinsic Mode Function's Noise contamination · Mode mixing · Selective accuracy of IMF

1 Introduction

Improved signal denoising method is defined by reducing the unwanted signals from the raw signal which are collected from the sensors that installed outside of the pipeline network [1]. The ultrasonic inspection and magnetic flux leakage inspection technique are the common methods for evaluating and monitoring the state of pipelines through the collected signals from the interior of the offshore pipeline [2]. The signal characteristics will have derived the anomalies, signal amplitudes, signal phases and signal frequency to ascertain actions to be taken ahead.

R. Silhavy (Ed.): CSOC 2018, AISC 764, pp. 241–255, 2019.
https://doi.org/10.1007/978-3-319-91189-2_24

Lacking techniques for evaluating and monitoring the state of pipelines through the collected signals from the interior of the pipeline leads to inaccurate corrosion detection. This happens due to high background noise and noise condition which will affect the actual signal. For example, the signals which can reveal the anomalies can be overlapped. Thus, the recovery of a signal from observed noisy data is important. The analysis of techniques of signal processing is essential to highlight the wanted information and attenuate the undesired one [3]. Butterworth low pass filter, wavelet-based thresholding filter (Wavelet Transform) and Hilbert Huang Transform (HHT) are few examples of complex decomposition signal breaking into finite components [4]. These existed denoising methods in signal processing still have limitations in the removal of noise reduction to achieve the accurate reading of signal signatures. In general, most often, signal processing methods are applied to solve the hidden results from the raw signal including removing noisy signals from the data collected through sensors which are installed outside of the pipeline network.

Empirical Mode Decomposition (EMD) algorithm in Hilbert Huang Transform (HHT) contains undesirable signal signatures. Undesirable signal signature contains anomalies such as noises interference, which leads to signal misinterpretation. Signal is decomposed into many IMFs by EMD in HHT. Based on the analysis on EMD algorithm, the mode mixing problems are encountered when signal contains intermittency. The mode mixing will affect the decomposition of signal into different IMFs on the time series as similar scale is residing in different IMF. Thus, this drawback creates a necessity for a proposal and development of an improved decomposition of signal using Ensemble Empirical Mode Decomposition (EEMD).

The problem found creates a necessity for a proposal and development of an improved method to decompose signals using Ensemble Empirical Mode Decomposition (EEMD). This is to improve the reconstruction of noise removed signal when the selection of the Intrinsic Mode Functions (IMFs) is relevant to the corresponding of the most important structures of the signal. The following research questions are extracted from the problem statement.

[RQ1]: What are the correct values for each parameter that are relevant with signal processing method using Ensemble Empirical Mode Decomposition (EEMD)?

[RQ2]: To what extent does EEMD that takes advantage of IMFs, able to determine accurate IMFs after the decomposition in time series?

[RQ3]: To what extent does the selection of the Intrinsic Mode Functions (IMFs), able to reconstruct the noise removed signal after the signal – filtering method been proposed?

The research aims are to improve the existing Empirical Mode Decomposition (EMD) algorithm using Hilbert Huang Transform (HHT). Through the preliminary research conducted, several research questions had been brought forward. To address the research questions arose; three research objectives have been devised:

[RO1]: To evaluate the correct values for each parameter that is relevant to Hilbert Huang Transform (HHT) using Ensemble Empirical Mode Decomposition (EEMD).

[RO2]: To compare and analyze signal energy of each Intrinsic Mode Functions (IMFs) in the Ensemble Empirical Mode Decomposition (EEMD) using Hilbert Huang Transform (HHT) in identifying the accurate IMFs.

[RO3]: To construct signal signature based on the selected principal Intrinsic Mode Functions (IMF) components to indicate the noise removed signal using data from oxidized carbon steel.

The study focuses on the existing EMD algorithm using HHT to recover a signal from the raw signals. The improved signal processing methods, which the extent of the EMD algorithm; are Ensemble Empirical Mode Decomposition (EEMD) algorithm are tested against the dataset obtained through field testing. The parameters tested for this research work are limited to the analysis of the EEMD algorithm which is many ensemble members, fixed sifting numbers and amplitude of the added white noise.

The algorithm used in this research is EEMD algorithm and for the representations in the frequency-time domain, HHT is used. The signal processing methods, EEMD will be improved to achieve higher accuracy based on the selection of the Intrinsic Mode Function (IMFs) leading to the noise removal signal for the oxidized carbon steel. High accuracy is defined as the ability of the signal processing algorithm, EEMD to the selection process of IMFs adaptive, and this selection IMFs achieves high Signal to Noise Ratio (SNR) while the Percentage of RMS Difference (PRD) and Max Error values are low.

However, for proof of concept, the existing algorithm without any improvements is used. The existing implementation of the EEMD algorithm is utilized to perform signal decomposition to enhance the thoroughness of denoising by fixing the parameters analysis of the EEMD. The algorithm is tested against two data sets obtained from the field testing. The study is limited to the signals that were corrupted by additive white Gaussian noise and is conducted based on extended numerical experiments.

2 Literature Review

2.1 Selection of Signal Processing Method

Table 1 describes the existing method in signal processing methods. For this research, EEMD by HHT is selected. The main reason of EEMD by HHT is selected due to the feature that it is data driven basis and suitable for analysis on the signal interpretation algorithm. It is important that the signal processing method can decomposed the signal into mono component functions. EEMD by HHT works through performing a time-adaptive decomposition operation.

Table 1. Comparison between signal processing methods

Name	Advantages	Disadvantages
Fourier transform by (Fast Fourier Transform) [5]	Fast Fourier Transform has accuracy in analytical form of smoothness signal of the underlying function [5]	Limitations for Fast Fourier Transform for pipeline having no localization of signal discontinuity [6]
Wavelet based thresholding filter by (Wavelet Transform) [7]	Wavelets have allowed filters to be constructed for signal localization for both time and frequency domain [7]	Limitations for the Wavelet Transform in signal generated for length of the basic wavelet function, wavelet base, scale, threshold function and optimal threshold values are fixed [8, 9]
Empirical Mode Decomposition by Hilbert Huang Transform (HHT) [10]	No pre-determine filter or wavelet function for denoising the signal [11]	Mode mixing occur in the Empirical Mode Decomposition (EMD) [11]
Ensemble Empirical Mode Decomposition by Hilbert Huang Transform (HHT) [10]	No pre-determine filter or wavelet function for denoising the signal [12]	Parameters analysis of the Ensemble Empirical Mode Decomposition (EEMD) is not fixed for the decomposition of the signal leading to the mode mixing [11]

By having the decomposition signal without requires no prior knowledge of the target signal, this will be more efficient in filtering accuracy. Even though EEMD can filter between noise and denoised signal, it still lacks in accuracy to filter between selections of the IMFs. As EEMD, the method still encounters the problem of mode mixing when the signal contains intermittency when the parameters of the EEMD is not set up. The parameter of the EEMD is based on the knowledge of the researchers [11–13].

Wavelet Transform method are more effective in performance compared with the EMD. Still this method involves major drawbacks when it comes to the selection of the wavelet based [12, 13]. The next subsection discusses in brief on EEMD algorithm.

2.2 Ensembles Empirical Mode Decomposition (EEMD)

EEMD, means from the numbers of IMF's been extracted from the adapted of EMD to the original signal. The ensemble mean is calculated with a different addition of white noise.

Figure 1 shows the process of EEMD algorithm. Given the signal, EEMD can be summarized as follows. By having many amplitudes of added white noise to the original signal resulted as noise-added signal. Next, the EMD operation will be implemented with the noise-added signal. Lastly, the ensemble means of each IMF component will be calculated.

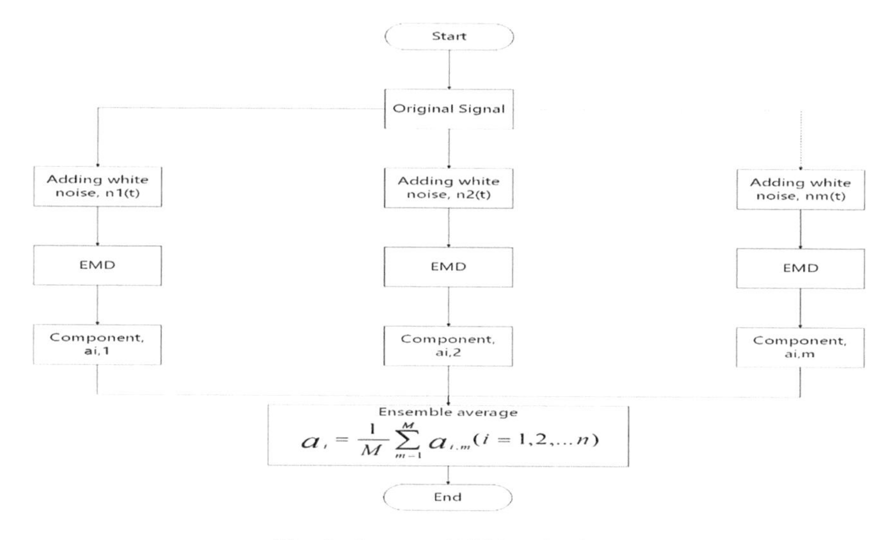

Fig. 1. Process of EEMD algorithm

Figure 2 shows an example of an IMF (green curve) with corresponding envelopes. A signal consists of higher-frequencies waves riding upon lower-frequencies carrying waves. The characteristic of time scales is used to find the intrinsic modes.

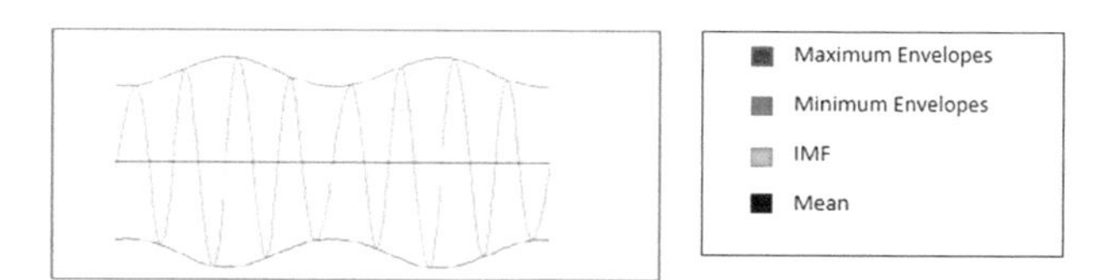

Fig. 2. Example of an Intrinsic Mode Function (IMF)

The main idea of having the sifting process which is the IMF is to separate the data into a slowly varying local mean part and a fast-varying symmetric oscillatory part. The latter part becoming the IMF represented by the green envelope and the local mean represented by the black line defining a residue [14]. This residue serves as input for further decomposition, with the process being repeated until no more oscillations can be obtained [14].

To overcome partially on the mode mixing phenomenon within the IMF, EEMD is proposed [15, 16]. This EEMD defines the true IMF components as the mean of an ensemble of trials, whereby each of the signals consist of the added white noise of finite amplitude. With this EEMD approach, the mode mixing scales in the IMF could be partially resolves based on the recent studies of the statistical properties of white noise [17]. The next section discusses existing works that utilize EEMD.

2.3 Existing Work for Ensembles Empirical Mode Decomposition (EEMD)

The content in the Table 2 explains the comparison between works implementing EEMD in decomposing the signal. Based on the existing works, all of them proved that EEMD could perform decomposition of signal. Either it is decomposition of signal in acoustic, ultrasonic guided waves, or any other types of signals, EEMD can decompose the signal into mono components to be more meaningful to be interpreted.

Table 2. Existing works of EEMD

Name	Type of signal	Findings	Way forward
Adnan et al. [8]	Acoustic signal	The signal processing is used to decompose the raw signal and show in time-frequency. Acoustic signal and HHT is the best method to detect leak in gas pipelines	This proposed approach could be applied for analysis of acoustic signal for the fibre optic sensor in pipeline
Camarena-Martinez et al. [18]	PQ disturbances	An approach based on EMD for analysis of PQ signals has been presented. The method consists of iterative down sampling stage that helps to extract the fundamental component as the first IMF	This proposed approach could be applied for analysis of acoustic signals for the fibre optic sensor in the pipeline
Xu et al. [19]	Acoustic signal	An adaptive EMD threshold denoising method for acoustic signal has been presented. The method is introduced into the threshold determination process of the denoising approach	This proposed approach could be applied for analysis of acoustic signals for the fibre optic sensor in the pipeline
Siracusano et al. [20]	Acoustic emission signal	An approach on a framework based on the Hilbert Huang Transform for the analysis of the acoustic emission data. By having the EMD features, the acoustic signal could isolate and extract the signal to be interpreted such as the amplitude to be more meaningful	This proposed approach could be applied for analysis of acoustic signals for the fibre optic sensor in the pipeline

However, most of the existing works explained that there is minimal work in analyzing signal energy of each IMF's in the EEMD using HHT in identifying the accurate signal reading in frequency-time domain. For this research, the question is whether the improvement EEMD algorithm indeed helps in reaching the goal of the data analysis in extracting the signals. Thereby, leading to the identification of t signal signature when the carbon steel is oxidized. Thus, for this research, the number of ensemble numbers, and the amplitude of the added noise, will be the parameters to be studied in reaching the goal of the research.

2.4 Testbed: Hilbert Huang Transform (HHT)

Having obtained the IMFs using the EEMD method, the IMFs will apply the HHT technique to each of the IMF component. HHT can be used to highlight the time–frequency representations of IMFs [5, 8, 14, 21]. Thus, this time-frequency, the representations of IMFs allow visualizing the modulation of each component of the IMFs.

Specifically, the EEMD is used to separate the modes and resulted in IMFs and this HHT is performed to compute the instantaneous amplitude and instantaneous frequency of these IMFs. From the time-frequency distribution and characteristic parameters such as instantaneous frequency and amplitudes are obtained. The target characteristics of frequency can be derived from the instantaneous frequency and amplitudes based on the energy distribution.

3 Methodology

There are several activities which are carried out for this research. For example, Fig. 3 depicts the activities conducted for this research.

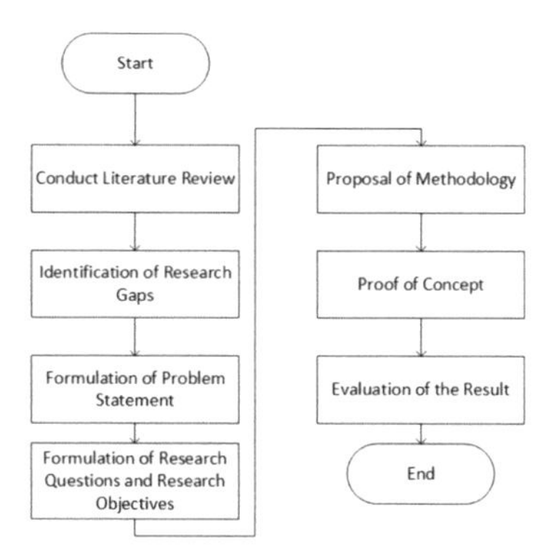

Fig. 3. Research activities

3.1 Solution Requirement

In this research, the parameter that is utilized is the accuracy. The parameter is selected based on the research gaps found from the existing implementations of the Ensemble Empirical Mode Decomposition (EEMD) algorithm. Accuracy is defined as context is defined as the closeness of the retrieve result to the intended result. Hence, accuracy of the algorithm is measured by the signal energy of each IMFs in the EEMD algorithm using HHT in identifying the accurate signal-reading in frequency–time domain.

Thus, from the signal energy of the IMFs in the EEMD algorithm, the selected of the IMFs, the signal will be reconstructed indicated the noise removed signal of oxidized carbon steel. Based on the critical analysis performed, EEMD is selected as the decomposition of signal for this research.

Figure 4 shows the signal processing flow. The procedure of the signal processing flow is as follows:

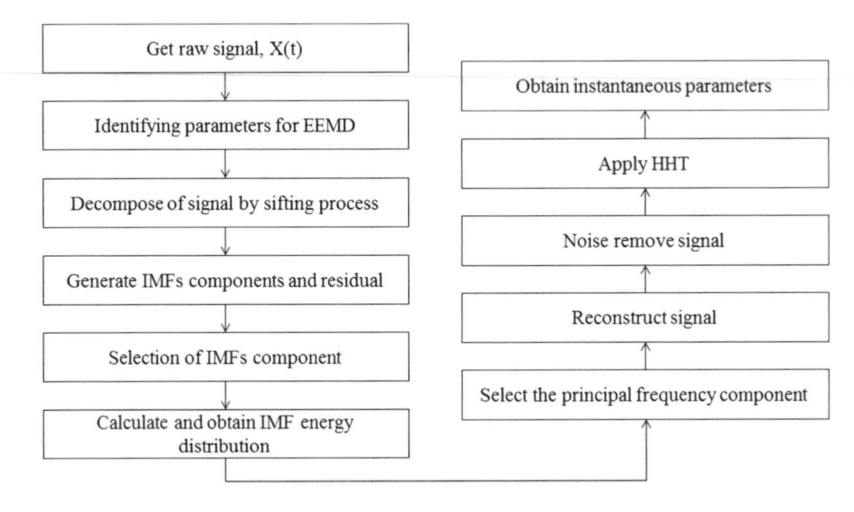

Fig. 4. Signal processing flow

The first step of the program will be to decompose the original signal by using the EEMD with the parameter that has been analyzed. The parameters that have been analyzed were the number of the ensemble numbers and the amplitude of the added noise. The decomposition of the signals by sifting process, IMF represents the series of stationary signals with different amplitude and frequency bands.

Frequency component, with different amplitude and frequency bands in each of the IMFs contained sampling of frequency. It resulted simple oscillatory modes as different frequency bands ranging from high to low. The parameters analysis of the EEMD such as the amplitude of white noise and the number of ensemble numbers is studies as it will be governed by the well-established statistical rule of:

$$\varepsilon_n = \frac{\varepsilon}{\sqrt{N}} \tag{1}$$

To select the IMFs which, contain most energy; the correlation between the instantaneous energy and frequency will be studied. Thus, from the relationship between instantaneous energy and frequency, the IMFs that contain most of energy will be selected. The justification on why the IMFs that contain most of energy will be selected is due to the approximately to the original signal. Next, the energy of the distribution of the IMF's will be calculated.

Proof of concept was done to prove that EMD and EEMD can decompose signals in the time domain and suitable to process the output signal from the fibre optic sensor as the output signal of the sensor is non-stationary.

The methodology that has been adopted is by utilizing EEMD algorithm for signal processing method purpose. EEMD is the right method for the problem at hand as it is one of the most accurate data analysis methods to reduce noise from the true signal in the signal processing compared with other algorithm based in the analysis from the literature review.

Through having an accurate signal processing algorithm, it will fulfil the objective for this research and address the problem of accurately selecting the adaptive IMFs of the EEMD. The method that has been adopted is also suitable to be used in the real dataset. It is because the method chosen will be able to decompose any type of signal into mono - components function, hence, making it suitable to be analyzed by the developed algorithm. Steps taken by the algorithm to decompose the signal in time domain have been illustrated and explained in clarity.

4 Proof of Concept

4.1 Objective of Proof of Concept

The objective is to prove that EMD and EEMD can decompose signals in time domain. Both methods are suitable to process the output signal from the fibre optic sensor as the output signal of the sensor is non-stationary and to determine the most suitable parameter for EEMD.

4.2 Data Preparation

The data are in the form of raw signal consists of time interval and mile voltage. The dataset will undergo pre-processing phase to ensure that it will be suitable for further analyses by EEMD. Details about the parameters of datasets are as follows in Table 3 used in the experiment. The data are in the form of raw signal consists of time interval and mile voltage. The dataset will undergo pre-processing phase to ensure that it will be suitable for further analyses by EEMD.

The constructed signal is composed by 2 different datasets from two sensors collected from experiments conducted during the field testing. Figure 5 shows the time representations for TEK 0098 while Fig. 6 for TEK 0099. The Y-axis represents voltage and the X axis represents time as depicted in Figs. 5 and 6 respectively.

Figures 5 and 6 shows the amplitude fluctuation in certain period, as many as 2500-time samples were recorded. Each spike represents the amplitude along the time interval. The amplitude shows major fluctuation in certain period compared with the Fig. 5. The average of the amplitude in TEK 0098 is −0.00049 mV per 2500 sampling. While for the TEK 0099 the average of the amplitude is 0.00099 mV per 2500 sampling.

Table 3. Parameters of data sets used in the experiment setup

		TEK 0098	TEK 0099
EMD	Source of data sampling	Field testing	Field testing
	Numbers of sampling	2500	2500
	Sifting number of each IMF	10	10
	Amplitude of noise	0	0
	Number of ensemble	1	1
EEMD	Sifting number of each IMF	10	10
	Amplitude of the added white noise	0.2	0.2
	Number of ensemble	100	100

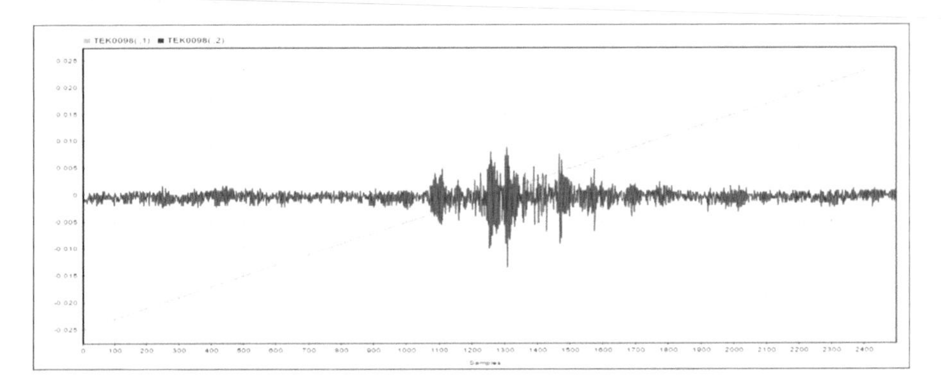

Fig. 5. Time domain representations for TEK 0098

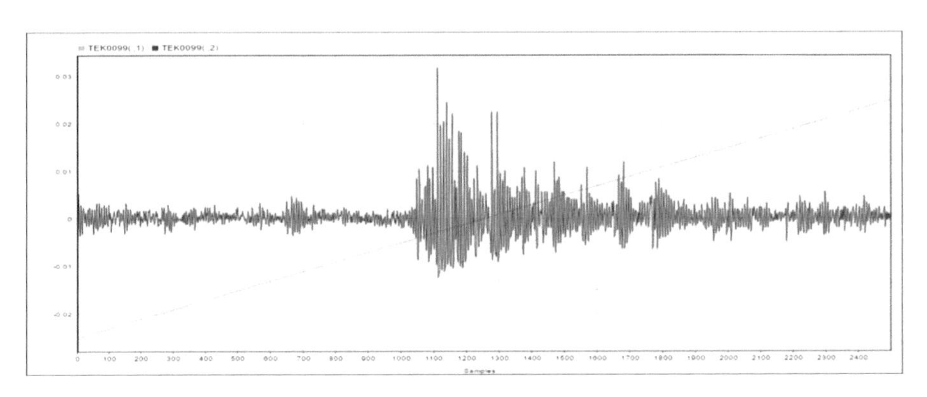

Fig. 6. Time domain representations for TEK 0099

The amplitude change shows anomalies captured by the sensors. At this stage, the noise is embedded within the raw signal. When there is a leak aperture in the fluid pipeline system, the fluid inside the pipe escapes to the external environment in high velocity.

5 Results and Discussions

5.1 EMD and Discussion

In Figs. 7 and 8, IMF1 to IMF3 are observing the noises coming from the experiment data. Comparing with the particle signals, they have higher frequency and smaller amplitude. As the plotted IMFs are using the same scale, IMF1 to IMF3 occurs mode mixing as similar scale is residing in different Intrinsic mode functions (IMF). During the sifting process, EMD acts as a filter bank. The property of EMD to behave as a dyadic filter bank resembling those involved in wavelets has been useful in signal denoising.

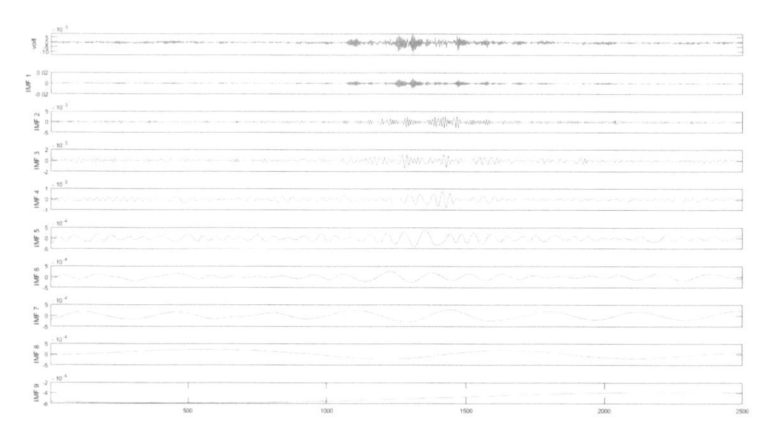

Fig. 7. EMD decomposition result of signal in TEK 0098

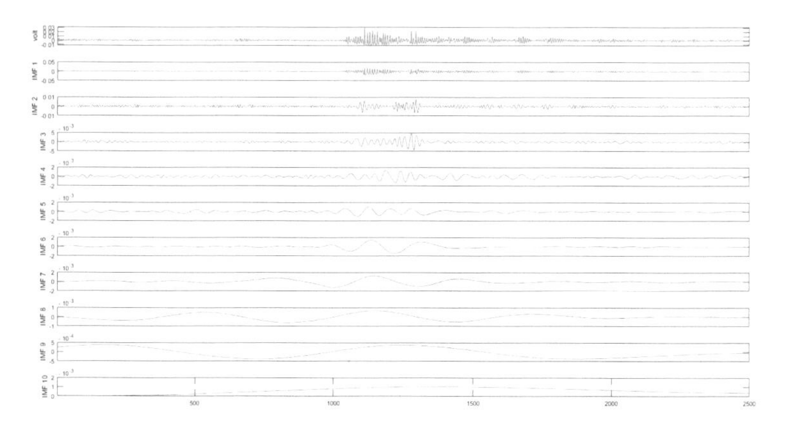

Fig. 8. EMD decomposition result of signal in TEK 0099

Mode mixing occurs in IMF1 until IMF3 due to noise interference. Thus, the EMD process may cause to generate the false signals. EMD method is stable and not sensitive to the noise. However, in some situation when the frequencies of useful signals are close, EMD process also can be disturbed by the change of noise. For this research, the EMD process have been disturbed throughout the background signal and noise interference as the leakage occurs from the fibre optics sensor.

In most case, the noise in a system is unpredictable and uncontrollable. For the implementation of EMD in this project, if the changes in the noise will totally distort the output of EMD, thus resulted in undesirable of signal that will make the EMD method is not suitable for the decomposition of the signal. To overcome this problem, an improved EMD method must be used to eliminate the influence of intermittency that causes mode mixing phenomenal. This is the reason the EEMD method is introduced and being implemented in this research purposes.

5.2 EEMD and Discussion

EEMD is not self-adapting as EMD. EEMD instead, can reduce the mode mixing occurrence. EEMD integrates white noise to improve EMD. Before the raw signals need to be process through EEMD, there are two parameters that need to be initialized, which are, the amplitude of added white noise and the number of ensembles.

Based on the critical analysis conducted from the literature reviews, the range of the standard deviation of amplitude when added white noise is between from 0.2 until 0.4 Hz [12, 14, 21]. As white noise is noise whose power spectral density is uniform over the frequency range. From the experiment analysis conducted, standard deviation of 0.2 was chosen [12, 14, 21]. The 0.2 standard deviation value for white noise amplitude is sufficient to decompose signal for the sifting process.

The second parameter is the number of ensembles, which means that how many ensemble operations need to be conducted in the decomposition process. In each ensemble operation, amplitude of added white noise will be added to the original data before performing the EEMD operation. Theoretically, to reduce the effect of the added white noise; a higher number of ensembles will be better for the EEMD operation. Unfortunately, the higher number of ensembles increases the computational time, thus will reduce the efficiency of EEMD operation. This will result in more time is needed to perform signal decomposition process. From the critical analysis been conducted, the range for the ensemble number in 100 to 200 are enough to remove most of the added white noise.

The increment of the number of ensembles has no more effect on the decomposition results when the standard deviation of the amplitude of the added white noise is representing the smaller difference between the input signal and the final reconstructed result of IMF components.

For this research purpose, the standard deviation of the amplitude of the added white noise is 0.2 and 100 is for the ensemble number are chosen for the EEMD decomposition of signal. The value setting had been chosen after the comparison of decomposition results with different preset parameters in the simulation and according to the analysis conducted from the literature reviews [12, 14, 21]. Figures 9 and 10

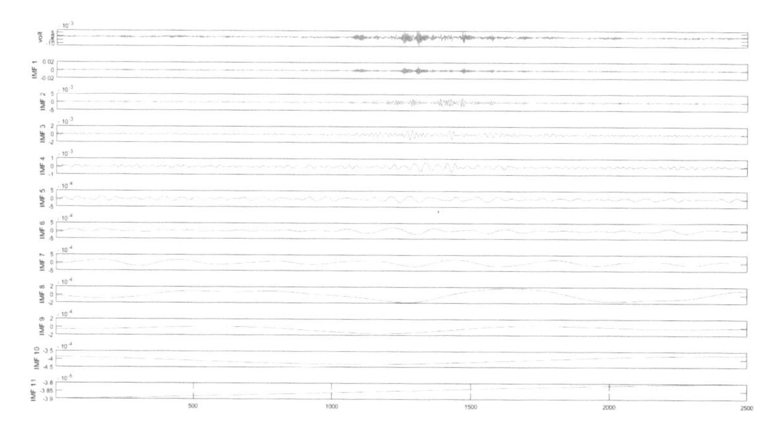

Fig. 9. EEMD decomposition result of signal in TEK 0098

Fig. 10. EEMD decomposition result of signal in TEK 0099

derived by the simulation carried out to proved that it is feasible to decompose the signal in time domain using EEMD.

The number of iterations is set up to 10 for this EEMD operation. This has generated 10 IMFs. Some dataset may contain high-frequency oscillations; hence more iteration may be needed. The higher in numbers of iterations are needed to generate new local extreme and it will distort the results of each IMFs. For this simulation, the number of iterations is set up to 10. The 10 iterations for this EEMD operation is symmetrically sufficient to keep the upper and lower envelopes of IMFs with respect to the zero mean.

From Figs. 9 and 10, the objective of the proof of concept was achieved. The parameters analysis of EEMD had been successfully been set up which are the amplitude of added white noise is 0.2, sifting number is 10, and the ensemble number is 100 based on extended numerical experiments.

Acknowledgments. This work was supported by Development of Intelligent Pipeline Integrity Management System (I-PIMS) Grant Scheme from Universiti Teknologi PETRONAS.

References

1. Underground pipeline corrosion
2. Shi, Y., Zhang, C., Li, R., Cai, M., Jia, G.: Theory and application of magnetic flux leakage pipeline detection. Sensors **15**(12), 31036–31055 (2015)
3. Gaci, S.: A new Ensemble Empirical Mode Decomposition (EEMD) denoising method for seismic signals. Energy Procedia **97**, 84–91 (2016)
4. Agarwal, M., Jain, R.: Ensemble empirical mode decomposition: an adaptive method for noise reduction. IOSR J. Electron. Commun. Eng **5**, 60–65 (2013)
5. Karkulali, P., Mishra, H., Ukil, A., Dauwels, J.: Leak detection in gas distribution pipelines using acoustic impact monitoring. In: 42nd Annual Conference of the IEEE Industrial Electronics Society, IECON 2016. IEEE (2016)
6. Datta, S., Sarkar, S.: A review on different pipeline fault detection methods. J. Loss Prev. Process Ind. **41**, 97–106 (2016)
7. Jiao, Y.-L., Shi, H., Wang, X.-H.: Lifting wavelet denoising algorithm for acoustic emission signal. In: 2016 International Conference on Robots and Intelligent System (ICRIS). IEEE (2016)
8. Adnan, N.F., Ghazali, M.F., Amin, M.M., Hamat, A.M.A.: Leak detection in gas pipeline by acoustic and signal processing-a review. In: IOP Conference Series: Materials Science and Engineering. IOP Publishing (2015)
9. Fang, Y.-M., Feng, H.-L., Li, J., Li, G.-H.: Stress wave signal denoising using ensemble empirical mode decomposition and instantaneous half period model. Sensors **11**(8), 7554–7567 (2011)
10. Yang, J., Wang, X., Feng, Z., Huang, G.: Research on pattern recognition method of blockage signal in pipeline based on LMD information entropy and ELM. In: Math. Probl. Eng. **2017** (2017)
11. Kevric, J., Subasi, A.: Comparison of signal decomposition methods in classification of EEG signals For motor-imagery BCI system. Biomed. Sig. Process. Control **31**, 398–406 (2017)
12. Rostami, J., Chen, J., Tse, P.W.: A signal processing approach with a smooth empirical mode decomposition to reveal hidden trace of corrosion in highly contaminated guided wave signals for concrete-covered pipes. Sensors **17**(2), 302 (2017)
13. Samadi, S., Shamsollahi, M.B.: ECG noise reduction using empirical mode decomposition based combination of instantaneous half period and soft-thresholding. In: 2014 Middle East Conference on Biomedical Engineering (MECBME). IEEE (2014)
14. Saeed, B.S.: De-noising seismic data by Empirical Mode Decomposition (2011)
15. Huang, Y., Wang, K., Zhou, Z., Zhou, X., Fang, J.: Stability evaluation of short-circuiting gas metal arc welding based on ensemble empirical mode decomposition. Meas. Sci. Technol. **28**(3), 035006 (2017)
16. Potty, G.R., Miller, J.H.: Acoustic and seismic time series analysis using ensemble empirical mode decomposition. J. Acoust. Soc. Am. **140**(4), 3423–3424 (2016)
17. Honório, B.C.Z., de Matos, M.C., Vidal, A.C.: Progress on empirical mode decomposition-based techniques and its impacts on seismic attribute analysis. Interpretation **5**(1), SC17–SC28 (2017)
18. Camarena-Martinez, D., et al.: Novel down sampling empirical mode decomposition approach for power Quality analysis. IEEE Trans. Ind. Electron. **63**(4), 2369–2378 (2016)

19. Xu, J., Wang, Z., Tan, C., Si, L., Liu, X.: A novel denoising method for an acoustic-based system through empirical mode decomposition and an improved fruit fly optimization algorithm. Appl. Sci. **7**(3), 215 (2017)
20. Siracusano, G., Lamonaca, F., Tomasello, R., Garescì, F., La Corte, A., Carnì, D.L., Carpentieri, M., Grimaldi, D., Finocchio, G.: A framework for the damage evaluation of acoustic emission signals through Hilbert-Huang transform. Mech. Syst. Sig. Process. **75**, 109–122 (2016)
21. Wu, Z., Huang, N.E.: Ensemble empirical mode decomposition: a noise-assisted data analysis method. Adv. Adapt. Data Anal. **1**(01), 1–41 (2009)

Hyper-heuristical Particle Swarm Method for MR Images Segmentation

Samer El-Khatib[1], Yuri Skobtsov[2(✉)], Sergey Rodzin[1], and Viacheslav Zelentsov[3]

[1] Southern Federal University, Rostov-on-Don, Russia
samer_elkhatib@mail.ru, srodzin@yandex.ru

[2] St. Petersburg State University of Aerospace Instrumentation, Saint Petersburg, Russia
ya_skobtsov@list.ru

[3] St. Petersburg Institute of Informatics and Automation, Russian Academy of Sciences (SPIIRAS), Saint Petersburg, Russia
v.a.zelentsov@gmail.com

Abstract. An important factor in the recognition of magnetic resonance images is not only the accuracy, but also the speed of the segmentation procedure. In some cases, the speed of the procedure is more important than the accuracy and the choice is made in favor of a less accurate, but faster procedure. This means that the segmentation method must be fully adaptive to different image models, that reduces its accuracy. These requirements are satisfied by developed hyper-heuristical particle swarm method for image segmentation. The main idea of the proposed hyper-heuristical method is the application of several heuristics, each of which has its strengths and weaknesses, and then their use depending on the current state of the solution. Hyper-heuristical particle swarm segmentation method is a management system, in the subordination of which there are three bioinspired heuristics: PSO-K-means, Modified Exponential PSO, Elitist Exponential PSO. Developed hyper-heuristical method was tested using the Ossirix benchmark with magnetic-resonance images (MRI) with various nature and different quality. The results of method's work and a comparison with competing segmentation methods are presented in the form of an accuracy chart and a time table of segmentation methods.

Keywords: MRI image segmentation · Particle swarm optimization k-means · Swarm intelligence · Segmentation · Bio-inspired methods

1 Introduction

The development of image recognition methods is one of the relevant and difficult tasks in artificial intelligence. When creating recognition systems, which contain high requirements for accuracy and performance, occurs needless to apply new methods to automate the procedure of image recognition. Despite the fact that the task of developing image recognition methods is well researched theoretically, however, there is no universal method for solving it, and the practical solution seems to be very difficult.

R. Silhavy (Ed.): CSOC 2018, AISC 764, pp. 256–264, 2019.
https://doi.org/10.1007/978-3-319-91189-2_25

At computer processing and recognition of images the wide range of problems is solved. One of the main stages of recognition is the process of dividing the image into non-overlapping areas (segments) that cover the entire image and are homogeneous by some criteria. Segmentation simplifies the analysis of homogeneous areas of the image, as well as brightness and geometric characteristics. The segmentation is implemented using special methods. Their goal is to separate the analyzed object, structure or area of interest from the surrounding background. This is a difficult task, the performance quality of which significantly affects the accuracy and the possibility of subsequent computer analysis of images, since there are difficulties associated with noise, blurring of images, etc.

To solve the image segmentation problem, there have been developed many methods based on the luminance, gradient and texture information of the image [1]. Results of image segmentation and recognition methods research are set forth in works of Woods and Gonsalez [1].

A suitable way to effectively solve the problem of image segmentation is to use mathematical transformations describing the collective behavior of a decentralized self-organized system that consists of a multiple agents interacting locally with each other and with the environment to achieve a predetermined goal. In nature, examples of such systems are swarm systems [2, 6, 7, 9, 10]. Each agent functions autonomously, using own rules. At the same time, the behavior of the entire system is surprisingly daunting [11].

This article is dedicated to development and evaluation for hyper-heuristical particle swarm segmentation method for MR images.

2 Hyper-heuristical Particle Swarm Method for Image Segmentation

Hyper-heuristic is a search procedure aimed to automate the process of selection, combination and adaptation of several simple heuristics to effectively solve a problem. The main idea of the proposed hyper-heuristic method is the application of several heuristics, each of which has its weak and strong places, and then their use depending on the current state of the solution.

Formulation of the problem is similar to the statement of the segmentation problem using the ant colony optimization algorithm [3, 4] and consists of the following: for the given source images in the form of a set of pixels with visual properties such as brightness, color, texture, and also a certain size, level of noise, contrast and quality, it is necessary, within the available time resources, to find the markup of image onto K segments, which provides acceptable accuracy and quality of image recognition.

Hyper-heuristic particle swarm segmentation method is a management system, in the subordination of which there are three bioinspired heuristics: PSO-K-means [6], Modified Exponential PSO [5], Elitist Exponential PSO [7]. Each of the hyper-heuristics is applied depending on the quality of the source images: good quality (no noise and other artifacts); with the presence of noise; contrast images, blurred images. The control scheme of the method is shown on Fig. 1.

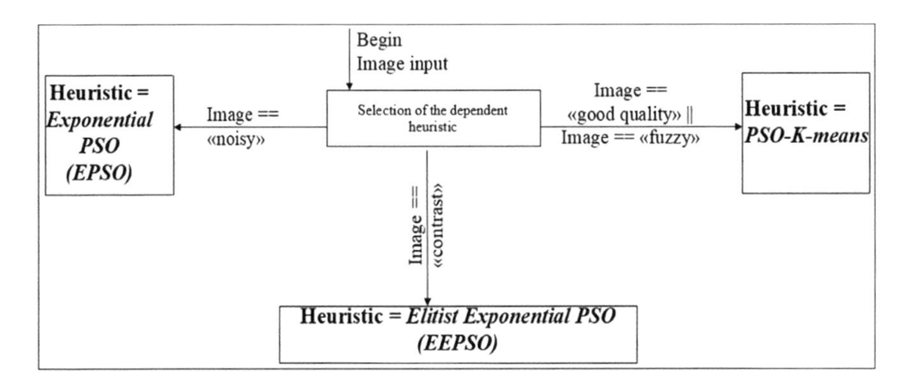

Fig. 1. Dependent hyper-heuristic management scheme

2.1 PSO-K-Means Segmentation Algorithm

When implementing the PSO-K-means heuristic, the n-dimensional search space is populated by a swarm of m particles (solution population). The coordinate of the i-th particle ($i \in [1{:}\ m]$) is described by the vector $x_i = (x_i^1, x_i^2, \ldots, x_i^n)$, that defines a set of optimization parameters: velocity $v_i(t)$ and position $x_i(t)$ at the moment of the time t. The position of the particle changes according to the following:

$$x_i(t+1) = x_i(t) + v_i(t+1) \tag{1}$$

where $x_i(0) \sim U(x_{\min},\ x_{\max})$.

For each position of the n-dimensional search space which the i-th particle has visited, the value of the fitness-function $f(x_i)$ is evaluated. In this case, the best value of the fitness-function is stored, as well as the position coordinates in the n-dimensional space corresponding to this value of the fitness-function.

Correction of each j-th coordinate of the velocity vector ($j \in [1{:}\ n]$) of i-th particle is performed according to the formula:

$$v_{ij}(t+1) = v_{ij}(t) + c_1 r_{1j}(t)[y_{ij}(t) - x_{ij}(t)] + c_2 r_{2j}(t)[\hat{y}_j(t) - x_{ij}(t)] \tag{2}$$

where $v_{ij}(t)$ – j-th velocity component ($j = 1,.., n_x$) of the i-th particle in the moment of time t; $x_{ij}(t)$ – j-th coordinate of the particle i; c_1, c_2 – constant acceleration coefficients, which define the effectiveness of PSO-K-means heuristic; $r_{1j}(t), r_{2j}(t) \sim U(0,1)$ – random values in the range of [0,1]; $y_{ij}(t)$ and $\hat{y}_j(t)$ – cognitive (the best position of i-th particle on coordinate j ($gbest$)) и and social components of the swarm.

The best position ($gbest$) at the moment of time (t + 1) is calculated as:

$$y_i(t+1) = \begin{cases} y_i(t) & if\, f(x_i(t+1) \geq f(y_i(t)) \\ x_i(t+1) & if\, f(x_i(t+1)) < f(y_i(t)) \end{cases} \tag{3}$$

The social component of the swarm $\hat{y}_j(t)$ at the moment of the time t is calculated as

$$\hat{y}(t) \in \{y_0(t), \ldots, y_{n_s}(t)\} | f(y(t)) = \min\{f(y_0(t), \ldots, f(y_{n_s}(t)))\} \quad (4)$$

where n_s – the common amount of the particles.

When correcting the velocity vector v_i the following modification of the formula (2) used:

$$v_{ij}(t+1) = \omega v_{ij}(t) + c_1 r_{1j}(t)[y_{ij}(t) - x_{ij}(t)] + c_2 r_{2j}(t)[\hat{y}_j(t) - x_{ij}(t)] \quad (5)$$

which includes the factor ω – inertia coefficient before j-th coordinate of velocity vector for i-th particle, so the speed changes more smoothly.

Each particle x_i represents the K centers as $x_i = (m_{i1}, \ldots, m_{ij}, \ldots, m_{iK})$, where m_{ij} represents the center of the cluster j for the particle i.

The fitness-function for each set of clusters is calculated as:

$$f(x_i, Z_i) = \omega_1 \overline{d}_{\max}(Z_i, x_i) + \omega_2(z_{\max} - d_{\min}(x_i)) \quad (6)$$

where $Z_{\max} = 2^s - 1$ for *s-bit* image; Z – the connection matrix, which represents connectivity between pixel and cluster for the particle i. Each element of the matrix z_{ip} represents, if the pixels z_p belongs to the cluster C_{ij} for the particle *i*. Constants ω_1 and ω_2 are user defined; $\overline{d}_{\max}$ – the maximum mean Euclidean distance from the particles to the associated clusters:

$$\overline{d}_{\max}(Z_i, x_i) = \max_{j=1..K}\left\{ \sum_{\forall Z_p \in C_{ij}} d(Z_p, m_{ij}) / |C_{ij}| \right\}, \quad (7)$$

$d_{\min}(x_i)$ – minimal Euclidean distance between pairs of cluster centers:

$$d_{\min}(x_i) = \min_{\forall j_1, j_2, j_1 \neq j_2} \{d(m_{ij_1}, m_{ij_2})\} \quad (8)$$

During the search, the particles exchange information among themselves, on the basis of which they change their location and speed of movement, using global and local solutions at the present moment. The global best solution is known for all particles, and in case of finding the best value, it is corrected.

The input data for PSO-K-means heuristic are: the number of clusters K; the number of the particles m, that directly perform the segmentation; the maximum number of iterations of the method n_{t0} to find a solution; acceleration coefficients c_1 and c_2 – personal and global components of the final particle velocity.

2.2 Modified Exponential PSO Segmentation Algorithm

Let's consider the following dependent heuristic *Exponential PSO(EPSO)*, included in the hyper-heuristic particle swarm method. Heuristic *EPSO* is preferred for noisy images segmentation. The input data is the same as in PSO-K-means heuristic. The EPSO algorithm steps is described below.

EPSO(images)
{
Initialization;
// Choice of the number of particles in a swarm m, personal and global acceleration coefficients c_1 and c_2, the maximum number of iterations of the method n_{t0}, the number of clusters K, determination of boundaries for the search parameters and parameters of fitness-function f under formula (6);
Creation an initial population of pixels (particles) distributed over the image area;
While(stop criteria is true)
{
Calculation pixel values based on the fitness-function (6);
If the current pixel (the current position of the particle) is better than the previous one,
than update it according to the formula (1);
Determination the best pixel (particle) in the population;
Pixel velocity correction according to formula (5).
Calculatation *pbest* (4) and *gbest* (3) – best local and global solutions for particles (pixels);
Moving the particle to a new position according to formula (1);
}

The peculiarity of EPSO heuristics is that, value of parameter ω is not constant. It decreases over each iteration from ω = 0,9 till ω_{min}= 0,4. The process becomes faster and less inert, if we modify the inertial weight ω and present in exponential form. Thus, the movement of particles becomes faster and less compact. ω calculated as

$$\omega = \omega e^{-\left(\frac{(\max - i)}{\max}\right)} + 0,4, \tag{9}$$

where *max* – the maximal number of iterations; i – current iteration.

2.3 Elitist Exponential PSO Segmentation Algorithm

The third dependent heuristic of the proposed particle swarm method is Elitist Exponential PSO (EEPSO). Its idea is as follows. In formula (2) r_1 and r_2 – are normally distributed random number. If inertia weight ω is small, there is possibility of premature process stagnation. Heuristic *EEPSO* uses grow rate coefficient β for each particle. If fitness value for particle on t-th iteration is greater than on (t-1)-th iteration than β increases. After the pbest value for all particles is calculated, the current best value of pbest is updated. The value of *gbest* is replaced to *pbest* value with the largest coefficient β.

The steps of the EEPSO heuristic are similar to the EPSO steps, except for the fitness-function:

$$f(x_i, Z_i, \beta_i) = \omega_1 \bar{d}_{\max}(Z_i, x_i) + \omega_2 \beta_{\max}(z_{\max} - d_{\min}(x_i)), \quad (10)$$

where $\beta_{\max}$ – maximal allowed rate of growth.

Input data is similar to input data used in *PSO-K-means* heuristic, but it is necessary to use additionally β_{max} – the parameter for maximal allowed rate of growth. Heuristic *EEPSO* can be used for the contrast MRI images segmentation.

3 Testing the Particle Swarm Segmentation Method

Developed hyper-heuristical particle swarm method for image segmentation was investigated. Investigation parameters are: Ossirix benchmark [8], set of 150 images with various initial conditions - good quality (no noise and other artifacts); with the presence of noise; contrast images, blurred images. Hyper-heuristical method was compared with k-means, Darwinian PSO [9] and FO Darwinian PSO [10] segmentation algorithms.

Results are presented by processing images with various initial conditions (Figs. 2, 3, 4 and 5).

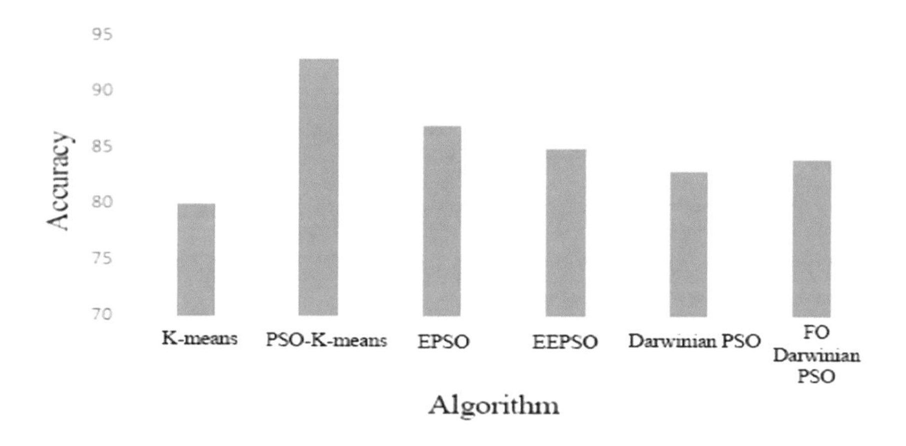

Fig. 2. Evaluation of segmentation accuracy for good quality images

There was established that each of the dependent heuristics in the hyper-heuristical particle swarm method shows the best results under certain initial conditions of the images. Use of dependent heuristics allows for all investigated types of images to show an average of 4.5% more accurate results than using Darwinian PSO, FO Darwinian PSO methods for MR images segmentation.

When comparing the operation time of segmentation methods (Table 1), it was established that the hyper-heuristic particle swarm method shows an average time of operation that is 2 times less than using the hybrid ant colony optimization method [3, 4], and the results of segmentation for images of good quality and blurred images are comparable.

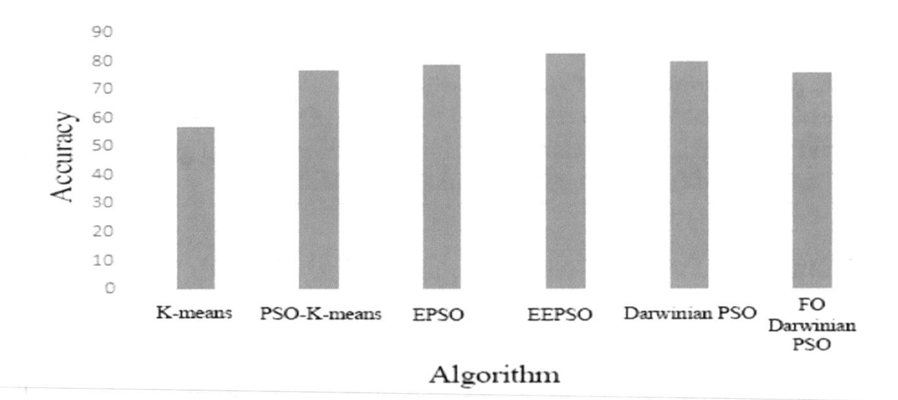

Fig. 3. Evaluation of segmentation accuracy for contrast images

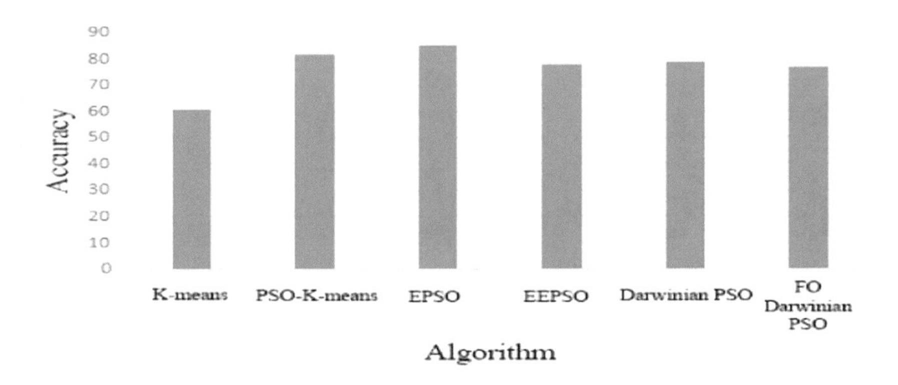

Fig. 4. Evaluation of segmentation accuracy for noisy images

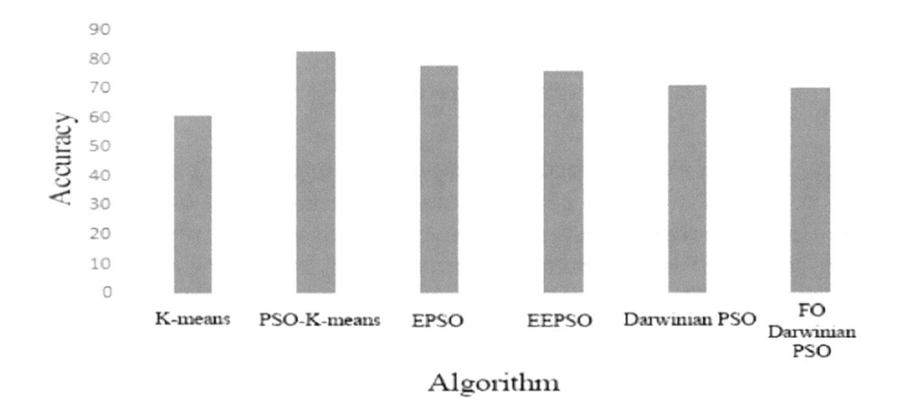

Fig. 5. Evaluation of segmentation accuracy for blurred images

Table 1. Comparative evaluation of segmentation methods operation time

Algorithm	k-means	Mean-shift	K-means PSO	Modified Exponential PSO	Elitist EPSO	Hybrid ACO	One cut	Fuzzy C-means	Grow cut	Random walker	Darwinian PSO	FO Darwinian PSO	JSEG
1	3,9	1,3	7,48	7,34	7,35	12,1	2,0	9,5	14,8	5,01	11,9	11,4	7,8
2	0,5	0,1	0,21	0,2	0,18	0,93	1,4	1,2	2,09	2,31	5,88	5,87	2,2
3	5,9	2,5	17,7	16,1	19,1	28,3	5,3	15	44,3	9,5	17,7	15,6	10
4	3,5	3,0	7,13	7,27	7,32	11,8	2,5	9,3	7,68	6,02	16,5	16,2	10
5	0,3	0,1	0,18	0,18	0,19	0,93	1,7	0,9	1,35	2,2	16,3	15,5	0,7

4 Conclusion

In the article was presented hyper-heuristic particle swarm method for MRI images segmentation, which is characterized by the choice, combination and adaptation in the process of searching the solution of several heuristics depending on the quality of the original images (without artifacts and distortions, noisy, blurred, contrasting), and uses dynamic inertia weight coefficient that allows to increase quality and reduce the time of image processing (on average by 9%). There were obtained comparative evaluations for the quality and processing time for the hyper-heuristic particle swarm segmentation method on different quality images from Ossirix benchmark comparing to more than ten competitive segmentation methods. Experimentally obtained scientific data indicates the following: a hyper-heuristic swarm method shows an average 2 times less working time than by using a hybrid ant method, and the results of segmentation for noisy and blurred images are comparable.

Acknowledgements. The research on creating the hyper-heuristic method for image segmentation was supported by the Russian Science Foundation (project 17-11-01254).

The research described in Section 2.3 of this article is partially supported by the state research 0073–2018–0003. Approbation of the results is partially supported by the Russian Science Foundation (project 16-07-00336).

References

1. Gonzalez, R.C., Woods, R.E.: Digital Image Processing, 3rd edn. Prentice-Hall, Upper Saddle River (2008)
2. Kennedy, J., Eberhart, R.C.: Particle swarm intelligence. In: Proceedings of the IEEE International Joint Conference on Neural Networks, pp. 1942–1948 (1995)
3. El-Khatib, S., Rodzin, S., Skobtcov, Y.: Investigation of optimal heuristical parameters for mixed ACO-k-means segmentation algorithm for MRI images. In: Proceedings of III International Scientific Conference on Information Technologies in Science, Management, Social Sphere and Medicine (ITSMSSM 2016). Part of series Advances in Computer Science Research, vol. 51, pp. 216–221. Published by Atlantis Press (2016). https://doi.org/10.2991/itsmssm-16.2016.72. ISBN (on-line): 978-94-6252-196-4
4. El-Khatib, S.A., Skobtcov, Y.A.: System of medical image segmentation using ant colony optimization. St. Petersburg State Polytech. Univ. J. Comput. Sci. Telecommun. Control Syst. **2**(217)–**3**(222), 9–19 (2015). https://doi.org/10.5862/jcstcs/1

5. El-Khatib, S.: Modified exponential particle swarm optimization algorithm for medical image segmentation. In: Proceedings of XIX International Conference on Soft Computing and Measurements (SCM 2016), St. Petersburg, 25–27 May 2016, vol. 1, pp. 513–516 (2016)
6. Saatchi, S., Hung, C.C.: Swarm intelligence and image segmentation swarm intelligence. ARS J. (2007)
7. Das, S., Abraham, A., Konar, A.: Automatic kernel clustering with a multi-elitist particle swarm optimization algorithm, science direct. Pattern Recogn. Lett. **29**, 688–699 (2008)
8. Ossirix image dataset. http://www.osirix-viewer.com/
9. Ghamisi, P., Couceiro, M.S., Ferreira, M.F., Kumar, L.: An efficient method for segmentation of remote sensing images based on darwinian particle swarm optimization. In: Proceedings of the IEEE International Geoscience and Remote Sensing Symposium – Remote Sensing for a Dynamic Earth (IGARSS 2012), pp. 20–28 (2012)
10. Ghamisi, P., Couceiro, M.S., Martins, M.L., Benediktsson, J.A.: Multilevel image segmentation based on fractional-order darwinian particle swarm optimization. IEEE Trans. Geosci. Remote Sens. **52**(5), 1–13 (2013)
11. Skobtsov, Y.A., Speransky, D.V.: Evolutionary Computation: Hand Book, 331 p. The National Open University “INTUIT”, Moscow (2015). (in Russian)

A Hybrid SAE and CNN Classifier for Motor Imagery EEG Classification

Xianlun Tang[1], Jiwei Yang[1](✉), and Hui Wan[2](✉)

[1] College of Automation, Chongqing University of Posts and Telecommunications, Chongqing 400065, China
849195277@qq.com

[2] College of Computer and Information Science, Chongqing Normal University, Chongqing 401331, China

Abstract. The research of EEG classification is of great significance to the application and development of brain-computer interface. The realization of brain-computer interface depends on the good accuracy and robustness of EEG classification. Because the brain electrical capacitance is susceptible to the interference of noise and other signal sources (EMG, EEG, ECG, etc.), EEG classifier is difficult to improve the accuracy and has very low generalization ability. A novel method based on sparse autoencoder (SAE) and convolutional neural network (CNN) is proposed for feature extraction and classification of motor imagery electroencephalogram (EEG) signals. The performance of the proposed method is evaluated with real EEG signals from different subjects. The experimental results show that the network structure can get better classification results than other classification algorithms.

Keywords: Electroencephalography (EEG) · Convolutional neural networks (CNN) · Motion imaging · Sparse autoencoder (SAE)

1 Introduction

Brain-computer interface (BCI) [1] systems do not depend on the peripheral nervous system of human body and muscle tissue, so BCI can provide a controlled means of communication for people with disabilities who are conscious but lose some of their bodily functions, helping them to actively interact with the outside world. Among noninvasive brain-computer, EEG is the most popular area of research, mainly due to its advantages of good time resolution, convenient use, good portability and low configuration cost. However, this technique is highly noise-sensitive so that it is difficult to filter out interference components of data. Before using EEG as a brain machine interface, a lot of training is needed to ensure the normal and effective work of the device. Therefore, the classification of EEG has been a key and difficult point in the development and design of brain-computer interfaces. Because of the obvious differences between EEG signals in different individuals and in different experiments, it is difficult to find features with good discriminability and representativeness. At the same time, it is very difficult to find a robust and accurate classification model because the

R. Silhavy (Ed.): CSOC 2018, AISC 764, pp. 265–278, 2019.
https://doi.org/10.1007/978-3-319-91189-2_26

EEG signals of a specific behavior are only a small fragment compared with the total sample and may also interfere with the interference caused by other behaviors. However, the classification problem has always been an area where deep learning is very good. Deep learning is distinguished from other machine learning by using the traditional manual extraction of features, but unsupervised or semi-supervised and step-by-step extraction. Therefore, deep learning modeling is not required a lot of prior knowledge. Multi-layer neural networks can not only represent the non-linear relationship between data, but also realize the level-by-level abstraction of complex data [2]. At present, deep learning has also been attracted a great deal of attention in the field of EEG classification research. Some people have tried to apply the deep learning method to EEG classification problems and have achieved good results [3].

At present, a lot of studies have been focused on the feature extraction and classification of EEG data such as auto regression (AR) [4]. There are some methods which observe the difference in power between the bilateral sides of hemisphere during the imagery. Some of these methods have good result but are too complex [5] or demand too much computing time [6], which is hard to apply in real-time applications. In addition, in Wu and Chen' study [7], they applied the common spatial pattern (CSP) to feature extraction and the linear discriminate analysis (LDA) to the classification of motor imagery EEG data and obtained an average recognition rate of 80% for two subjects. For BCI application, we need to recognize the fact that different subjects may have different mental conditions and hence, a general classification model is not suitable for all subjects. As a result, we need to find out a desirable system for the higher accuracy at hand and that is what we use here.

In this study, we propose a new network structure that mixes SAE and CNN to extract and classify the features of EEG. The model uses an unsupervised sparse autoencoder to train the convolution kernel of CNN. Sparse autoencoder is a three-layer fully connected neural network, and the training method of the encoder is unsupervised because label is not needed during training. The purpose of encoder training is to learn better representations of the input data so that it can be used to train better classifiers. In this paper, ZCA whitening is used to eliminate the ingenious correlation between adjacent pixels, and make the nodes of the hidden layer in the network much more than the input layer. During training, the number of activation nodes in the hidden layer is limited, and the sparse features are learned. After training the convolution kernel, we classify it by using logistic regression, then add a subsampling layer and a logistic regression layer to the convolution layer, which forms the network structure used in this paper.

2 An Overview of the SAE and CNN

2.1 An Overview of SAE Network Structure

Sparse autoencoder (SAE) [8] is a more commonly used method in deep learning. Its specific idea is to automatically obtain features through sparse coding, and then to train the acquired features in multi-level artificial neural networks. Because of its automatic processing, it does not require a very strong prior knowledge. Figure 1 is a typical sparse autoencoder network model.

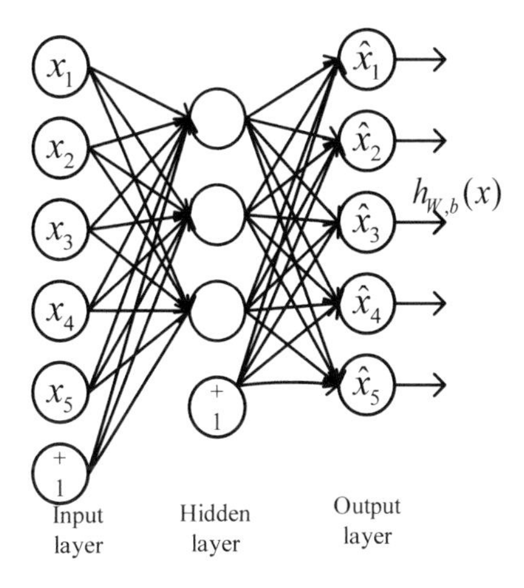

Fig. 1. Basic structure of SAE

SAE is a three-layer neural network [9], $\{x_1, x_2, \ldots x_5\}$ is the input vector, $\{\hat{x}_1, \hat{x}_2, \ldots \hat{x}_5\}$ is the output vector. The network has a hidden layer and $h_{w,b}(x)$ is the mapping of network from input to output. When the encoder is trained, output target value y is equal to the input vector x, making the training of the encoder can still use the traditional back propagation (BP) technology.

The autoencoder attempts to learn a mapping function $h_{w,b}(x)$ from input to output and makes $h_{w,b}(x) = x$. That is, the network tries to learn an identity function that makes $\hat{x} = x$. By using nonlinear activation functions in hidden layers and limiting the number of nodes in hidden layers, the network can learn interesting structures in the input data. If the input data has some rules, then the network can learn the better representation of the input data. For example, if you make the number of hidden layer nodes lower than the input layer, the network can learn a compressed representation of the input data. If the number of nodes in the hidden layer is much larger than that in the input layer, you may learn the sparse features of the input data.

2.2 An Overview of CNN

In 1960s, Hubel and Wiesel found that their unique network structure can effectively reduce the complexity of feedback neural network when studying neurons in the cortex of cat for local sensitivity and directional selection, and then proposed convolution neural network (CNN) [10]. As a multi-layer sensor, it is characterized by not needing a lot of manual processing, which has been widely used in image and video recognition. CNN is also suitable for EEG classification, especially for P300 wave classification of EEG signals [11]. Figure 2 shows a typical CNN structure.

The training process can be broadly divided into the following steps:

1. The original image of 32 * 32 is convoluted with 6 different convolution kernels and the size of convolution kernel is 5 * 5.

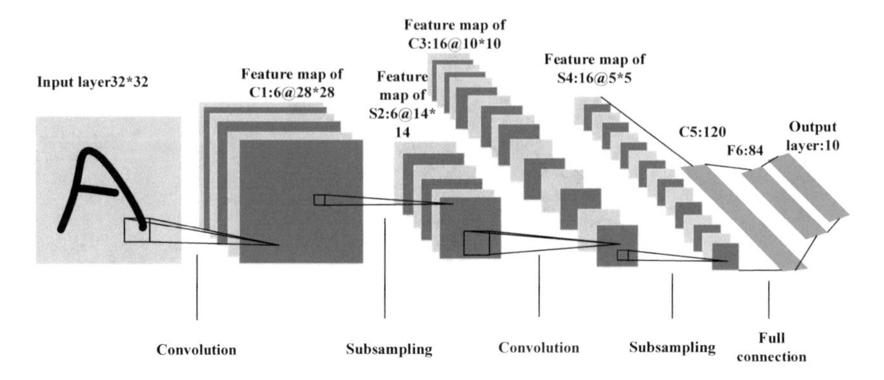

Fig. 2. A typical CNN structure.

2. The six feature maps of the first convolution layer C1 are respectively processed by pooling and the size of pooling template is 2 * 2. That is, 4 pixels (2 * 2) on the feature map in C1 are mapped to one pixel in the pooling layer S2, so the size of feature map is changed from 28 * 28 to 14 * 14 after pooling.
3. The six feature maps of the pooling layer S2 are further subjected to convolution processing, resulting in the production of the second convolution layer C3. This convolution uses a convolution kernel size of 5 * 5, and introduces 16 convolution kernels, resulting in 16 feature maps in the convolutional layer C3. The size of each feature map is 10 * 10.
4. The 16 feature maps of the second convolution layer C3 are processed by pooling and the size of pooling template is 2 * 2, which generates second pooling layer S4. S4 contains 16 feature maps, and the size of each feature map is 5 * 5.
5. The second pooling layer S4 is rasterized, and all feature maps are expanded and combined into a column vector to form a fully connected layer, which is involved in the final data training classification.

The specific process of convolution and subsampling can refer to Fig. 3.

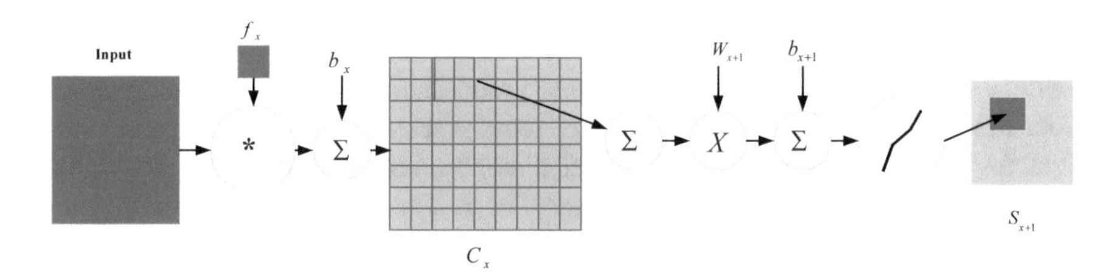

Fig. 3. The process of convolution and subsampling

3 A Framework that Combines SAE and CNN

Firstly, we trained sparse autoencoder for feature extraction. The weight matrix between the input layer and the intermediate hidden layer in the sparse autoencoder can be regarded as a feature extractor, which is also a convolution kernel in a convolutional neural network. In this paper, the convolution kernel is trained in a pre-training way, and the training process is supervised by backward propagation technique.

3.1 Preprocessing by ZCA Whitening

Autoencoder is a three-layer neural network, and the network structure of autoencoder in this paper is also the same. The network structure is shown in Fig. 1.

The sample sequence is randomly disrupted, and then a rectangular window of $5 * 5$ is used to capture 100 matrices for each sample as training input. In this paper, the matrix is pretreated by ZCA whitening [12]. Whitening can reduce the redundancy of the input. More precisely, the input of the learning algorithm has the following properties by whitening the data:

- The correlation between features is weak.
- All features have the same variance.

The process of preprocessing is shown in Fig. 4.

Fig. 4. The process of preprocessing

In order to use PCA, the data is firstly processed with zero mean:

$$\mu = \frac{1}{m}\sum_{j=1}^{m} x_j \tag{1}$$

$$x_j = x_j - \mu, j = 1, 2, .., m \tag{2}$$

The covariance matrix of the data is obtained by the use of the next formula:

$$\sum = \frac{1}{m}\sum_{j=1}^{m} x_j x_j^T \tag{3}$$

Find the eigenvectors and eigenvalues of the covariance matrix. According to the eigenvalues corresponding to the eigenvectors, the eigenvectors are composed of a feature vector matrix in the form of columns:

$$U = [u_1 \, u_2 \ldots u_n] \tag{4}$$

Assume that the eigenvalues from large to small are $\lambda_1, \lambda_2, .., \lambda_n$. U is used as a base coordinate, and x is expressed as a vector in the U space, and the vectors that are not related to each dimension are obtained:

$$x_{rot} = U^T x \tag{5}$$

In order to make each dimension of x_{rot} a standard deviation, the x_{rot} is transformed into a vector $x_{PCAwhite}$ with PCA whitening:

$$x_{PCAwhite,i} = \frac{x_{rot,i}}{\sqrt{\lambda_i + \varepsilon}} \tag{6}$$

ε is a regularized item which prevent eigenvalues from being too small. At present, each dimension of $x_{PCAwhite}$ has neither relevant nor same variance, which meets the requirements of our preprocessing, but we use ZCA whitening here:

$$x_{ZCAwhite} = U x_{PCAwhite} \tag{7}$$

In this paper, we do not want to reduce the data dimension in the process of preprocessing, so we use ZCA whitening which provides a basis for better access to the feature templates we want to obtain.

3.2 The Network Structure of Autoencoder

Autoencoder is a three-layer neural network, and the network structure of autoencoder in this paper is also the same. The network structure is shown in Fig. 1. The input of the network is the data after ZCA whitening. The number of hidden layer nodes in this article is about two times than the number of input nodes. The hidden layer nodes use the sigmoid activation function, the output layer uses the identity function, and the specific process and symbol are defined as follows:

Define the input as a vector x, the weight matrix between the input layers and hidden layers is W^1, the weight matrix between the hidden layers and output layers is W^2, and the sigmoid function is f, the output of the hidden layer is:

$$a^2 = f(W^1 x) \tag{8}$$

The output of the output layer is:

$$a^3 = W^2 a^2 \tag{9}$$

3.3 Definition of Loss Function of SAE

We take the mean square deviation of the difference between the input data and the output data as the main part of the loss function. Here the training set is $\{x^1, x^2, .., x^m\}$, and the function from the input to the output is mapped to $h_{w,b}(x)$. For the input x, the mean square deviation is defined as:

$$J(W, b; x) = \frac{1}{2} \left\| h_{W,b}(x) - x \right\|^2 \tag{10}$$

Based on the upper formula, the regular term of the weight value is added, and the following loss function is obtained.

$$J(W, b) = \left[\frac{1}{m} \sum_{i=1}^{m} J\left(W, b; x^i\right) \right] + \frac{\lambda}{2} \sum_{l=1}^{n_l - 1} \sum_{i=1}^{s_l} \sum_{j=1}^{s_{l+1}} \left(W_{ji}^l\right)^2 \tag{11}$$

Where n_l represents the number of layers of the network, here is 3, and s_l represents the number of nodes in the l layer. λ is a weight attenuation coefficient, which is a compromise between the weight loss and the mean square deviation loss.

For each input, if a node in the hidden layer is activated, we can approximately assume that this input has the characteristics represented by this node. For a given input, if the feature point is larger than the data dimension, most of the hidden layer nodes should be in an inactive state. SAE adds a sparse restriction based on the existing loss function in order to learn sparse features.

In this paper, we use $a_j^2(x)$ to represent the output value of the jth hidden layer node when inputting data x. This article defines a sparse parameter ρ and uses the average activation value of the hidden node to fit sparse parameters:

$$\hat{\rho}_j = \frac{1}{m} \sum_{i=1}^{m} \left[a_j^2\left(x^i\right) \right] \tag{12}$$

So the sparse constraint that we want to add is:

$$\hat{\rho}_j = \rho \tag{13}$$

For this reason, this paper adds an item to the loss function to punish the degree of deviation of $\hat{\rho}_j$ and ρ:

$$\sum_{j=1}^{s_2} \left[\rho \log \frac{\rho}{\hat{\rho}_j} + (1 - \rho) \log \frac{1 - \rho}{1 - \hat{\rho}_j} \right] \tag{14}$$

The loss function after adding sparse constraint is as follows:

$$J(W,b) = \left[\frac{1}{m}\sum_{i=1}^{m} J(W,b;x^i)\right] + \frac{\lambda}{2}\sum_{l=1}^{n_l-1}\sum_{i=1}^{s_l}\sum_{j=1}^{s_{l+1}}\left(W_{ji}^l\right)^2 + \beta\sum_{j=1}^{s_2} KL(\rho\|\hat{\rho}_j) \tag{15}$$

β is a sparse coefficient which is used to adjust the importance of sparse constraint. The updated gradient formula is as follows:

$$\frac{\partial}{\partial W_{ij}^l} J(W,b) = \left[\frac{1}{m}\sum_{i=i}^{m}\frac{\partial}{\partial W_{ij}^l} J(W,b,x^i)\right] + \lambda W_{ij}^l \tag{16}$$

In order to speed up the training, the LBFGS training method is used in this paper. In the training process, we set the sparse coefficient ρ as 0.03 based on experience, λ as 3e−3 for weight reduction and β as 5 for adjusting the importance of sparse constraints.

3.4 Construction of a Convolution Neural Network Classifier

Now that we have a convolution layer for feature extraction, we add a subsampling layer and a logistic regression layer to the convolution layer. A logistic regression classifier is trained for each label, and the classification results of each classifier are combined as the final classification result.

The network structure of this article is shown in Fig. 5.

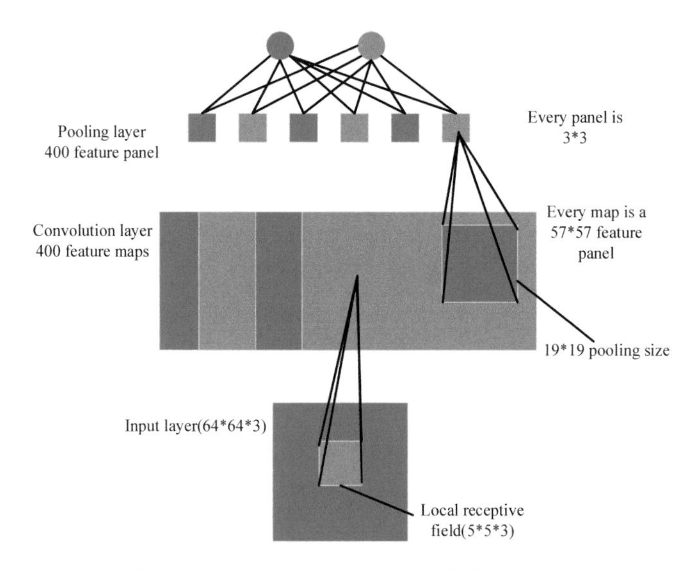

Fig. 5. The network structure of CNN classifier

The input size of network is 64 * 64. Next is a convolutional layer with 400 convolution kernels, each of which extracts features from a 5 * 5 sub-region of the input and the step of sub-region slides from the upper left corner to the lower right corner is 1. Each convolution kernel extracts 57 * 57 eigenvalues. After convolution, 400 57 * 57 feature maps are obtained. The sample area of the subsampling layer is 19 * 19, so every feature map is sampled by the feature map whose size is 3 * 3, and the way of sampling is mean sampling. The feature after sampling is the feature extracted from the EEG data at the end of this article, which is 400 3 * 3 feature maps. The feature maps are expressed as a vector, and the 400 vectors are combined to form a feature vector with a size of 3600. The vector dimension is large, but the features are sparse, most of the vector element values are close to zero.

4 Experimental Results and Analyses

4.1 Collection Process of EEG Dataset

In this paper we have used two data sets: one is from the BCI competition and the other is from our own experiment. Emotiv brain signal acquisition system [13] mainly uses noninvasive electroencephalograph technology to sense and learn each user's brain neuron signal mode, interprets their ideas, feelings and emotions, and then wirelessly transmits to the computer. Whether people are bored or excited, and whether they are focused on working tasks or in a relaxed state of mind, Emotiv EEG signal acquisition instrument can detect and sense people's emotions. At the same time, the device can detect the state of human muscles through the brain, so it can find a smile or frown, and some corresponding reaction. The most remarkable feature of the system is that EEG signals can be acquired, and users can quickly master the software system to understand different patterns of brain waves.

As can be seen in Fig. 6, the EEG signal acquisition device mainly includes electrode cap, electrode, electrode box, Emotiv wireless USB receiver, tampon and conductive liquid. The electrodes are equipped with cotton plugs, and a proper amount of conductive liquid is injected on the cotton stopper to ensure that the electrodes can induce the EEG signal. The electrode is placed in accordance with the international 10–20 standard electrode placement method, its placement is shown in Fig. 7. The location of the 14 electrodes of the device is AF3, F7, F3, FC5, T7, P7, O1, O2, P8, T8, FC6, F4, F8, AF4. Reference electrodes "CMS" and "DRL" are at P3 and P4 respectively. The sampling frequency of the EEG acquisition instrument is 128 Hz.

The experimental data were collected in a relatively quiet environment and the subjects conducted a number of experiments. The single experimental process shown in Fig. 8, the specific process is as follows: at the beginning of the experiment ($t = 0$), the subjects sat in a chair and remained relaxed. When $t = 2$ s, the subject heard a left-hand, right-hand or closed-eye sound, and then participants would perform the corresponding task of imagining left hand, right hand movement or closing eyes; When $t = 4$ s, the subject would hear a stop sound before the subjects stopped the experiment and prepared for the next experiment after a short break. During the whole experiment, EEG signals were collected from each subject for 120 experiments for each type of experimental task. the EEG signals from 3S to 4S were selected as the experimental samples, that was, each group had only 128 samples.

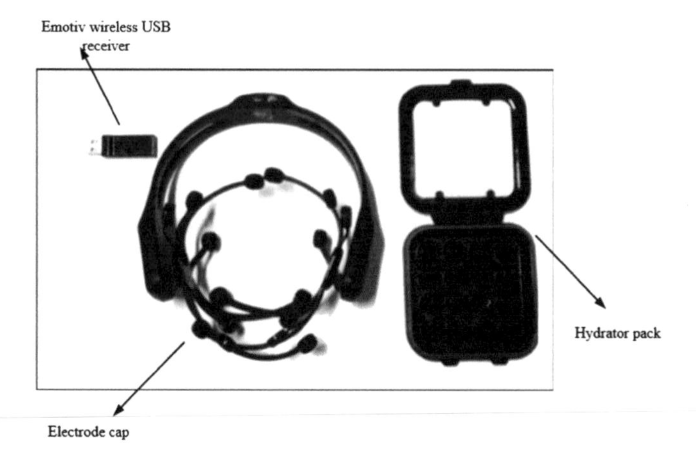

Fig. 6. Emotiv EEG signal acquisition device.

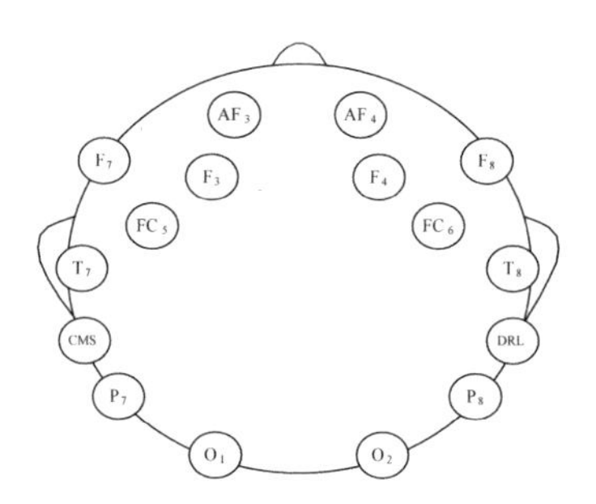

Fig. 7. The location of electrodes of the device

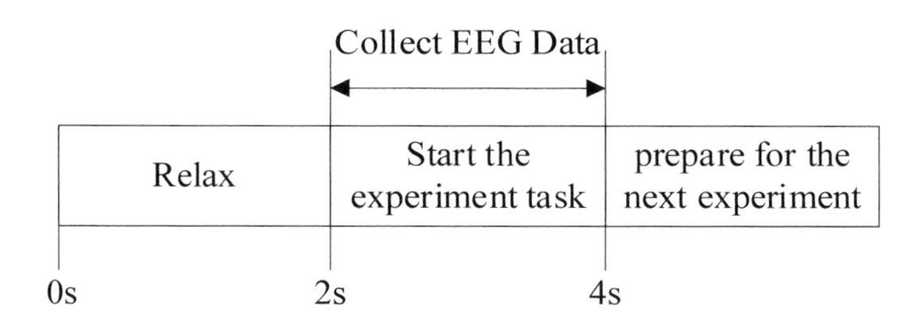

Fig. 8. The process of single experiment

4.2 Experimental Results and Analyses

In this study we use two datasets. The first data sets is the data sets 1 from the BCI competition IV [14], which were recorded from 4 human subjects performing motor imagery tasks. We consider the four human subjects that are referred to as "a" "b" "c" and "d". For each subject two classes of motor imagery were selected from the three classes left hand and right hand. The other data set, dataset 2, is recorded in our laboratory from 4 subjects, male, right-handed. There are 4 sessions in one experiment. Each session has 40 trials. The EEG data were recorded with a 32-channels Ag/AgCl electrode cap with 10–20 system. The sample rate is 1 k Hz, which is then down-sampled to 100 Hz.

It shows the average of the mean squared error with respect to the target vector over the four subjects, and individual mean square error for each of the four subjects to evaluate the performance. The last row shows the results obtained by our method on the BCI data sets 1.

The BCI competition results report the performance of different systems for the four human subjects: “a”, “b”, “c” and “d”. The BCI Competition results show the average mean squared error with respect to the target vector of four subjects, and the individual mean square error for each of the four human subjects to evaluate the performance of different methods. In particular, we include performance of the methods with ranks 1–5 (in column 1 of Table 1, we show these ranks). The last row depicts the performance of our method. The mean square errors obtained by the proposed method for subjects “a”, “b”, “c”, “d” are 0.56, 0.52, 0.59 and 0.48, respectively and the overall MSE over the four subjects is 0.575. We note here that we have considered only two classes (excluded the relaxed state). The algorithm is also applied to data on four subjects generated by our experiments and the average mean square error obtained is 0.57. The maximum classification accuracy rate that we have obtained for our data is 91%. The results discussed above suggest that the proposed method is a simple but effective method.

Table 1. Part of the BCI competition result from “Competition IV”.

Rank	a	b	c	d	mse
1	0.46	0.45	0.45	0.39	0.466
2	0.48	0.44	0.48	0.35	0.472
3	0.36	0.53	0.57	0.44	0.483
4	0.54	0.46	0.58	0.35	0.484
5	0.47	0.51	0.51	0.42	0.418
SAE+CNN	0.56	0.52	0.59	0.48	0.575

In the following, we analyze the decrease of minimum mean square error of SAE +CNN during training. In this experiment, we use the data of 4 channels, 12 channels and 16 channels to train respectively (Fig. 9).

The 16-channel classification accuracy is still better than the less-channel-selective data. We can see that although the overall trend tends to be flat after 30 iterations and the local fluctuations are still quite significant.

Finally, we compare the advantages and disadvantages of the three methods of SAE +CNN and SVM and shallow softmax neural network, and find that SAE+CNN has certain advantages in classification accuracy (Fig. 10).

The abscissa in the figure above represents 4 channels, 12 channels and 16 channels respectively. From the figure we can see, SVM generalization ability is stronger than

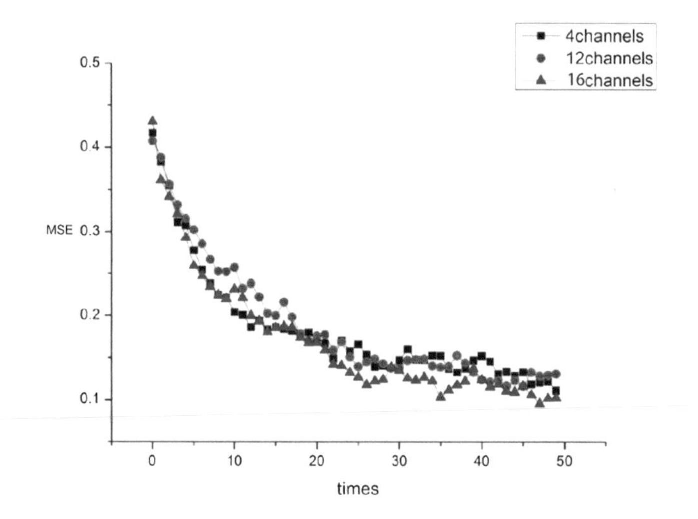

Fig. 9. The mean square error of different channels declines

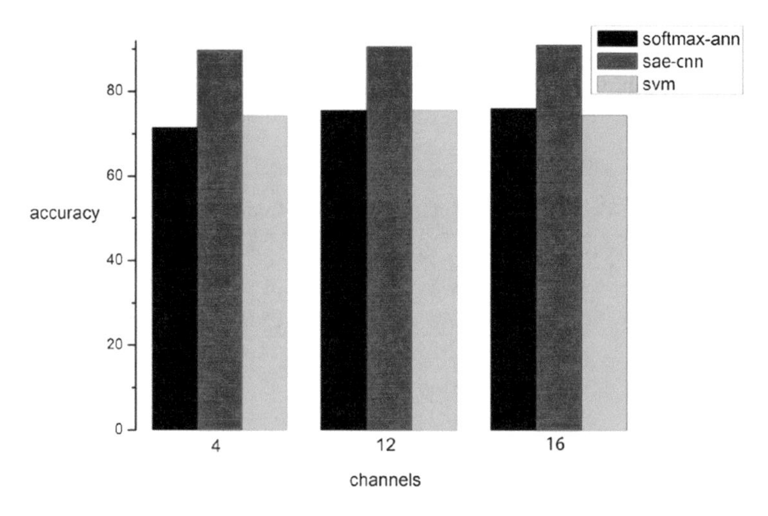

Fig. 10. Comparison of classification accuracy of different classifiers

Softmax neural network. When using less channel data, SVM classification is better than Softmax. While Softmax can better fit the data. When the number of channels increases, the accuracy of Softmax increases, while the accuracy of SVM has no obvious upward trend. Because the SAE+CNN model automatically obtains the more representative features, the accuracy is obviously improved, and its accuracy is positively correlated with the number of channels. The accuracy of the 4-channel was 89.8%, that of the 12-channel was 90.6%, and that of the 16-channel was 92%.

5 Conclusion

In this study, we have used a new way of hybrid SAE and CNN classifier for motor imagery classification based on EEG signals. A subject may not be able to do the motor imaginary task immediately after the visual cue is shown and actual time to start the motor imaginary action after the cue is shown could be different for different subjects. We have used a two-level cross-validation scheme to choose the most useful time interval that is related to the motor imaginary task for a given subject. We have used two data sets, one from the BCI competition and the other data set is generated in our laboratory. The experimental results show that the proposed study has some advantages for BCI applications. The method is simple, computationally efficient and yields good classification rates. Thus, the classification method developed here may be used to assess human imagery state in real-life applications.

Funding. This work was supported by the National Nature Science Foundation of China under Project 61673079, 61703068 and the Chongqing Basic Science and Advanced Technology Research under Project cstc2016jcyjA1919.

References

1. Vrushali, R.: Survey of brain computer interaction. Int. J. Adv. Res. Electr. Electron. Instrum. Eng. **2**(4), 1647–1652 (2013)
2. LeCun, Y., Bengio, Y., Hinton, G.: Deep learning. Nature **521**(7553), 436–444 (2015)
3. An, X., Kuang, D., et al.: A deep learning method for classification of EEG data based on motor imagery. In: Huang, D.S., Han, K., Gromiha, M. (eds.) Intelligent Computing in Bioinformatics, ICIC 2014, pp. 203–210. Springer, Cham (2014)
4. Robinson, D.A.: A method of measuring eye movement using a scleral search coil in a magnetic field. IEEE Trans. Biomed. Eng. **10**, 137–145 (1963)
5. Huang, H., Chuang, Y., Chen, C.: Multiple kernel fuzzy clustering. IEEE Trans. Fuzzy Syst. **20**, 120–134 (2012)
6. Johns, M.W., Tucker, A., Chapman, R.J., Crowley, K.E., Michael, N.: Monitoring eye and eyelid movements by infrared reflectance oculography to measure drowsiness in drivers, pp. 234–242 (2007)
7. Wu, S., Wu, W.: Common spatial pattern and linear discriminant analysis for motor imagery classification. In: 2013 IEEE Symposium on Computational Intelligence Cognitive Algorithms, Mind, and Brain (CCMB), pp. 146–151 (2013)
8. Su, Y., Li, J., et al.: Nonnegative sparse autoencoder for robust endmember extraction from remotely sensed hyperspectral images. In: IGARSS 2017 – 2017 IEEE International Geoscience and Remote Sensing Symposium, pp. 205–208 (2017)
9. Tabar, Y.R., Halici, U.: A novel deep learning approach for classification of EEG motor imagery signals. J. Neural Eng. **14**(1), 016003 (2016)
10. Lecun, Y., Boser, B., Denker, J.S., et al.: Backpropagation applied to handwritten zip code recognition. Neural Comput. **1**(4), 541–551 (1989)
11. Cecotti, H., Graser, A.: Convolutional neural networks for P300 detection with application to brain-computer interfaces. IEEE Trans. Pattern Anal. Mach. Intell. **33**(3), 433–445 (2011)

12. Hyvärinen, A., Hoyer, P.: Emergence of phase and shift invariant features by decomposition of natural images into independent feature subspaces. Neural Comput. **12**(7), 1705–1720 (2007)
13. Trigui, O., Zouch, W., Messaoud, M.B.: A comparison study of SSVEP detection methods using the Emotiv Epoc headset. In: Sciences and Techniques of Automatic Control and Computer Engineering, 21–23 December 2015
14. Francois, J., Rouck, A., Verriest, G.: Electrooculography in essential hemeralopia, vol. 204, pp. 1035–1045 (1971)

Semantic Bookmark System for Dynamic Modeling of Users Browsing Preferences

Syed Khurram Ali Shah, Shah Khusro(✉), Irfan Ullah, and Muhammad Abid Khan

Department of Computer Science, University of Peshawar, Peshawar 25120, Pakistan
khurram.ali28@gmail.com,
{khusro,cs.irfan,mabid}@uop.edu.pk

Abstract. Web search engines are successful at finding relevant resources; however, the search engine results page (SERP) list contains so many results that the intended ones are difficult to identify, therefore, requires manual user inspection and selection. A user may equally likely visit an already visited web page, which requires them to repeat the same search process. To deal with this issue, several re-visitation approaches are used, which include history, bookmarking and URL auto-completion. Among these, bookmarking is the most effective and user-friendly approach. Today, bookmarking is available in almost all the frequently used web browsers. However, they provide static and unstructured bookmarks. This makes it difficult for the end-user to easily manage, organize and maintain the hierarchical structure of bookmarks especially when their number exceeds a certain limit. Besides their hierarchical structure, the available bookmarking systems provide keyword-based searching with no exploitation of the semantics of the visited resources making it difficult to re-visit a required resource without reference to its context. We exploit Semantic Web technologies to devise a more effective, accurate and precise bookmarking service so that the cognitive efforts of the users could be reduced to a significant level. The proposed solution uses an extension that uses ontology to generate semantic bookmarks by tracking the user browsing activities, resulting into a more user-friendly re-visitation experience.

Keywords: Re-visitation · Bookmarks · Ontology · Web browser Web page

1 Introduction

The Web has become one of the most important platforms for providing access to information regarding almost any aspect of life. In response to search query, web search engines employ efficient Information Retrieval algorithms to bring the most relevant resources, which the users select and browse. Sometimes, a user may recall the memory through several cues, but none of them is enough to locate the previously visited web pages. Therefore, most often users bookmark web pages using the built-in bookmarks functionality of web browsers. According to Papadakis et al. [1], more than 45% of web pages are revisited by the same user. Therefore, these selected web pages,

R. Silhavy (Ed.): CSOC 2018, AISC 764, pp. 279–287, 2019.
https://doi.org/10.1007/978-3-319-91189-2_27

if bookmarked automatically or semi-automatically, will not only facilitate users in browsing but also constitute their personal information space [2, 14]. In addition, the manual bookmarking requires an extra cognitive effort from the user, which interrupts the interaction flow of web page and requires users to think whether they will visit the page again [3]. It is therefore, required that the user activities during Web navigation are tracked and exploited in automatic or semi-automatic bookmarking of web pages having greater probability of re-visitation [4]. This research work aims to develop a semantic bookmarking system in the form of a Chrome browser extension that tracks the user activities and exploits the metadata associated with the web pages using our own web page ontology [11] so that bookmarks could be automatically generated or recommended. Section 2, covers related work, Sect. 3 covers the proposed solution, development considerations, and experimental settings, Sect. 4 presents results, and Sect. 5 concludes our discussion and presents some future work.

2 Related Work

To reduce the cognitive overload in using current bookmarking tools, several revisitation tools and features have been incorporated into web browsers. These include the back/forward button; history list; and bookmarking. Among these, according to [5], bookmarking is the most popular revisitation technique.

The possible re-visitation of a page depends on several factors including its usefulness, quality, usefulness in the near future, user personal interest in it, term frequency, etc., [6]. Reasons of using bookmarks include reducing the cognitive overload in managing URLs; developing information lists for personal use; and facilitation in re-visitation. However, sometimes, users miss or forget to bookmark the candidate web pages, which is difficult to search in the history list especially when number of visited pages is too many. Therefore, some automatic ways are required for bookmarking such web pages.

Several bookmark management solutions are used as third-party tools to automate bookmarking. HIBO [7] bookmarks manager is used locally to organize the bookmarks. It downloads the URLs, parses and summarizes the page using the lexical chaining technique to extract the thematic words from the web page URL. It maps these words with the hierarchy of categories by following a fixed hierarchy structure to fit the page category. However, users are unable to create additional categories.

Bookmarking tool [8] stores the hyperlinks in user directory. Its extension displays bookmarks in arc, timeline and tree views. In the arc view, URLs of the bookmarked pages are displayed in the order of their frequency of visits. These distributions are divided into five arcs on screen and most frequently visited bookmarks are presented in larger shape. In timeline view, the bookmarks are shown in squares. It shows the bookmarks in most recent month-wise along with the activity performed on the bookmarks. In tree view, the bookmarks are represented by circles while folders are represented using squares with rounded corners.

Smart bookmark [9] is a Firefox extension that bookmarks web pages using their URL and provides revisitation using script commands. It consists of graphical visualization combining textual information and screenshots. It tracks the user actions along with the description of the page. The description part includes the text labels, location of DOM elements (by using XPath); and the screenshots. It detects the overall change [10] between the existing page P and the re-fetched page R and calculates the total weight of the difference between P and R by using W_P and W_R. The total difference is computed as the cost of operations required for transforming P to R divided by $W_P + W_R$ [10]. If the change is greater than 10%, then R is concluded as new page, otherwise both are treated same [9].

At present, users use either the web browser's built-in bookmarking function or social bookmarking sites including Delicious, Diigo, etc., for re-visitation proposes. The existing browsers including Google Chrome, Mozilla Firefox, Microsoft Edge, etc., allow users to manually bookmark web pages and assign different tags [5]. However, manually bookmarking web pages introduces cognitive overload and prevents the users to exploit fully revisitation as most of the time, important web pages are missed to be bookmarked while the user is browsing the Web. Some automatic ways have been introduced, however, their limited capturing and exploiting of the associated metadata and user activities, make these tools less useful. Therefore, these automatic methods should be augmented with semantics-aware bookmarking features. This demands the capturing the user behavior by recording the metadata including title, meta keywords, headings, bold letters and paragraphs etc., of the visited web pages. The authors [15] did this in clustering user behaviors by adopting an unsupervised click-stream analysis considering several factors including e.g., most frequently visited web pages, longer session on a web page, copying a text etc. [15]. One possible solution can be developing a web browser extension to capture all the relevant metadata about a page visited by the user to model semantically the process of revisitation using a web page ontology, which we have developed in [11]. In the next Section, we propose the development of such a system in the form of an open-source extension for Chrome web browser.

3 Methodology

The proposed solution consists of an extension, a bookmarks repository for storing implicit and explicit bookmarks, an inference engine for extracting user interests and preferences from the recorded activities, and Web Page Ontology [11] for storage, generating and recommending bookmarks semantically. Figure 1 graphically illustrates bookmarks generation process. In the following sub-sections, we present briefly its working, experimental setting, and evaluation.

3.1 Parsing Web Pages

Parsing involves extracting useful content from the web pages and the associated metadata that have been frequently visited by the user or on which the user spends much of their time. In the Chrome web browser, parsing can be performed after the page has been loaded, where it is loaded to Document Object Model (DOM). The DOM tree helps in parsing the page content to populate the ontology with the extracted useful information.

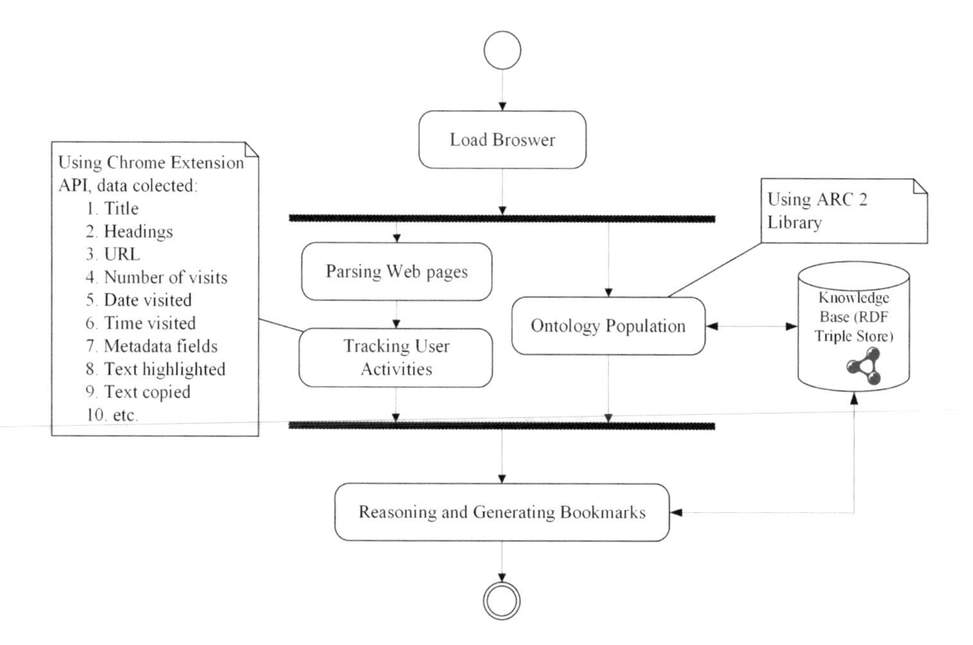

Fig. 1. The proposed solution

3.2 Tracking User Activities

We developed the extension to record user activities while browsing and capturing important metadata in order to use it in inferencing, generating, and recommending bookmarks semantically. The recorded data includes clicks on link, form submission, text copy, etc. We used a JavaScript event handler to record opening web pages, navigation between tabs, time and number of opening and closing web pages, form filling, time spent on a certain page and window, choosing items from drop-down lists, checking checkboxes, and selecting radio buttons, etc. The extension is capable of representing user actions and associating appropriate labels in order to populate the ontology. Table 1 lists some of the labels that the extension records and associates with web page elements in order to detect, record, and store the user activities in the ontology. Using DOM tree, we produce a simple file to store information about the candidate web page in the special directory within the Chrome folder.

3.3 Knowledge Base

After processing, the recorded data populates the ontology (Web Page Ontology [11]), which covers all the web page DOM elements and user activities.

3.4 Reasoning and Generating Semantic Bookmarks

To generate semantic bookmarks, we need reasoners that could make inferences on the stored knowledge using SPARQL queries. This component is coded using the ARC2,

Table 1. Summary of labels tracked by the extension

Element type	Label to be associated with/Process triggered
Hyperlinks in the page	The text of a web link
Button	The content of button or value associated with the button
Drop-down select list box	The text content with the list box
Image within the web page	ALT or title associated with the image tag
Radio button or checkbox	The text node associated with the radio button or checkbox
Text box	The extension will look all the text which is enter with the text box
Meta data	The text within the metadata element
Title of web page	The content within the title tag of web page
Heading within the web page	The text within the heading tag of html web page, e.g. <h1>, etc.
Paragraph text	The text within the paragraph tag of html web page, e.g. <p>

which is built-in PHP library to parse Web Ontology Language (OWL) file of the populated Web Page Ontology. ARC2 is a flexible RDF framework, which is used for the Semantic Web retrieval process. It also provides the SPARQL query and parsing the file for WAMP system [15]. Figure 2 depicts the snippet of the source code to query the knowledge base using SPARQL and ARC2 framework. The reasoners also exploit the user profile, which the extension generates, maintains, and updates continuously so that together with the data in the knowledge base, semantic bookmarks could be generated and recommended/presented to the user at the end of the session if the user misses it to explicitly bookmark it.

3.5 Experimental Settings

Following are some details about experimental settings and evaluation metrics used.

Data Collection. The extension, after adding to a Google Chrome browser, recorded browsing records while browsing the Web through Google Web Search engine. For every information need, we first formulated a query (keyword-based), which we run on the search engine and browse the resulting relevant pages so that the extension could associate it to that keyword. The recorded data includes the URL of the visited web page, which is also known as a candidate bookmark and the time stamp when user visits the web page. This data is also helpful in understanding the user navigation behavior [12]. In addition, the extension also recorded other user activities and associated metadata implicitly. The Web Page Ontology exploited this data to generate semantic bookmarks to add them to a given category. To evaluate and confirm the accuracy of the automatically generated semantic bookmarks, we searched the bookmarked pages against the corresponding keyword to see whether they have been accurately bookmarked in relation to that keyword and assigned the appropriate category. Currently, we have manually defined certain categories using some closed-World assumptions. However, we are working on it and soon will be able to dynamically generate and recommend categories based on the content and navigation

```
$array_one = comma_separated_to_array($bookmark);

$length = count(array_unique($array_one));


$a ='';
$result3 = $store->query('
PREFIX rdf: <http://www.w3.org/1999/02/22-rdf-syntax-ns#>
PREFIX owl: <http://www.w3.org/2002/07/owl#>
PREFIX xsd: <http://www.w3.org/2001/XMLSchema#>
PREFIX rdfs: <http://www.w3.org/2000/01/rdf-schema#>

SELECT  ?link ?label ?metadata ?heading ?catLabel
    WHERE { ?URL rdfs:label ?label.
            ?URL rdfs:externalLink ?link.
            ?URL rdfs:hasMetaData ?metadata.
            ?URL rdfs:hasHeading ?heading.
            ?URL rdfs:hasCatLabel ?catLabel.
    }
');

$result4 = $store->query('
PREFIX rdf: <http://www.w3.org/1999/02/22-rdf-syntax-ns#>
PREFIX owl: <http://www.w3.org/2002/07/owl#>
PREFIX xsd: <http://www.w3.org/2001/XMLSchema#>
PREFIX rdfs: <http://www.w3.org/2000/01/rdf-schema#>

SELECT ?catLabel
    WHERE { ?URL rdfs:hasCatLabel ?catLabel.
    }
');
```

Fig. 2. Querying the ontology using ARC2 and SPARQL

details of the user. These results are presented in the next Section, in terms of precision and recall.

Evaluation Metrics. The selection of the metrics depends on the nature of data and the problem at hand [13]. For evaluation of the proposed system, we used precision and recall as evaluation metrics. The reason behind the selection of these metrics is their wide applicability and use in similar problems in the state-of-the-art literature. To understand precision and recall, consider A as the number of relevant URLs bookmarked; B as the total number of relevant URLs that should be bookmarked; and C as total number of URLs bookmarked by the proposed solution. Equation (1), gives the precision (P), whereas Eq. (2) gives the recall (R):

$$P = \frac{A}{C} \tag{1}$$

$$R = \frac{A}{B} \tag{2}$$

4 Results and Discussion

Table 2 reports the recorded data for keyword used and the corresponding A, B, C, P, and R computed from the available navigation records.

Table 2. The summary of the labels (Keyword ID) tracked by the extension with their corresponding precision and recall

Keyword ID	A	C	B	P	R
1	23	70	30	0.328571429	0.766666667
2	26	66	32	0.393939394	0.8125
3	29	87	39	0.333333333	0.743589744
4	36	58	39	0.620689655	0.923076923
…					
55	24	83	26	0.289156627	0.923076923
56	21	65	24	0.323076923	0.875
57	28	96	33	0.291666667	0.848484848
58	22	79	31	0.278481013	0.709677419
59	27	80	33	0.3375	0.818181818
60	28	76	34	0.368421053	0.823529412

After sorting the values of both precision and recall and plotting precision against recall, we get Fig. 3, in which the almost straight line shows precision of semantically bookmarked URLs at different levels of recall. The figure clearly demonstrates moderate improvements in precession as the value of recall increases. However, for a clearer picture of the scenario, it is necessary to increase the volume of dataset. We plan to increase the dataset as our future work.

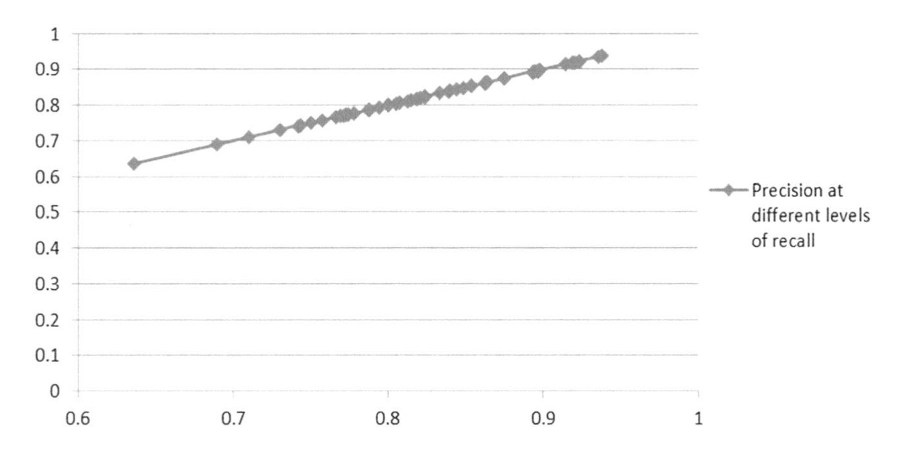

Fig. 3. Precision of the proposed solution at different levels of recall

5 Conclusion and Future Work

In this research work, we presented the design and development of a semantic bookmarking system that generates and recommends bookmarks semantically by exploiting the page content, associated metadata and user navigation behavior. With the help of a web browser extension, it tracks and records all the browsing activities of the user, populates the associated web page ontology, and makes inferences for possible bookmarks. The work is one of the first steps, to the best of our knowledge, towards exploiting user navigation behavior ontologically and semantically for a more enhanced re-visitation experience based on bookmarks. This is a work in progress, where much more is to be explored and added to the proposed semantic bookmark solution including the automatic generation and recommendation of categories so that users could be assisted in efficiently categorizing their bookmarks in a more managed manner and associating the generated bookmarks to their appropriate category. In addition, the prototypical implementation will be extended to other open-source browsers so that more users could benefit from it. We also plan to work in ranking the generated semantic bookmarks as well as searching bookmarks especially when bookmarking data increases with the passage of time.

References

1. Papadakis, G., Kawase, R., Herder, E., Nejdl, W.: Methods for web revisitation prediction: survey and experimentation. User Model. User-Adap. Inter. **25**(4), 331–369 (2015). https://doi.org/10.1007/s11257-015-9161-7
2. Staff, C.: Boosting Bookmark Category Web Page Classification Accuracy using Multiple Clustering Approaches (2018). https://www.researchgate.net/publication/228897005_Boosting_Bookmark_Category_Web_Page_Classification_Accuracy_using_Multiple_Clustering_Approaches
3. Abrams, D., Baecker, R., Chignell, M.: Information archiving with bookmarks: personal web space construction and organization. In: Proceedings of the CHI 1998 Conference on Human Factors in Computing Systems, Los Angeles, California, pp. 41–48 (1998)
4. Wang, G., Zhang, X., Tang, S., Zheng, H., Zhao, B.Y.: Unsupervised clickstream clustering for user behavior analysis. In: Proceedings of the 2016 CHI Conference on Human Factors in Computing Systems, San Jose, California, pp. 225–236 (2016)
5. Al-Mamun, A., Noori, S.R.H.: Vismarkmap – a web search visualization technique through visual bookmarking approach with mind map method. Int. J. Adv. Comput. Sci. Appl. **7**(11) (2016). http://dx.doi.org/10.14569/IJACSA.2016.071103
6. Toda, M.: Web refinding support system based on process recollective activity. Int. J. Inf. Technol. Comput. Sci. **4**(5), 1–7 (2012)
7. Kokosis, P., Stamou, S.: HiBO: a system for automatically organizing bookmarks. In: Proceedings of ACM/IEEE Joint Conference on Digital Libraries, Denver, USA, pp. 155–156 (2005)
8. Bry, F., Wagner, H.: Collaborative Categorization on the Web: Approach, Prototype, and Experience Report, pp. 1–14 (2003)
9. Hupp, D., Miller, R.C.: Smart bookmarks: automatic retroactive macro recording on the web. In: Proceedings of the 20th Annual ACM Symposium on User Interface Software and Technology, Newport, Rhode Island, pp. 81–90 (2007)

10. Hu, S., Wen, J., Dou, Z.: Following the dynamic block on the web. J. World Wide Web **19** (1), 1–25 (2015)
11. Ali, S., Khusro, S.: POEM: Practical ontology engineering model for semantic web ontologies. Cogent Eng. **7**(1) (2016)
12. Kellar, M., Watters, C., Shepherd, M.: A field study characterizing web-based information-seeking tasks. J. Am. Soc. Inform. Sci. Technol. **58**(7), 999–1018 (2007)
13. Fenstermacher, K.D., Ginsburg, M.: A lightweight framework for cross-application user monitoring. Computer (Long. Beach. Calif) **35**(3), 51–59 (2002)
14. Wang, G., Zhang, X., Tang, S., Zheng, H., Zhao, B.: Unsupervised clickstream clustering for user behavior analysis. In: Proceedings of the 2016 CHI Conference on Human Factors in Computing Systems, pp. 225–236 (2016)
15. Raufi, B., Ismaili, F., Ajdari, J., Ferati, M., Zenuni, X.: Semantic resource adaptation based on generic ontology models. In: 2014 9th International Conference on Software Paradigm Trends (ICSOFT-PT), Vienna, Austria, pp. 103–108 (2014)

Models, Algorithms and Monitoring System of the Technical Condition of the Launch Vehicle "Soyuz-2" at All Stages of Its Life Cycle

Aleksey D. Bakhmut[1(✉)], Kljucharjov A. Alexander[1], Aleksey V. Krylov[2], Michael Yu. Okhtilev[2], Pavel A. Okhtilev[2], Anton V. Ustinov[1], and Alexander E. Zyanchurin[2]

[1] St. Petersburg State University of Aerospace Instrumentation, Saint-Petersburg, Russia
adbakhmut@gmail.com

[2] St. Petersburg Institute for Informatics and Automation of the Russian Academy of Sciences (SPIIRAS), Saint-Petersburg, Russia
pavel.oxt@mail.ru

Abstract. A model complex and algorithms for assessing the technical condition (TC) and reliability of the launch vehicle (LV) "Soyuz-2" with the decision support (DS) for managing its life cycle (LC) is considered in the article. On the basis of the analysis of modern problems and the requirements for the efficiency, quality and reliability of the assessment of the LV and the reliability of LV, it was concluded that it is necessary to use the new intelligent information technology (IIT), presented in the article, when designing the automated monitoring systems of the condition and DS for managing the LC LV "Soyuz-2". As a theoretical basis for this technology, the modification of the generalized computational model (GCM) as a knowledge representation model, allowing to build simulation-analytical model-based complexes for monitoring conditions and managing complex organizational and technical objects (COTO), is considered.

Keywords: Condition monitoring · Complex organizational and technical object · Artificial intelligence system · Control theory structural dynamics · Parallel distributed computing · Distributed computing · Intelligent user interface · Cognitive image

1 Introduction

The current stage in the development of the LV and space industry is characterized by an increasing the complexity of space complexes (SC) and its management processes. So, decision makers face difficulties associated with low operational efficiency, quality and reliability in assessing TC and reliability of LV in managing its LC [1].

LV and organizations, associated with it, belong to the class of COTO in the considered subject area (SA). It is characterized by extremely large volumes of

R. Silhavy (Ed.): CSOC 2018, AISC 764, pp. 288–297, 2019.
https://doi.org/10.1007/978-3-319-91189-2_28

information resources of various nature, incomplete and undetermined information about the condition, the presence of various types of structures that change over time: organizational structures of enterprises, product structures, and so on [2, 3].

Nowadays there is no solution of this problem if not to use effective intelligent complex tools for automating the processes of monitoring the condition and the DS for the management of the LC of COTO, capable for solving a wide range of problems.

When dealing with SC, automated systems for monitoring condition and DS, according to GOST 1410-002-2010, should function within the system of information of TC and reliability of SC and the products included in their composition [4].

There is a need to organize distributed computing as part of the monitoring system because of the territorial remoteness of organizations, in which information is generated about the condition of the LV "Soyuz-2" at all stages of its LC.

The decision maker is forced to make decisions considering the large amount of information for analyzing. So, it is required to solve the problem of increasing the effectiveness of human-machine interaction through the development of an intelligent user interface for the monitoring system.

2 Modeling of the Subject Area

The object can be represented by a set of semantically related entities [2]. Then the change of one structure leads to change in others dynamically. The account of mutual influence is possible only on the basis of polymodel description of the SA. In this case, each structure is represented by its model [2].

From the position of the artificial intelligence (AI) theory one of the ways of constructing the information model of the SA is the formation of a conceptual polymodel description of the SA on the basis of facts, rules, and data composition and structure of this SA given by an expert non-programmer. Such a conceptual description due to the use of some data and knowledge representation models (DRM and KRM) that allow to form the corresponding database (DB) and knowledge base (KB) can be interpreted in the automatic mode as a finite set of parameters and computational algorithms in program schemes [5, 6]. There are a number of approaches that offer synthesis of software solutions on the basis of a model description of an SA [7–9].

3 Data Representation Model

DRM is a conceptual semantical model that provides internal interpretability and structured information about TC and reliability of COTO. Many processes within SOTO are associated with the formation of documents as a basis of organizational and technical information. The documental model can be represented as a frame, the attributes of which are metadata, and the slots define relations between it [10].

In the proposed SA, the DRM can be represented as follows:

$$M_D = \; <Id, Product, X^{Cl}, X^A, X^V, X^{St}, Cl, A, V, St> \tag{1}$$

where $Id = \{id_i | i \in I_D\}, Id \in \mathrm{N}$ – finite set of document identifiers; $Product = \{product_s | s \in I_{Product}\}$ – set of product elements; $X^{Cl} = \{x_j^{Cl} | j \in I_{X^{Cl}}\}$ – finite non empty set of document classes; $X^A = \{x_k^A | k \in I_{X^A}\}$ – finite non empty set of document attributes; $X^V = \{x_n^V | n \in I_{X^V}\}$ – finite set of attribute values, values can be of any nature; $X^{St} = \{x_m^{St} | m \in I_{X^{St}}\}$ – finite set of document statuses; $Cl : id_i \rightarrow x_j^{Cl}$ – mapping function that maps each document to a specific class; $A : x_j^{Cl} \rightarrow X_k^A, X_k^A \in X^A$ – mapping function that maps each class to certain subset of attributes X_k^A, X_k^A of different classes can intersect; $V : x_k^A \rightarrow X_n^V, X_n^V \in X^V$ – mapping function that maps each attribute x_k^A to a certain subset of its values X_n^V, X_n^V of different attributes can intersect; $St : <x_k^A, X_n^V> \;\rightarrow x_m^{St}, x_m^{St} \in X^{St}$ – mapping function that maps attribute and a subset of its values to a certain status x_m^{St}.

According to the described DRM of SA, a DB scheme can be formed on the basis of one of the notable notation for constructing “entity-relationship” models [11, 12]. Following the concept of “programming without programming” [7], it should be noted that there are approaches that allow to synthesize a DB on the basis of DRM of SA. So, both the DB structure and the application programs for data processing within the corresponding software complex (SwC), interacting with it, can be synthesized automatically.

For the interaction of KRM and DRM, it is necessary to specify some query language that allows the expert to form expressions (models) interpreted as schemes of application programs for receiving, adding and changing data to the DB, regardless of the selected DRM (relational, hierarchical, etc.) and technical features of the DB.

One of the most common DRM is the relational model, based on the Codd Relational Algebra [16]. Structured Query Language (SQL) is most often used to interact with relational DB in practice (and its dialects, depending on the implementation of a particular DB management system).

We can give the following example for the problem. The following expression $A : x_j^{Cl} \rightarrow X_k^A, X_k^A \in X^A$ can be constructed for a mapping function that maps each document class to a subset of attributes using the SQL:

```
SELECT attribute.name FROM class, attribute,
class2attribute
WHERE class.name = x[j, Cl] and class.name =
class2attribute.class_name
  AND attribute.name = class2attribute.attribute_name.
```

Execution of the query involves converting the specified expression into some internal representation, optimal for processing on a computer. As a result, syntactical constructions that approximate the expression to natural language from consideration are completely excluded [13, 14]. The query is represented by the Codd Algebra as a relational expression:

$$\pi_{name}\left(\left(\sigma_{name=x_j^{Cl}}(Class)\right) >\!\!<_{name=class\_name} class2attribute\right) >\!\!<_{attribute\_name=name} Attribute. \quad (2)$$

Then the syntactical tree of the request is constructed, which defines the relations between relational operators in an explicit form [13]. This tree is a schema of the query execution program (Fig. 1).

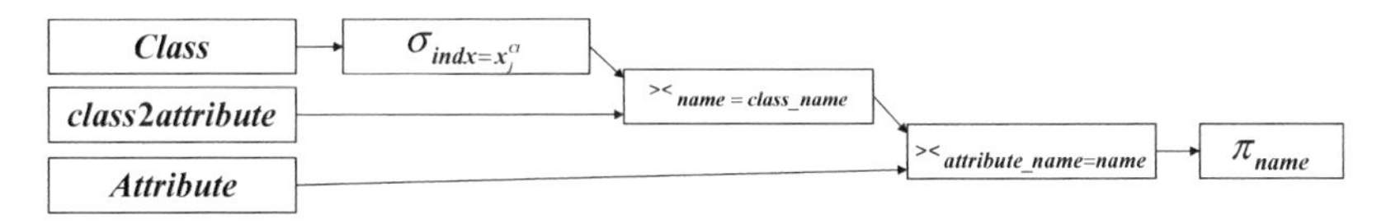

Fig. 1. Example of the schema of the query program for a set of attributes by the document class.

In relational DRM, only binary relationships can be established between entities. This fact leads to the conclusion that the schema of the query execution program is a binary tree.

Relational DRM is a special case of graph as well as aggregate DRM [15]. Consequently, the definition of some formal grammar and the corresponding query language on graph DRM provides a basic possibility of constructing schemes for executing queries for any DB, regardless of the type of DRM. Such an opportunity is essential for the synthesis of monitoring systems of TC and the reliability of COTO, where elements and subsystems of COTO can be represented by arbitrarily complex structures. This determines the need to use a complex of different DRM, and hence, an universal language for describing the models of interaction between KRM and DRM.

4 Condition Assessment Model of a COTO Element

The approach proposed in this paper is based on the application of a G-model complex, which is a kind of GCM [7]. The application of this approach allows us to synthesize program schemes in the corresponding SC for assessing the COTO condition using methods of the Pattern Recognition Theory in the generated G-models as a KRM of the relevant SA.

Here is the condition assessment model of a product element (Fig. 2). The model allows to assess the condition of the COTO element associated with the received electronic document. The expert should set additional information within the KB to implement calculations for a given model: (see (1)): possible ranges of attribute values X_n^V and associated document statuses x_m^{St}, possible sets of document statuses and associated generalized document class statuses $s_{Cl}^{St} = \{M_{Cl}^{St}, st_{Cl}\}$, possible sets of generalized statuses of document classes and associated estimates of product condition parameters $m_{Cl}^{St^c} = \{St^c, s\}$.

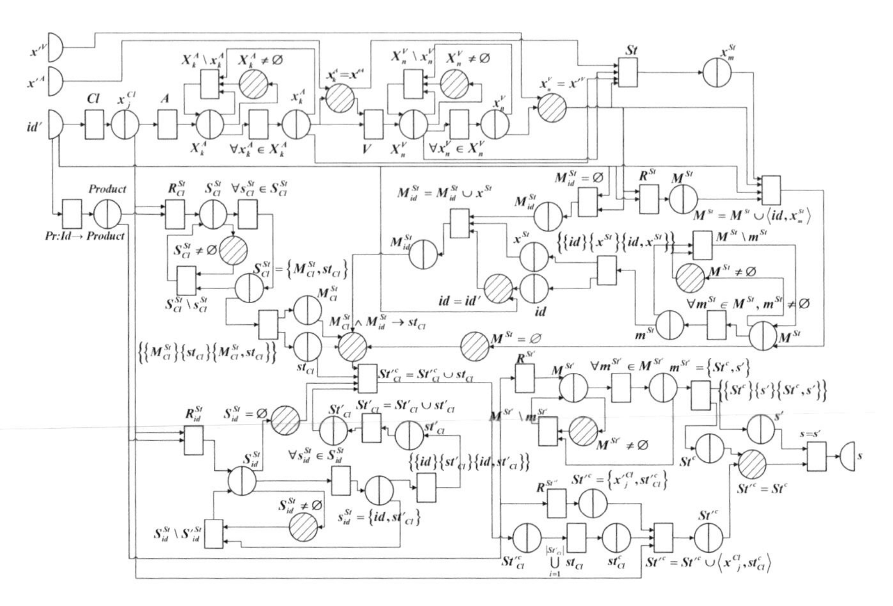

Fig. 2. Model state estimation of an element of a COTO

The following cortege [7] describes the G-model proposed for consideration:

$$M_G = \; <X_M, R_M, P_M, \Delta_M> \;, \tag{3}$$

where the set X_M is represented $Id, Product, X^{Cl}, X^A, X^V, X^{St}$ (see (1)) and X^R - set of intermediate auxiliary parameters (see Fig. 2); R_M is represented Cl, A, V, St (see (1)) and R^{St} - set of relations, among which the mapping functions that put their documents in correspondence to the elements of the product, as well as known operations on sets.

The presented model needs to be explained and especially the processes simulated in it. Entering a document into the system is accompanied by setting the identifier id' of this document and the values x'^V of the attributes x'^A. The document status x_m^{St} for each value is determined. As a result, the document acquires a multitude of statuses M_{id}^{St}. Further, the generalized status of the document st_{Cl} is determined as a result of the search for such a set of statuses M_{Cl}^{St} from the sets specified by the expert, which would coincide with M_{id}^{St}. It is possible when several documents are entered for each class. In this case, st_{Cl} will be added to the set $S_{Cl}^{\prime c}$ of all generalized statuses of the documents of this class. As a result of the operation of combining the elements of the set $S_{Cl}^{\prime c}$ a generalized state of the class of documents st_{Cl}^c will be computed. In this case, if in a particular case the range of admissible values st_{Cl} is represented only by $\{a, b\}$, then st_{Cl}^c will be represented by a single value from the pair $\{a, b\}$. Further, for the totality of all generalized statuses of document classes St'^c, it is necessary to find such a set St^c and the associated assessment of the condition parameter of the COTO s element from the sets specified by the expert, that $St'^c = St^c$. As a result of processing and analysis of document metadata, the status of the COTO element can be comprehensively assessed.

It is possible to synthesize the program scheme (G-net) according to the corresponding generated G-model in automatic mode [7]. The G-net is verified if it has the properties of completeness, closure, and consistency. In the SA, the model will have the completeness property if the set of specified intervals of attribute values X_n^V covers all possible values X^V for all attributes. The model will have the property of closure if for each interval of attribute values X_n^V the status x_m^{St} in the final set of document statuses X^{St} is uniquely determined. The model will have the property of consistency if in constructing the DNF (in the case of zero-order logic or propositional calculus) over the set of implicants of any predicate functions, there are not found two concurrently defined contradictory predicates that have the values *true* and *false*. Thus, the verification of a given model is ensured by the correctness of the expert's knowledge embedded in the model.

5 The Organization of Distributed Computing as Part of the Monitoring System of Technical Condition and Reliability of COTO

A computational program that implements a certain algorithm for assessing the condition of COTO is given by the G-net [7]. On this basis, a computational program execution can be represented as a process of entering chips (parameters) in a position (inputs of operators) and triggering transitions (execution of operators). At the same time, the non-determinism of classical Petri nets behavior provides a possibility of flexible assignment of available resources to select the optimal route for calculation program execution (the sequence of triggering transitions) [16, 17]. Thus, a computing system model can be represented as a certain Petri net. The approach is that the computation program and the computing system model are defined using the same formalism [18].

Since the computing system is distributed, we consider it as a set of independent processes (executable software components) interacting by means of messages transmission for data exchange and actions synchronization (control of the calculation process) [19]. Messages in a broad sense can have a complex structure and contain various parameters and attributes. Processes running within one or more computers will be called nodes of a distributed computing system [20]. Each node can receive, transmit and accumulate parameters, as well as execute operators. The distributed computer system (DCS) model can be constructed using a Petri net. In this model node behavior is described by means of transitions and also nodes and messages state is described by means of positions. As an example, consider the process of sending a message from one DCS node to another. Thus, DCS can be represented as a set of nodes:

$$DCS = \{N_1, N_2, \ldots, N_m\} \tag{4}$$

Each node is responsible for the set of operators according to a calculation program. Operator execution goes after required parameters set coming in operators input.

After an operator implementation, the node sends a message to another node (if the operators are located on different DCS nodes). Thus, there is a set of messages:

$$MS = \{(s, r)|s, r \in DCS^{\wedge}s \neq r\} \quad (5)$$

where s – sender, r – recipient, s and r are located on different DCS nodes. The message has a following appearance:

$$MS = \sum_{r \in DCS - \{s\}} (s, r) \quad (6)$$

where the sum is taken over set of elements by adding of multisets:

$$\{s, r\}|r \in DCS - \{s\} \quad (7)$$

each of one contains one element. The result set contains one message for each receiver r that has s as a sender.

Due to the fact that the scheme of the computation program and the model of the computing system are defined using the same formalism. This gives many advantages of the Petri net to be used, not only in modeling, but also in the execution of the computation program. This approach makes it possible to easily scale the DCS and, at the same time, preserve the natural parallelism of the SA, which contributes to the improvement of the quality and operability of assessment the COTO condition.

6 Intelligent User Interface of the COTO Condition Monitoring System as a Solution of Human-Machine Interaction Problems

Another important aspect of the design, development and operation of monitoring system is the development of an intelligent user interface [1].

An interface element has a type and a set of attributes. Attributes are associated with the values of those parameters that characterize the SA (the set of subject variables of the G-model).

Based on the current state of affairs in the field of human-machine interaction [5, 21], we can conclude that this area needs to form new approaches to the design, development and interacting with user interfaces. Thus, we can talk about the formation of a cognitive image, which means a set of methods and methods of figuring the conditions of the problem, which allows either to immediately see the solution or to get a clue to find it [22, 23]. The cognitive image can be described by a triplet:

$$KO = \langle S_{KO}, I_{KO}, F_{KO} \rangle \quad (8)$$

$S_{KO} = \{M_G | G = 1, 2, 3 \ldots n\}$ is a finite non empty set of G-models describing the SA; $I_{KO} = \{E_I | I = 1, 2, 3 \ldots n\}$ is a finite non empty set of interface elements that provide user-defined functionality; F_{KO} is a mapping of elements of the set S_{KO} in I_{KO}

that associates the subject variables of the G-models to the elements of the interface. Thus, the transitivity is formed "the state of the original object - the state of the computing process - the state of the adaptive user interface" [7].

The interface elements are set as follows:

$$E_I = \langle ID_E, T_E, A_E, F_E, HND_E \rangle \tag{9}$$

$ID_E = \{1, 2, 3 \ldots n\}$ is an interface element identifier; $T_E = \{Button, Form\ldots\}$ is an interface element type; $F_E : T_E \to A_E$ a mapping that maps a type of an interface element to a set of attributes; $HND_E : A_E \to X_G$ a mapping that maps a set of attributes into a set of objective variables of the G-model.

The attributes of the interface element are:

$$A_E = \langle L_A, SZ_A, PAR_A, ECT_A \rangle \tag{10}$$

L_A the location described by the given metric; SZ_A the size described by the given metric; $PAR_A = \{E_I | 1\}$ the parent of a specific interface element; ECT_A set allowing to expand the list of attributes under the conditions of tasks, which can be empty. For example, the description of the color of the interface element in RGBA format is {r, g, b, a} (the last parameter specifies the transparency of the interface element).

One of the actual tasks is the description of the interactive user interface as an integral part of the monitoring system for technical condition, reliability and DS for the management of the LC of COTO.

For example, in the problem considered in this article, the interface of the subsystem for assessing the condition of the product of the LV "Soyuz-2" according to its documents, can be described as follows:

I_{KO} = {(Electronic structure of the product, E_1), (Interactive intelligent 3D model, E_2), (Block for handling contingencies, E_3), (Registration card, E_4), (Organizational-staff structure, E_5), (Interactive intellectual technological schedule, E_6)}.

Let's describe the interface element "Registration card". This element is designed to enter information about the document. As was told above the document is specified by the following parameters: id' is the document identifier, $x^{\prime A}$ is the document attribute and $v^{\prime A}$ is the attribute value. In addition, you need to extend the set of attributes with the color parameter. Color is given in RGBA format $A_{E4} = \{L_{A4}, SZ_{A4}, PAR_{A4}, \{r, g, b, a\}_{A4}\}$. Thus, the registration card is described:

$$E_4 = \{4, Form, \{(x, y), (w, h), (I_{KO}), (r, g, b, a)\}, f(Form) = \{L_4, SZ_4, PAR_4, COL_4, \varnothing\} \\ id' \to A_{A4}\} \tag{11}$$

7 Conclusion

The considered approach of assessing TC and reliability with the DS for managing the LC of COTO as part of a system of information of the TC and reliability allows us to say that the use of this new IIT in the system allows to make an integrated assessment of the TC and reliability of COTO, which meets the goals and objectives of decision makers. The G-model as a kind of the GCM has a number of advantages for modeling, estimating and predicting the structural dynamics of COTO.

The parallelism in the processing of information in a DCS can significantly accelerate the computational process and shorten the time for decision-making.

The formation of the cognitive image with the intelligent interface allows the operator to make a qualitative decisions in the conditions of receipt of fuzzy, inaccurate, incorrect information on the current situation.

Based on the results of the study, a SwC was developed for monitoring the TC and reliability of COTO using the new IIT.

Acknowledgments. The research described in this paper is partially supported by the Russian Foundation for Basic Research (grants 16-07-00779, 16-08-00510, 16-08-01277, 16-29-09482-ofi-i, 17-08-00797, 17-06-00108, 17-01-00139, 17-29-07073-ofi-i), grant 074-U01 (ITMO University), project 6.1.1 (Peter the Great St. Petersburg Politechnic University) supported by Government of Russian Federation, Program STC of Union State "Technology-SG" (project 1.3.3.3.1), state order of the Ministry of Education and Science of the Russian Federation № 2.3135.2017/4.6, state order of the Ministry of Education and Science of the Russian Federation № 2.3135.2017/4.6, state research 0073–2014–0009, 0073–2015–0007, International project ERASMUS +, Capacity building in higher education, № 73751-EPP-1-2016-1-DE-EPPKA2-CBHE-JP, International project KS1309 InnoForestView Innovative information technologies for analyses of negative impact on the cross-border region forests.

References

1. Avtamonov, P.N., Bahmut, A.D., Krylov, A.V., Okhtilev, M.Yu., Okhtilev, P.A., Sokolov, B.V.: Application of technology for decision support at various stages of the life cycle of space assets as part of the information system on technical condition and reliability. In: Actual Problems of Rocket and Space Technology (V Kozlov readings). Samara Publ., pp. 222–233 (2017). ISBN 978-5-93424-798-1. (in Russian)
2. Okhtilev, M.Yu., Sokolov, B.V., Yusupov, R.M.: Intellectual technologies for monitoring the state and managing the structural dynamics of complex technical objects. Nauka Publ., Moscow, 410 p. (2006). (Informatics: unlimited possibilities and possible limitations). ISBN 5-02-033789-7. (in Russian)
3. Soloviev, I.V.: General principles of managing a complex organizational and technical system. Perspect. Sci. Educ. **2**(8), 21–27 (2014). https://pnojournal.wordpress.com. ISSN 2307-2447. (in Russian)
4. Avtamonov, P.N., Okhtilev, M.Yu., Sokolov, B.V., Yusupov, R.M.: Actual scientific and technical problems of development and implementation of the interconnected complex of unified integrated decision support systems (DSS) in the ACS by the objects of military-state administration. Izvestia of the Southern Federal University. Technical science. Issue № 3 (152), pp. 14–27. Southern Federal University Publ., Rostov-on-Don (2014). (in Russian)

5. Pospelov, D.A.: Artificial intelligence. In: 3 kN. Book 2. Models and Methods: Handbook, 304 p. Radio and Communication Publ., Moscow (1990). (in Russian)
6. Gushchin, A.N.: Fundamentals of Knowledge Representation: Textbook, Handbook. Score. State. Tech. Un-t, 31 p. St. Petersburg Publ. (2007). ISBN 5-85546-285-4. (in Russian)
7. Okhtilev, M.Yu.: Basics of the theory of automated analysis of measurement information in real time. In: Synthesis of the Analysis System, 161 p. VIKU them. Mozhaisky Publ., St. Petersburg (1999). (in Russian)
8. Narinyani, A.S.: Model or Algorithm: A New Paradigm of Information Technology. Information Technology Publ. № 4, pp. 11–16 (1997). (in Russian)
9. Tyugu, E.H.: Conceptual Programming. Problems of Artificial Intelligence, 255 p. Nauka Publ., Moscow (1984). (in Russian)
10. Larin, M.V.: Electronic documents: theory and practice. Vestnik RGGU. Serija: dokumentovedenie i arhivovedenie. Informatika. Zashhita informacii i informacionnaja bezopasnost'. Izd. RGGU № 2(145), pp. 53–63 (2015). (in Russian)
11. Chen, P.P.-S.: The entity-relationship model—toward a unified view of data. ACM Trans. Database Syst. (TODS). **1**(1), 9–36 (1976). Special issue: papers from the international conference on very large data bases
12. Kuznecov, S.D.: Database. Introductory Course, 28 March 2017. http://citforum.ru. (in Russian)
13. Date, C.J.: An Introduction to Database Systems, 8th edn., 1005 p. Pearson/Addison Wesley Publ., Boston (2004). ISBN 5-8459-0788-8
14. Gray, P.M.D.: Logic, Algebra, and Databases, 294 p. Wiley, New York (1984). ISBN 0-321-18956-6
15. Pramod, J., Sadalage, M.F.: NoSQL Distilled, 192 p. Addison-Wesley Publ., Reading (2012). ISBN-10 0-321-82662-0
16. Peterson, J.L.: Petri Net Theory and the Modeling of Systems. Prentice Hall Publ., Upper Saddle River (1981). 290 c. ISBN 0-136-61983-5
17. Kotov, V.E.: Petri nets. Science, 160 p. (1984). (in Russian)
18. Zajcev, D.A.: The paradigm of computing on petri nets. Avtomatika i telemehanika № 8. Science, pp. 19–36 (2014). (in Russian)
19. Tanenbaum, J., van Steen, M.: Distributed Systems. Principles and Paradigms, 877 p. Piter, St. Petersburg (2003). (in Russian)
20. Kosjakov, M.S.: Introduction to Distributed Computing, 155 p. NIU ITMO, St. Petersburg (2014). (in Russian)
21. Cooper, A., Reimann, R., Cronin, D.: About Face: The Essentials of Interaction Design, 720 p. Wiley, Indianapolis (2014). ISBN 1-118-76657-1
22. Pospelov, D.A.: Cognitive graphics - a window into a new world. Programmnye produkty i sistemy (Softw. Syst.) **2**, 4–6 (1992). (in Russian)
23. Fu, K.: Structural methods in pattern recognition. Mir, 320 p. (1977). (in Russian)

Proactive Management of Complex Objects Using Precedent Methodology

Aleksey D. Bakhmut[1], Aleksey V. Krylov[2], Margaret A. Krylova[3], Michael Yu. Okhtilev[2], Pavel A. Okhtilev[2(✉)], and Boris V. Sokolov[2,4]

[1] St. Petersburg State University of Aerospace Instrumentation, Saint-Petersburg, Russia
adbakhmut@gmail.com
[2] St. Petersburg Institute for Informatics and Automation of the Russian Academy of Sciences (SPIIRAS), Saint-Petersburg, Russia
pavel.oxt@mail.ru, sokolov_boris@mail.ru
[3] Herzen State Pedagogical University of Russia, Saint-Petersburg, Russia
m.a.krylova@bk.ru
[4] ITMO University, Saint-Petersburg, Russia

Abstract. The article describes the approach to designing proactive decision support systems (DSS) based on the modification of the generalized computational model as a knowledge representation model (KRM). It is proposed to use a precedent methodology for making management decisions. It was found that this approach allows to predict the situations and ensure the model's self-learning by forming a precedent knowledge base.

Keywords: Decision support systems
Complex organizational and technical object · G-model · Proactive system
Precedent methodology

1 Introduction

One of the directions in the development of decision-making automated systems is related to the modeling of human thought processes and is studied in the artificial intelligence theory [1–4]. One of the most important components of this theory are expert intellectual systems (ES) that represent a class of computer information systems based on work with knowledge.

ES are used in DSS to improve the quality and validity of decisions and reduce the time for its making.

The basis of the ES is the knowledge base (KB), which is formed and filled by the experts of subject area. A KB can be formed using the "if … then …" constructions with the use of some KRM [5]. On the basis of such a KRM, decision making, associated with the processes of complex organizational and technical object (COTO), can be modeled in the form of a decision tree [19] expanded due to the ability of frames to constructively describe complex analytical processes. This approach allows to specify a declarative-procedural description of the functioning of a COTO, where

R. Silhavy (Ed.): CSOC 2018, AISC 764, pp. 298–307, 2019.
https://doi.org/10.1007/978-3-319-91189-2_29

knowledge becomes an active control element. It is essential that in such a KRM, the model is of primary importance as a means of describing the subject area. Algorithms become only invariant automatic blocks in such models [6, 7]. Nowadays an example can be given of one of the variants of constructing such a KRM, for the description of complex systems. It is called the "G-model" [6] and has both a theoretical basis, and is already actively used for really existing complex systems.

2 Knowledge Representation Model

The theory of intellectual analysis based on G-models is a unified methodology for the automated assessment of the condition of complex systems and the synthesis of an appropriate analysis program. This methodology of intelligent interactive monitoring and management of COTO has received wide and successful implementation in various subject areas (astronautics, nuclear energy, ecology, logistics). "Interesting prospects are opened by using this technology and corresponding systems when creating a network of situational decision support centers for monitoring and managing facilities in the military, political, communication and social spheres" [8–12].

The G-model makes it possible to create simulation-analytical polymodel complexes and appeared as a result of generalization of existing approaches in such sections of the artificial intelligence theory [6] as:

- conceptual programming;
- constraint programming;
- multi-agent modeling, etc.

The constructing a KRM based on G-models in conceptual programming is the following:

- introduce the concept of computational models as a frame describing the computing environment;
- knowledge of complex situations are represented by sets of attributes (parameters), actions, and also conditions of their application;
- the slots of the frame are filled in with the specification of the knowledge, when the frame is used in connection with a certain specific situation;
- KRM is the form of a semantic net, complemented by a description of actions and a set of rules to perform these actions;
- at the nodes of the semantic network are the denotations of the subject area represented by the measured and calculated parameters characterizing the subject area;
- the relationship between the denotations are presented in the form of maps, limited to the set of computable functions. In programming terms, these functional relations can be called operators;
- network objects, which are variables linked by relations, called compute models. In advanced computational models operators conditions and loops are also defined;
- synthesis of the program of calculation occurs after the axiomatization and interpretation of KRM, which consists in the formation for each operator network of a certain computable relation.

G-model is a kind of generalized computational model (GCM), defined with the constraint programming and is described by the following cortege:

$$M_G = \langle X_M, R_M, P_M, \Delta_M \rangle \quad (1)$$

where X_M is set of analyzed object parameters, R_M is set of relations between parameters, P_M is set of predicates, subject variables in which are elements from X_M, Δ_M is mapping from R_M to P_M.

The G-model can be represented in the form of its operator scheme and simply the scheme (Fig. 1).

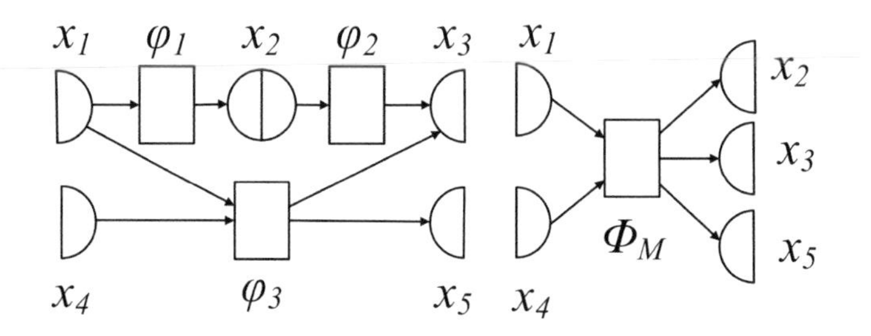

Fig. 1. Operator scheme (left) and scheme (right) of the G-model

When modeling a real system with the help of the G-model, a polymodel complex is formed where each model is represented by its own operator scheme. Thus, a multiagent approach is implemented, where each operator scheme is an autonomous semantic analyzer (or an agent). The complex forms a set of intellectual agents, where each of them performs its role [6].

A systematic analysis of the classical KRM and combination of production-frame G-model in some significant for problem indicators was carried out [13] (Table 1).

As a result of the analysis it was revealed that the G-model has a number of advantages in comparison with other KRM when solving the problem of automated monitoring of structural states of COTO.

3 A Precedent Methodology

In the conditions of incomplete or underdetermined dynamically changing information, when using simulation-analytical models, it becomes difficult to form proactive conclusions for making decisions for the management of COTO. A precedent methodology for solving this problem can be used, and therefore it is necessary to use a relevance relation that is not taken into account in such models, as well as the inductive inference method necessary to generate conclusions on certain semantic metrics when comparing the analyzed condition and precedents.

The reasoning based on precedents is based on the accumulation of experience and the subsequent adaptation of the solution of a known problem to the solution of a new problem. The precedent approach allows to simplify the decision making process under

Table 1. System analysis of classical KRM and G-net

Kind of model	Declarative-procedural modeling	Structural relationships	Functional relations	Causal relationship	Semantical relationships	Relationship of relevance	Modeling of superficial knowledge	Modeling deep knowledge	Modeling of hard knowledge	Modeling of soft knowledge	Method of deductive inference	Method of inductive inference	Total (12):
Causal model	+	-	-	+	-	-	+	-	+	+	+	-	6
Formal Logical Model	-	-	+	+	-	-	+	-	+	-	+	-	5
Semantic net (net model)	+	+	-	-	+	+	-	+	+	+	-	-	7
Frame Model	+	+	+	-	-	+	-	+	+	+	-	-	7
Causal-frame model (G-model)	+	-	+	+	+	-	+	+	+	+	+	-	9

time constraints and in the presence of various kinds of uncertainty in the initial data and expert knowledge, as well as in the case of various emergency situations.

Models that use the precedent base (PB) for their reasoning are called precedent models.

The reasoning methods based on precedents include four main stages, forming the so-called CBR-cycle, presented on Fig. 2 [15].

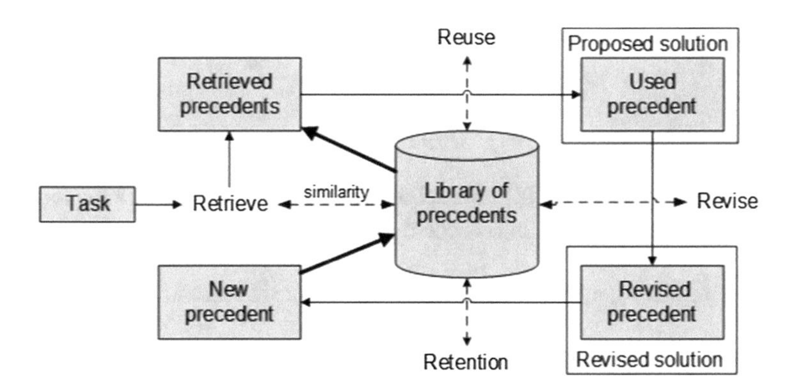

Fig. 2. The process of using precedent in the solution of new tasks

According to Fig. 2, the process of using precedents is as follows:

1. In the event of a specific task, the most appropriate (similar) precedent for the current situation is retrieved from the PB by a software tool based on the methods of finding solutions in the used PB. Depending on the subject area and the method used to find solutions in the used PB, different metrics are used to compare the current situation with the situations from the PB.

2. Reuse of the extracted precedent for the solution of the current task (problem) is carried out under the conditions of similarity.
3. In case of partial failure of the similarity conditions, the decision is reviewed and adapted according to the chosen precedent (or set of precedents) according to the conditions of the current task (problem).
4. A new precedent is formed from the new conditions and the solution of an existing precedent (precedents) adapted for them, which is stored in the PB.

When using the precedent methodology, some problems arise in interpreting data and knowledge stored in the used PB. There is a number of methods for finding solutions based on precedents (extracting precedents) and their modifications [15]. Examples include: the method of the nearest neighbor; method of extracting precedents based on decision trees; method of extracting precedents based on knowledge; method of extraction, taking into account the applicability of the precedent.

The most difficult moments in the precedent approach are the choice of a suitable precedent and the adaptation of its solution to the current situation.

In the management problems of COTO, the current situation is represented by a set of features that in the computational process is a vector of parameter values.

$$X\{x_i \in X\} \tag{2}$$

where (2) is set of values of parameters of the current situation.

From this point of view, precedents, also represented by sets of parameter values, serve as alternatives for which it is necessary to decide is it optimal or not.

$$P\{P^j | P^j \subset P\} \tag{3}$$

$$X \leftrightarrow P^j, |X| = \left|P^j\right|, \forall P^j \subset P \tag{4}$$

$$P^j\{p_i^j | p_i^j \in P^j, d^j\} \tag{5}$$

where (3) is set of precedents, (4) - the set of parameters of the current situation is equal to the set j precedent, (5) is set of precedent parameter values and solution d^j associated with it.

The task of finding a suitable precedent can be considered as a multicriteria choice problem.

Thus, the task of formalizing the reasoning by precedents is considered as the search for the minimum distance over the set of values of the precedent parameters and the current situation.

$$\left|x_i - p_i^j\right| \to \min_{X,P}, \; i = \overline{1:n}, j = \overline{1:m} \tag{6}$$

where (6) is task of finding the precedent closest to the current situation on the set of values of the parameters characterizing it.

In the case of multi-criteria the problem of finding the minimum (maximum) is unsolvable with classical mathematical theory, since more often in practice this or that alternative will be better in one parameter and worse in the other.

However, when considering the problem as a multicriteria choice problem, first of all a set of non dominated alternatives can be found by simply comparing the values of the parameters.

$$P_{нд} = \{ p_i^j \mid p_i^j \in P^j \wedge (\left|x_i - p_i^j\right|, \left|x_i - p_i^k\right|) \notin R_{dom} \forall p_i^k \in P^k, k \neq j \} \quad (7)$$

where (7) is the set of non dominated alternatives of precedents.

If the preference for at least one criterion is at variance with the preference for the other, then such alternatives are considered incomparable or non dominated.

Then we can use the method of analyzing hierarchies. More precisely, its part - the method of paired comparisons of Thomas Saaty:

1. A paired comparison matrix is constructed

$$A = (a_{ij})_{n \times n} = \begin{pmatrix} a_{11} & a_{12} & \dots & a_{1n} \\ a_{21} & a_{22} & \dots & a_{2n} \\ \dots & \dots & \dots & \dots \\ a_{n1} & a_{n2} & \dots & a_{nn} \end{pmatrix} \quad (8)$$

When constructing the matrix a scale of the relative importance of T. Saaty is used. It allows to specify a relationship between each pair of parameters in order of importance from 0 to 9:

- the scale should be able to capture the difference in the sensations of experts when they make comparisons, to distinguish as many of their shades as possible;
- the expert must be confident in all gradations of his judgments simultaneously.

The matrix A is inversely symmetric, i.e. its elements, located symmetrically with respect to the main diagonal, are inverse to each other. The diagonal consists of 1.

The task is to find the vector w - the normalized weight vector of the precedent parameters.

2. For this, it is necessary to find the maximum of the proper value of the matrix by solving the characteristic equation:

$$D = \begin{pmatrix} a_{11} - \lambda & a_{12} & \dots & a_{1n} \\ a_{21} & a_{22} - \lambda & \dots & a_{2n} \\ \dots & \dots & \dots & \dots \\ a_{n1} & a_{n2} & \dots & a_{nn} - \lambda \end{pmatrix} = 0 \quad (9)$$

3. A system of linear equations is solved:

$$(A - \lambda_{max} E) w = 0 \quad (10)$$

4. The vector w is normalized.

5. There is a minimum of a linear additive convolution:

$$V^d = \min_{x_i \in X, p_i^j \in P^j} \sum_{i=1}^{n} w_i \left| x_i - p_i^j \right| \quad (11)$$

Thus there is a precedent closest to the current situation.

It should be noted that the solution of the problem of searching difference $\left| x_i - p_i^j \right|$ can be found a valid vector of deviations V^d between the current situation and precedent.

To reduce the number of precedents, stored in the database, a precedent matrix is associated with the permissible deviation matrix. Each precedent in a pair with deviations defines a certain state space with which a single solution is associated. It should be noted that there can be several variants of such deviations in the set of parameters of the situation. Columns of the matrix should be a set of non dominated alternatives to allowable deviations:

$$V\{V^j | V^j \subset V\} \quad (12)$$

$$V^d \leftrightarrow V^j, \left| V^d \right| = \left| V^j \right|, \forall V^j \subset V \quad (13)$$

$$V^j\{v_i^j | v_i^j \in V^j\}, v_i^j : [0, 1] \quad (14)$$

Multiplying the vector of the values of the precedent parameters and the deviation vectors for each variant of the deviations, the set of vectors of maximum deviations from the precedent will be obtained:

$$V_{max}^j = P^j \times V^j \quad (15)$$

The task is to find such a variant of deviations from the precedent, in which the deviation will be minimal:

$$\left(v_i^d - v_{max\ i}^j \right) \to \min_{V^d, V_{max}}, \; i = \overline{1:n}, j = \overline{1:m} \quad (16)$$

Then again the method of paired comparisons of T. Saati is applied, where the indicators of relative importance are given for variants of deviations.

The weights of the deviations will be obtained, and the most preferable deviation for the current selected precedent can be found.

The process can be iterative to find such a precedent, the deviation of which will be minimal:

$$\min_{v_i^d \in X, v_i^j \in V^j} \sum_{i=1}^{n} w_i \left(v_i^d - v_{max\ i}^j \right) \quad (17)$$

As a result, one of two options can be applied to the current situation:

$$\begin{cases} v_i^d - v_{max\ i}^j \leq 0 \rightarrow d^j \\ v_i^d - v_{max\ i}^j > 0 \rightarrow X, P^j, v_i^d \end{cases} \tag{18}$$

In the case of a difference that is less than or equal to zero over the whole set of parameter values, we fall into the allowable values of the deviation vector, so a decision related to the precedent can be given, in another case - the precedent is not applicable for its rejection, but the percentage of discrepancy between the closest precedent and the current situation, the decision related to the nearest precedent and a suggestion to bring a new solution to the PB can be shown.

The search for optimal methods for adapting the found precedent to the current situation requires more detailed consideration in further research.

The result of the search and adaptation is the solution of the current problem. To close the problem solving cycle and fill the knowledge of the system, it is necessary to keep the current precedent in the PB. Based on the results of the implementation of the decision on the new precedent, an assessment of its quality is carried out, which is also attached to the description of the precedent for further accounting. Training of the expert system on precedents is carried out on the basis of monitoring the response in the implementation of the decision on the precedent. Negative results have the same weight for a precedent expert system, as precedents with positive quality solutions. The system should understand that when making wrong decisions, a negative result is obtained.

4 Conclusion

The use of precedent methodology in G-models allows, when building expert management systems for complex objects in any field, to introduce additional opportunities in increasing the efficiency of decision making, their adequacy, and predicting. Such systems, regardless of the scope of their application (technical, economic, political, social, including their combinations) provide ample opportunities for the development of management decisions, drawing on the practical experience gained in the PB. Of course, it is necessary to have large PB of a particular subject area for the adequacy of issued solutions in the current situation, which requires a sufficiently large time resource for their development.

Acknowledgments. The research described in this paper is partially supported by the Russian Foundation for Basic Research (grants 16-07-00779, 16-08-00510, 16-08-01277, 16-29-09482-ofi-i, 17-08-00797, 17-06-00108, 17-01-00139, 17-29-07073-ofi-i), grant 074-U01 (ITMO University), project 6.1.1 (Peter the Great St. Petersburg Politechnic University) supported by Government of Russian Federation, Program STC of Union State "Technology-SG" (project 1.3.3.3.1), state order of the Ministry of Education and Science of the Russian Federation № 2.3135.2017/4.6, state order of the Ministry of Education and Science of the Russian Federation № 2.3135.2017/4.6, state research 0073–2014–0009, 0073–2015–0007, International project ERASMUS +, Capacity building in higher education, № 73751-EPP-1-2016-1-DE-

EPPKA2-CBHE-JP, International project KS1309 InnoForestView Innovative information technologies for analyses of negative impact on the cross-border region forests.

References

1. Pospelov, D.A.: Artificial Intelligence. In: 3 kN, Book 2, Models and Methods: Handbook - M., 304 p. Radio and Communication Publ. (1990). (in Russian)
2. Gaaze-Rapoport, M.G., Pospelov, D.A.: Structure of research in the field of artificial intelligence. In: Explanatory Dictionary of Artificial Intelligence – Moskow, Radio i svjaz', pp. 5–20 (1992). (in Russian)
3. Gushhin, A.N.: Basics of Knowledge Representation: A Tutorial. Ball. gos. tehn. un-t, 31 p. St. Petersburg. (2017). ISBN 5-85546-285-4. (in Russian)
4. Kalinin, V.N.: Theoretical foundations of systemic research tutorial, 293 p. VIKU them. Mozhaisky Publ., St. Petersburg (2016). (in Russian)
5. Expert Systems, 20 November 2017. http://www.studfiles.ru/preview/400100
6. Okhtilev, M.Y.: Basics of the theory of automated analysis of measurement information in real time. Synthesis of the analysis system, 161 p. VIKU them. Mozhaisky Publ., St. Petersburg (1999). (in Russian)
7. Okhtilev, M.Y., Sokolov, B.V., Yusupov, R.M.: Intellectual technologies for monitoring the state and managing the structural dynamics of complex technical objects, (Informatics: Unlimited Possibilities and Possible Limitations), 410 p. Nauka Publ., Moscow (2006). ISBN 5-02-033789-7. (in Russian)
8. Avtamonov, P.N., Okhtilev, M.Y., Sokolov, B.V., Yusupov, R.M.: Actual scientific and technical problems of development and implementation of the interconnected complex of unified integrated decision support systems (DSS) in the ACS by the objects of military-state administration. Technical science, Issue № 3 (152) - Izvestia of the Southern Federal University, Rostov-on-Don: Southern Federal University Publ., pp. 14–27 (2014). (in Russian)
9. Sokolov, B.V., Jusupov, R.M.: Complex modeling of risks in the development of management decisions in complex organizational and technical systems. Probl. Manag. Inf., №. 1, pp. 1–22 (2006) (in Russian)
10. Gurban, I.A., Pecherkina, M.S., Lykov, I.A.: Automation of analysis of regions' stability to social and economic crises. Bull. Ural Fed. Univ. Ser. Econ. Manag. **15**(6), 906–925 (2016). (in Russian)
11. Budaev, J.V.: Conceptual metaphor in the service of US intelligence. Polit. Anal. **2**, 12–16 (2015). (in Russian)
12. Bykov, I.A.: Empirical policy research using network analysis methods. Polit. Anal. **11**, 49–59 (2011). (in Russian)
13. Okhtilev, P.A., Bahmut, A.D., Krylov, A.V.: Review and application of knowledge representation models in the intellectual system for monitoring structural states of complex organizational and technical objects. Scientific Session of the SUAI, № (2). Technical Science, pp. 185–191. SUAI, St. Petersburg (2017). ISBN 978-5-8088-1218-5. (in Russian)
14. Khaing, Z.L., M'o, A.K., Varshavskij, P.R., Alehin, R.V.: Implementation of the precedent module for intelligent systems. Softw. Syst. Prod. **2**, 26–31 (2015) (in Russian)
15. Varshavskij, P.R., Alehin, R.V.: The method of finding solutions in intelligent decision support systems based on use cases. Int. J. Inf. Mod. Anal. **2**(4), 385–392 (2013)
16. Maslennikova, N.J., Slinkova, O.K.: The concept and essence of monitoring from a system approach point of view. Sci. Time **6**(6), 110–121 (2014)

17. Solov'ev, I.V.: General principles of managing a complex organizational and technical system. Perspect. Sci. Educ. **2**(8), 21–27 (2014). ISSN 2307-2447. (in Russian)
18. Daneev, A.V., Vorob'ev, A.A., Lebedev, D.M.: Investigation of the behavior dynamics of complex organizational and technical systems under the influence of unfavorable factors. Bull. Voronezh Inst. Min. Intern. Aff. Russ. **2**, 163–171 (2010). (in Russian)
19. Murthy, S.: Automatic Construction of Decision Trees from Data: A Multidisciplinary Survey. Data Mining and Knowledge Discovery, 49 p. Kluwer Academic Publishers, Boston (1998)

Artificial Intelligence and Algorithms in Intelligent Systems

Advanced Analytics: Moving Forward Artificial Intelligence (AI), Algorithm Intelligent Systems (AIS) and General Impressions from the Field

Carla Sofia R. Silva[1,2(✉)] and Jose Manuel Fonseca[1,2]

[1] Nova University, Lisbon, Portugal
silvacarla.uab@gmail.com, jmf@uninova.pt
[2] CTS – Center of Technology and Systems, FCT Campus, 2829-516 Caparica, Portugal

Abstract. The possibility of creating thinking systems discusses issues that may arise in the near future of AI. However this outlines challenges to ensure that AI operates safely as it approaches humans in its intelligence from Algorithms Intelligent Systems. To understand how progress may proceed we need to understand how existing algorithms are developed and improve, differentiating the concepts between data analytics and data algorithmic decision making. This article reviews the literature on AI and AIS and presents some general guidelines and a brief summary of research progress and open research questions. The first section reviews the basic foundation of Artificial Intelligence to provide a common basis for further discussions and the second section of this paper suggests the development of Algorithm Intelligence Systems models including: learning algorithms such as learning from observations, learning in Neural and Belief Networks and reinforcement learning.

Keywords: Artificial intelligence · Learning algorithms
Decision making and data analytics

1 Artificial Intelligence

We believe that understanding intelligence involves understanding how knowledge is acquired, represented and stored, how is it generated and learned, how motivates and how emotions are developed and used. AI has always been interesting philosophers and scientists in understanding how can machines think. Turing in 1950 answer this problem rephrasing that depends on how we define machine and think [1]. To many people the word "machine" evokes images of steel parts and steam hissing, however the computer has expand this notion to a growing understanding biological mechanisms. In to a logical explain of the development and function of molecules, scientists explains the machines through a complex system of humanlike thought. This comes closer to understand the word "think" in machine learning and it requires interaction between machine and the environment leading to what is called emergent behavior

R. Silhavy (Ed.): CSOC 2018, AISC 764, pp. 308–317, 2019.
https://doi.org/10.1007/978-3-319-91189-2_30

liked a dynamic environment [2]. These systems, inspired by biological models, are interesting for the ability to learn. Human Intelligence encompasses many abilities, including the ability to perceive and analyze a visual scene and the ability to understand and generate language.

There are common literatures in the field of AI and there are many definitions of Artificial Intelligence, in the early 1960s Marvin Minsky indicated that "*artificial intelligence is the science of making machines do things that would require intelligence if done by men.*" [3]; and from the 90's when Bransford discuss

> "*What is artificial intelligence? It is often difficult to construct a definition of a discipline that is satisfying to all of its practitioners. AI research encompasses a spectrum of related topics. Broadly, AI is the computer-based exploration of methods for solving challenging tasks that have traditionally depended on people for solution. Such tasks include complex logical inference, diagnosis, visual recognition, comprehension of natural language, game playing, explanation, and planning*" [4].

The study of measurement of intelligence has long histories and many ideas and techniques relevant to mechanizing intelligence have been developed by AI researchers. Gardner [5] and Stenberg [6] developed a definition of Intelligence like a combination of the ability to learn. This includes all kinds of informal and formal learning via any combination of experience, education, and training; to pose problems. This includes recognizing problem situations and transforming them into more clearly defined problems; and to solve problems. This includes solving problems, accomplishing tasks, and fashioning products.

AI research develops on the problem of connecting symbolic processes in to the camp of robotics in physical environments, creating a dynamic field and giving a new emphasis on integrated, autonomous systems – robots and software [7] agents that roam the internet that will motivate and guide artificial intelligence in this research field.

1.1 Pluralization of Intelligence

The Turing Test, proposed by Alan Turing, was designed to provide a satisfactory operational definition of intelligence. Turing [8] defined intelligent behavior as the ability to achieve human-level performance in all cognitive tasks. The test he proposed is that the computer should be interrogated by a human via a teletype, and passes the test if the interrogator cannot tell if there is a computer or a human at the other end. Therefore if we are going to discuss some that some programs think like humans, we must know how humans think and interact. We must get inside the human mind through introspection and psychological achievements. From this point of view if the inputs and outputs programs and human behavior matches this can be evidence how human does operate. Ideally, we would like to know the relations between inputs and outputs for algorithms progress.

According to Simon [8] evaluating the success of an artificial intelligence research effort can be relatively simple or it can be complex. When the Logic Theorist (LT) demonstrated that a rather primitive heuristic search, with a modest capability for selectivity, could find proofs for many theorems in Principia, a basic work on logic, that fact alone told us a great deal about intelligence. Simon [8] argues that significance of the result depended on the task being nontrivial for humans. It depended also on the fact that the program required modest amounts of computation, but amounts comparable to what we might think a human brain could provide. It depended on the fact that LT's heuristics, though simple, made its search highly selective as compared with brute-force search. Perhaps the most profound impact of AI will be the effects that Artificial Intelligences will have on our understanding of ourselves. Copernicus and other astronomers moved us from the center of the Universe to a small planet in one of the galaxies. Later Darwin moved us from the center of creation to our present place among DNA life forms. These changes were not well accepted at the time, these changes of perspective were difficult for mankind, so what changes are waiting for us? However the difference that has given the Humans a dominant advantage over the other species is not strength or even speed, but intelligence. Just as human intelligence has allowed us to develop tools and strategies for controlling our environment, a superintelligent system would likely be capable of developing its own tools and strategies for exerting control [9]. Smarter-than-human systems must be trusted to make good decisions, but what does it mean for a decision to be good? Formally, given a description of an environment and an agent embedded within, how is the best action identified respecting some preferences? This question addresses to the Decision Theory, which has been studied extensively by philosophers. Much of the power of decision analysis lies in its ability to effectively integrate the many factors that commonly affect a decision [10]. Such an integrating capacity makes decision analysis a very successful means of facilitating the decision-making process. There is a useful role distinction that can be made between the decision analyst and decision maker. The decision analysts implement an interactive process known as the decision analysis cycle (see Fig. 1) and the decision maker controls the valuable resources to be allocated and defines the situation within which the decision is to be made.

The decision analysis cycle is divided into four distinct phases [11]. The first phase – Basis Development – encompasses the formal definition of the decision maker's preference and a quantitative description of that part considerer important. This description contains variables of two types – controllable as decisions and fully modelled and aleatory as uncertainties and incompletely modelled. This phase is used as the input to the deterministic analysis phase, whose purpose is to reduce the size of the original model eliminating unimportant variables. The model variables are then ordered by their deterministic sensitivity to reduce the size of the original decision model. Without this the model will be too large to be subject of probabilistic – Decision Theory – analysis. The third phase – probabilistic analysis consists in three tasks: risk attitude assessment – which is more meaningful after the decision problem has been well defined, calculation of the optical policy – is the central task in the probabilistic analysis phase and evaluation of the risk sensitivities – involving stochastic sensitivity analysis. The last phase of the decision analysis cycle is the basis appraisal that consists of reviewing the decision basis and the corresponding recommendation resulting from the probabilistic analysis phase. The decision analysis is a dynamic process whose aim is not to produce a correct answer but to produce insight where is lacking. Decision analysis helps us to deal with our limited abilities when we are making complex choices. A more complex detail of this cycle can be found in the research of Matheson and Howard [13] and Howard [12].

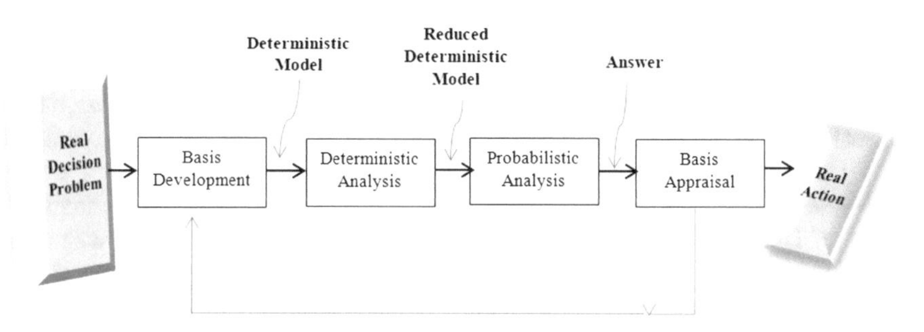

Fig. 1. Decision analysis cycle [11]

Once we have a suitable mathematical model for our problem, we can attempt to find a solution in terms of that model. The solution is finding an algorithm, which is a sequence of instructions, each of which as a clear meaning and it can be performed. [14]. Until a few years ago, the area of AI was seen as theoretical area, with applications only in small, curious, challenging but of little practical value. Most of the practical problems that required computation were solved by coding, in some programming language. From the 1970s, there was a greater dissemination of the use of AI-based computing techniques to solve real problems [15]. Having the strategy, we now decide on the tactics. In this stage we must select the specific method to be used for searching patterns, or if we want to use precision versus understandability, it is better with neural networks, while the latter is better with decision trees. This approach attempts to understand the conditions under which a data mining algorithm is most appropriate [16]. Therefore we can collect and interrelated data and a set of software programs to manage and access the data [17]. Relational data can be accessed by database queries written in a relational query language. This expert system, together with a powerful facility for manipulating and evaluating decision theoretic models constitutes an intelligent decision system.

1.1.1 From Problems to Programs

In certain aspects of the problem it can be expressed in terms of a formal model, than we must research for solutions in terms of a precise model and determine whether a program already exists to solve that problem or we can discover what is known about the problem to be solved to help construct a good solution.

By combining predictive with descriptive analytics, we are able to develop new applications. Taking qualitative decisions is however not solely the result of a data-analysis: an efficient decision support system is indispensable. A decision is supported by data, information, objectives, decision criteria and a decision-supportive model that gives clear insights in all possible alternatives and restrictions of the problem [18].

Formal decisions methods, such us mathematical programming and decision theory are solution techniques and doesn't addresses the problem issue. But the algorithms need to be complemented with a methodology concerned with formulating decision rather than solving decisions. Making decisions means designing and committing to a strategy in allocating resources. Although when we speak about decision we must imply the nature of discourse, than we can distinguish very different domains in decision. Therefore when we find decisions for some problems asked, we must assume that the action is subject to the outcome and a good decision do not guarantee good outcomes. Because of this, arguments must be made indicating that in decision making a measure of uncertain must obey the axioms of a probability measure [19]. This expert system, along with a powerful facility for evaluate and endorse decision theoretic models constitutes an intelligent decision system.

2 Data Structures and Algorithms

AI algorithms play an increasingly large role in modern society and it will become increasingly important to develop AI algorithms that are not just powerful and scalable, but also transparent to inspection. Here we must take into account the differentiation between predictive models and descriptive models highlighting some techniques and algorithms that have been used and developed in science (see Table 1 - summarizing some aspects not to be exclusively an assumption of all review).

The researcher after determine the research question must approach the exact appropriate model to be used if the all underlying assumptions are valid, making assumptions about predict or descriptive models to be used.

In the use of algorithms in real problems the knowledge that has of the application to the real is very important. Although a single algorithm is chosen it may be necessary to make adjustments to its free parameters, which leads to the obtaining of multiple models for the same data. The evaluation of any proposed technique can be performed according to several aspects, such as comprehensibility of knowledge, learning and storage of the model. Considering the predictive models as measures related to the performance obtained from the predictions made and the descriptive models referring and identifying structures in a set of data.

Table 1. Classification scheme for research items in research application

Research item	Output	Parametric Model	Research_Application [algorithm]
Pattern recognition	Continues	Regression	Correlation analysis using χ^2
Two group and multiple group classification	Discrete	LOGIT Discriminant function	
Asses underline structure	Continues	Principal component analysis Factor analysis	
	Discrete	Cluster analysis	k-*means* centroid clustering algorithm PAM - k-*medoids* representative clustering algorithm Expectation Maximization Algorithm *Chameleon* – Hierarchical Algorithm *DBSCAN* algorithm δ-Cluster Algorithm
Significance group differences	Continues Discrete	ANOVA PROBIT	
Predictive Models	Decision Tree algorithm Neural networks algorithms Graphic search algorithm Naive Bayes algorithm		ID3(Iterative Dichtomiser), C4.5, CART Backpropagation algorithm, OET algorithm Predict class membership probabilities - Sequential covering algorithm.
	Bagging algorithm *AdaBoost.* Algorithm *Sequential covering* Algorithm		Create an ensemble of classification models for a learning scheme where each model gives an equally weighted prediction. Create an ensemble of classifiers. Each one gives a weighted vote. Learn a set of IF-THEN rules for classification
Descriptive Models	Genetic algorithm		Attempt to incorporate ideas of natural evolution. + *Fuzzy* set approaches
	Data computation algorithm		*Star-Cubing* algorithm, BUC algorithm
	Apriori Algorithm		The algorithm uses prior knowledge of frequent item set properties
	FP-*Grown* Algorithm		Mine frequent item sets using an FP-tree by pattern fragment growth.

2.1 Predictive and Descriptive Models

One of the important modeling techniques to be use is the artificial neural networks. With this technique it's detect the underlying functional relationships within a set of data and perform such tasks as pattern recognition, classification, evaluation, modeling, prediction and control [20] and can be used as parametric models.

Scientists speculate that the brain may have one fundamental algorithm for intelligence – this idea grows with the concept of brain's plasticity. Artificial Neural Network can be used to classify any kind of input but it doesn't imply an algorithm to all functions.

Artificial Neural Networks with no hidden layers is similar to a generalized linear model. Neural Networks with hidden layers are a subset of the larger class of non-linear regression and discrimination models. However as non-parametric models, Artificial Neural Networks can easily incorporate multiple sources of evidence without simplifying assumptions concerning the functional form of the relationship between output and predictor variables. In parametric models the outliers in data produce influence on the size of correlation coefficient and the univariate outliers are very easy to detect.

2.1.1 *Apriori* Algorithm and FP-*Grown* Algorithm

The algorithm introduced by Agrawal and Srikant [21], was the first algorithm proposed for the extension of *itemsets* and association rules. It is based on the principle that any subset of frequent *itemsets* must be therefore a frequent *itemset*. Uses a wide area search with a test algorithm, and the degree of interest is represented by the trust of the rules, given by Eq. 1 and, which consists of the probability of a set of terms occurring given that another set occurred:

$$Confidence\,(A \rightarrow B) = \frac{P(AUB)}{P(A)} = P(B/A) = \frac{Support\,account(AUB)}{support\,account(A)} \tag{1}$$

The strategy of this algorithm involves using databases several times to calculate the frequent candidate itemsets. Alternatively, algorithms were used and developed to minimize the recurrent need to use databases. These algorithms seek to generate sets of itemsets by adopting a strategy in depth. One of the developed algorithms was the FP - growth (Frequent Pattern Grown) created by Han et al. [22]. This algorithm (see Table 2) is used in two phases, in the first one builds a data structure, the FP-tree, traversing the databases twice, used for association rules. This association is first created by a set of frequent items, creating the root node, then generating a set of nodes and then a common prefix. In this way, the representation of databases is as small as possible and it can be represented us follows:

There is no clear criteria regarding which functions to be used, but the best guidance is to know our data regarding the functions to be used in to a correct model, and determine which architecture is appropriate for our problem.

Tomasik [23] argues that Artificial intelligence (AI) will transform the world later this century. He expects this transition will be a “soft takeoff” in which many sectors of society update together in response to incremental AI developments, though the possibility of a harder takeoff in which a single AI project “goes boom” shouldn't be ruled out.

Decision-theoretic regression is a general concept whose applicability is not restricted to decision-tree representations of transition functions. The same principles apply to any structured representation as long as one can develop a suitable regression operator for that representation.

Table 2. Mine frequent itemsets using an FP-tree by pattern fragment growth.

Input

D, a transaction database;
min_sup, the minimum support count threshold.

Output: The complete set of frequent patterns.

Method:

1. The FP-tree is constructed in the following steps:

(a) Scan the transaction database *D* once. Collect F, the set of frequent items, and their support counts. Sort *F* in support count descending order as *L*, the list of frequent items.
(b) Create the root of an FP-tree, and label it as "null." For each transaction Trans in *D* do the following.
Select and sort the frequent items in *Trans* according to the order of *L*. Let the sorted frequent item list in *Trans* be [p\P], where *p* is the first element and *P* is the remaining list. Call insert_tree ([p\P], T), which is performed as follows. If *T* has a child *N* such that *N.item-name* = *p.item-name*, then increment *N's* count by 1; else create a new node *N*, and let its count be 1, its parent link be linked to *T*, and its node-link to the
nodes with the same item-name via the node-link structure. If *P* is nonempty, call insert_tree (P, N) recursively.

2. The FP-tree is mined by calling FP_growth (*FP_tree, null*/, which is implemented as follows.

procedure FP_growth(Tree,α)

(1) **if** *Tree* contains a single path *P* **then**
(2) **for each** combination (denoted as β) of the nodes in the path *P*
(3) generate pattern β U α with *support_count = minimum support count of nodes in* β;
(4) **else for each** a_i in the header of *Tree* {
(5) generate pattern $\beta = a_i$ U α with *support_count* = a_i*.support_count*;
(6) construct β's conditional pattern base and then β's conditional FP_tree $Tree_{\beta}$;
(7) **if** $Tree_{\beta} \neq \emptyset$; **then**
(8) call FP growth(*Tree*_ , _); g

Concluding Remarks

There are many interesting directions in which the work described here can be extended. We have attempted to cover some algorithms in the broader context to interconnect with methodological predictive and descriptive models. There is a brief review analysis of application algoritms in research questions related to research models and techniques. The algorithms we present aren't representative of the full knowledge in Intelligent Systems but the complexity of them are important to understand the progressively design programs and analysis that can be achieve with this development. We also introduce some ideas of Decision Theory making use of the most recent sophisticated information mathematical technologies for improve business intelligence, medical issues and knowledge management, highlighting the fact that Intelligent System are transversal to all subject science.

We anticipate major advances in Artificial Intelligence and decision science analysis based on develop of interact between disciplines.

Acknowledgments. Thanks to our colleagues for the discussion and comments on various aspects of this research. We also want to thank to Center of Technology and systems – CTS in UNINOVA in Nova University in Lisbon for supporting this research.

References

1. Nilsson, N.J.: Artificial Intelligence: A New Synthesis. Morgan Kaufmann Publishers Inc., San Francisco (1998)
2. Carter, M.: Minds and computers: an introduction to the philosophy of artificial intelligence. Hist. Philos. Log. **30**(3), 306–308 (2009)
3. Feigenbaum, J., Feldman, E.A.: Computers and Thought. McGraw Hill, New York (1963)
4. Bransford, J.D., Brown, A.L., Cocking, R.R.: How People Learn: Brain, Mind, Experience, and School. National Academy Press, Washington (1999)
5. Gardner, H.: Multiple Intelligences: The Theory in Practice. McGraw Hill, New York (1993)
6. Sternberg, R.: The Triarchic Mind: A New Theory of Human Intelligence. Penguin Books, New York (1988)
7. Etzioni, D., Weld, O.: A softbot-based interface to the internet. Commun. ACM **37**(7), 72–76 (2005)
8. Turing, A.M.: Computing machinery and intelligence. Mind **59**, 433–460 (1950)
9. Salamon, A., Muehlhauser, L.: Intelligence explosion: evidence and import. In: Eden, A., Moor, J., Søraker, J., Steinhart, E. (eds.) Singularity Hypotheses. The Frontiers Collection. Springer, Heidelberg (2012)
10. Soares, N., Fallenstein, B.: Agent foundations for aligning machine intelligence with human interests: a technical research agenda. Technol. Singul. Manag. Journey, pp. 1–14 (2017)
11. Holtzman, S.: Intelligent Decision Systems. Addison-Wesley Publishing Company, New York (1989)
12. Howard, R.A.: The foundations of decision analysis. IEEE Trans. Syst. Man Cybern. Part C Appl. Rev. **4**, 211–219 (1968)
13. Matheson, R.A., Howard, J.E.: Readings on the Principles and Applications of Decision Analysis, pp. 445–475. Strategic Decisions Group, California (1967)
14. Aho, A.V., Hopcroft, J.E., Ullman, J.D.: Data Structures and Algorithms. Addison-Wesley Publishing Company, California (1983)
15. Gama, J., Carvalho, A.P.L., Faceli, K., Lorena, A.C., Oliveira, M.: Extracçao de conhecimento de dados, 2nd edn. Silabo, Lisboa (2015)
16. Maimon, O., Rokach, L.: The Data Mining and Knowledge Discovery Handbook. TEL-AVIV University of Israel, Israel (2005)
17. Han, J., Kamber, M., Pei, J.: Data Mining Concepts and Techiques. Elsevier, New York (2012)
18. De Coninck, N.: The relationship between big data analytics and operations research. Universiteit Gent (2017)

19. Kolmogoroff, A.: Foundations of the Theory of Probability. Chelsea Publishing Co., New York (1959)
20. Lawrence, P., Andriola, J.: Three-step method evaluates neural networks for your application. In: EDN, pp. 93–100 (1992)
21. Agrawal, R., Srikant, R.: Fast algorithms for mining association rules in large databases. In: Proceedings of the 20th International Conference on Very Large Data Bases, pp. 487–499 (1994)
22. Han, R., Pei, J., Yin, J., Mao, Y.: Mining frequents patterns without candidate generation. Data Min. Knowl. Disc. **8**, 53–87 (2004)
23. Tomasik, B.: Artificial Intelligence and its implications for future suffering (2016)

Hierarchical System for Evaluating Professional Competencies Using Takagi-Sugeno Rules

Ondrej Pektor[1], Bogdan Walek[1(✉)], Ivo Martinik[2], and Michal Jaluvka[1]

[1] Department of Informatics and Computers, University of Ostrava, 30. dubna 22, 701 03 Ostrava, Czech Republic
{ondrej.pektor, bogdan.walek, michal.jaluvka}@osu.cz

[2] Department of Computer Science, VSB-Technical University of Ostrava, Sokolska trida 33, Ostrava, Czech Republic
ivo.martinik@vsb.cz

Abstract. The article deals with a way of automated evaluation of competent persons evaluated as a percentage. In the introduction, the need of evaluating professional competencies of new or existing employees is outlined. We define the problem of overall evaluating various professional competencies. Subsequently, we propose a hierarchical system that automatically generates aggregate evaluation of any number of evaluating computations including continuous visualisation of the results. As a calculating unit, we use the Takagi-Sugeno expert system. This solution can be extended with other attributes that can be evaluated for competencies. The proposed hierarchical system is experimentally verified.

Keywords: Competency model · Fuzzy expert system · Takagi-Sugeno Hierarchical system · Professional competencies · Job position competencies

1 Introduction

In companies, it is emphasised that job positions are occupied by competent employees. Competences are defined as knowledge, skills and attributes that distinguish an expert from a layman. Another definition is that competency is a construction which helps define the level of skill and knowledge [1, 2]. The person competent to take up a particular job must meet the given competences to a minimum. This information is called the competency model. It can be successfully used to define which employee skills are crucial for the job position [3, 4].

For instance, based on the evaluation of the job applicant competencies, the overall job position suitability can be determined. This article emphasises professional competencies of an employee as a generalisation of the characteristics of the candidates from the previous article [5].

Evaluating the employee's competences can take place in various situations. For instance, in decision-making and recruitment of job applicants or in regular evaluation

R. Silhavy (Ed.): CSOC 2018, AISC 764, pp. 318–325, 2019.
https://doi.org/10.1007/978-3-319-91189-2_31

of existing staff. The assessment of each competence can take the form of a language expression, years of practice or percentage. In the case of the number of years of practice, this rating can be converted to percentages according to the given reference value (maximum, minimum, etc.) and in the case of a language expression, the conversion to percent is easy if the number and order of language expressions are fixed.

This article does not deal with a way of collecting information about people or how to evaluate their individual competencies.

One of the parts of the evaluation process is an algorithm that calculates the overall evaluation based on the evaluation of any number of the rated competencies in percent. For example, a multi-criteria analysis [5] or a weighted sum method can be used successfully. These methods, however, do not provide a simple tool for tracking the calculation process, they do not allow the adjustment of the equalisation algorithm and, last but not least, they only work with value and importance. Therefore, it is not possible to include additional attributes in the evaluation process.

There are a few approaches for evaluating competences:

One of them evaluates competences of employees using computerised tests [6]. Another approach selects appropriate employees to construction projects using a fuzzy adaptive decision-making model [7]. Another approach uses a fuzzy model for competency-based employee evaluation [8].

The aim of this article is to propose mechanisms and procedures that would be useful for transparent and variable calculation of the overall evaluation of competences of an employee based on the percentage assessment of individual competencies. Another feature of this solution will be the extensibility of any other attributes that can be listed and evaluated for competencies.

2 Methodology

The hierarchicity of the solution process is used because it helps use possibly an unlimited number of competencies. If we use a classical fuzzy expert system, the size of the knowledge base would depend on the number of input competencies.

For evaluation of competencies, we use the Takagi-Sugeno fuzzy expert system [9, 10]. The main difference between the Mamdani and Takagi-Sugeno fuzzy system is in the fuzzy rule consequent. Mamdani fuzzy systems use fuzzy sets as rule consequent. Takagi-Sugeno fuzzy systems uses linear functions of input variables as rule consequent [11, 12]. This important property of Takagi-Sugeno is suitable for our proposed approach. The hierarchical system is universal to be used with any evaluating algorithm.

2.1 Hierarchical Evaluation Process

The proposed evaluation procedure consists of evaluating competence pairs using a computing unit. Subsequently, the pairs of intercalculations are evaluated. After the necessary number of steps, which depends on the number of evaluated competencies, the result is a final evaluation in percentages. Generally, we can use n-tuplet of input competences.

In Fig. 1, there is an example of four competencies. Each competence has its name, value (rating) given in percent, and importance. This importance is a value from the interval [1, 5], which means that a higher number represents more important competency.

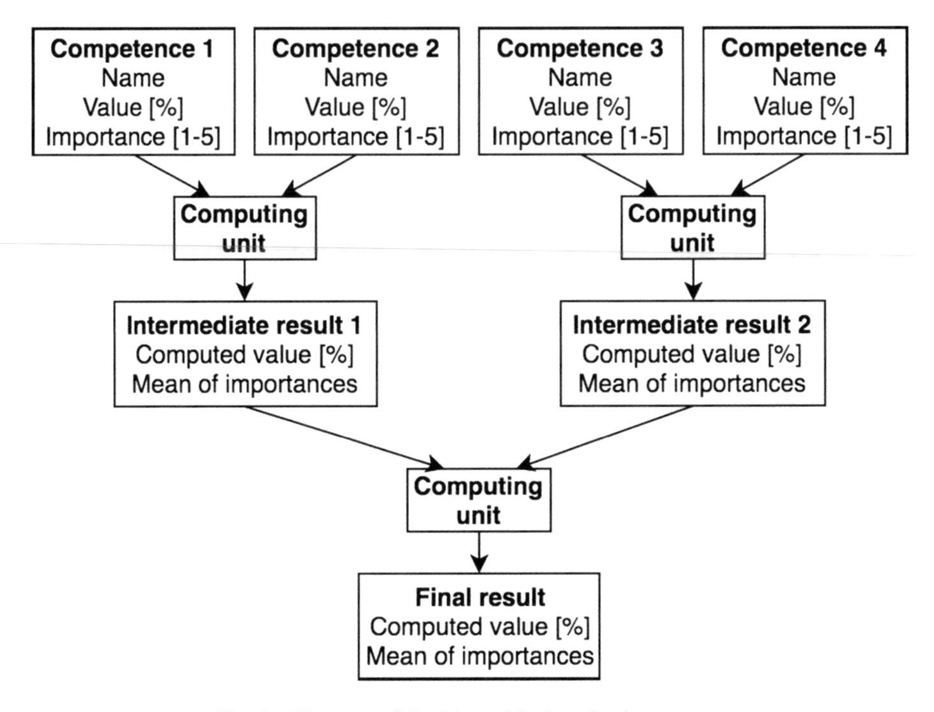

Fig. 1. Diagram of the hierarchical evaluating system

There are also computational units which is a function that calculates the total score of two (or generally any predefined number) competencies. The output of these computational units is an intermediate result, which is a competence that has its value and importance calculated from the competences that entered the previous computing unit.

These operations are performed in the indicated hierarchy until the intermediate result is only one. This last intermediate result becomes the final evaluation of all input competencies.

2.2 Computing Unit

The computing unit is shown in Fig. 1 as a box between two input competencies and one output competence.

As a computing unit, there can be used any function that is capable of processing all attributes of the input n-tuplet of competences and based on them, generating a cumulative evaluation that is used as an input to another computing unit, or it is the final evaluation of all inputs.

As an example, a Takagi-Sugeno inference system based computing unit is proposed here. This system has three inputs and one output.

The first two inputs (inputA and inputB) are the percentage rating of competences divided by two, so the values are from the interval [0, 50]. Another importance is the combined importance of entry competencies. The value is from the interval [−5, 5]. In the case of a negative input, the first input is more important, in the case of a positive input, the second input is more important. In the case of 0 as input, the inputs are equally important.

2.3 Fuzzification of Importance

Importance is divided into three fuzzy sets according to the used language expressions, which determine which of the inputs has a higher importance, see Fig. 2.

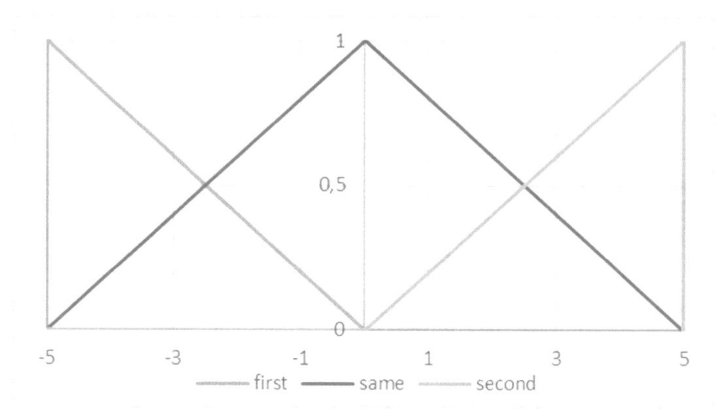

Fig. 2. Importance fuzzy sets

The first fuzzy set called "first" is used when the first input is more important. Its possible value is from [−5, 0]. The second fuzzy set "same" is used when both inputs are equally important. It has a possible value from the interval [−5, 5] and the last fuzzy set is used in case the second input is more important and it has a possible value from the interval [0, 5].

2.4 Knowledge Base of Computing Unit

This knowledge base is used for the inference mechanism of the computing unit.

The knowledge base consists of three IF-THEN rules of the Takagi-Sugeno type. In the antecedent, there is only the input variable "importance" and based on its value, the linear functions used in the consequent change.

```
IF importance IS first THEN suitability IS inputA * 1.5 +
inputB * 0.5
IF importance IS same THEN suitability IS inputA + inputB
IF importance IS second THEN suitability IS inputA * 0.5
+ inputB * 1.5
```

Coefficients used in the consequents can be set either for the entire hierarchical system globally or for each evaluation separately. Due to the use of Takagi-Sugeno rules, no defuzzification method is needed.

3 Verification

The proposed approach has been tested on experimental data using a Java application. Experimental data was selected from a job position called "Web app developer" in a cooperating real company. Experimental data represents a list of required competences for this job position. The results of the calculations were exported and converted to a scheme, see Fig. 3.

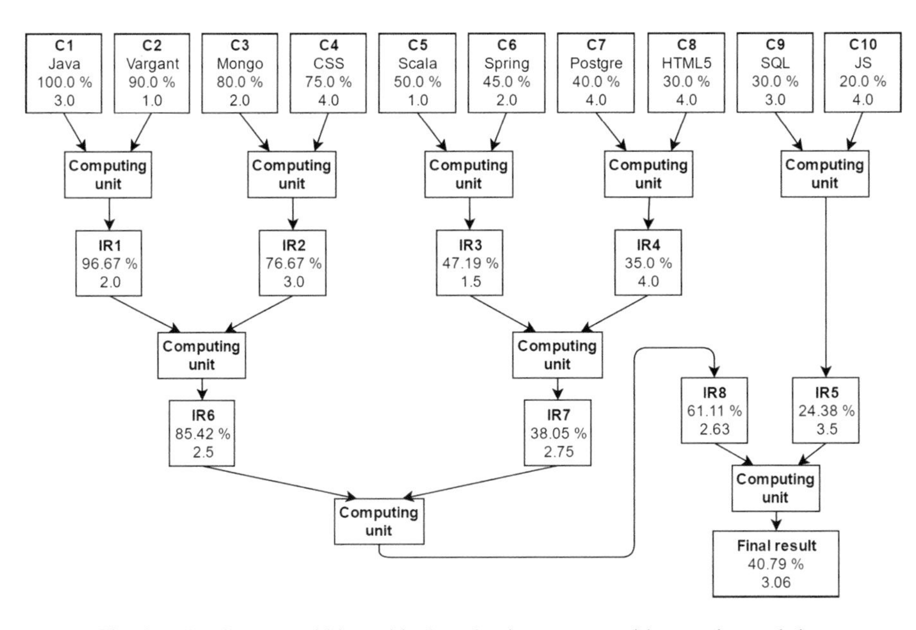

Fig. 3. The diagram of hierarchical evaluating system with experimental data

3.1 Determination of Experimental Inputs

To test the proposed solution, it was necessary to select a set of competences, their importance for the given job position and their evaluation for a given employee. These competences were selected from the hard skills listed in the previous article [5].

The assessment of these competencies is shown in percentages for the purposes of the calculation. The importance is then a number from the interval [1, 5], which determines the importance of the competence.

The first column of Table 1 provides the name of the competence, the second provides the percentage evaluation of competence (how much the given person meets the competence), and the sub-column is the importance of the given competence. These competencies are already sorted by value. More information about this step is described in Sect. 3.3.

Table 1. Experimental competences

Name	Value [%]	Importance [1, 5]
Java	100	3
Vargant	90	1
Mongo	80	2
CSS	75	4
Scala	50	1
Spring	45	2
Postgre	40	4
HTML5	30	4
SQL	30	3
JavaScript	20	4

In general, it is possible to add any other attributes to the competencies that can enter the evaluation. However, the computing unit must be ready for these inputs and must define the evaluation by using them.

3.2 Testing on Experimental Data

Figure 3 is a graphical representation of the entire algorithm procedure for testing with the data. At the top of the list, all ten input competences are ranked according to the percentage rating. The other levels of the hierarchy are the computing units listed in Sect. 2.2. If there is a case where an odd number of competencies or intermediate results appear at some level, the last of them is included in the next level of the hierarchy.

3.3 Comparing the Selected Attribute Sorting Method

The order in which the competences enter the hierarchical system has a provable effect on the course and outcome of the evaluation. Hence, the competencies of entering the hierarchical system are sorted according to their percentage rating. We tested a variation on the test data sample where attributes were ranked by importance, resulting in a final value of 56.02%, but the importance of this value was only 2.13. This approach provides results with lower credibility for these experimental inputs. Therefore, the ordering was pro-rated.

3.4 Comparison with the Multi-criteria Analysis Mentioned in the Previous Article

The proposed approach is an alternative to using multi-criteria analysis, because it also has an arbitrary number of inputs and one total output. However, it provides greater variability through the ability to modify the process of evaluating the proposed Takagi-Sugeno fuzzy expert system or using another algorithm. But the greatest benefit is that the algorithm can be extended with any number of additional attributes for each competence. This can be done by extending the computing unit with more attributes which will be included in the computation [13].

4 Conclusion

The article deals with a way of automated evaluation of competent persons evaluated as a percentage.

In the introduction, the need of evaluating professional competencies is outlined. Whether recruiting new employees or evaluating existing ones. We define the problem of overall evaluating various professional competencies.

Subsequently we propose a hierarchical system that automatically generates aggregate evaluation of any number of computations including continuous visualisation of results. As a calculating unit, we use the Takagi-Sugeno fuzzy expert system. This solution can be extended with other attributes that can be evaluated for competencies.

Another possible improvement is to extend the whole system with subcategories of competencies that will be evaluated separately and then to enter into further hierarchical or expert evaluation to determine the competence of an employee with regard to the job position or function in the organisation.

The proposed hierarchical system was experimentally verified.

Acknowledgment. This work was created during the completion of a Student Grant called "Application of fuzzy methods for evaluating competencies and developing an adaptive web" with student participation, supported by the Czech Ministry of Education, Youth and Sports.

References

1. Shippmann, J.S., et al.: The practice of competency modeling. Pers. Psychol. **53**(3), 703–740 (2000)
2. Harzallah, M., Vernadat, F.: IT-based competency modeling and management: from theory to practice in enterprise engineering and operations. Comput. Ind. **48**(2), 157–179 (2002)

3. Campion, M.A., et al.: Doing competencies well: best practices in competency modeling. Pers. Psychol. **64**(1), 225–262 (2011)
4. García-Barriocanal, E., Sicilia, M.A., Sánchez-Alonso, S.: Computing with competencies: modelling organizational capacities. Expert Syst. Appl. **39**(16), 12310–12318 (2012)
5. Walek, B., Pektor, O., Farana, R.: Proposal of the web application for selection of suitable job applicants using expert system. In: Software Engineering Perspectives and Application in Intelligent Systems, pp. 363–373. Springer, Cham (2016)
6. Zandbergs, U., Judrups, J.: Evaluating competences with computerized tests. In: Engineering for Rural Development, Jelgava, pp. 625–630 (2015)
7. Shahhosseini, V., Sebt, M.H.: Competency-based selection and assignment of human resources to construction projects. Sci. Iranica **18**(2), 163–180 (2011)
8. Golec, A., Kahya, E.: A fuzzy model for competency-based employee evaluation and selection. Comput. Ind. Eng. **52**(1), 143–161 (2007)
9. Lilly, J.H.: Takagi–Sugeno fuzzy systems. In: Fuzzy Control and Identification, pp. 88–105 (2010)
10. Scherer, R.: Takagi-Sugeno fuzzy systems. In: Multiple Fuzzy Classification Systems, pp. 73–79 (2012)
11. Takagi, T., Sugeno, M.: Fuzzy identification of systems and its applications to modeling and control. IEEE Trans. Syst. Man Cybern. **1**, 116–132 (1985)
12. Ying, H.: Sufficient conditions on uniform approximation of multivariate functions by general Takagi-Sugeno fuzzy systems with linear rule consequent. IEEE Trans. Syst. Man Cybern. Part A Syst. Hum. **28**(4), 515–520 (1998)
13. Zopounidis, C., Pardalos, P.M.: Handbook of Multicriteria Analysis, vol. 103. Springer Science & Business Media, Heidelberg (2010)

Discovering Association Rules of Information Dissemination About Geoinformatics University Study

Zdena Dobesova(✉)

Department of Geoinformatics, Faculty of Science, Palacky University, 17. listopadu 50, 779 00 Olomouc, Czech Republic
zdena.dobesova@upol.cz

Abstract. The article presents the data mining of dataset about spreading information among the university study applicants. The data were collected during admission procedure of applicants for bachelor study branch Geoinformatics and Geography at Palacky University in Olomouc (Czech Republic). Answers were received by questionnaire in two years, 2016 and 2017. Data collecting and processing aimed to discover the dissemination of first information about this specific specialization among graduates at secondary schools. Statistics and data mining techniques, namely finding association rule were used. Data mining discovered some unexpected relation and association in the data. Interesting results bring feedback about the impact of various presentation activities like Open Day, GIS Day or publishing information on the Internet. Results will also be reflected in future advertisement strategies of the study branch Geoinformatics to assure increasing interest of the potential applicants.

Keywords: Data mining · Associations rules · Questionnaire · Geoinformatics Classification · Advertisement · University applicant

1 Introduction

The right choice for future profession and specialization is a fundamental decision in the life of young people. This decision is mainly connected with finishing common secondary school and continuing study at university. The information dissemination about interesting study branches is important both for applicants both for universities. The importance for the universities is the fact that the number of potential students in the Czech Republic has been decreased due to demography evolution. Recently, some study branches have not filled up the declared quota of students. The best result is when personal interest of student corresponds with the offer of study to assure his professional success in career.

The Department of Geoinformatics has been collecting data about spreading information to applicants. The reason was to explore the penetration and influence of advertisement and other activities to the potential university students of geoinformatics. The generation of association rules, one of data mining technique, was used [1, 2].

R. Silhavy (Ed.): CSOC 2018, AISC 764, pp. 326–335, 2019.
https://doi.org/10.1007/978-3-319-91189-2_32

The advertisement activities and collection of data are described in Sect. 2. The processing data by association rules is described in Sect. 3.

2 Information About University Study

Delivering information about offered study branches at universities has several various ways. Some of them are common, and others are specific for the Czech Republic or Palacky University in Olomouc.

2.1 Advertisement Strategies About Study Branches at University

Palacky University issued printed leaflets and booklets about a list of offered study branches. The materials are delivered to the secondary schools, especially to the teachers and students. Detail study information is also presented in digital format on web pages of universities to describe the structure and aim of study braches. Description about specializations is supplemented by information about admission procedure, deadlines for application, fees and opportunities to forgive enrol exams.

Sometimes, the students actively try to find interesting study branch and information about the structure of the study. They ask their teachers at secondary schools for information about study opportunities at universities. Students also debate with their parents and siblings about the future profession. The recommendation and experiences of friends and schoolmates are very influential for their final decision. Students from the Czech Republic actively attend special study fair Gaudeamus that is organized in Brno and Prague in the fall and winter of each year [3]. Moreover, the Gaudeamus fair takes place in Nitra and Bratislava in the Slovak Republic. All national and abroad universities present their various study programs at these fairs.

Moreover, universities organize their form of public presentation of studies. Palacky University organizes annually two Open Days, one in November (Friday) and second in January (Saturday) [4]. Applicants could attend the university to personally speak with future teachers and receive information about admission procedure. Many excursions are arranged to present departments, classrooms, laboratories, equipment and research results of scientific teams. The excursions are arranged both during Open Days and optionally during the whole academic year.

2.2 Information About Study of Geoinformatics

Department of Geoinformatics at Palacky University issued a unique collection of printed leaflets about bachelor study branch Geoinformatics and Geography and master study Geoinformatics. The aim is to spread information to future potential students to the secondary schools and the public. Leaflets contain a description of the aim of study and list of subjects with highlighted topics. One leaflet "How live geoinformatics professionals in practice?" brings interviews with geoinformatics graduates about their different jobs. Several other leaflets exist about department and scientific research. They describe research projects, published books, map and atlas production (Fig. 1).

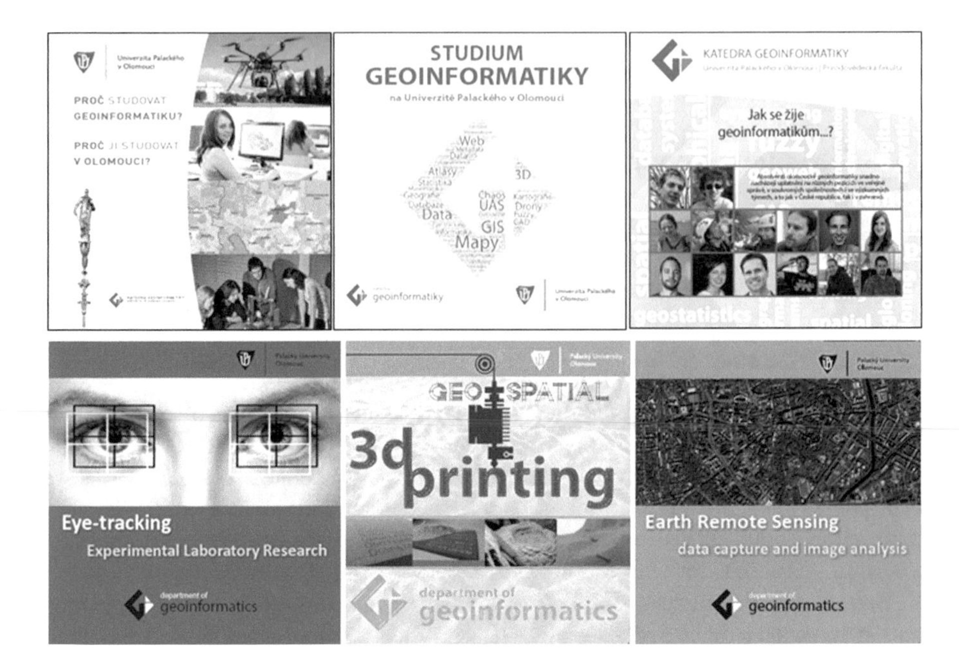

Fig. 1. Collection of leaflets about Geoinformatics study and research

Study branch Geoinformatics is presented at all presentation days like Open Days and Gaudeamus Fairs. Moreover, the study is presented at special action named GIS Day. GIS Day is an annual worldwide action that presented geographic information systems (GIS) technology. It is regularly organized at third Wednesday in November. Schools, firms, and many organizations prepare many presentations for the public to discover and explore the benefits of GIS. GIS Day a good initiative for people to learn about geography and the uses of GIS [5]. More than one thousand participants take part at this event around the whole world. The Department of Geoinformatics regularly organizes a presentation for a student from secondary schools from Olomouc and other neighbouring cities. Groups of students with their teachers of geography attend computer laboratories at the department and try to use GIS software, prepare maps or solve a geographical task. They make an idea about utilization of GIS technology, and they receive basic information about opportunities to study Geoinformatics discipline. Present students of master study Geoinformatics also attend secondary schools with presentation lectures about study opportunities. These activities run during the whole school year in various cities and schools.

Besides these all presentations activities, other publicity channels exist. The Teacher News (www.ucitelskenoviny.cz) is issued in the Czech Republic. The autumn issue regularly brings a complete list of Czech Universities and all study branches offered for next academic year. Next information channel is radio and TV. Local and national radio and television broadcasting (like Český rozhlas, Radiožurnál, ČT1, ČT24) bring several reportages about Palacky University and interviews with famous researchers and teachers. These entire information channels could be a source of first information or a hint about particular study branch.

It is very difficult to evaluate the influence of all mentioned activities to the attraction of potential applicants. The research task was to find the impact of presentation days, leaflets and other channels of advertisement to the spreading of information to the future applicant of study branch Geoinformatics.

3 Data Acquisition and Data Mining

The data have been collected from applicants for evaluation of the advertisement impact. The Department of Geoinformatics prepared a questionnaire for applicants to imagine the way of spreading the first information about the study. The statistical methods and data mining methods were used for processing data from the questionnaire. Data mining is the process of discovering and extracting hidden patterns from different data types to guide decision makers and making decisions [6]. Data mining performs different methods including classifications, clustering, regression, association rule discovery, decision tree, and pattern recognition. The method of association rules generation was chosen because the received transactional data had a suitable form for this mining method. The rules could be stored and reused as knowledge in intelligent and expert systems [7]. Clustering and classification of data, as frequent data mining methods, was used as next step of data analysis [8].

3.1 Questionnaire and Data Matrix Creation

The data were collected during admission periods in two years 2016 and 2017. The data was gathered by questionnaire from April to September each year. It is hard to meet all applicants at the same time. Some applicants only send the application in February without starting the study and any other connections to the university.

Some attended applicants were first asked to fill the questionnaire in April on information weekend named "GIS weekend" organized by Department of Geoinformatics. This meeting with applicants was an introductory weekend about enrolment tests, information about the structure of the study. The presence was optional for applicants, so the number of received questionnaires was approximately 50%. Subsequently, the questionnaire was delivered during enrol exams in June to the remains students. Finally, the last group of the student was questioned in September (about 8%) at the beginning of the study. These students had forgiven enrol exams, and they did not attend GIS weekend in April. The groups of students were punctually identified to prevent duplicated in filled questionnaires. This way of the questionnaire spreading covered nearly all applicants except them that they did not attend exams and did not start the study of Geoinformatics (about 10%). They only sent the application in enrolment period without any other interests.

The structure of questionnaire was straightforward. The necessary information like name of the student, secondary school, and the city was filled with the introduction section. The main question was:

"How did you find out information about studying Geoinformatics at Palacky University?"

There were 12 possible answers:

- Teacher of Geography at secondary school *(TeacherG)*
- Teacher of Information science at secondary school *(TeacherI)*
- Lecture at secondary school provided by Department of Geoinformatics *(Lecture)*
- Attending of GIS Day at Department of Geoinformatics *(GIS Day)*
- Open Day at Palacky University (November and January) *(Open Day)*
- Friends, schoolmates, siblings *(Friend)*
- Parents and grandparents *(Parent)*
- Gaudeamus Fair in Prague or Brno *(Gaudeamus)*
- Leaflets about study branch Geoinformatics *(Leaflets)*
- TV and Newspapers *(TV News)*
- Internet by your search *(Internet)*
- Teacher News *(T News)*

Each type of answer was assigned with the code. The codes of answers are in Italics in brackets above. The codes are used in the graph in Tables 1, 2 and Fig. 2.

Table 1. Input transactional data example.

ID	Items
1	TeacherG, Open Day, Internet
2	GIS Day, Gaudeamus, Leaflets, Internet
3	Friend, Internet
4	Parents, Gaudeamus, Leaflets
5	Lecture, Open Day, Internet

Table 2. Input matrix example for association rules.

ID	Teacher G	Lecture	Open Day	GIS Day	Friend	Parents	Gaudeamus	Leaflets	Internet
1	1	0	1	0	0	0	0	0	1
2	0	0	0	1	0	0	1	1	1
3	0	0	0	0	1	0	0	0	1
4	0	0	0	0	0	1	1	1	0
5	0	1	1	0	0	0	0	0	1

Some student's responses contain one or more answers. The maximum was six answers by one applicant; it means six various sources of information. The most frequent was two or three sources of information about the study. The answers from the questionnaire were collected like transactions in the table. The term of the transaction is used in Market Basket Analysis (MBA) to discover association rules in data mining process [1, 2, 9]. Students were assigned by identification number ID. The example of input data for data mining is in Table 1.

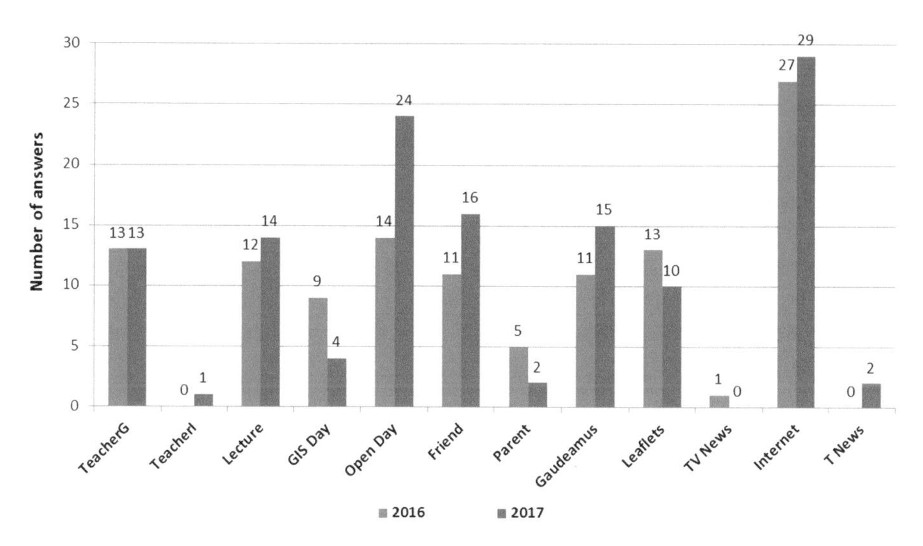

Fig. 2. Graph with numbers of responses for each source of information about the study.

The transactional table data were converted to the matrix where items are binary attributes. Attribute value 1 represents the presence of answer; the 0 represent the absence of an answer. The resulting matrix is very sparse matrix due to the low frequency of answers (a lot of zero values). The example of the data matrix is in Table 2. The data in Table 2 correspond to the transactional data in Table 1. There are not all 12 possible answers (attributes) to reduce the size of the table in this article. The conversion of input data to the matrix was source data for statistics and association rules generations.

Finally, the data was supplemented by the information, if students started or did not start the study of Geoinformatics in September. It was taken from evidence about new students in the first grade. It is also interesting the association of information dissemination about geoinformatics study and the real start of the study. Moreover, information about attending the GIS weekend was added to the data form questionnaire.

3.2 Data Analyzing

Firstly the basic statistics were prepared. A total number of respondents was 46 in the year 2016 and 55 in 2017. The number of answers for each type of information for two years is displayed in Fig. 2. The highest number of answers has the source form Internet and web pages. The answers are expectable because of the primary information about bachelor study, an overview of subjects, an organization of study and examples of enrolling test (mathematics and geography) are placed on web pages Department of Geoinformatics (www.geoinformatics.upol.cz). Also, faculty web pages contain necessary information about admission procedure, conditions, and dates (www.prf.upol.cz). Also, other sources from the Internet could be assumed like Facebook. The particular web pages were not inquired in the questionnaire because the answers would not be reliable. This answer could be assumed to be biased and over evaluated by the time delay between received first information and time of survey (sometimes more than six months). Also, there is a bias about the primer information about the existence of study branch Geoinformatics on the Internet. Primer information could be outside of the Internet, but applicants subsequently looked for detailed information on the Internet. It is hard to remember for applicants after two or six months to distinguish if the information form Internet was solely primary information. However, the answers verified that the Internet is also the most frequent information source about the study. Surprisingly, a minimal number of answers got: teacher of information science subject, TV and newspapers, Teacher News. Very influential are Open Day, Gaudeamus, Lectures at secondary schools and Leaflets. Fiends and siblings have more answers than recommendations by parents.

3.3 Association Rules

Association rule discovery task is a typical example of unsupervised learning [6, 9] where the aim is to discover the correlation among attributes in the transaction data like presence-absence data received by questionnaire. Based on converted data to the matrix, association rules find relations among items. The association rules are displayed on the next form (1), where the antecedent is a precondition of rule and succedent is the result of the rule. The interpretation of the rule is: When the precondition is valid than also the result is valid [9]. The inference made by an association rule does not necessarily imply causality. Instead, it suggests a strong co-occurrence relationship between items in the antecedent and consequent [1].

$$Antecedent => Succedent \tag{1}$$

The evaluation of the validity of association rules depends mainly on three important values called support, confidence and lift. Support represents the frequency of the items in datasets, in other words, objects that fulfill both antecedent and succedent. The confidence indicates how often the rule has been found to be true in itemsets that represent this rule. Lift is the ratio of the probability when antecedent and succedent occur together to the two multiple individual probabilities of antecedent and succedent [1]. Antecedent and succedent are independent when the value of the lift is 1. The higher value of lift means that existence of antecedent and succedent is not just a random occurrence, but because of some relationship between them.

In the process of the rules generation is user-set two thresholds: minimum support and minimum confidence. The process of generating association rules consists of two steps. Firstly, to find all frequent itemsets and secondly generate the association rules from frequent itemsets. The recommendation for minimum confidence is about 70% for data obtained by questionnaire [2], and minimal support 5%. The Apriori algorithm is the most well-known algorithm for finding the frequent itemsets and Boolean association rules [10]. The software WEKA v 3.8 was used for data mining of presented questionnaire data. The Apriori algorithm is implemented in WEKA.

Both the matrix with one-zero (presence-absence) information, both matrix only one-one (presence) information was processed. The matrix with presence-absence data produces a lot of strong rules (high support and high confidence) but only with absence items. The sparse matrix did not produce a lot of frequent positive, strong rules.

The extracted presence-absence rules are briefly described like this:

- ***Parent = 0 TeachN = 0 => TeacherI = 0 TVnews = 0*** (*support 92%, confidence* 99%)
  This rule means that many students do not receive any information from Parents, Teacher News, and Teachers of Informatics science. All these channels are a weak source of information with high occurrence and high confidence. It is the strong rule.

The rules with lower support and some presence-absence items are like these:

- *TeacherG = 0* ***Gaudeamus = 1*** *TeachN = 0 => Lecture = 0 GIS_Day = 0* ***Open-Day = 1*** *Friend = 0 (support 17%, confidence 59%, lift 2.7)*
  This rule means that visiting Gaudeamus Fair and visiting Open Day has high co-occurrence without other information. The rule covers a group of students that have no information from teachers of geography, information lectures, friends and GIS Day, so they very often attend the public presentation, and they prefer the personal receiving of information by humans. They travel to the fair and the university to find some information.
- *TeacherG = 0 Lecture = 0* ***Gaudeamus = 1*** *TeachN = 0 => TeacherI = 0 GIS_-Day = 0* ***OpenDay = 1*** *Friend = 0 (support 14%, confidence 71%, lift 2.67)*
  This rule is near the previous one. Missing information from teachers of geography, lecture at secondary school, Teacher News and friends forces the students to attend the Gaudeamus and Open Day personally.
- *Gaudeamus = 0 and Open Day = 0 =>* ***Internet = 1*** *(support 29%, confidence 52%, lift 1,02)*
  This rule expresses that student without any information from campaign use namely the Internet. The absence at the Gaudeamus Fair and Open Day (no personal attendance) and searching information on the Internet, on the other hand, are an independent phenomenon according to the lift.

At the second step, only presence matrix of items was used. The presence of an item in a transaction is considered more important than its absence in that case. The list of interesting rules is:

- ***Friend = 1 Leaflet = 1 => Internet = 1****(support 7%, confidence 86%, lift 1.55)*
  This rule could be considered as a searched exception. It has low support and high confidence. Students that have information from friends and read the leaflets also use the Internet. These students did not attend public presentation like Fair.
- ***TeacherG = 1 => GIS_Day = 1*** *(support 26%, confidence 27%, lift 2.09)*
  This rule expresses: If a student has information from Teacher of Geography at secondary school they also attend (in 27% of cases) GIS Day. According to the lift 2.09, the information from the teacher has high co-occurrence with GIS Day.
- ***Friend = 1 => OpenDay = 1*** *(support 27%, confidence 30%, lift 0.79)*
  This rule expresses that student with information from friends also attends Open Day. It is not a frequent rule but the preferable rule.

Finally, the generation of association rules from the data with the supplemented information about the start of study brings some interesting rules. There is no strong rule with the very high support and confidence. Discovered rules are:

- ***OpenDay = 1 Internet = 1 => Study = 1*** *(support 20%, confidence 75%, lift 1.33)* That means that students who attend Open Day and have information on the Internet they start the study.
- ***Friend = 1 and Internet = 1 => Study = 1*** *(support 13%, confidence 85%, lift 1.5)* It is the rule with the low support and high confidence in case of started study. It could be assumed as an exception that student starts study based only the recommendation of friends and the Internet.
- ***Internet = 1 => Study = 0*** *(support 13%, confidence 85%, lift 1.5)* This rule revealed situation when a student had information only from Internet he did not start the study. The Internet as a sole channel of information is very weak, and students are not interested in the study of Geoinformatics.

4 Results

The evaluation of the information dissemination about study brings some interesting information. The Apriori algorithm in WEKA software was used for the generation of association rules. Gaudeamus Fair, University Open Day and Internet are the most influenced activities to the applicants. If students attended Gaudemus Fair, they also attend Open Day and use the Internet. Personal meeting of the applicant with teachers of the university was very influential. Fair and Open Day are frequently used way of collecting information by secondary school graduates. These ways of personal dissemination of information form the first group of active applicants. The second group of applicants used the only Internet and they did not attend Gaudeamus and Open Day. The addition of the information about the start of the study revealed that students with information based only on Internet very often did not start study (support 13% and confidence 85%). High support 92% were discovered in rules with the absence of information from teachers at secondary schools, Teacher News, TV news and parents. These channels of information are very weak.

Surprisingly, the recommendation and experiences of friends and schoolmates are a very influential way of receiving information and final decision to start the study. The combination of friend, leaflets, and the Internet is interesting rule exception with small support but high confidence. This rule describes a small third group of students that prefer private searching of information. Also, high confidence has a combination of Open Day and Internet source to start the study. The Internet as a sole source of information did not assure the start of the study. Detection of this isolated source affirms that more channel of information is necessary to assure interest of graduates and successful start of the study.

The discovered rules help improve and intensify some forms of spreading information about the study. The most important is active and punctual delivering information during Gaudeamus Fair and Open Day about the study branch Geoinformatics. Other activities like GIS Day, lectures at secondary school are less frequent but could be an indirect source for friends, schoolmates and teachers of geography. These people potentially hand over the information to the target group of applicants. Finally, the actual and complete university web pages about the study are assumed as the necessary essential base of information.

Application association rules as one of data mining techniques is a new approach in processing questionnaire survey about dissemination information about the university study. The comparison with other study branches and universities is impossible due to lack of reliable information. This area opens new opportunities for retrieving interesting information about future applicants.

Acknowledgment. This article has been created with the support of the Operational Program Education for Competitiveness – European Social Fund (project CZ.1.07/2.3.00/20.0170 Ministry of Education, Youth, and Sports of the Czech Republic).

References

1. Tan, P.-N., Steinbach, M., Kumar, V.: Introduction to Data Mining, 1st edn. Addison-Wesley Longman Publishing Co., Inc., Boston (2005)
2. Šarmanová, J.: Methods of data analysis (Metody analýzy dat). Technical University of Ostrava, Faculty of Mining and Geology, Ostrava (2012). (In Czech)
3. MP-Soft, Gaudeamus Fair. https://gaudeamus.cz/
4. Palacky University: Why study at Palacky University? (Proč studovat na Univerzitě Palackého?). www.studuj.upol.cz. (In Czech)
5. Esri, GIS day. http://www.gisday.com/
6. Witten, I.H.: Data Mining. Morgan Kaufmann, Burlington (2011)
7. Brus, J., Dobesova, Z., Kanok, J., Pechanec, V.: Design of intelligent system in cartography. In: Proceedings of 9th Roedunet International Conference (RoEduNet) 2010, pp. 112–117 (2010)
8. Petr, P., Krupka, J., Provaznikova, R.: Statistical approach to analysis of the regions. In: Fujita, H., Sasaki, J. (eds.) Selected Topics in Applied Computer Science, pp. 280–285. World Scientific and Engineering Acad and Soc, Athens (2010)
9. Pavel, P.: Methods of data mining (Metody Data Miningu), part 2. University of Pardubice, Faculty of Economics and Administration, Pardubice (2014). (In Czech)
10. Agrawal, R., Srikant, R.: Fast algorithms for mining association rules in large databases. In: Proceedings of 20th International Conference on Very Large Data Bases, pp. 478–499 (1994)

Predicting User Age by Keystroke Dynamics

Avar Pentel(✉)

School of Digital Technologies, Tallinn University, Tallinn, Estonia
pentel@tlu.ee

Abstract. Keystroke dynamics is investigated over 30 years because of its biometric properties, but most of the studies are focusing on identification. In current study our goal is to predict user age by keystroke data. We collected keystroke data through different real life online systems during 2011 and 2018. Data logs were labeled with user age, gender and in some cases with other available information. We analyzed 2.3 million keystrokes, from 7119 keystroke data logs, produced by ca 1000 individual subjects, presenting six different age groups. All these data logs are also made available to research community, and the web address is provided in the paper. We carried out binary and multiclass classification using supervised machine-learning methods. Binary classification results were all over the baseline, best f-score over 0.92 and lowest 0.82. Multiclass classification distinguished all groups over baseline. Analyzing distinguishing features, we found overlap with text-mining features from previous studies.

Keywords: Keystroke dynamics · Age prediction · Multiclass classification Supervised machine learning

1 Introduction

When interacting with other people, the age is important attribute that we are taking into account. The same action, the same sentence is interpreted differently depending on author's age. Thus, systems ability to detect automatically user's age is crucial component of intelligent system that serves humans. As humans, we do not recommend the same things to adults and children, and we can expect similar behavior from automatic recommendation systems. With widespread of social media, people can register accounts with false age information about themselves. Younger people might pretend to be older to get access to restricted content, older people might pretend to be younger in order to communicate with youngster. As we can imagine, this kind of false information might lead to serious threats, as for instance pedophilia or other criminal activities.

Many studies are done on user age prediction based on text-mining approaches. There are differences in vocabulary between age groups [1–4]. Similarly, there are stylistic differences between age groups [3, 5, 6]. The shortcomings of text-mining approaches are language dependency and the need for longer texts. But besides content and style, there are physical patterns of typing, keystroke dynamics, that can reveal new valuable information, and allow making better predictions about user age on shorter input.

R. Silhavy (Ed.): CSOC 2018, AISC 764, pp. 336–343, 2019.
https://doi.org/10.1007/978-3-319-91189-2_33

Keystroke dynamic is now studied over 30 years [7] and is well accepted as biometric property allowing user identification. As byproduct of those identification studies, there are reports, that some user properties, including age, influences identification quality [8]. Among other properties, gender and affective sate were mentioned. In recent years, there is growing interest to use keystroke dynamics in user profiling. There are some studies on emotion detection [9, 10], gender and age [11, 12] prediction. However, the number of studies on age prediction is still small, and there is lack of available age and keystroke datasets.

In this study we present our experimentations on dataset that is collected between 2011–2018. Some of the same data is used in our other studies and reported separately [10, 12, 15]. Combining keystroke data from wide variety of contexts, we put together all data to a single dataset not depending on if it was typed on desktop, laptop, tablet or smartphone. Our intention in doing so was to set a baseline with relatively diverse data and simple set of features, so we could in further studies step-vise exclude data with some common properties and to test if it improves the models. Similarly we can test new feature types, and classification methods. All data logs used in this study are made available to research community, and can be downloaded from following web address:

http://www.tlu.ee/~pentel/keystroke_age/.

2 Methods

2.1 Description of Data Sources

Keystroke data collected 2011–2018 from six different data sources as described in [11], but in this study we use about 10 times more instances of data, and we classify between more age groups. These data sources represented online content management systems, feedback questionnaires, testing environments, and controlled experiments. Most of the users were typing in Estonian language, there were handful of logs in Russian and some logs in English and German, but typed by native Estonians or Russians. As data collection time was relatively long there are number of data instances created by the same users, but representing different age groups. As our data sources had very different context, starting from content management systems, where users created long documents, and ending with short comments and feedback, then number of keystrokes per user varied significantly between groups and in groups. Distribution of the final dataset by age groups and average number of key presses per instance is presented in Table 1.

2.2 Data Collection Methods

Data collection was carried out in real life online environments, where in many cases we had no control over user input technology. We still collected data about screen resolution, a *user agent* string to specify used web browser, operating system and, in our case, most importantly, if the used device was desktop computer or laptop, tablet or smartphone.

Table 1. Distribution of keystroke data across age groups

Age group	Number of instances in group	Average number of key presses per instance
<=15	376	308.41
16–19	258	3217.42
20–29	357	176.11
30–39	1327	170.58
40–49	3213	241.28
>=50	1588	176.93
Total	7119	321.88

Our experimental results show, that there are significant differences in typing patterns between virtual keyboards on smart devices and physical keyboards. Namely, the key is hold down significantly longer on physical keyboards (ca 7 times longer), but the time between last key release and next key press takes longer about the same magnitude on virtual keyboards. There is no such significant difference between typing patterns between laptop and desktop, furthermore, there is no straightforward method to detect if web user is using desktop or laptop. Despite differences in typing patterns, in this study we did not exclude keystroke data collected from phones and tablets.

For data collection, we implemented JavaScript code [9] for logging keystroke data, and server backend in PHP, and integrated it in to all of our data collection systems. By keystroke data we mean three numbers that characterizing each key press:

(1) keyCode - numeric representation of a key,
(2) keyDown time - time, when the key was pressed,
(3) keyUp time - the time when the key is released.

Out of those triples, we calculated different variables that characterizing particular user.

2.3 Preprocessing and Feature Extraction

We used in this study only 4 types of features. All used feature types were time-based: key hold time, key seek time and n-graph latencies. The descriptions of these feature types are given in Table 2. On figure below the table (Fig. 1), the visual scheme of keystroke features is presented as example of two-key sequence. For each instance, we calculated mean time for each key or n-key combination, and used it as variable in feature vector.

Because the logs lengths varied significantly between and in the age groups, we calculated frequency of each feature in the set of all instances, and excluded all features with frequency below 500 occurrences in set. In this way we got 134 features to extract from the each log.

Another important issue with keystroke dynamics is the problem of missing values. When in text mining we have n-gram based features, they are calculated by relative frequency, and having zero frequency for some feature is normal. In keystroke

Table 2. Description of used keystroke features

Type	Explanation	Total	Used
Hold	Mean time between key press and release for particular key	100	29
Seek	Mean time between previous key release and current key press for particular key	100	29
n-graph latencies	Mean time of 2–4 consecutive key presses, starting from first key press and ending with last key release for particular n-graph	335 418 118	63 5 2
Pauses between words	Mean time of n-graphs that includes space character	6	6

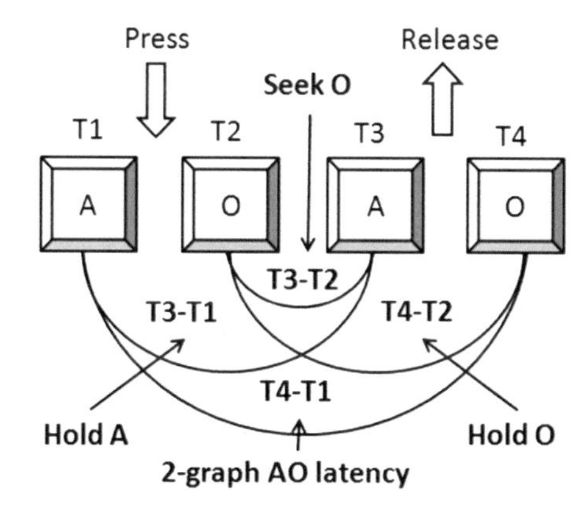

Fig. 1. Visual scheme of used keystroke features

dynamics, when time based features are in use, zero means that particular key or combination of keys are typed instantly, or, in case of standardized data, that the particular keys are typed as average. While latter is better option, it does not characterize the situation adequately. If in question is some key combination that is physically difficult to type or is it so rare that people are not used to it, then the probability that it is typed slower and the probability that it is missing value, are both high. If all instances were standardized then it would rather make sense to substitute missing values with feature mean, or to use imputation algorithm [13] for substitution missing values. In current study we substituted all missing values with feature mean.

2.4 Machine Learning

We performed binary classification between each age group and multiclass classification for all classes together. Before generating models, we balanced our classes by undersampling the majority class or classes in case of multiclass classification. This means that in case of multiclass classification, all classes were undersampled to 258 instances, like the smallest age group 16–19, and in case of binary classification the number of instances per class varied from 258 to 1588, depending on sizes of binary classes (Table 1). The models were generated using Support Vector Machines, Random Forest, C4.5, and Logistic Regression. Java implementations of listed algorithms that are available in the Weka [14] library were used. 10-fold cross validation was used for evaluation of all models. It means, we partitioned our data into 10 even sized and random parts, and then using one part for validation and another 9 as training dataset. We did so 10 times and then averaged validation results. Presenting our results here, we use a single f-score value that is an average of both classes' f-score values.

3 Results

In all cases of binary classification, Random Forest yield to best results. In Table 3, the f-scores of binary classification between each age group are presented. Results are weighted average f-scores of both classes. As all classes were balanced then the baseline in binary classification was 0.5.

Table 3. Binary classification between each age group.

	<=15	16–19	20–29	30–39	40–49
16–19	0.82				
20–29	0.87	0.83			
30–39	0.89	0.9	0.86		
40–49	0.9	0.9	0.87	0.87	
>=50	0.89	0.92	0.86	0.87	0.86

As mentioned above, we tested four different algorithms. We do not present here the results of each algorithm separately, because Random Forest classifiers were only slightly better than those generated by C4.5 and SVM. We speak here about differences as small as 0.9 vs. 0.88, for example. Differences between models generated by Logistic Regression and Random Forest were bigger, but Logistic regression still yield to results over 0.7 in each classification.

The best results with multiclass classification were also made by Random Forest. We present the results in form of confusion matrics, f-scores and ROC area in following Table 4.

Table 4. Multiclass classification on balanced dataset using Random Forest. Confusion matrics.

Classed as	<=15	16–19	20–29	30–39	40–49	>=50	F-score	ROC Area
<=15	**166**	24	27	16	8	17	0.69	0.9
16–19	15	**189**	34	5	7	8	0.73	0.93
20–29	6	31	**154**	27	11	29	0.55	0.87
30–39	13	4	24	**157**	23	37	0.59	0.86
40–49	12	6	30	28	**149**	33	0.61	0.87
>=50	15	3	31	38	33	**138**	0.53	0.85
Weighted average (baseline F-score is 0.167):							**0.62**	**0.88**

As we can see in the case of multiclass classification, all the results are far over baseline 0.167, but also number of false positives and false negatives are in all cases smaller than the number of true positives. Similar pattern emerged also with other tested algorithms.

Features with the greatest info gain are presented in Table 5. We compared these 20 features, with text-mining frequency based features [6] extracted from different data but the same language, and we found that there is only one (4. KS) feature that is not presented in the top of frequency based n-graphs and n-grams. The finding indicates that the frequency of some character combinations used in particular age group is correlated to typing speed of those characters.

Table 5. Top 20 features in multiclass classification

1.	I{space}	6.	D{space}	11.	{space}O	16.	SI
2.	AL	7.	Seek Ä	12.	LE	17.	EL
3.	S{space}	8.	AK	13.	MA	18.	E{space}
4.	KS	9.	Seek I	14.	IN	19.	JA
5.	EI	10.	L{space}	15.	Seek L	20.	LA

We could not use current keystroke dataset for n-graph frequency analysis, because this includes cases where the user types someone else's text and in that case n-graph frequencies would characterize rather the original author of the texts and not the typist.

4 Conclusion

Keystroke dynamics data collection is unobtrusive for users and can be easily implemented in to real life web environments. Opportunity to guess the age of anonymous user right on the way by analyzing typing patterns can have many useful applications starting from automatic recommendation systems and ending with cyber crime detection.

We demonstrated that keystroke data that is collected through different environments and contexts, and produced by different devices - desktops, laptops, tablets and smartphones, can still provide enough information to distinguish between different age groups. Our main intention with this study was to set a baseline for further studies, where we can exclude data with some specific properties and to classify it separately. Building different models for smartphones and tablets, desktops and laptops will improve the prediction accuracy.

Similarly we can test step-vise if some new feature types will improve our models; for instance these based on frequency of correction keys or others calculated by larger groups of keys that are grouped by their physical location or by frequency they are represented in language. There are groups of key combinations that are more difficult to type than others. These difficulties might be the result of the inconvenient physical locations of the keys or rareness of the key combination in the user language. Focusing on these difficulties, we can get some cues about user's vocabulary, and we can also get information about physical properties of the user. As we experienced in our handedness detection study [15], creating combined features that are based on physical location of the keys, can significantly improve classification results. Both cues - physical and vocabulary - help to make better predictions about user's age group.

While we had some good profiling results in this study, our main contribution is a successful attempt to profile users on very diverse keystroke data, which is collected mainly in real life situations and with large number of individual subjects. We also made this dataset available on the web, and provided link to it in the end of the introduction section of this paper. Our results are still preliminary, but it is good starting point for further studies, and we anticipate a more systematic and in-depth study in the near future.

References

1. Schwartz, H.A., et al.: Personality, gender, and age in the language of social media: the open-vocabulary approach. PLoS ONE **8**(9) (2013). Ed. Tobias Preis
2. Zhang, J., et al.: Your age is no secret: Inferring microbloggers' ages via content and interaction analysis. In: Proceedings of the 10th International Conference on Web and Social Media, ICWSM 2016, p. 476–485. AAAI Press (2016)
3. Rosenthal, S., McKeown, K.: Age prediction in blogs: a study of style, content, and online behavior in pre - and post - social media generations. In: Proceedings of the 49th Annual Meeting of the Association for Computational Linguistics: Human Language Technologies, vol. 1 (2011)
4. Morgan-Lopez, A.A., et al.: Predicting age groups of Twitter users based on language and metadata features. PLoS ONE **12**(8) (2017)
5. Flekova, L., Preotiuc-Pietro, D., Ungar, L.H.: Exploring stylistic variation with age and income on Twitter. In: Proceedings of the 54th Annual Meeting of the Association for Computational Linguistics, vol. 2 (2016)
6. Pentel, A.: Effect of different feature types on age based classification of short texts. In: Proceedings of 6th International Conference on Information, Intelligence, Systems and Applications. IEEE Digital Library (2015)
7. Garcia, J.: Personal identification apparatus. US Patent Office, 4621334 (1986)
8. Monrose, F., Rubin, A.D.: Keystroke dynamics as a biometric for authentication. Future Gener. Comput. Syst. **16**, 351–359 (2000)
9. Vizer, L.M.: Detecting cognitive and physical stress through typing behavior. In: Proceedings CHI EA 2009 CHI 2009 Extended Abstracts on Human Factors in Computing Systems, pp. 3113–3116 (2009)
10. Pentel, A.: Emotions and user interactions with Keyboard and Mouse. In: Proceedings of 8th International Conference on Information, Intelligence, Systems and Applications. IEEE Digital library (2017)
11. Fairhurst, M., Costa-Abreu, M.D.: Using keystroke dynamics for gender identification in social network environment. In: 4th International Conference on Imaging for Crime Detection and Prevention (ICDP) 2011
12. Pentel, A.: Predicting age and gender by keystroke dynamics and mouse patterns. In: UMAP 2017 Adjunct Publication of the 25th Conference on User Modeling, Adaptation and Personalization. pp 381–385. ACM Digital Library (2017)

13. Schafer, J.L.: Analysis of Incomplete Multivariate Data. Chapman and Hall, New York (1997)
14. Hall, M., et al.: The WEKA data mining software: an update. SIGKDD Explor. **11**, 1 (2009)
15. Pentel, A.: High precision handedness detection based on short input keystroke dynamics. In: Proceedings of 8th International Conference on Information, Intelligence, Systems and Applications. IEEE Digital library (2017)

Salary Increment Model Based on Fuzzy Logic

Atia Mobasshera, Kamrun Naher, T. M. Rezoan Tamal, and Rashedur M. Rahman(✉)

Department of Electrical and Computer Engineering, North South University, Plot-15, Block-B, Bashundhara Residential Area, Dhaka, Bangladesh
{atia.mobasshera, rezoan.tamal, rashedur.rahman}@northsouth.edu, kamruntithy@gmail.com

Abstract. It is quite challenging for the human resource management professionals to deal with the employees' salary expectations compared to their work during the salary increment period. Employees' salary is raised based on some major factors which include their mastery, responsibility, workload factors and many more. For the salary increment, the employee's performance is estimated based on crisp values. This estimation leads to uncertainty and vagueness. In order to handle this situation of uncertainty, we have proposed the use of fuzzy logic in salary increment model. Our data set consists of the salary increment factors of the employees along with the increased salary percentage of 100 employees. The implementation of Adaptive Neuro Fuzzy inference system (ANFIS) on salary increment model is described in this paper. We have approached the Sugeno fuzzy inference model to generate the fuzzy rules and membership functions of input (salary increment factors) and output (salary increment percentages) data.

Keywords: Salary increment model · ANFIS · FIS · Fuzzy logic
Human resource management

1 Introduction

Salary is one of the major concerns of the employees. It is considered to be the justification of one's skill as well as performance. Salary depends on uncertain properties and different organizations use different methods to analyze the performance level of their employees. However, often some conventions are followed to maintain the justification of salary standards.

Point method is widely used for performance evaluation of the employees due to its high level of accuracy in the result [1, 2]. In the point method, performance evaluation factors are rated in points and the total point is calculated for the overall performance evaluation which is used for the salary increment structure [3]. In a generalized view, if someone knows about the factors that increase the salary and how these factors impact the increment process, then he/she can focus on improving his/her performance in a better way. This will encourage both employers and employees to effectively judge their performance as well as salary. As salary increment is proportional to the

R. Silhavy (Ed.): CSOC 2018, AISC 764, pp. 344–353, 2019.
https://doi.org/10.1007/978-3-319-91189-2_34

performance of the employees, being able to predict it can help them a lot to raise their level of confidence.

In this paper, we have proposed a system that demonstrates some major factors regarding salary increment. We have built this system using fuzzy logic. Fuzzy logic is a problem solving technique that was introduced by Lotfi Zadeh in [4] to deal with vague or imprecise problems [5, 6]. Fuzzy logic works on uncertainty. In fuzzy logic we do not specify the exact value of the variables; rather we express our variables or factors using range or intervals. A variable can be associated with any value from an interval and the association is defined by a membership function. In fuzzy logic this membership function lies from 0 to 1 and the variables can take any value from this range.

We have collected data from a private organization named Delta Group Limited in Bangladesh. It is a textile and ready-made garment company. They have twenty six hundred employees. Our dataset includes all the factors that the company considers while making their salary increment model. Employees are classified into three categories: management, junior management and non-management. We have taken data of 100 employees from the non-management section. Different departments have different factors to evaluate the performance of their employees. Our dataset includes the annual salary raise of each employee along with the point achieved in the increment factors. Name, address and designation of the employees are not mentioned in the data sheet due to privacy issues. It can be seen that this company analyzes employees' performance over some variables by using crisp values ranging from 1 to 5. After that they follow some rules to calculate their salary increment percentage. In real life, we express the variables linguistically, such as low, medium and high in general. We cannot apply any crisp value to these variables, For example- if we say an employee has a low work standard, we cannot define how low it is. If someone gets 5 points, then it is low then again if he/she gets 1 point, it is also called low. So we have to express our variables using interval where the variables can take any value from a given range. This has led us to consider fuzzy logic in developing our system. In our application, we have evaluated the factors and variables using intervals. These variables can have any values or degree of membership from these intervals. We have considered the degree of membership within the range of 0 to 1.

2 Related Works

In some of the researches, fuzzy logic has been used for job evaluation, which is a key factor for salary increment. In papers [1–3], reasonable internal compensation structures have been proposed using the process of point factor job evaluation approach. Point factor approach is a widely used method for job evaluation throughout the world. It evaluates each position's value through some weighing factors such as labor amount, and quality which is then expressed in a certain number of points for the factor weighing. In [4, 5], Fuzzy Analytic Hierarchy process (FAHP) has been used for job evaluation systems. Fuzzy pairwise comparisons are used in judgment matrices to obtain the fuzziness in the evaluation process of each job. This method results in obtaining a more accurate weight for factor scoring. In the paper [6], the authors have discussed multi-criteria disaggregation–aggregation approach to deal with job evaluation in large organizations. Job evaluation is one of the most important functions of human resource management. It also enables the design and improvement of human resources. But the authors put forward a few inhibiting components such as the existence of multiple factors in the evaluation process, fuzziness in the data, and many more. In [7], the authors described the successful applications of FIS in automatic control, data classification, decision analysis and expert system. The application of Fuzzy Inference Systems (FIS) in multiple disciplines are known by different names. For example, simple and vague, fuzzy systems, fuzzy rule based systems which mean systems are set based on fuzzy rules. In paper [8], we see that companies use factors such as mastery, responsibility, workloads, and physical and working condition as main challenges and these factors have also divided into sub factors. The paper demonstrates that employees' performance is evaluated using point scaling and based on their total points, salary is calculated. But it is not very convenient, as different factors have different contribution towards salary increment. So salary prediction is done by using fuzzy theory because fuzzy property has more effective control over various factors or variable intervals. Based on these factors, companies evaluate the salary scale for employees. These variables/factors are uncertain and vague. So FIS is used to determine and evaluate these factors [8]. FIS is important in this scenario as it can provide crisp output for uncertain data or fuzzy numbers or values. Fuzzy inference system is generally based on two methods, e.g., Sugeno and Mamdani methods. Fuzzy inference system uses input data to deliver output in fuzzy logic [9]. It uses membership functions and fuzzy logic operations (if/then rules). As the factors are uncertain and decision making is done by using human intuition as well as experience; fuzzy logic is better suited for this situation. In decision making process data and information are explained in a more brief way and linguistically [9]. In paper [10] inflation and profit are the major factors and certain rules have been applied according to their linguistic value (such as low, medium and high) and membership functions. These rules and membership functions are under Mamdani method and defuzzification has been used to determine the expected salary. Mamdani method uses min operation and variables or factors have linguistic values, and membership functions. So, specific rules are formed according to the number of variable and their linguistic value combination. If we give any numerical value in the input variable, then expected/approximate salary can be predicted. This expected salary is gained based on the rules that were previously stored as Mamdani rules [10]. Every possible outcome for any given input was generated by verifying the Mamdani rules and the approximate result was calculated through the defuzzification process [10].

3 Fuzzy Inference System (FIS) and Adaptive Neuro Fuzzy Inference System (ANFIS)

We have used FIS and ANFIS methodology in our research; these techniques are elaborated in the next two sections.

3.1 Fuzzy Inference System (FIS)

Fuzzy Inference Systems (FIS) have been applied successfully in many applications in decision analysis, expert systems, data classification, and computer vision [7]. Let us consider our scenario where we could also apply the ideas of fuzzy logic. For instance, "If the work standard is low and work habit is very low then salary increment will be low". This kind of example is known as propositional logic because the final result depends on the condition. The antecedent part is known as the premise and the resulting part is known as a consequence. Here we see that, if the condition is true, then the result will be true. For every truth value of the premises the consequence has to be true and vice versa. There can be one or multiple input variables and one or more output variables. We connect these variables using logical AND/OR operations. In the above example, we have used two input variables and one output variable. Work standard and work habit are linguistic variables as we express our input variables with the help of real life observations, for example- very low, low, average, high and very high. We do not express it using any numerical value or data. Now the question is how we can express our real life observation using fuzzy terms, and notations and how fuzzy logic can help us to explain our application of human intuition.

Suppose, x = work standard, y = work habit z = salary increment Ax = low By = very low, Cz = low

We can say, if x is Ax AND y is By then z is Cz.

We cannot define any exact numerical data that could be considered as low or very low or average or high. So we express a range from 0 to 1 and define the term low that can be any value from this interval. We do not declare low simply as crisp 0 or 1, rather we express our uncertainty using an interval. So this becomes an example of fuzzy number. When we say the range can be from 0 to 1, we are talking about the membership function. In FIS we set both input variables and output variables which are mostly obtained from linguistic variables and express the variables using membership functions. This membership function can be of Mamdani type and Sugeno type. For our research we have used Sugeno type membership function. After setting input and output variables, the rules are formed using a combination of input and output variables and their membership functions. Suppose, performance of employee has 3 input variables that have linguistic values such as low, average and high. Output salary has also 3 linguistic values and they are low, average and high. So the combinations will be: (i) If the performance is low, then salary increment will be low, (ii) If performance is average then salary increment will be average, (iii) If performance is high, then salary increment will be high. Linguistic value low, average and high can have membership values ranging from 0 to 1.

3.2 Adaptive Neuro Fuzzy Inference System (ANFIS)

Adaptive Neuro-Fuzzy Inference System (ANFIS) uses linguistic variables and their membership functions. Given a crisp dataset as input, ANFIS generates linguistic input and output variables and membership functions and creates conditions or rules. ANFIS works through multiple layers. The first layer is input variables, the second layer consists of input variables, and membership functions, the third layer has multiple outputs and the final layer generates only one output through defuzzification (crisp value). Though we provide crisp data, but ANFIS converts them into fuzzy variables and membership functions and numbers.

4 Proposed Methodology

We have worked with the data of the non-management sector of the company. Non-management employees' major salary increment factors are work standard, work habit, attitude, and resourcefulness. The company uses a crisp rating from 1 to 5 to evaluate the performance of the related employees. Table 1 shows the increment factors which are the input variables with membership functions.

Table 1. Input variables and membership functions

Input variables	Membership function		
Work standard	Low	Medium	High
Work habit	Low	Medium	High
Attitude	Low	Medium	High
Resourcefulness	Low	Medium	High

In our application, we consider total mark range 0–5 as very low increment, 6–10 as low increment, 11–16 as average increment and 17–20 as high increment. For example-if any employee gets total 11 to 16 points his salary will be incremented by 10% and so on. These are based on real life data and final salary increment. This distribution is also maintained in other departments, though the factors are different. Next we demonstrate our salary increment model using ANFIS tool. Figure 1 shows the ANFIS structure. We have used the following steps-

1. *Import the file in ANFIS as training data and generate FIS using gaussmf membership functions where every variable has 3 linguistic values- low, average, high.*
2. *Train the data by setting error tolerance level 0 and epochs to 40.*
3. *Take another 30 employees' data as test data and record performance.*

From our data sheet we see that, if any employee gets-

1. *6 to 10 mark, salary increment will be 5% (low increment)*
2. *11 to 16 mark, salary increment will be 10% (average increment)*
3. *17 to 20 mark, salary increment will be 15% (high increment)*

Here increment percentage is crisp data. After generating FIS, numerical input data is converted into fuzzy input variables where range is from 1 to 5 and every variable has membership function for low, average and high. During this time, FIS rules are built, as there are 4 input variables and 3 membership functions for each variables, we have total 81 rules. 1st layer in ANFIS consists of crisp input data. 2nd and 3rd layer works on fuzzification whereas 4th layer generates the crisp output that is salary increment percentage. In our rule view, if we change the input variables' crisp number from 1 to 5 then we clearly see the fluctuation in the output (Fig. 2). These outputs are very close to our actual salary increment.

5 Experimental Result and Analysis

The analysis of salary increment model using fuzzy logic is done by taking The 100 employees' records which are divided into two groups. 70 employees' data for training purpose and 30 employees' data for testing purpose. There are 4 input variables with 3 membership functions for each variable for the training purpose. The surface graph of the rules is shown here for better understanding. The inputs which are the salary increment factors are taken in the x axis and y axis and the output (salary increment percentage) is shown in the z axis. The surface graphs are shown in Fig. 3.

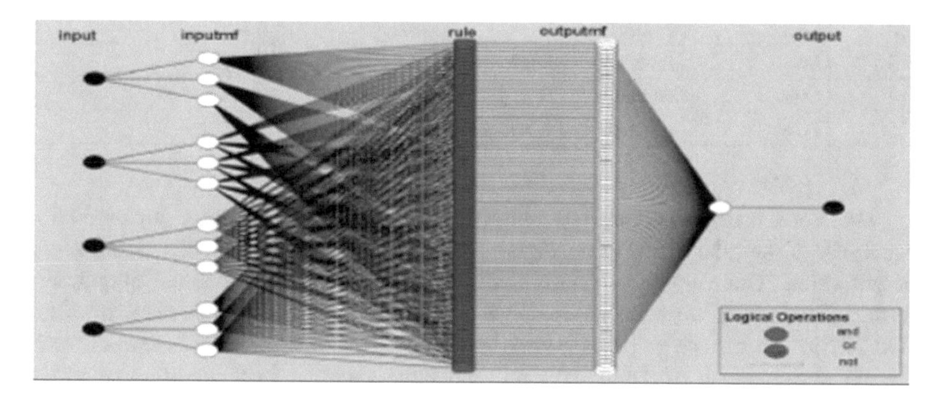

Fig. 1. ANFIS structure of salary increment model

It can be seen from the Fig. 4 that the majority of the training data matches with the FIS output. The average training error is (1.9506e–06). It can be depicted from Fig. 4 that the testing data does not match with FIS output as much as the training data. The average testing error is relatively high and that is (1.6579e–05). RMSE is 0.0005. The increased average error of the testing data could be due to small number of testing data which is 30. In Table 3, the salary increment factors of the employees are illustrated. From the 30 employees' data used for testing, 6 test cases are shown. The factors are rated with points ranging from 1 to 5 and the employees' salary increment is measured in percentage. From the test case 3 it can be seen that, the employee has achieved 5 points that is the maximum in both work standard and work habit but he has achieved 4 points in attitude and resourcefulness. According to ANFIS generation his salary increment is supposed to be 14.4% but in reality it was 15%. It can be inferred that the variation in the salary increment between ANFIS and actual is minimal. Table 2 depicts the situation.

Table 2. 15 rules out of 81 rules generated by ANFIS

Rules	Work standard	Work habit	Attitude	Resourcefulness	Output (salary increment %)
1.	High	High	High	Low	5.5
2.	High	High	High	Average	15
3.	High	High	High	High	16.4
4.	High	High	Average	Low	9.21
5.	High	High	Average	Average	9.3
6.	High	High	Average	High	10
7.	High	High	Low	Low	6.31
8.	High	High	Low	Average	8.92
9.	Low	Low	Low	Low	3.26
10.	Low	Average	Average	High	10.3
11.	Low	Average	High	Low	1.52
12.	Low	Average	High	Average	1.53
13.	Low	Average	High	High	4.23
14.	Low	Average	Average	Average	4.83
15.	Low	Average	Average	Low	3.38

Therefore, it can be concluded that ANFIS generated output has a fair prediction capability of the salary increment. Figure 2 and Table 3 show the predicted and actual output value. There is slight variation recorded in predicted and actual output. For further clarification we have compared the surface graph that is generated from the ANFIS with the surface graph of the original data. TeraPlot software is used to generate the 3D surface graph of the actual data. The same variables in *x, y* and *z* axis of the ANFIS generated surface graph is used for the surface graph generation of the original data. In the Fig. 5 work standard is plotted along *x* axis and work habit, attitude and resourcefulness is plotted in *y* axis separately and the salary increment percentage is plotted along *z* axis. The similarity in the shape of the two surface graphs-ANFIS

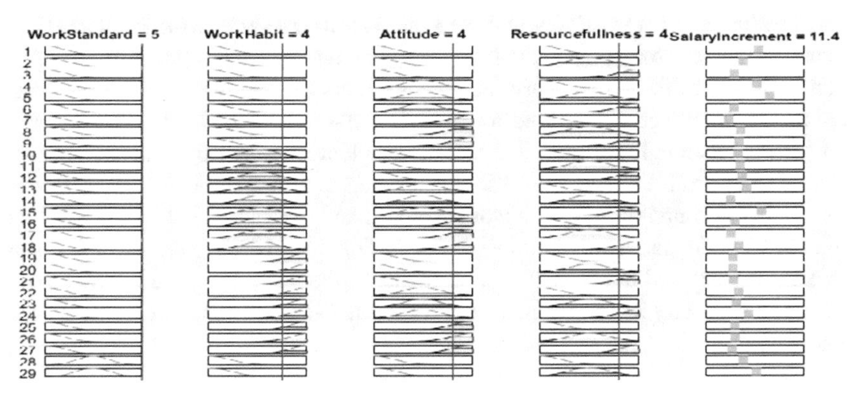

Fig. 2. Partial view of the Rule viewer generated by ANFIS

generated and actual data wise plotted surface graph is clearly visible. However, there are slight variations in the two graphs which could be due to low number of input data. The results could be more precise if we could use more number of employees, i.e., 500 or more. It can be inferred from our result that fuzzy logic is a useful tool than can be used for salary increment.

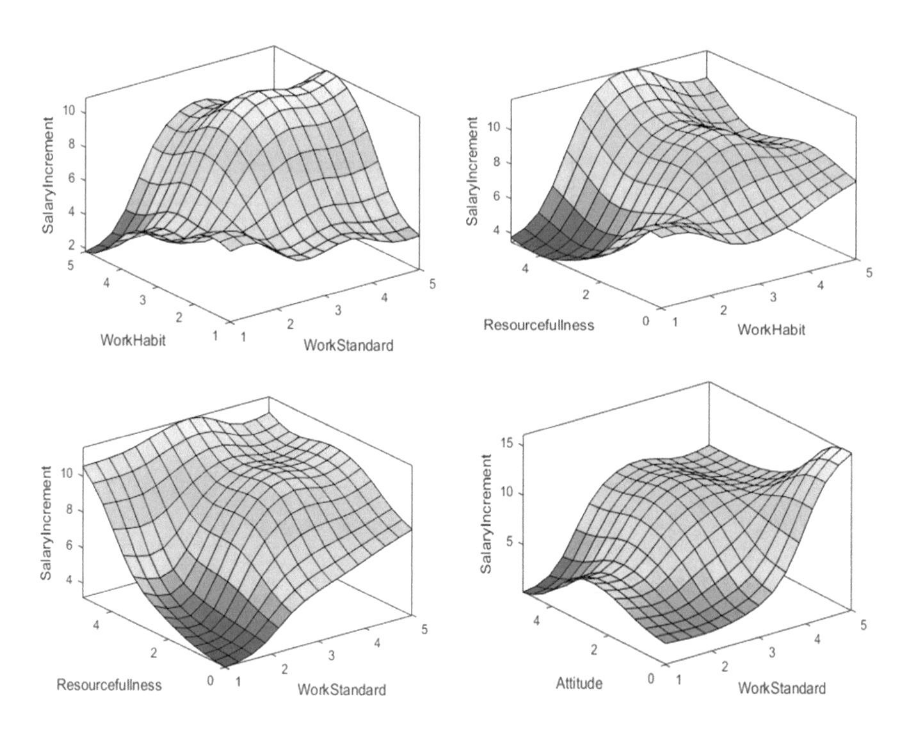

Fig. 3. Surface view of the ANFIS generated rules

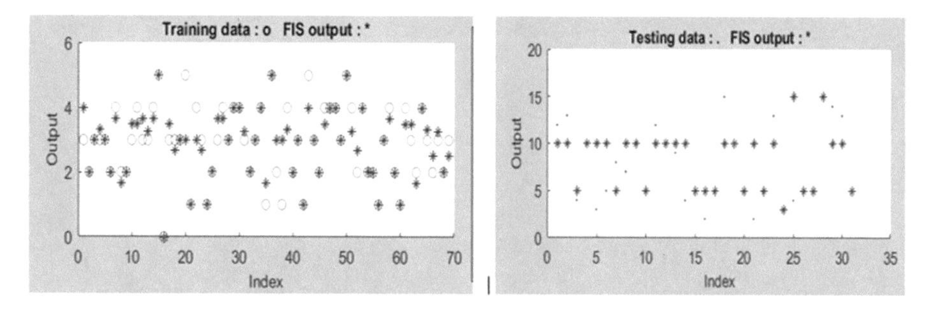

Fig. 4. Comparison of actual and predicted value of training and testing data

Table 3. Sample 6 test cases to show the variation in between ACTUAL and ANFIS generated output

Test case	Input variables				Output source	Salary increment% (Output)
	Work standard	Work habit	Attitude	Resourcefulness		
1	5	5	5	5	ANFIS	17%
					ACTUAL	15%
2	3	2	3	3	ANFIS	6.15%
					ACTUAL	10%
3	5	5	4	4	ANFIS	14.4%
					ACTUAL	15%
4	2	5	2	3	ANFIS	5.01%
					ACTUAL	5%
5	5	5	2	2	ANFIS	8.56%
					ACTUAL	10%
6	4	4	2	2	ANFIS	9.71%
					ACTUAL	10%

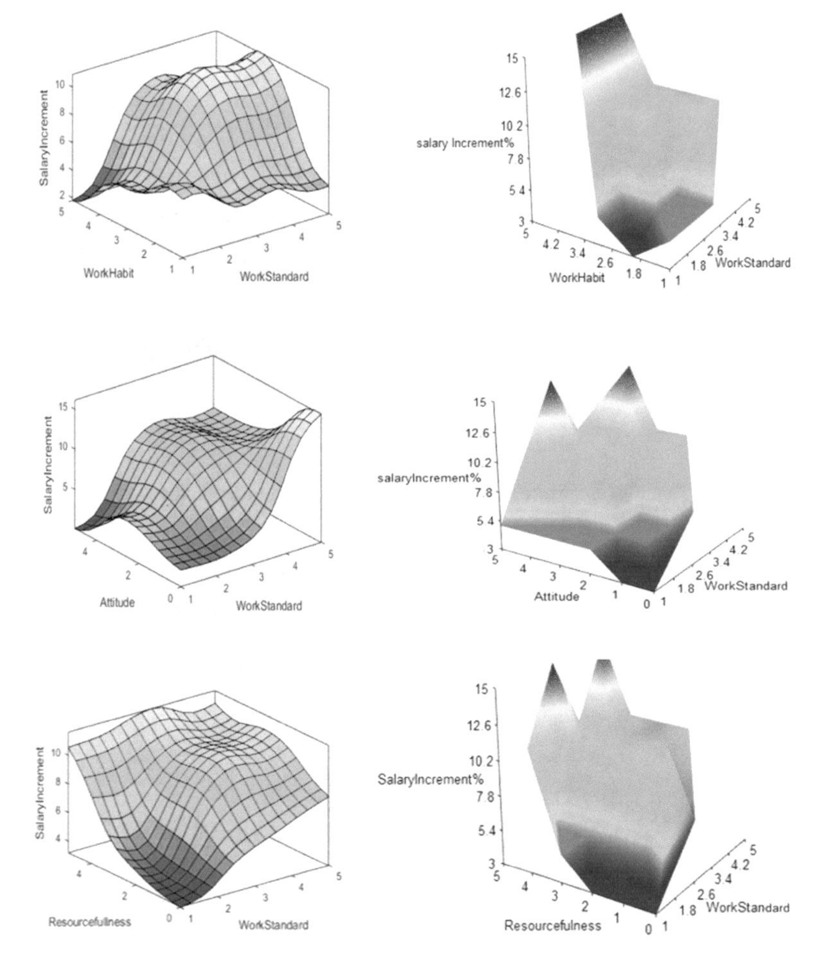

Fig. 5. ANFIS generated surface graph in left and surface graphs from original data generated by TeraPlot software in the right

6 Conclusions and Future Work

In our paper, we have identified multiple factors that have major impact on predicting the salary increment of the employees. Even though there are small percentage of errors in the generated prediction, the results indicate that the salary increment suggested by fuzzy logic is very close to actual salary increment. Despite the fact that we have considered four major factors in performance evaluation to generate our prediction model, there might be some other factors that could improve the accuracy of it. In near future we want to apply the concept of fuzzy logic in the respective company which follows these factors to set their employees' salary scale. Besides, we can always modify these factors, according to industries' demand and gradually the rule base can be extended to make the system more standard.

References

1. McCormick, E.M.: Job Analysis: Methods and Application. AMACOM, New York (1979)
2. Sun, X., Luo, N.: Study on the Effectiveness of Point-Factor Job Evaluation System in Operation Position. Academicpub.org (2017)
3. Doverspike, D., Carlisi, A.M., Garrett, G.V., Alexander, R.A.: Generalizability analysis of a point-method job evaluation instrument. J. Appl. Psychol. **68**, 476–483 (1983)
4. Dweiri, F.T., Kablan, M.M.: Using fuzzy decision making for the evaluation of the project management internal efficiency. Decis. Support Syst. **42**(2), 712–726 (2006)
5. Shuliang, L.: The development of a hybrid intelligent system for developing marketing strategy. Decis. Support Syst. **27**(4), 395–409 (2000)
6. Spyridakos, A., Siskos, Y., Yannacopoulos, D., Skouris, A.: Multicriteria job evaluation for large organizations. Eur. J. Oper. Res. **130**, 375–387 (2009)
7. Ajzen, I., Fishbein, M.: Understanding Attitudes and Predicting Social Behavior. Prentice-Hall, Englewood Cliffs (1990)
8. Eraslan, E., Atalay, K.: A new approach for wage management system using fuzzy brackets in industry. Math. Prob. Eng. **2013**, 1–11 (2013)
9. Nobari, S., Jabrailova, Z., Nobari, A.: Using fuzzy decision support systems in human resource management. In: International Conference on Innovation and Information Management (ICIIM 2012), IPCSIT, vol. 36, pp. 204–207 (2012)
10. Oderanti, F., De Wilde, P.: Wage increment problems and fuzzy inference approach. In: 2011 Annual Meeting of the North American Fuzzy Information Processing Society (2011)

Fuzzy Logic Based Weight Balancing

Md Sakib Ibne Farhad, Ahmed Masud Chowdhury,
Md. Ehtesham Adnan, Jebun Nahar Moni, Rajiv Rahman Arif,
M. Arabi Hasan Sakib, and Rashedur M. Rahman(✉)

Department of Electrical Engineering and Computer Science,
North South University, Plot-15, Block-B, Bashundhara,
Dhaka 1229, Bangladesh
{sakib.farhad, ehtesham.adnan, nahar.moni, arabi.hasan,
rashedur.rahman}@northsouth.edu,
ahmedmasudchowdhury@gmail.com, rajivarif@yahoo.com

Abstract. Human race is affected by a silent but dangerous disease known as obesity. This is the result of the difference between the high calorie intake and low calorie burn. Though most of the people do not realize that they are slowly being engulfed by this disease which can be easily prevented by some simple measures. Along with obesity, underweight is also an issue. To avoid these issues a calorie based diet is the best solution. To recover from this obesity, people need to lose weight and to recover from underweight, people need to gain some weight, both depend on calorie intake and calorie burn. So, to assist with this weight-loss and weight-gain a fuzzy logic based weight loss/gain training program is proposed in this research.

Keywords: Fuzzy logic · Fuzzify · Defuzzify · Weight-loss · Weight-gain
Calorie deficit

1 Introduction

Obesity is turning into an epidemic of significant scale all around the globe. The World Health Organization (WHO) media center states that in 2014 more than 1.9 billion adults were overweight, among whom over 600 million were obese [1]. This sums up to the fact that almost one-third of the world is either overweight or obese. The indicator of a person's weight-related wellbeing is called the Body Mass Index (BMI). Different ranges of BMI indicate different stages of a person's weight-related wellbeing. Table 1 illustrates the WHO BMI chart [2]. Besides an unfit body, overweight and obesity can cause cardiovascular diseases, diabetes and even terminal disease, like cancer.

According to the WHO, the fundamental reason for overweight and obesity is an imbalance created by consumption of high energy and low physical activities [1]. When the calorie consumption of an individual is greater than the expended calorie, then the person starts to gain weight. If the process could be reversed then the consequence would be reversed as well, i.e., the person would start reducing weight. The similar but opposite issues occur in case of underweight. This paper offers an approach

R. Silhavy (Ed.): CSOC 2018, AISC 764, pp. 354–363, 2019.
https://doi.org/10.1007/978-3-319-91189-2_35

Table 1. BMI chart

Classification	BMI (kg/m^2)	
	Principal cut-off points	Additional cut-off point
Underweight	**<18.50**	**<18.50**
Severe Thickness	<16.00	<16.00
Moderate Thickness	16.00–16.99	16.00–16.99
Mild Thickness	17.00–18.49	17.00–18.49
Normal range	**18.50–24.99**	**18.50–22.99**
		23.00–24.99
Overweight	**≥ 25.00**	**≥ 25.00**
Pre-Obese	25.00–29.99	25.00–27.49
		27.50–29.99
Obese	**≥ 30.00**	**≥ 30.00**
Obese class I	30.00–34.99	30.00–32.49
		32.50–34.99
Obese class II	35.00–39.99	35.00–37.49
		37.50–39.99
Obese class III	≥ 40.00	≥ 40.00

to re-balance the calorie burn and calorie intake, of an individual, by the application of a fuzzy logic based weight-loss/gain training program.

This paper is based on the basic theory of calorie difference i.e. the difference of calorie intake and the calorie burn of an individual, discussed in the Theory section. The Methodology section illustrates the application of fuzzy logic on the calorie intake, calorie burn and calorie difference which goes through fuzzy inference and will eventually result in a crisp output, after defuzzification. Depending on an individual's BMI the system may suggest for a weight loss or gain. The ultimate result is delivered to the user along with some food suggestions. To make this whole process user friendly an application is developed which takes input from the user and delivers the output.

2 Theory

It is widely believed that a Calorie difference (difference between Calorie intake and Calorie burn) of 7700 kcal will result into a weight loss/gain of 1 kg and vice versa [3]. To stay at a deficit, one has to crunch the Calorie intake and increase his/her activities to facilitate the Calorie burn. In case of weight gain one needs more Calorie intake than burn. The mathematical relationship among Calorie deficit (ΔC), Calorie intake (calorieIn) and Calorie Burn (calorieBurn) can be stated as shown in Eq. (1)

$$\Delta C = calorieIn - calorieBurn \quad (1)$$

The minimum amount of intake that one needs to maintain, without any loss of muscle mass, during a diet plan is 1200 kcal [4]. According to the Center of Disease

Control and Prevention, the rate at which an individual can safely lose body weight is approximately 1 to 2 lbs (0.454 kg to 0.907 kg) a week [5].

The program is intended for weight loss/gain plans. The fuzzy model presented takes the conversion of 7700 kcal = 1 kg into consideration and suggests the user a calorie intake and calorie burn, provided the users feed in the rate at which they wish to lose/gain their weight.

Another important terminology used in this paper is Basal Metabolic Rate (BMR). It is the amount of energy spent per unit time by endothermic animals in sedentary state [6]. BMR is the energy needed to run the internal functions of a human body, which includes breathing, maintaining body temperature, blood circulation, heartbeat, cerebral functions etc.; the unit for BMR is calorie. Most of the calories of a human body are burnt through the basal metabolism. The resting metabolism accounts for 60% to 80% of total energy expenditure [7]; hence, the consideration of BMR is vital in the calculation of weight gain/loss. For this paper, the Harris-Benedict equation for BMR, revised by Mifflin and St Jeor in 1990 [8], is used. The following Eqs. (2) and (3) show the BMR for men and women, with respect to their height, weight and age.

BMR for men is given in Eq. (2):

$$BMR = (10 \times \text{Weight(kg)}) + (6.25 \times \text{height(cm)}) - (5 \times \text{age(years)}) + 5 \quad (2)$$

BMR for women given in Eq. (3):

$$BMR = (10 \times \text{Weight(kg)}) + (6.25 \times \text{height(cm)}) - (5 \times \text{age(years)}) - 161 \quad (3)$$

3 Methodology

As discussed in the Theory section, a net deficit or gain of 7700 kcal will result in a loss or gain of 1 kg weight respectively. So, 7700 kcal distributed over a period of 7 days makes it an 1100 kcal each day. Hence, the weight loss/gain program, if followed properly, should decrease/increase approximately 1 kg of an individual's weight per week. The program takes the user's weight, height, age and sex as input in the first phase to calculate BMI and BMR in order to give the user an overview of his/her current situation. In the second phase, for the case of weight loss, two user inputs, namely Loss Per Week (LPW) and Calorie Intake (calorieIn) are taken. LPW is the amount of weight he/she wishes to lose per week. CalorieIn is the daily calorie intake of the user and calorieBurn is the amount of calorie the user should burn to lose weight. In case of weight gain the program takes Gain Per Week (GPW) and BMR as inputs and gives out calorieIn. All LPW, GPW, calorieIn, BMR and calorieBurn are Fuzzified using Triangular Membership Functions in Fuzzy Logic Toolbox MATLAB. The linguistic variables used for every parameter (i.e. LPW, GPW, calorieIn, BMR, and calorieBurn) are Very Low (VL), Low (L), Optimum (O), High (H), and Very High (VH).

3.1 Weight Loss Program

The following Fig. 1 shows the membership functions of LPW (input for weight loss)/GPW (input for weight gain).

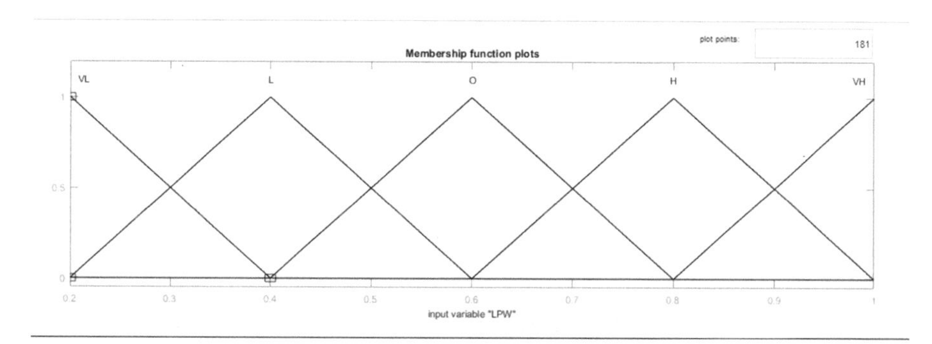

Fig. 1. Membership function for LPW/GPW

Table 2 shows the ranges of the different membership functions of LPW (input for weight loss)/GPW (input for weight gain) and the corresponding points where the membership reaches unity.

Table 2. Ranges and unity membership points of LPW (input for weight loss)/GPW (input for weight gain)

	Range	Unity membership point
VL	0.2–0.4	0.2
L	0.2–0.6	0.4
O	0.4–0.8	0.6
H	0.6–1.0	0.8
VH	0.8–1.0	1.0

The following Fig. 2 shows the membership functions of calorieIn (input for weight loss)/BMR (input for weight gain).

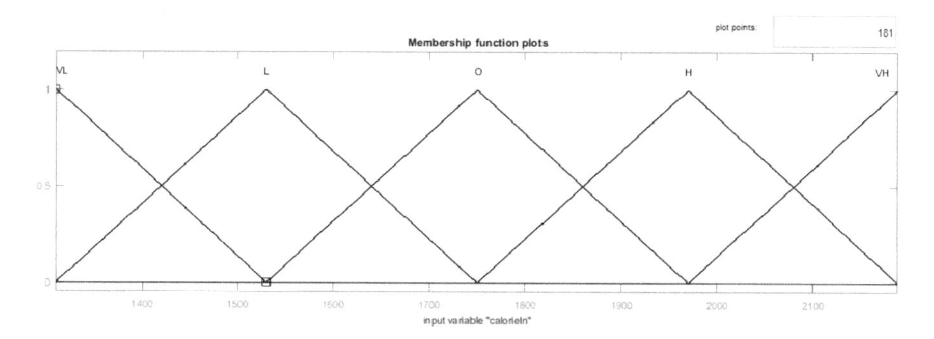

Fig. 2. Membership function of calorieIn/BMR

Table 3 shows the ranges of the different membership functions of calorieIn (input for weight loss)/BMR (input for weight gain) and the corresponding points where the membership reaches unity.

Table 3. Ranges and unity membership points of calorieIn (input for weight loss)/BMR (input for weight gain)

	Range	Unity membership point
VL	1310–1530	1310
L	1310–1750	1530
O	1530–1970	1750
H	1750–2190	1970
VH	1970–2190	2190

The Fig. 3 illustrates the membership functions of calorieBurn (which is the output for weight loss)/calorieIn (which is the output for weight gain).

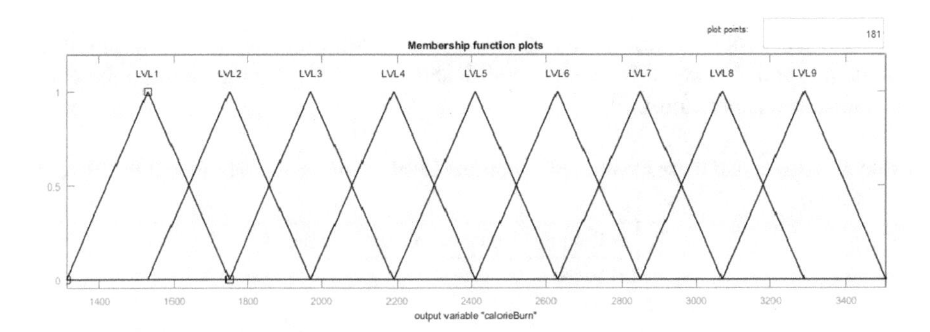

Fig. 3. Membership function of calorieBurn/calorieIn

Table 4 shows the ranges of the different membership functions of calorieBurn (output for weight loss)/calorieIn (output for weight gain) and the corresponding points where the membership reaches unity.

Table 4. Ranges and unity membership points of calorieBurn (output for weight loss)/calorieIn (output for weight gain).

	Range	Unity membership point
Level 1	1310–1750	1530
Level 2	1530–1970	1750
Level 3	1750–2190	1970
Level 4	1970–2410	2190
Level 5	2190–2630	2410
Level 6	2410–2850	2630
Level 7	2630–3070	2850
Level 8	2850–3290	3070
Level 9	3070–3510	3190

A sample output for the weight loss program of the fuzzy system for an input of LPW = 0.6 is shown in Fig. 4.

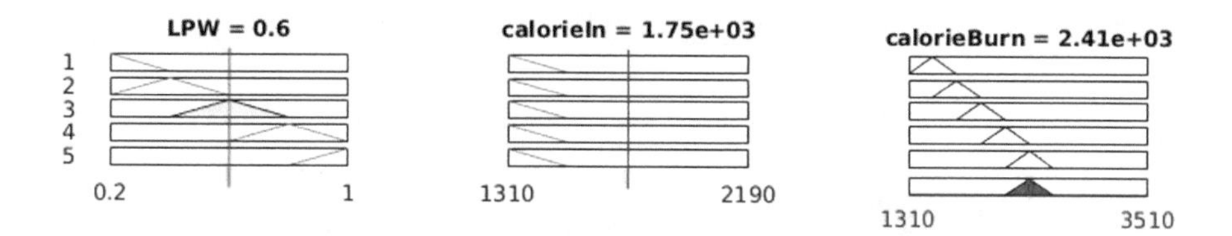

Fig. 4. A sample output for the weight loss program

In Fig. 4, an input of 0.6 at LPW and 1750 at calorieIn is given in the fuzzy system, meaning the user wants a weight loss of 0.6 kg per week. For the user to achieve this weight loss, the Fuzzy system suggested the user a daily burn of 2410 kcal depending on his daily calorieIn input. If the user maintains this set of Calorie intake and burn, then the user should achieve a weight loss of approximately 0.6 kg.

3.2 Weight Gain Program

The membership function of GPW is the same as the illustration provided on Fig. 1. Table 2 shows the ranges of the different membership functions of GPW and the corresponding points where the membership reaches unity.

The membership function of BMR is the same as the illustration provided on Fig. 2. Table 3, shows the ranges of the different membership functions of BMR and the corresponding points where the membership reaches unity.

The membership function of calorieIn is the same as the illustration provided on Fig. 3. Table 4 shows the ranges of the different membership functions of calorieIn and the corresponding points where the membership reaches unity.

A sample output for the weight gain program of the fuzzy system for an input of GPW = 0.6 is shown in Fig. 5.

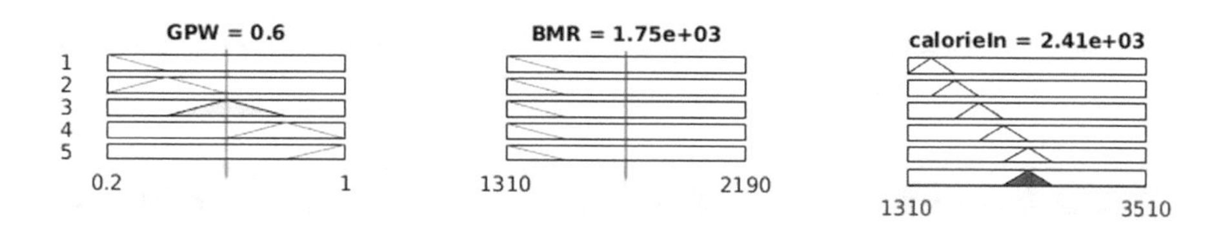

Fig. 5. A sample output for the weight gain program

In Fig. 5, an input of 0.6 at GPW and 1750 at BMR was given in the fuzzy system, meaning the user wants a weight gain of 0.6 kg per week. For the user to achieve this weight gain, the Fuzzy system suggested the user a daily intake of 2410 kcal depending on his inputs. If the user maintains this set of Calorie intake, then the user should achieve a weight gain of approximately 0.6 kg.

The following Table 5 illustrates the rules of the fuzzy system for both weight loss and gain.

Table 5. Rules of the fuzzy system for both weight loss and gain

CalorieIn/BMR LPW/GPW	VL	L	O	H	VH
VL	Level 1	Level 2	Level 3	Level 4	Level 5
L	Level 2	Level 3	Level 4	Level 5	Level 6
O	Level 3	Level 4	Level 5	Level 6	Level 7
H	Level 4	Level 5	Level 6	Level 7	Level 8
VH	Level 5	Level 6	Level 7	Level 8	Level 9

The whole process is illustrated on the flowchart shown in Fig. 6.

The application is developed using JAVA and the IDE was NetBeans. This application is divided into three phases. In the first phase the user gives Sex, Age, Weight and Height as inputs.

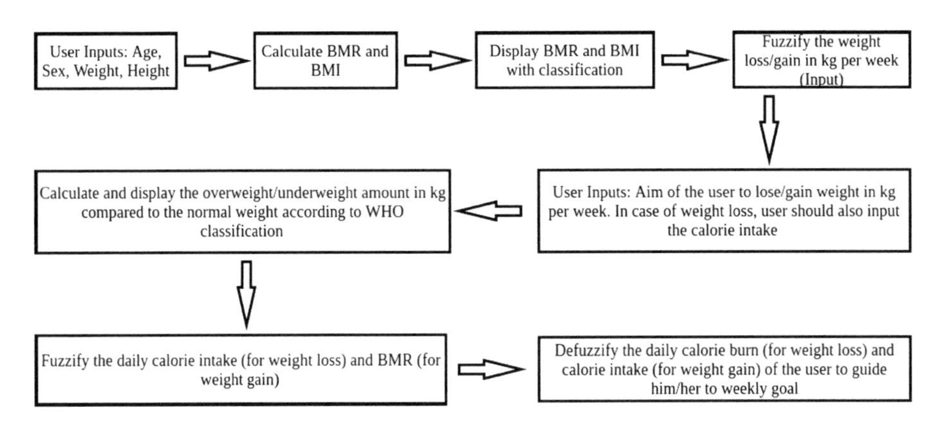

Fig. 6. The flowchart for the entire process

Figure 7 shows the first phase with some user inputs and the corresponding calculated results will be shown in the second phase.

Fig. 7. First phase: collecting primary user information (sex, age, weight, height)

The second phase calculates user's BMI, BMI Category, Normal Weight, Over or Under Weight and BMR based on the inputs given in the first phase. These details will help the user to judge his/her current physical condition, pertaining to his/her weight. In this phase user will also put in two inputs. Input 1 is the expected amount of weight loss or weight gain in Kg of the user and in Input 2 the user puts in the amount of daily Calorie Intake. Considering these two inputs, an output will be generated. If the user wants to plan a diet based on calorie, he/she can go to third phase by clicking on the Food Habit button.

Figure 8 shows the second phase with some user input and the corresponding result.

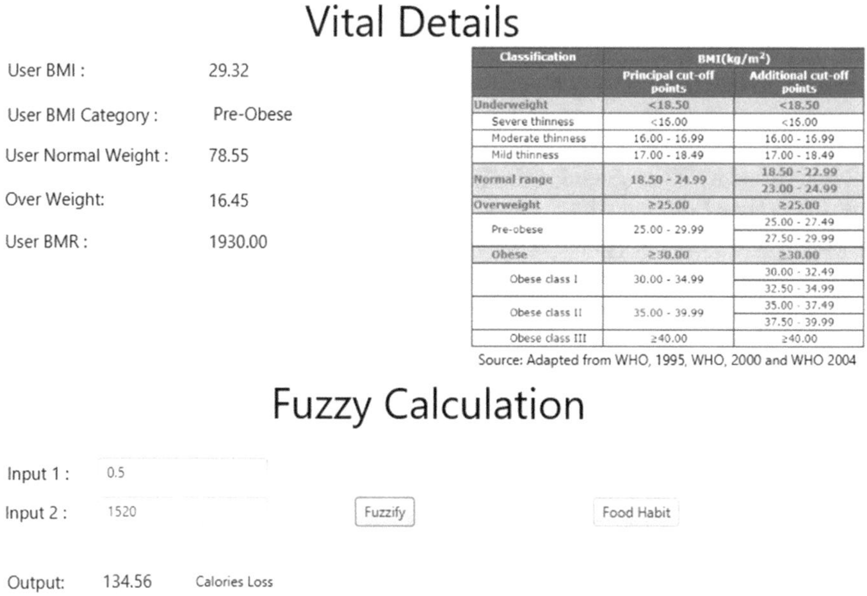

Classification	BMI(kg/m²)	
	Principal cut-off points	Additional cut-off points
Underweight	<18.50	<18.50
Severe thinness	<16.00	<16.00
Moderate thinness	16.00 - 16.99	16.00 - 16.99
Mild thinness	17.00 - 18.49	17.00 - 18.49
Normal range	18.50 - 24.99	18.50 - 22.99
		23.00 - 24.99
Overweight	≥25.00	≥25.00
Pre-obese	25.00 - 29.99	25.00 - 27.49
		27.50 - 29.99
Obese	≥30.00	≥30.00
Obese class I	30.00 - 34.99	30.00 - 32.49
		32.50 - 34.99
Obese class II	35.00 - 39.99	35.00 - 37.49
		37.50 - 39.99
Obese class III	≥40.00	≥40.00

Source: Adapted from WHO, 1995, WHO, 2000 and WHO 2004

Fig. 8. Second phase: user's physical condition assessment and weight gain/loss plan

The third phase allows user to plan a food diet based on his/her calorie need. In this paper only six food items are listed. In this phase, a user will input the amount of food in grams and can adjust the amounts to get the required calorie intake. If the total calculated calorie does not match with the desired calorie intake, a recommendation will be shown.

Figure 9 shows the third phase with some user input and the corresponding result.

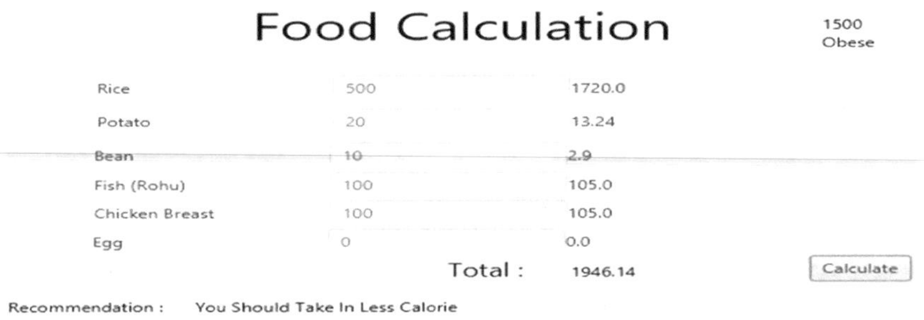

Food Calculation

1500
Obese

Rice	500	1720.0
Potato	20	13.24
Bean	10	2.9
Fish (Rohu)	100	105.0
Chicken Breast	100	105.0
Egg	0	0.0
	Total :	1946.14

Calculate

Recommendation : You Should Take In Less Calorie

Fig. 9. Third phase: food intake suggestions

4 Related Works

Many attempts of different natures have been taken in achieving similar goal of weight gain or loss. A fuzzy logic based physical activity analysis and management system, using Smartphone Embedded Sensors is proposed by Liu and Chan [9]. The analysis and management is delivered by a prompting mechanism, namely, Accumulated Activity Effective Index Reminder (AAEI Reminder). The AAEI Reminder is walking and running dependent. Fuzzy logic is used for the decision of prompt delivery. There are two stages of decision; first, if the situation requires a prompt; second, the amount exercise to be prompted. In Lee et al. [10], for diet assessment, a type-2 fuzzy ontology and fuzzy markup language is used. For the diet assessment, the authors collected the nutrition facts of different food items from Internet and constructed a type-2 fuzzy ontology; then different people input the food they have eaten, after which the diet assessment agent displays its diet assessment of the food consumed. Another approach of a mathematical diet model is used by Toumasis [3]. In [3], a mathematical model is presented taking the fact that carbohydrates burn to produce glucose, glycogen and fats, which in turn helps us sustain our body temperature of 37° C, heartbeat, walk, talk and even breathe. Toumasis [3] suggests even Calculus can be used for the common objective of modelling weight gain or loss. In [11] Wing and Phelan suggested that generally negligible amount of people succeeds in long term maintenance of weight loss, which, in turn, became the motivation for this paper.

5 Conclusion

A fuzzy system is developed based on different facts and figures (e.g. BMI), provided by relevant organizations like WHO, for the purpose of suggesting a plan for weight gain/loss. The system is further developed into a software with a User Interface. The output of the fuzzy system does indeed show a similar pattern to that of a crisp system. The values, however, are close approximation of what is expected for a normal person without any physical issues. More inputs like blood pressure levels, cholesterol level or any other relevant physical conditions could be associated, in the future, with the system for a more comprehensive output. The software needs to be fed with data by users manually. Hence, the system could be further improved for a more rapid prompt, where the data could be fed automatically when a user exercises. In the future, the software could be developed as a mobile phone application, which could take direct input (e.g. Calorie burned) from an individual's workout, by the means of the sensors associated with the phone, and suggest him/her a diet plan for the day, in real time for weight gain/loss.

References

1. Obesity and overweight, June 2016. http://www.who.int/mediacentre/factsheets/fs311/en/
2. BMI classification. http://apps.who.int/bmi/index.jsp?introPage=intro3.htm
3. Toumasis, C.: A mathematical diet model. Teach. Math. Appl. **23**(4), 165–171 (2004)
4. Cespedes, A.: Non-Starving, 1200-calorie diet. 18 July 2017. http://www.livestrong.com/article/312891-calories-per-day-to-lose-3-lbs-perweek/
5. Michalek, A.: What is the maximum weight loss per day? 18 July 2017. http://www.livestrong.com/article/405585-what-is-the-maximumweight-loss-per-day/
6. McNab, B.K.: On the utility of uniformity in the definition of basal rate of metabolism. Physiol. Zoology **70**(06), 718–720 (1997). https://doi.org/10.1086/5158817
7. Manini, T.M.: Energy expenditure and aging. Ageing Res. Rev. **9**, 1–11 (2010)
8. Mifflin, M.D., St Jeor, S.T., Hill, L.A., Scott, B.J., Daugherty, S.A., Koh, Y.O.: A new predictive equation for resting energy expenditure in healthy individuals. Am. J. Clin. Nutr. **51**(2), 241–247 (1990). PMID: 2305711
9. Liu, C.-T., Chan, C.-T.: A fuzzy logic prompting mechanism based on pattern recognition and accumulated activity effective index using a smartphone embedded sensor. Sensors (Basel, Switzerland) **16**(8), 1322 (2016). http://doi.org/10.3390/s16081322
10. Lee, C.-S., Wang, M.-H., Acampora, G., Hsu, C.-Y., Hagras, H.: Diet assessment based on type-2 fuzzy ontology and fuzzy markup language. Int. J. Intell. Syst. **25**, 1187–1216 (2010). https://doi.org/10.1002/int.20449
11. Wing, R.R., Phelan, S.: Long-term weight loss maintenance. Am. J. Clin. Nutr. **82**(1 Suppl), 222S–225S (2005). PMID: 16002825

Analysis of Spatial Data and Time Series for Predicting Magnitude of Seismic Zones in Bangladesh

Sarker Md Tanzim, Sadia Yeasmin, Muhammad Abrar Hussain, T. M. Rezoan Tamal, Rashidul Hasan, Tanjir Rahman, and Rashedur M. Rahman(✉)

Department of Electrical and Computer Engineering, North South University, Plot-15, Block-B, Bashundhara Residential Area, Dhaka, Bangladesh
{sarker.tanzim, sadia.yeasmin, abrar.hussain, rezoan.tamal, rashidul.hasan, Tanjir.rahman, rashedur.rahman}@northsouth.edu

Abstract. The paper demonstrates the use of clustering to find different sensitive seismic zones and time series for the earthquake hazard prediction. Anticipating seismic activities using previous history data is obtained by applying hierarchical, *k*-means and density based clustering. Data is collected first and then clustered. Finally, the clustered data is used to obtain the different seismic zones on map. On the top of that data is used in linear regression to build a predictive model for forecasting upcoming earthquakes' magnitudes for different regions in and nearby areas of Bangladesh.

Keywords: Earthquakes · Magnitudes · Data mining
Clustering · Frequent pattern · Regression · WEKA · Forecasting

1 Introduction

Earthquakes are ground motions caused by the sudden release of elastic energy stored in the rocks along period of time [1]. They have much complex spatio-temporal distribution. As seismological data is multidimensional and we use the spatial dataset; so the whole process needs special data mining techniques.

Bangladesh is situated at the south of the Himalayas, surrounded by India and Myanmar. Dhaka, the capital city of Bangladesh, is currently holding around 16 million people and an average of 52,000 people is living in per square feet. Many buildings, bridges and other constructions are not built up to tolerate earthquake impacts. Currently Bangladesh is on three tectonic plates - Indian plate, Eurasian plate and Burmese plate which are moving slowly in separate directions and causing collisions. Scientists predict a major earthquake in this region and considering the type of soil and infrastructure, Bangladesh is not prepared for that at all. Being one of the most crowded country, a seismic event on this area will make a devastation surely.

In this study, three clustering techniques have been used to identify separate zones of Bangladesh that are prone to seismic events. We have compared hierarchical,

R. Silhavy (Ed.): CSOC 2018, AISC 764, pp. 364–373, 2019.
https://doi.org/10.1007/978-3-319-91189-2_36

DBSCAN and *k*-means clustering to identify the suitable clusters for our work. Upon the clustering three seismic zones are identified that are most likely to have earthquake events. Among them two zones are found to be more seismic prone from our acquired result. The efficiency of the selected cluster is also evaluated using decision tree, *k*-nearest neighbor and support vector machine (SVM) and upon the result the most accurate evaluation model is identified. Then a time series analysis is performed to forecast future patterns of seismic events. Simple linear regression, Gaussian process, Multilayer perceptron and SMOreg are used for predicting the model which forecasts when, where and with what magnitude a seismic event could be occurred. The overall model will give us a better prediction of seismic hazards so that we can take precautionary steps which is expected to cause less harm.

2 Background

Abser et al. [1] used a software that can estimate the probability of earthquake. They have included distance from epicenter, communication links, population density, development and severity as determinant factors of earthquake. This analysis will certainly give much accuracy on calculating the risk of a certain area. On the other hand, this study is depended heavily on user input which is inefficient to give an overall earthquake risk status. Moreover, there is no data mining tool used in this research which is essential to determine the pattern and trend of earthquake possibility in an area. In [2] a method is suggested to get benefit from hierarchical clustering technique. Authors selected five attributes for three clustering. At first they considered only non-spatial attributes, then the spatial attribute (location) and finally they used all of them. Three different supervised training models were applied on the obtained clusters, namely, decision tree, *k*-nearest neighbors (KNN), and SVM. After analyzing the RMSE, the authors suggested that K-NN performed very poorly and decision tree performed slightly better than SVM. In another work [8], comparison of different methods is used to predict earthquake. K-nearest neighbor graph considers all the distributions of the training data. According to the research, for predicting earthquake the spatial, temporal and the magnitude prediction are important issues. The spatial prediction is better handled by data mining and clustering but neural networks are better for the temporal and magnitude prediction. Different computational intelligence techniques were used to predict earthquakes in [5].

Some preprocessing techniques were used to include 28 years of data on 33 seismic events in [3]. The authors experimented these data with seven classification algorithms like *k*-NN, SVM, ANN, and J48 etc. Among many algorithms, they obtained most accuracy by Multi-Objective Info-Fuzzy Network (M-IFN) algorithm. In [4] Multi-linear regression was applied to find earthquake magnitude by specifying spatial attributes as independent variables. They showed comparison of predicted magnitudes by WEKA and SPSS tool and found higher coefficient of determination in WEKA. A prediction is made for earthquakes by analyzing the pattern of time and date of occurrence using quantitative association rules and regression in [6]. K-means clustering was used in [7] on magnitude of seismic activities to organize homogeneous data together. Given a year and a quadrant of earth, these rules were able to predict the number of seismic vents with about 90% accuracy. Spatial data mining was applied to locate seismic zones based on real life datasets. Spatial data mining is not as same as the conventional one because it additionally considers spatial data to find patterns of interest. After the application of normalization and discretization on spatial dataset, clustering methods were applied to group co-related spatial data together as they effectively translate to same seismic zone.

3 Data Collection

In this study we are considering the earthquakes generated in between 20.35° N to 27.33° N Latitude and 85.913° E to 94.087° E Longitude for the last sixty years (1951–2017). The area we have chosen covers the area of Bangladesh. The earthquake data is taken from United States Geological Survey (USGS) [9]. The data contains the depth of the places, magnitude of the earthquakes, time of the event occurred and some other related data. We have found 689 earthquakes occurred in that concerning area. The description of each attributes of our dataset is given in Table 1.

Table 1. Data Attributes

Attribute	Description
Date and Time	Year, month, day and hour, minute, second
Longitude	Decimal degrees of longitude
Latitude	Decimal degrees of latitude
Depth	Distance from the surface to earthquake's origin in km
Magnitude	The magnitude of the earthquake

4 Data Pre-processing

To deal with problems such as missing and imprecise location information, relevant features selection etc. regarding spatial data, we have applied some pre-processing techniques on our dataset. The spatial attributes that are used here– spatial locations along with latitude, longitude and depth.

We have added two other attributes in our dataset named as "N-Days" (occurred days of earthquakes passed since 1951) and "Gap Days" (days passed since last earthquake occurred) as used in [2]. By arranging the dataset into ascending order based on "N-Days", "Gap Days" are calculated by using the Eq. 1 where IP is the immediate predecessor of Passed Days.

$$\text{Gap Days} = \text{N-Days} - \text{IP} \tag{1}$$

Before clustering the data points or building the predictive model an important step is normalizing the variables to avoid outliers and extreme values. For that mean of zero and standard deviation method is used. Equation 2 shows the normalization of the dataset. Here m' = mean, s = standard deviation and N = normalized data.

$$N = (m - m')/s \tag{2}$$

5 Methodology

By using the spatial data our first goal is to analyze previous seismic events so that we can determine the areas where earthquake occurred frequently. This is done using three clustering methods. Then we evaluate the clusters using 3 different algorithms to find the accuracy of those clusters. Finally, we predict future seismic events by using time series analysis. Fig. 1 depicts the workflow of our system.

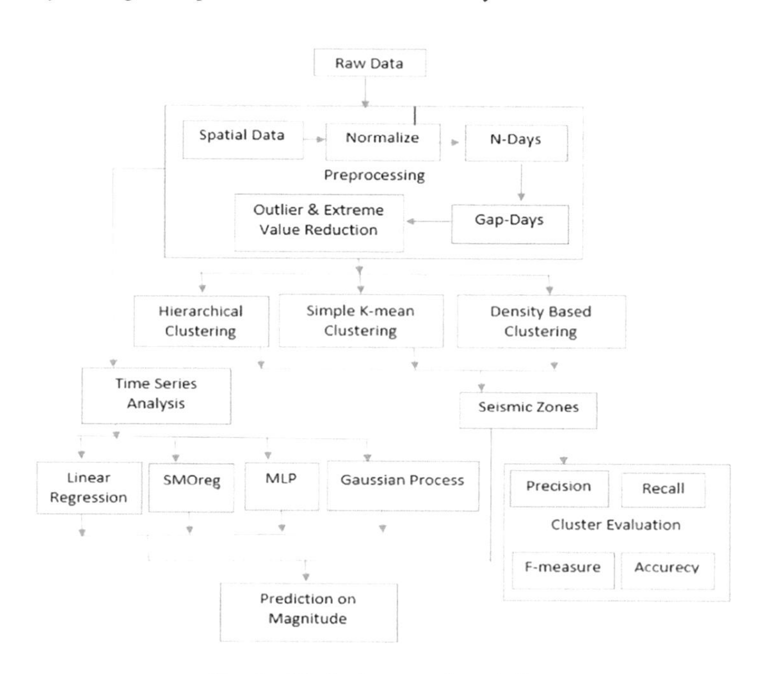

Fig. 1. Block diagram of work flow

5.1 Clustering

We cluster the data based on longitude and latitude to determine earthquake prone zones in and around Bangladesh. Then we divide the data based on magnitude, depth and time to determine how frequently significant earthquakes occurred in these areas. We use three different clustering methods for this purpose.

Hierarchical Clustering: We have done hierarchical clustering using average-link method as it is less sensitive to outliers. In this method the distance between clusters are calculated using the following formula.

$$\frac{1}{|A|.|B|}\sum\nolimits_{x \in A}\sum\nolimits_{y \in B} d(x, y) \tag{3}$$

Hierarchical clustering algorithms do not require the number of clusters to be predetermined and are mostly deterministic.

Density based Clustering: Density-Based Spatial Clustering of Applications with Noise (DBSCAN) provides the flexibility of not requiring the number of clusters as a parameter. Given a set of points from any dataset, it groups closely packed points together which leads to the discovery of clusters of arbitrary shape. Thus it infers the optimal number of clusters on its own.

K-means Clustering: We have tested with different values of *k, e.g., 2, 3, 4, and 5* for *k*-means clustering and got the least Sum of Squared Error (SSE) value for $k = 3$. We have calculated the mean of each cluster which has given us the centroids and then calculated the dissimilarity between a data point and its nearby centroids. Figure 2 shows this.

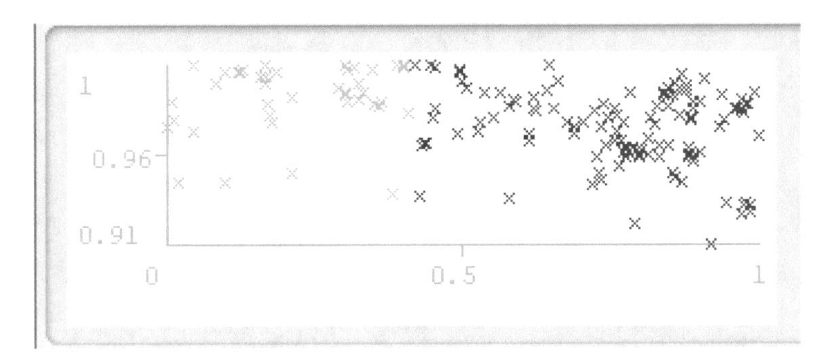

Fig. 2. Clustering using K means algorithms

We have compared the distance between the point and the centroids and assigned the point to a cluster with the closest centroid to the point. Then we have re-calculated the mean of that cluster which gave us the updated centroid [7]. We have limited the number of iterations to 9 which is enough to assign all the points to a cluster. We have observed that the SSE is 5.02 which is the lowest of the three clustering algorithm we used.

5.2 Cluster Evaluation

Evaluating the clusters is very important but a challenging one. Instead of validating the cluster itself, we have first divided the data set based on number of clusters. Next we have evaluated the accuracy of the clusters using SVM, KNN and Decision tree.

5.3 Time Series Analysis

To forecast time dependent events such as earthquake time series analysis is an excellent tool. Among many algorithms we have used linear regression, Gaussian process, and SMOreg and Multilayer perceptron to predict future earthquakes. We have compared their Mean Absolute Error (MAE) & Root Mean Squared Error (RMSE) to determine which of these algorithm is more efficient and accurate.

6 Findings

6.1 Evaluation of Clusters

To evaluate clustering results, recall, precision and F-measure [10] are calculated over the pair of clustering points. We have used Decision tree, SVM and K-NN to evaluate cluster assignment. We found 617 correctly classified instances out of 689 total instances with 89.61% accuracy using decision tree whereas SVM and K-NN gives 71.16% and 79.04% accuracy respectively. Table 2 shows different metrics used for cluster evaluation. Cluster evaluation comparisons by three methods for K-means clustering are listed in Table 3.

Table 2. Confusion Matrix of Classfiers

Output class	Target class	
	Negative	Positive
Classifier for negative	tn	fn
Classifier for positive	fp	tp

Here,

$$\text{Recall} = \text{tp} \,/\, (\text{tp} + \text{fn}) \tag{4}$$

$$\text{Precision} = \text{tp} \,/\, (\text{tp} + \text{fp}) \tag{5}$$

$$\text{Accuracy} = \left(\sum \text{mf} - \sum \text{af} \,/\, \sum \text{af}\right) * 100 \tag{6}$$

Where, mf = measured instances, af = accepted instances

Table 3. Accuracy of Classifiers

Algorithms	Precision	Recall	F-Measure	RMSE	Accuracy
Decision tree	0.905	0.864	0.884	0.1531	89.6068%
SVM	0.758	0.803	0.749	0.3541	71.1579%
K-NN	0.821	0.81	0.697	0.3121	79.04%

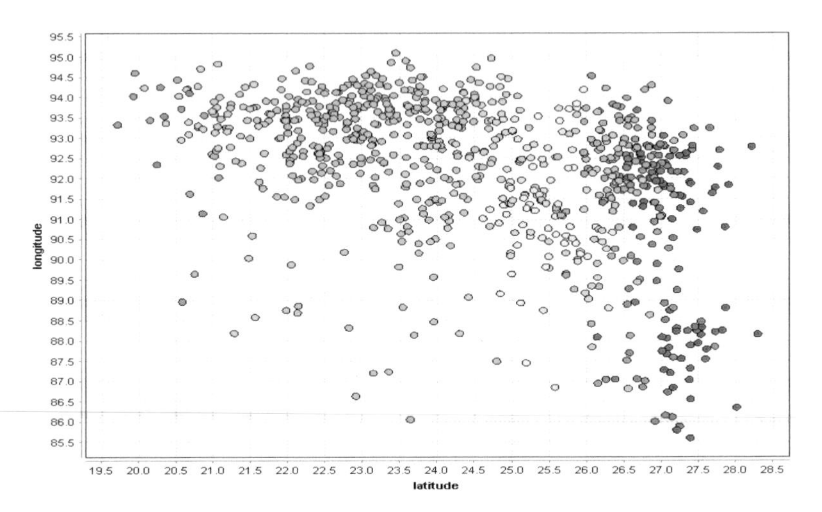

Fig. 3. Magnitude Clustering based on longitude and latitude

After applying three clusters on mapper we have finally got the model of magnitudes based on attributes of longitude and latitude. This map clearly gives us the visualization of low, medium and highly density based seismic zones of Bangladesh and its nearby areas which is shown in Fig. 3. The x axis represents latitude and y axis represents longitude. So every dot represents a seismic event. As we can see for longitude 90–95 and for latitude 21–27 the dots are very dense which are the north and north-west region of Bangladesh. From this we have estimated that these two regions are more prone as seismic zones compared to others.

6.2 Forecasting Future Patterns

In order to predict future values, data forecasting can be done based on previous events. Time series analysis is applied to model the series of data points in the dataset which are time-dependent. Among many, four algorithms that are used here are simple linear regression, Gaussian process, sequence minimum optimization (SMOreg) and multilayer perceptron. Our seismic event data is used to forecast future magnitude of earthquakes using each of the four algorithms [7]. A comparison of these algorithms based on acquired results are analyzed and from there a best prediction algorithm is selected for the Bangladeshi seismic data.

In every method time is set to be the independent variable and earthquake magnitude is set as dependent output variable. Prediction on magnitude is tested at step 1, 5 and 13 in WEKA Explorer. The purpose for lagging the data value is to remove a time varying mean. This will lead to find a slightly different pattern compared to the pattern derived from actual data as the variable under consideration is time shifted. Test data is evaluated by folding 5 data on a single instance [8, 9]. This steps are chosen randomly and in the testing phase predicted value of dependent attribute is calculated by folding number of steps of times instances (1, 5 and 13 fold instance in this case). This ahead-step representations are shown graphically to visualize the accuracy of predicted data

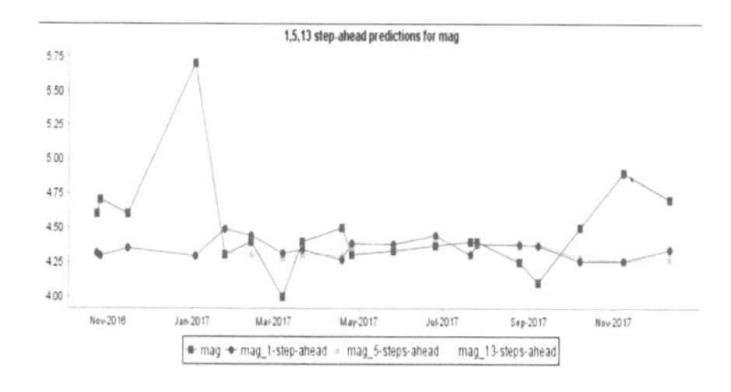

Fig. 4. Test prediction on Gaussian Process

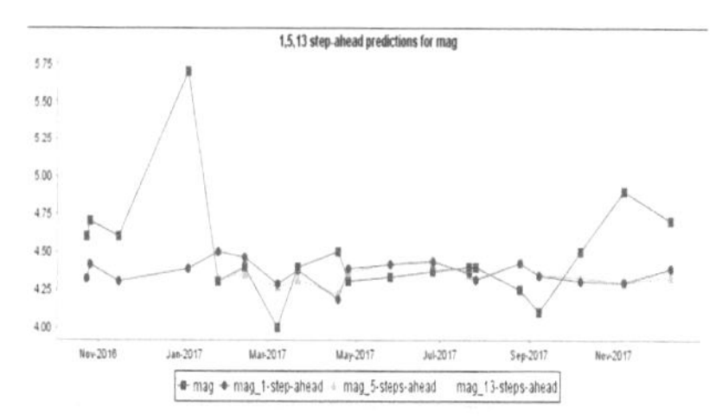

Fig. 5. Test prediction on Simple Linear Regression

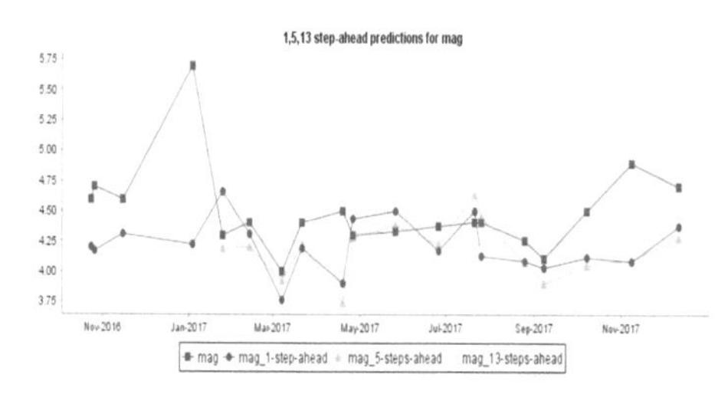

Fig. 6. Test prediction on Multilayer Perception

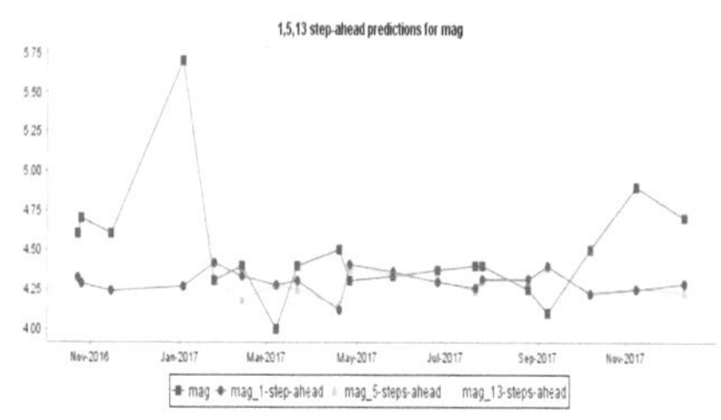

Fig. 7. Test prediction on SMOreg

with actual value. We have allowed to forecast different magnitude predictions in each of the four algorithm methods. 1, 5, 13 step ahead test predictions for all four algorithms are shown in Figs. 4, 5, 6 and 7 respectively.

We can visualize predicted value with 1, 5, 13 step-ahead and actual value representation in all four time series methods. We can easily estimate multilayer perception (MLP) to be the most accurate method to predict earthquake magnitude values with minimum error.

MAE and RMSE to Evaluate Model: Mean Absolute Error (MAE) is used to measure the magnitude of errors. By assigning a linear score and ignoring the directions, equal weights are given to every individual differences in MAE [8]. It can be formulated as:

$$MAE = \frac{1}{n}\sum\nolimits_{i}^{n} |e_i| \tag{7}$$

Mainly average magnitude of errors are measured by Root Mean Squared Error (RMSE). The calculation is done over the dataset by squaring the predicted and actual value. After that averaging the sample dataset is done. This two processes are done to highlight and avoid larger errors mainly.

$$RMSE = \sqrt{\frac{1}{n}\sum\nolimits_{i=1}^{n} e_i^2} \tag{8}$$

Both train and test data are evaluated at 1, 5 and 13 step-ahead by measuring MAE and RMSE in all four time series algorithms in WEKA to evaluate the final model. Test data is acquired by 5-fold technique. Comparison of MSE and RMSE values for four Time Series Algorithms is shown in Table 4 (data training phase) and Table 5 (data testing phase).

Table 4. Error in training data

Algorithms	Train data								
	1-step ahead			5-step ahead			13-step ahead		
	MAE	RMSE	Instances	MAE	RMSE	Instances	MAE	RMSE	Instances
Linear regression	0.327	0.432	619	0.328	0.434	615	0.329	0.436	607
Gaussian	0.328	0.434		0.328	0.435		0.328	0.435	
SMOreg	0.289	0.388		0.343	0.494		0.419	0.775	
MLP	0.318	0.439		0.319	0.441		0.324	0.444	

Table 5. Error in testing data

Algorithms	Test data								
	1-step ahead			5-step ahead			13-step ahead		
	MAE	RMSE	Instances	MAE	RMSE	Instances	MAE	RMSE	Instances
Linear regression	0.26	0.385	19	0.175	0.233	15	0.223	0.282	7
Gaussian	0.271	0.411		0.178	0.243		0.241	0.303	
SMOreg	0.361	0.485		0.259	0.352		0.317	0.394	
MLP	0.294	0.429		0.209	0.272		0.256	0.325	

Root mean squared error is minimum in multilayer perception (Tables 4 and 5) and thus seismic event forecast accuracy is better in MLP algorithm.

7 Conclusion and Future Work

Earthquake is an unpredictable event. It is very difficult to predict a natural phenomenon like sudden movement of earth's crust precisely, especially for countries like Bangladesh where earthquake does not occur frequently. But still there were many major seismic events that occur near Bangladesh, which could affect this country too. So, we have taken these events into consideration and using different data mining techniques we have predicted such events with a high degree of accuracy. Analyzing the data with proposed models we will be able to locate the earth quake prone areas near Bangladesh. With time series analysis we will be also able to predict future occurrences of seismic events that may affect Bangladesh. This prediction, if done more accurately, will help the people to take precautions for earthquakes which will minimize the damage.

We had many limitations while doing our work. There were not enough data and many of the data attributes were not well defied. In future we plan to work with a larger dataset with more attributes. We believe this will certainly improve the accuracy of our model.

References

1. Absar, N., Shoma, S.N., Chowdhury, A.A.: Estimating the occurrence probability of earthquake in Bangladesh. Int. J. Sci. Eng. Res. **8**(2) (2017). ISSN 2229-5518
2. Hashemi, M., Karimi, H.: Seismic source modeling by clustering earthquakes and predicting earthquake magnitudes. In: Smart City 360°, pp. 468–478 (2016)
3. Last, M., Rabinowitz, N., Leonard, G.: Predicting the maximum earthquake magnitude from seismic data in Israel and its neighboring countries. PLOS ONE **11**(1), e0146101 (2016). https://doi.org/10.1371/journal.pone.0146101
4. Dutta, P., Naskar, M., Mishra, O.P.: South Asia earthquake catalog magnitude data regression analysis. Int. J. Stat. Anal. **1**(2), 161–170 (2011). ISSN 2248-9959
5. Martínez-Álvarez, F., Troncoso, A., Morales-Esteban, A., Riquelme, J.C.: Computational intelligence techniques for predicting earthquakes. In: Corchado, E., Kurzyński, M., Woźniak, M. (eds.) Hybrid Artificial Intelligent Systems, HAIS 2011. Lecture Notes in Computer Science, vol. 6679. Springer, Heidelberg (2011)
6. Nivedhitha, U.S., Krishna, A.: Development of a predictive system for anticipating earthquakes using data mining techniques. Indian J. Sci. Technol. **9**(48) (2016). https://doi.org/10.17485/ijst/2016/v9i48/107976
7. Hoque, S., Istyaq, S., Riaz, M.M.: A clustering method for seismic zone identification and spatial data mining. Int. J. Adv. Res. Comput. Sci. Eng. Inf. Technol. **1**(2) (2013). ISSN 2321-3337
8. Kulkarni, A.D., More, A.: Analysis of the effect of cell phone radiation on the human brain using electroencephalogram. Orient. J. Comput. Sci. Technol. **9**(3) (2015). http://dx.doi.org/10.13005/ojcst/09.03.07
9. United States Geological Survey (USGS) (2015). http://earthquake.usgs.gov/earthquakes/search/
10. Zaidi, F., Archambault, D., Melançon, G.: Evaluating the quality of clustering algorithms using cluster path lengths. In: Perner, P. (ed.) Advances in Data Mining. Applications and Theoretical Aspects, ICDM 2010. Lecture Notes in Computer Science, vol. 6171. Springer, Heidelberg (2010). https://doi.org/10.1007/978-3-642-14400-4_4

Determination of the Data Model for Heterogeneous Data Processing Based on Cost Estimation

Jianping Zhang[1,2], Hui Li[1,2](✉), Xiaoping Zhang[1], Mei Chen[1,2], Zhenyu Dai[1,2], and Ming Zhu[3]

[1] College of Computer Science and Technology, Guizhou University, Guiyang, People's Republic of China
22zjpzjp22@gmail.com, {cse.HuiLi,gychm}@gzu.edu.cn
[2] Guizhou Engineer Lab of ACMIS, Guizhou University, Guiyang, People's Republic of China
[3] National Astronomical Observatories, Chinese Academy of Sciences, Beijing, People's Republic of China
mz@nao.cas.cn

Abstract. In heterogeneous data processing, various data model often make analytic task too hard to achieve optimal performance, it is necessary to unify heterogeneous data into the same data model. How to determine the proper intermediate data model and unify the involved heterogeneous data models for the analytical task is an urgent problem need to be solved. In this paper, we proposed a model determination method based on cost estimation. It evaluates the execution cost of query tasks on different data models, which taken as the criterion to measure the data model, and chooses a data model with the least cost as the intermediate representation during data processing. The experimental results of BigBench datasets showed that the proposed cost estimation based method could appropriately determine the data model, which made heterogeneous data processing efficiently.

Keywords: Heterogeneous data analysis · Data model determination Cost estimation · BigBench datasets

1 Introduction

In heterogeneous data processing [1], various model of relational and non-relational data often involved together for data fusion and analysis. Typical non-relational data includes graph data, textual data and the key-value data. Data heterogeneity makes the query processing, data maintenance and management exceptionally difficult. Therefore, it is common to transform data into a common model to make data processing easier [2, 3]. During the data fusion, XML data [4–6] usually used as the intermediate conversion model [7]. However, the operation of XML data mainly use XPath or XQuery query language, which often to be inefficient and not suitable for large-scale data processing.

To address this issue, we work on determining the corresponding optimal data model in heterogeneous data analysis task, and unify the heterogeneous data into the

R. Silhavy (Ed.): CSOC 2018, AISC 764, pp. 374–383, 2019.
https://doi.org/10.1007/978-3-319-91189-2_37

optimal model. There are exist some NoSQL database systems [8] support the flexible conversion between relational and non-relational data, which provides a basis for the implementation of our idea. In this work, we proposed a method for determining the optimal data model in various heterogeneous data model based on cost estimation of the query tasks. We introduced the cost model approaches [9, 10] in our work, which transforms the query statements into physical query plans, aggregates the cost of each physical operator in the query plan, and then determines the optimal intermediate data representation model based on the cheapest cost.

The rest of the paper is organized as follows. Section 2 describes the design and implementation of the cost model. Section 3 evaluates the accuracy of the cost model and its effectiveness in data model selection through experiments. Finally, Sect. 4 presents the summary of this work.

2 Construction of Cost Model

The cost model we utilized to estimate the execution cost of query statements mainly consist of the cost of CPU operation and I/O operation. Equation (1) expresses the cost model.

$$C_{total} = C_{cpu} + C_{IO} + C_{st} \tag{1}$$

Where C_{total} represents the total cost of the query, C_{cpu} represents the CPU cost, C_{IO} represents the IO cost, and C_{st} represents the cost of other operations prior to the CPU and I/O operations, which is recorded as the initial cost. Such as the cost of load raw data in a given operation.

The design idea of the cost model is to take the physical operator as the evaluation object, and obtain the total cost of the whole query statement by estimating the cost of all the physical operators. First, it transforms the predicate in the query statement into the corresponding physical operator, and obtains the physical query plan of the query statement according to the execution of the query. Then, the cost model estimates the query cost of each physical operator, including CPU cost and IO cost. Finally, the entire cost of the SQL statement can be obtained according to Eq. 1.

2.1 Physical Query Plan

The first step of cost estimation is to parse the SQL statement and transform it into the corresponding physical operators. In this paper, we implement our approach in OrientDB [11] system.

Standard Query. OrientDB uses SQL as its query language, supporting standard SQL syntax. As shown in Fig. 1(a).

During the parsing of the query statement, it transforms the *select* into the Project operator, the *from* into the ClassScan operator, the ratio operation and the *like* into the Filter operator, the *group* into the GroupBy operator, the *sum* aggregation operation into the Agg operator, the *order* into the Sort operator, and the *limit* into the Limit

```
select t1.c1,t1.c2,sum(t1.c3)
from t1 where t1.c1>5 and t1.c2
like “%ab%” group by t1.c1,t1.c2
order by t1.c1 limit 5,2
```

(a) Standard Query

```
select * from
(traverse * from #77:2 maxdepth 3)
where ss_item_sk=641
```

(b) Traverse Query

Fig. 1. Query statement

operator. According to the execution process, it obtains the physical query plan, as shown in Fig. 2(a).

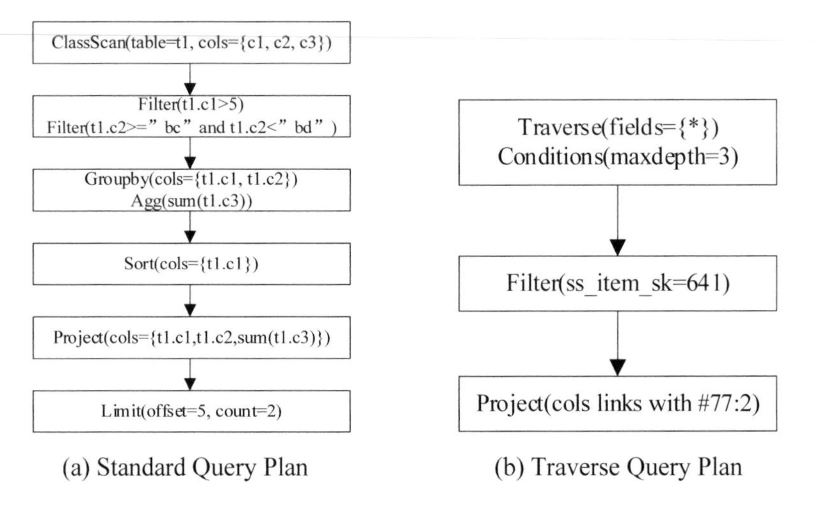

(a) Standard Query Plan (b) Traverse Query Plan

Fig. 2. Query plan

Graph Query. OrientDB supports the *Traverse* operation, which traverses the graph. The Fig. 1(b) display the query statement.

During the parsing of the query statement, it transforms the *select, traverse, equivalent* into Project, Traverse and Filter operator respectively. The Fig. 2(b) shows the resulting physical query plan.

2.2 Cost Estimation of Physical Operator

In order to get the total cost of the entire query plan according to its physical plan, we used the Profiler tool in the Database to analyze some simple query statements, and assigned the model parameters and their values according to the measurement results. Table 1 shows these parameters and how they are obtained.

The main operator and its cost calculation formula is as follows.

ClassScan. A scan operation is the most basic physical operator that scans a specified Class and reads the data from the storage. It includes I/O operations that read data and CPU operations that parse tuples. Equation (2) shows how to obtain its IO cost, where

Table 1. Cost model parameters

Parameters	Meaning of parameter	Method of obtaining
C_{seq_page}	The cost of sequential reading of a page	Profiler measurement
C_{cpu_tuple}	The CPU cost of parsing each record	Profiler measurement
C_{cpu_pro}	The CPU cost of a single project operation	Profiler measurement
C_{cpu_op}	The CPU cost of performing operation once	Profiler measurement
C_{comp}	The cost of performing a sort comparison	Computation of parameters
C_{ran_page}	The cost of random reading of a page	Profiler measurement
C_{cpu_bool}	The CPU cost of executing a decision	Profiler measurement

N_t represents the number of records to be accessed, W_t represents the average width of tuple, S_b represents the disk page size.

$$C_{scan\_io} = C_{seq\_page} * (N_t * W_t/S_b) \tag{2}$$

Equation (3) shows how to obtain the CPU operation cost of the operator.

$$C_{\text{scan}\_cpu} = C_{cpu\_tuple} * N_t \tag{3}$$

Project. Projection operation reads some columns for each record from the intermediate result after processing, which only involved in CPU operation. It is the cost calculation as Eq. (4) shows, where N_c represents the number of columns.

$$C_{\text{pro}} = C_{cpu\_pro} * N_c \tag{4}$$

GroupBy. The GroupBy operator aggregates the input data according to the specified columns. It need to evaluate the CPU cost of processing each record on each group column. The Eq. (5) shows the cost calculation, where N_{gc} represents the number of grouping columns.

$$C_{\text{group}} = C_{cpu\_op} * N_t * N_{gc} \tag{5}$$

The query will execute an aggregation operation after group operation. Equation (6) shows the cost computation, including the starting cost and the aggregation computing cost, in which N_g represents the number of groups obtained after grouping.

$$C_{agg} = C_{cpu\_tuple} * \left(N_t + N_g\right) + C_{cpu\_op} * N_t * N_{gc} \tag{6}$$

Sort. Sorting operation needs to consider two situations: internal sort and external sort. If the volume of input data is less than the sorted memory capacity, it takes the internal sort, no I/O operations occurring. On the other hand, it takes the external sort and considers the cost of I/O operations. Equation (7) is the cost calculation of CPU operations.

$$C_{sort\_cpu} = C_{comp} * N_t * \log_2 N_t \tag{7}$$

When it takes the external sort, the query will calculate the cost of I/O operations. As shown in Eq. (8), N_{p} represents the number of disk pages.

$$C_{sort\_io} = N_p * \left(C_{seq\_page} * 0.75 + C_{ran\_page} * 0.25\right) \tag{8}$$

Limit. In OrientDB, query usually use the *limit* in conjunction with *skip*, which returns resulting records in a count range. This operation needs to consider the starting cost for inputting raw data. Equation (9) shows the cost formula, where P_s represents the starting position, P_l represents the number of records is limited.

$$C_{\mathrm{limit}} = C_{cpu\_op} * (P_s + P_l) + C_{cpu\_tuple} * N_t \tag{9}$$

Filter. The Filter operator chooses records according to the conditions, which includes compared operation, equivalent operation, as well as the operations such as *like*, *between*, *contains* and so on. The cost estimation is mainly the total cost of performing predicate comparison for each record. As shown in Eq. (10), N_{pre} represents the number of predicates.

$$C_{\mathrm{filter}} = C_{cpu\_bool} * N_t * N_{pre} \tag{10}$$

Distinct. The query use the *distinct* to eliminate duplicate records in the input data. It sorts the input data with all columns, and eliminates duplicate rows by determining whether adjacent rows are equal. Therefore, the cost model transforms this operation into the sorting and comparison operation, and the cost estimation of the operator is the sum of the Sort operator and the Filter operator.

Traverse. This operator traverses the nodes connected by edge, each of which is a data record. OrientDB uses Class to store graph data, and divides it into one or more Clusters, which consists of multiple disk pages. It transforms the foreign key in relational data into edge in graph, so traversal operations executed across Classes. The cost of the operator consists of the cost of processing records and parsing edges. In Eq. (11), N_d represents the maximum depth of traversal.

$$C_{\mathrm{trav}} = \left(C_{ran\_page} + C_{cpu\_tuple}\right) * N_d \tag{11}$$

2.3 Data Model Determination Based on Cost Estimation

Now, we use the cost model to estimate the execution cost of the query on graph data. In the PostgreSQL database, it uses the *explain* command to estimate the execution cost of the query on relational data. By comparing its estimation with our approach, we choose the data model with the least cost, and unify the data into the chosen model, which achieve higher efficiency of data analysis. This method avoids the actual execution of query tasks, which reduces resource consumption and saves time. We can directly predict the optimal data model for the query according to the cost estimation.

3 Experiments

We performed two sets of experiments to evaluate performance of our method. The first set of experiments verifies the accuracy of the cost model by comparing the estimated cost with the actual execution cost. The second set measures the effectiveness of the cost model, which compares the execution cost on graph and relational data.

3.1 Datasets

The experiment used the datasets generated by BigBench [12, 13] as the experimental data. It involves structured data, semi-structured data and unstructured data, where the semi-structured data includes the log data of customer clicking in retail websites, the unstructured data consists of online commodity review data. Our experiments involves five tables, as shown in Table 2.

Table 2. Experiment datasets

Table name	Number of records (million)
customer	0.01/0.05/0.1/0.2/0.5
customer_demo	0.1/0.5/1/2
date_dim	0.01/0.05/0.1
item	0.01/0.05/0.1
store_sales	0.1/0.5/1/5/10/20

3.2 Query Cases

The query cases in experiments include single table query, join query of two tables and join query of multiple tables. The original BigBench query designed for running over Hadoop ecosystem [14], we rewrite those query cases to run in PostgreSQL and OrientDB before evaluation.

Q1 is a single table query that looks up the table *customer*. The filter condition is that the last name matches "To", and the born year is between 1980 and 1990. Lastly, it sorts the last name and limits 100 pieces of data to output. The Fig. 3 displays the query statement of Q1.

```
select c_last_name, c_first_name from customer where c_first_name like 'To%'
and c_birth_year between 1990 and 2000 order by c_first_name limit 20;
```

Fig. 3. Q1 Query statement

Q2 is a classification that clusters customers into book buddies or club groups based on in store book purchasing histories, involving join operation of two table. The Fig. 4 (a) shows the query statement of Q2.

```
select ss_customer_sk as cid,
count(case when i_class_id=1 then 1 else NULL end) as id1,
count(case when i_class_id=2 then 1 else NULL end) as id2,
count(case when i_class_id=3 then 1 else NULL end) as id3,
count(case when i_class_id=4 then 1 else NULL end) as id4,
count(case when i_class_id=5 then 1 else NULL end) as id5
from store_sales inner join item on ss_item_sk = i_item_sk
and i_category in ('Books') and ss_customer_sk is not NULL
group by ss_customer_sk having count(ss_item_sk) > 5
order by cid;
```

(a) Q2 Query

```
select sum(ss_quantity)
from store_sales, date_dim, customer_demographics
where ss_sold_date_sk=d_date_sk and d_year=1998
and cd_demo_sk=ss_cdemo_sk and (
(cd_marital_status='M' and cd_education_status='2 yr Degreee'
and ss_sales_price BETWEEN 100.00 and 150.00) OR
(cd_marital_status='M' and cd_education_status='4 yr Degree'
and ss_sales_price BETWEEN 50.00 and 100.00) OR
(cd_marital_status='M' and cd_education_status='4 yr Degree'
and ss_sales_price BETWEEN 150.00 and 200.00) )
```

(b) Q3 Query

Fig. 4. Query statement

Q3 is a statistical analysis that calculates the total sales by different types of customers (e.g., based on marital status, education status), sales price and different combinations of state and sales profit. The Fig. 4(b) is the query of Q3.

3.3 Analysis of Experimental Results

This section presents the detailed experimental evaluation.

Accuracy Verification of Cost Model. This experiment compares the estimated cost with the actual execution cost using the Q1 query. As shown in Fig. 5(a), the horizontal axis represents the data size of table *customer*, the vertical axis represents the query cost, which is a special unit to measure the query cost. The "gm" represents the estimation cost of query on graph model. The "ac" represents the actual execution cost of query.

In Fig. 5(a), the estimation cost is consistent with the actual execution cost. Although there is a gap, the trend of growth shows that both of them have the same growth rate. Therefore, we can conclude that the query cost estimated by the cost model can be very close to the actual execution cost. It means that the cost model is effective for determine the data model.

Verification of Data Model Determination Based on Cost Estimation. This part completes the selection of data model based on cost model.

The first experiment also used Q1 to compare the query cost on graph data with the cost on relational data, as shown in Fig. 5(b). The "rm" represents the query cost of relational data. The query cost on graph data is always smaller than that on relational data. When the data size reaches 0.5 million, the query cost of graph data is close to the relational data, and the growth rate exceeds that of relational data. Therefore, we will choose the graph data as the optimal data model when the data size is less than 0.5 million. Conversely, we choose the relational data.

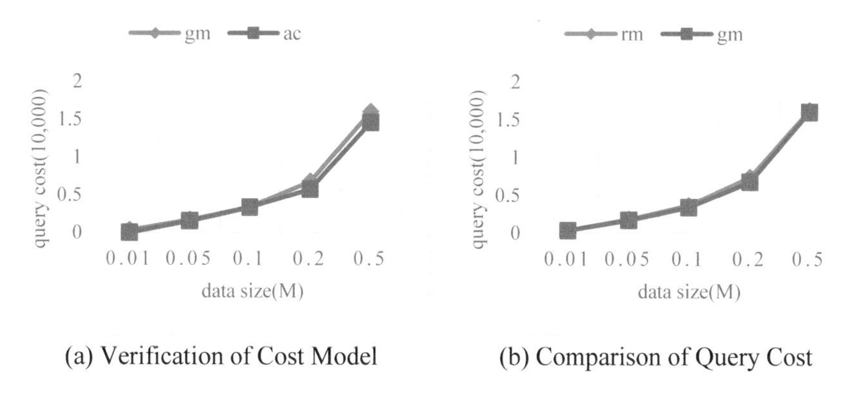

(a) Verification of Cost Model (b) Comparison of Query Cost

Fig. 5. Q1 Query cost

The second experiment involves join query of table *store_sales* and *item*. Towards graph data, the join operation is replaced by link association [15, 16], which only scanned table *store_sales*. The result shown in Fig. 6, where the horizontal axis represents the data volume in the table *store_sales*, and the "rm0.01", "rm0.05", "rm0.1" represents the query cost of relational data with different data size in table *item* respectively.

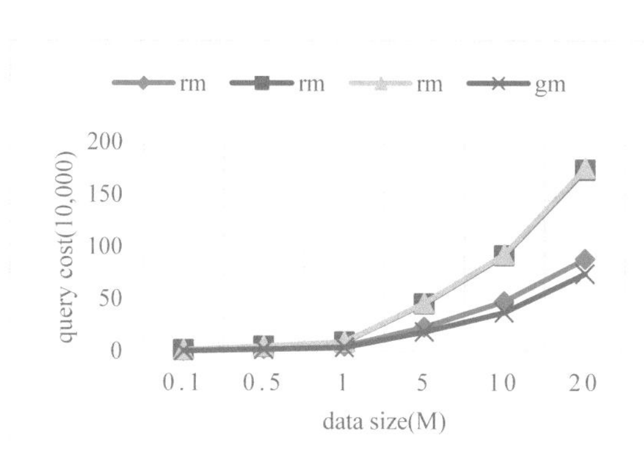

Fig. 6. Q2 Query cost

In Fig. 6, the query cost of relational data is generally greater than that of graph data. Therefore, we take the graph data as the optimal data model of the query task. From the result of "rm0.01", the difference of the query cost between relational data and graph data is very small. However, with the data size increasing of table *store_sales*, the query cost of graph data will exceed that of relational data, so we choose the relational data as the optimal data model.

The third experiment involves join query of multiple tables. First, we considers the change of data size in table *store_sales* and *customer_demo*, and the table *date_dim* is fixed. The result shows in Fig. 7(a), where the horizontal axis represents the change of data volume in the table *store_sales*, and the "rm0.1", "rm0.5", "rm1", "rm2" represents the query cost of relational data with different data size in table *customer_demo* respectively.

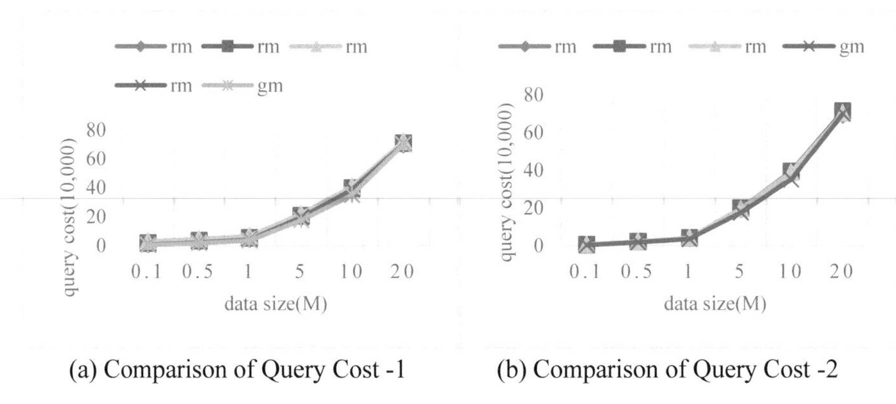

(a) Comparison of Query Cost -1 (b) Comparison of Query Cost -2

Fig. 7. Q3 Query cost

Then the experiment considers the different data size of table *store_sales* and *date_dim*, and the table *customer_demo* is changeless. From the Fig. 7(b), the "rm0.01", "rm0.05", "rm0.1" represents the query cost of relational data with different data size in table *date_dim* respectively.

The results are similar in Fig. 7(a) and (b), where the query cost of the graph data is lower than that of relational data. Therefore, we can choose the graph data as the optimal data model of the query. When the data volume of table *store_sales* reaches 20 million, their cost values become close. As the data continues to grow, we will select the relational data as the optimal data model.

4 Conclusion and Future Work

In this paper, we address the problem of low efficiency of the heterogeneity of data model in data analysis. We proposed a method for data model determination based on cost estimation. It can be used to choose the optimal data model in heterogeneous data processing. The experimental results show that the proposed method can effectively determine the optimal data model between relational data model and graph data model. In the future, we plan to incorporate more data models (e.g., key-value model, and document model) into the approach to make it has more widely application scenarios.

Acknowledgements. This work was supported by the Fund by The National Natural Science Foundation of China (Grant No. 61462012, No. 61562010, No. U1531246), Guizhou University Graduate Innovation Fund (Grant No. 2017078) and the Innovation Team of the Data Analysis and Cloud Service of Guizhou Province (Grant No. [2015]53).

References

1. Wang, L.: Heterogeneous data and big data analytics. Autom. Control Inf. Sci. **3**(1), 8–15 (2017)
2. Xuexian, C., Huiting, J.: Research on heterogeneous data sources integration. Comput. Eng. Sci. **30**(8) (2008)
3. Xiangjiang, K., Yupeng, M., Yingfan, L.: Data type's conversion at heterogeneous database systems. Appl. Res. Comput. (1), 217–218 (2006)
4. Jingling, Y., Lili, X., Lianchao, M.: Research on virtual approach about heterogeneous data integration based on XML. Appl. Res. Comput. **26**(1) (2009)
5. Weiwei, W., Qinghong, S.: Integrated platform of distributed heterogeneous data based on XML. J. SE Univ. **36**(5) (2006)
6. Castro, E., Cuadra, D., Velasco, M.: From XML to relational model. INFORMATICA **21**(4), 505–519 (2010)
7. Hui, H., Fu, S., Sui, L.: Data conversion model of heterogeneous database. Comput. Eng. Des. **26**(9) (2005)
8. Derong, S., Ge, Y., Xite, W.: Survey on NoSQL for management of big data. J. Softw. **24**(8) (2013)
9. Shengli, W., Nengbin, W.: Estimation of query cost in object-oriented database systems. Comput. Res. Dev. **35**(1) (1998)
10. Xunbo, S., Xiangguang, Z.: Cost model of query plan in distribute database system. Comput. Syst. Appl. (10) (2007)
11. OrientDB Document. http://orientdb.com/docs/. Accessed 28 Jan 2018
12. Ghazal, A., Rabl, T., Hu, M.: BigBench: towards an industry standard benchmark for big data analytics. In: SIGMOD (2013)
13. Baru, C., Bhandarkar, M., Danisch, M., et al.: Discussion of BigBench: a proposed industry standard performance benchmark for big data. In: TPCTC (2014)
14. Chowdhury, B., Rabl, T., Saadatpanah, P., et al.: A BigBench implementation in the Hadoop ecosystem. In: WBDB 2013. LNCS, vol. 8585, pp. 3–18 (2014)
15. Petermann, A., Junghanns, M., Muller, R., et al.: Graph-based data integration and business intelligence with BIIIG. VLDB **7**(13) (2014)
16. De Virgilio, R., Maccioni, A., Torlone, R., et al.: Converting relational to graph databases. In: GRADES (2013)

Aspects of Using Elman Neural Network for Controlling Game Object Movements in Simplified Game World

Dmitriy Kuznetsov(✉) and Natalya Plotnikova

National Research Mordovia State University, Saransk, Russia
kuznetsov.da@list.ru, linsierra@yandex.ru

Abstract. This paper describes architecture of an artificial intelligence system based on the Elman neural network. Simple training algorithms and neural network models are not able to solve such a complex problem as movements in the conditions of an independent game world environment, so a combination of a base neural network training algorithm and Q-learning agent approach is used as part of a player behavior control model. The paper also includes results of experiments with different values of model and game world characteristics and shows efficiency of the described approach.

Keywords: Elman neural network · RMS propagation · Q-learning Machine learning

1 Introduction

Nowadays artificial intelligent systems, in particular, artificial neural networks, are a very popular research area. There are even several AI competitions, such as Russian AI Cup or Code Game Challenge. The main idea of these competitions is implementation of simplified artificial intelligence for controlling some game object (usually a player), which can move, get bonuses, attack other players, lose or gain health and do other things according to the rules of the competition game world. The main purpose of each player is to get as many scores as possible and lose as little health as possible. The rules of calculating finale scores depend on the implementation of the game world.

There are a lot of techniques of creating such an AI. But one of the most promising approaches is using artificial neural networks (ANN). Neural networks are quite a wide tool for solving AI-problems. The main idea of using ANN is to control movements of a player by sending information about the player environment to neural network inputs and getting information about the next best movement from neural network outputs.

The main purpose of this research is to develop tools for creating and testing such a neural network, and to investigate player behavior quality correlation with neural network architecture, training algorithm characteristics and artificial intelligence model.

For this, we have used methods of system analysis, artificial neural networks theory, optimization theory, geometry modeling, and experiment planning theory.

First of all, it is necessary to describe the artificial intelligence model according to the game world model and rules.

R. Silhavy (Ed.): CSOC 2018, AISC 764, pp. 384–393, 2019.
https://doi.org/10.1007/978-3-319-91189-2_38

2 Game World and Artificial Intelligence Model

In our research, the game world is a rectangle map of a fixed size with some game objects inside. Each game object model is a circle of some predefined radius. The main game objects are players and bonuses. A bonus is a static game object which has the following characteristics: coordinates in the game world, positive or negative scores and lifetime. A player is a moveable game object which can move in some direction during one game world simulation tick and eat bonuses on its way. The player has the following characteristics: coordinates in the game world, speed, health, fullness. Health of the player decreases in case of eating bonus with negative scores. Fullness of the player increases in case of eating bonus with positive scores. Finale scores of the player after simulation is the sum of its health and fullness. Usually there are several players in the game world during one limited simulation. And finally, the player with the greatest scores wins.

A schematic model of the researched game world is presented in Fig. 1.

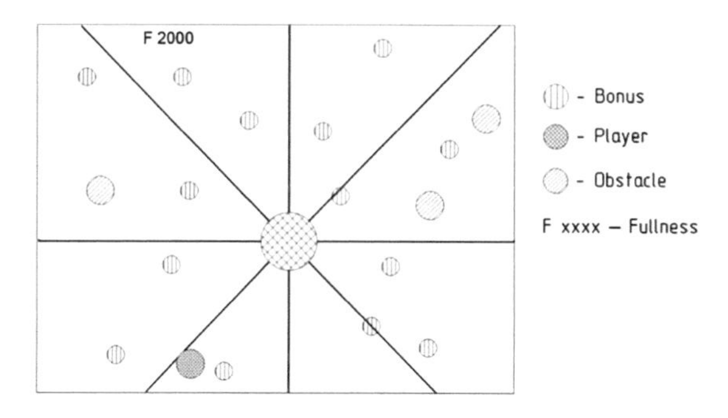

Fig. 1. Game world model

According to the introduced game world model, it is important for the player to get as many bonuses as possible, so to have information about bonuses positions at each moment of simulation time and to take a decision of a movement direction based on this information. The amount of bonuses in the game world may be great, and the amount of possible directions may be great too. So, we need to limit the range of available decisions to make modeling more efficient. For this purpose, so called sensors have been introduced. It is a property of the player object. Each sensor is a sector with center coincided with the player center and limited by the game world borders. A player object with 8 sensors is presented in Fig. 1. We can use information about bonuses in these sectors to determine profitability of the movement in the corresponding direction.

For this, we need to introduce a neural network model.

3 Neural Network and Training Algorithm

The most appropriate neural network types for solving this problem are multilayer perceptron and Elman neural network [1–3]. In our research, we use Elman neural network [4] because of its' time sequence memorization ability (it is important for us to take into account results of the previous movements). The architecture of the Elman neural network is presented in Fig. 2.

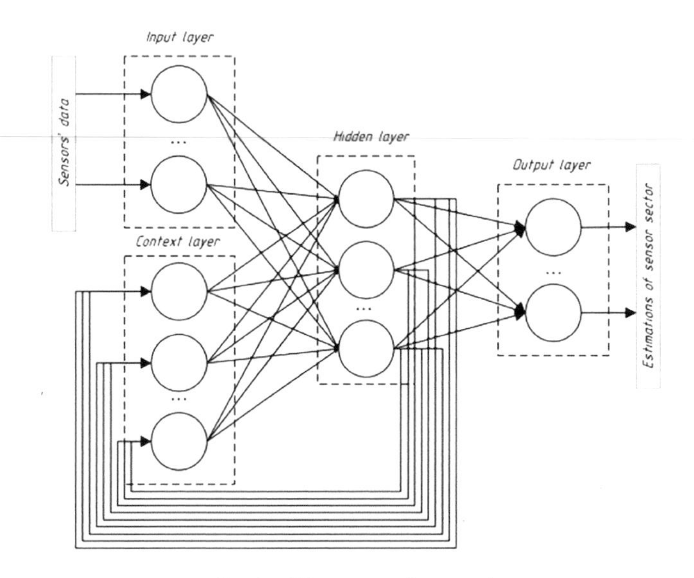

Fig. 2. Elman neural network

Information from sensors for a player is sent to neural network inputs. Information from one sensor is represented by a distance to and a type of the nearest bonus in the appropriate sector. Outputs of neural network are estimations of profitability of moving into the direction of each sensor sector. For example, for the player model with 8 sectors neural network should include 16 inputs and 8 outputs. The decision about a movement direction for each simulation tick is taken according to the best (maximum) estimation of the neural network [5].

Logic of the movement routine is presented in Fig. 3.

The training algorithm is part of the movement schema and should be applied for each simulation tick. For our purposes, we choose RMS-propagation training algorithm as a base in combination with Q-learning algorithm (Fig. 4).

RMS propagation is used to calculate and correct weights and biases of the neural network. Q-learning is responsible for the logic of the movement estimation calculation.

According to the steps of the training algorithm, we need to introduce reward and/or penalty (negative reward) for a movement:

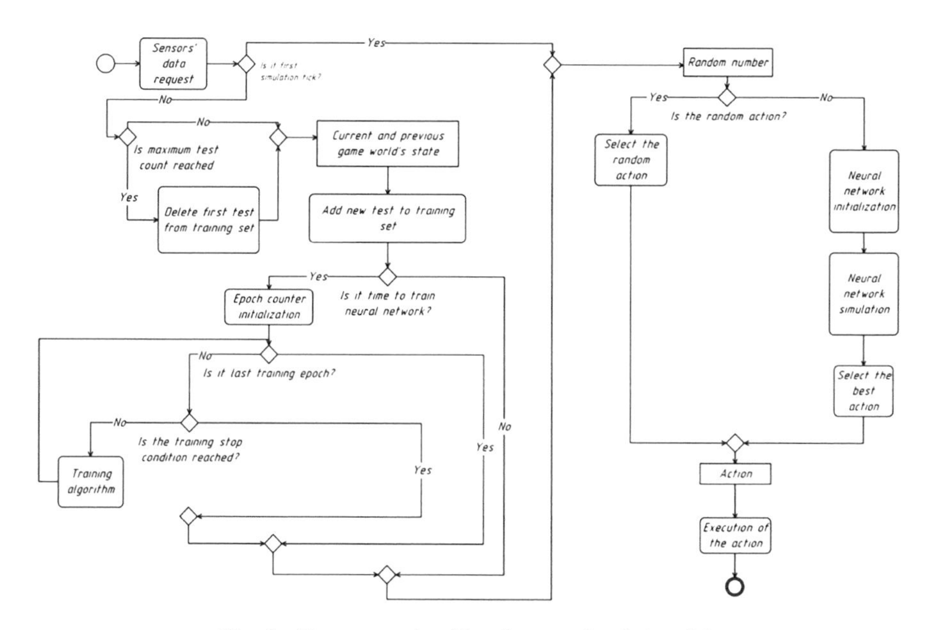

Fig. 3. Movement algorithm for one simulation tick

$$Reward_t = Fullness_t - Fullness_{t-1} + WallPenalty_t \tag{1}$$

$$WallPenalty_t = -0.9 * exp(-DistanceToWall_t / PermitedDistance) \tag{2}$$

The value of the reward represents profitability of the player movement and allows taking into account desirable and undesirable consequences of the movement. For example, in our case we want to increase fullness of the player and to avoid a lot of movements near the game world borders.

The described model consists of quite complex algorithmic components with many characteristics. And it is necessary to investigate the influence of these characteristics on the quality of the player behavior. For this purpose, we need a convenient research tool.

4 Training and Modeling System

A special training system has been developed to research the approach described below. The structure of this system is presented in Fig. 5.

Game World – a C++ object oriented implementation of the game world. It includes classes for the main objects and routines for providing the game world logic and rendering.

Neural Network – an implementation of the Elman neural network and "RMS-propagation and Q-learning" training approach [7, 8]. The Neural Network component has been developed on C++ programming language using an object oriented paradigm.

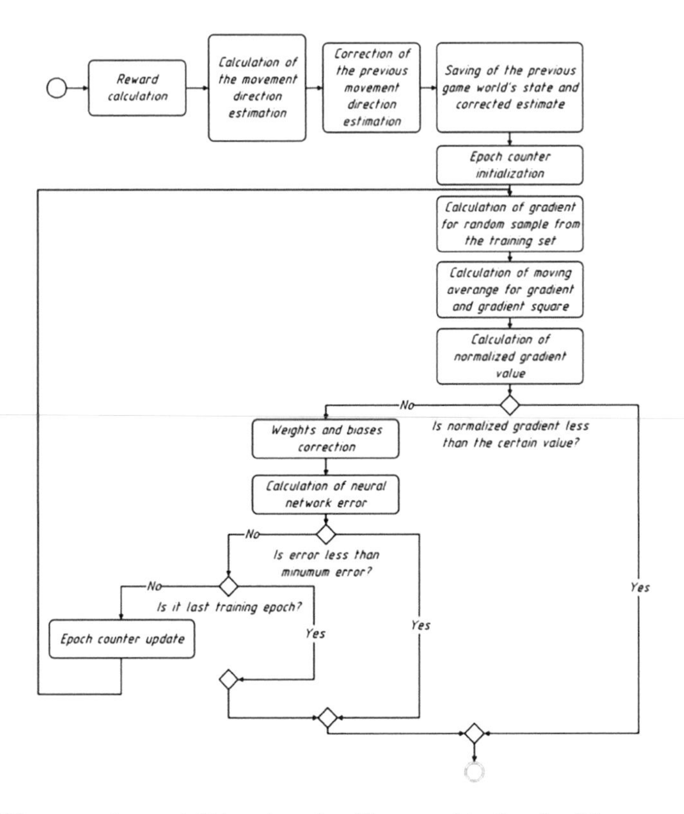

Fig. 4. RMS propagation and Q-learning algorithms combination for Elman neural network training

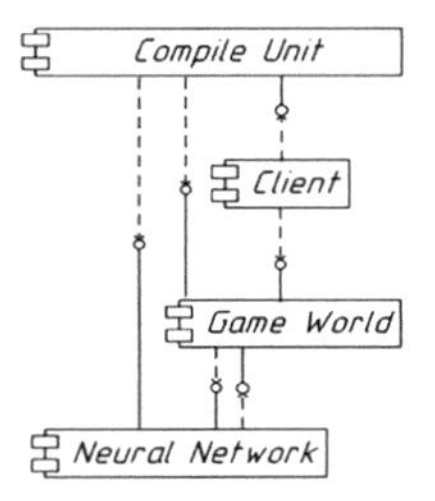

Fig. 5. Structure of the training system

Client – an application with GUI used to tune neural network settings, training algorithm parameters, and to compile and run the combination of the game world component and one or more player control solutions (including the neural network solution). It allows to simplify research process. This component has been developed on C# programming language. Interconnection between the client and the game world has been implemented using C++/CLI.

Compile Unit – an implementation of the compilation tool which allows compiling of any modification of neural network on the fly.

Also, the system includes some logging and data processing routines to simplify the investigation process.

5 Experiment Conditions and Results

All experiments and tests have been carried out on the Elman neural network with the following architecture. The number of layers is equal to three (input, hidden and output). According to the Elman neural network general architecture, there is also a context layer with the same neuron count as for the hidden layer. Hidden neurons use a sigmoid or hyperbolic tangent activation function. Output neurons use a linear activation function. Table 1 includes base characteristics of the neural network and training algorithm.

Table 1. Base Elman neural network and training algorithm characteristics

Characteristic	Value
Sensor count	8
Hidden neuron count	25
Trainset size for each training epoch	25
Full trainset size	5000
Training epoch count	10
Hidden layer activation function	Hyperbolic tangent
Based training algorithm	RMS Propagation
RMS Propagation training speed	0.0001
Q-Learning discount factor	0.9

Experiments have been conducted as a set of game world simulations. The investigated characteristic has been varied in some range and the other characteristics have been fixed according to the default values from Table 1. All the results are represented as charts reflecting dependency of player scores on the simulation tick count. This dependency allows showing efficiency of the player object behavior according to the game world rules and influence degree of some characteristics on the result of the game. The following characteristics have been investigated:

1. The base training algorithm. To be sure that the chosen base algorithm (RMS propagation [9]) is efficient, the comparison of RMS propagation and Resilent propagation [6] training algorithms has been performed. According to the chart in Fig. 6, RMS propagation allows to achieve better results.
2. The hidden layer neuron count has been varied in range from 5 to 300 neurons to investigate the optimal size of the neural network, which can be critical from the system performance point of view. And as presented in Fig. 7, the neural network with the greater neuron count shows better results as expected. So, it is naturally to use as many neurons as possible in the technical conditions of the experiment.

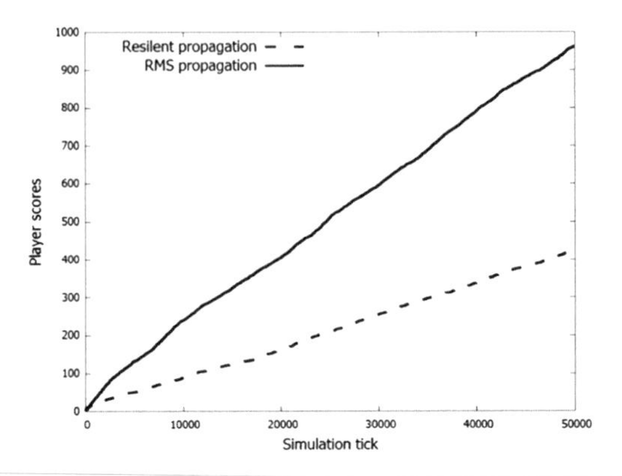

Fig. 6. Experiment results for different base training algorithms

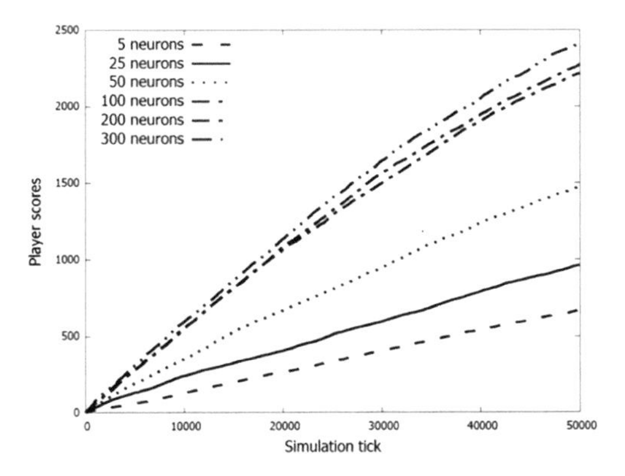

Fig. 7. Experiment results for different hidden layer neuron count

3. The hidden layer activation function. Two types of the activation function of the hidden layer neurons have been researched. According to the chart in Fig. 8, better results for short simulation period (approximately less than 13000 simulation ticks) are showed by the neural networks with the hyperbolic tangent activation function. But for long simulation period, the sigmoid activation function demonstrates stable increase of the efficiency.
4. Q-learning discount factor is a measure of trust to the previous neural network predictions of the player movements. The discount factor has been varied in range from 0 to 1.5. According to the chart in Fig. 9, the best result is shown by the neural network with Q-learning discount factor 0.9. For Q-learning discount factor equal to or greater than 1, the neural network demonstrates worse results. The main conclusion is that trust to the past decisions should be presented in the system sufficiently but over trust can lower system decision efficiency and quality of the player behavior.

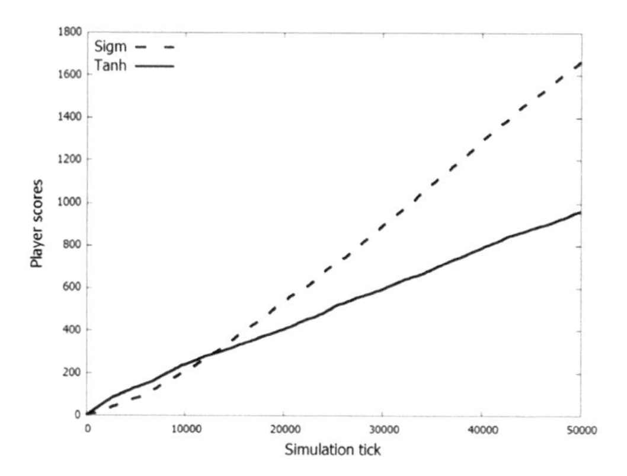

Fig. 8. Experiment results for different hidden layer activation functions

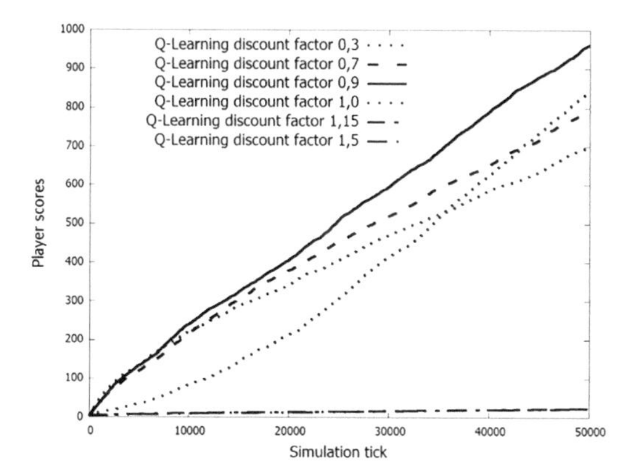

Fig. 9. Experiment results for different Q-learning discount factor values

5. RMS propagation training speed is the value of the gradient descent step. The influence of this parameter is presented as chart in Fig. 10. Greater values of the training speed show better simulation results. The main reason of that is low epoch count, so with the greater speed the algorithm moves forward faster and converges faster to the neural network error global minimum.
6. The sector count defines ability of the player to "see" objects around as described in the beginning of the paper. The more objects player can identify at the moment, the more accurate profitability of the movement in each direction can be calculated. This conclusion is reflected in Fig. 11 - the best simulation results are shown by the player model with 64 sectors.

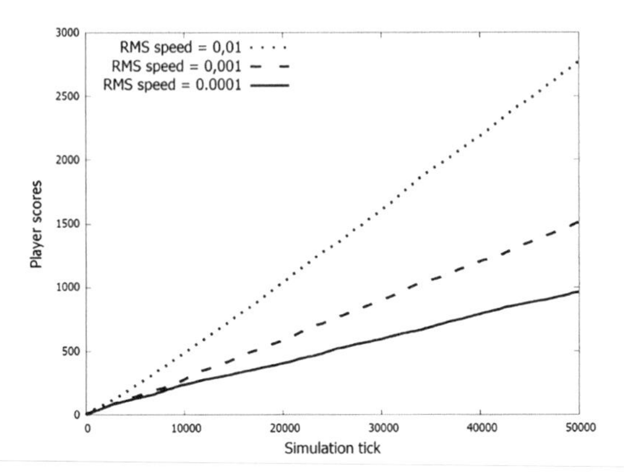

Fig. 10. Experiment results for different RMS propagation training speed values

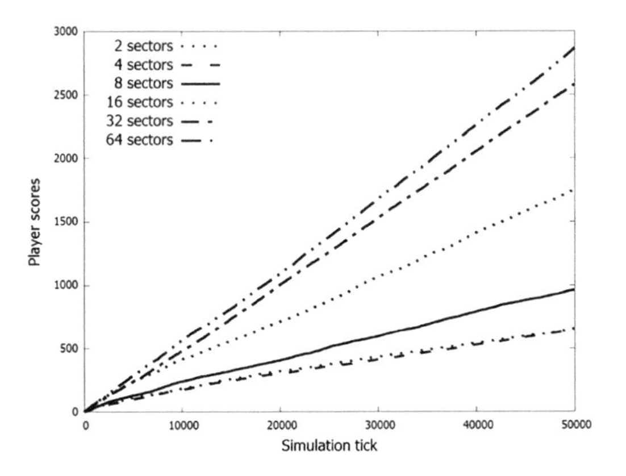

Fig. 11. Experiment results for different sector count

6 Conclusion

In this paper, we have described an approach to creating a simplified artificial intelligence model for controlling player movements in a game world environment. The main idea of the approach is using Elman neural network trained with a combination of Q-learning and RMS propagation algorithms to take a decision about the best movement direction. The value of movement quality depends on the preferable behavior of the player, for example in our research the player should move in such a way to get as more bonuses as possible.

A special training system has been developed to research the influence of the most important neural network, training algorithm and player model characteristics on the efficiency of the proposed player control model.

According to the conducted experiments, Elman neural network model has showed good learning ability and flexibility in taking into account different aspects and characteristics of the game world environment.

References

1. Hornik, K., Stinchcombe, M., White, H.: Multilayer feed forward networks are universal approximators. Neural Netw. **2**(5), 359–366 (1989)
2. Chen, T.P., Chen, H.: Universal approximation to nonlinear operators by neural networks with arbitrary activation functions and its application to dynamical systems. IEEE Trans. Neural Netw. **6**(4), 911–917 (1995)
3. Leshno, M., Lin, V.Y., Pinkus, A., Schocken, S.: Multilayer feed forward networks with a nonpolynomial activation function can approximate any function. Neural Netw. **6**(6), 861–867 (1993)
4. Wilamowski, B.M.: Neural networks and fuzzy systems for nonlinear applications. In: Proceedings of 11th INES 2007–11th International Conference Intelligent Engineering Systems, Budapest, Hungary, June 29–July 1 2007, pp. 13–19 (2007)
5. Macal, C.M., North, M.J.: Tutorial on agent-based modeling and simulation part 2: how to model with agents. In: Proceedings of Winter Simulation Conference WSC 2006, 3–6 December 2006, pp. 73–83 (2006)
6. Riedmiller, M., Braun, H.: A direct adaptive method for faster back propagation learning: the RPROP algorithm. In: Proceedings of International Conference Neural Networks, San Francisco, CA, pp. 586–591 (1993)
7. Demystifying Deep Reinforcement Learning. http://neuro.cs.ut.ee/demystifying-deep-reinforcement-learning/
8. Q-Learning. http://mnemstudio.org/path-finding-q-learning-tutorial.htm
9. Neural Networks for Machine Learning. https://www.cs.toronto.edu/~tijmen/csc321/slides/lecture_slides_lec6.pdf
10. Stroustrup, B.: Programming: Principles and Practice Using C++, 2nd edn. Addison-Wesley, Upper Saddle River (2014)

Intrinsic Evaluation of Lithuanian Word Embeddings Using WordNet

Jurgita Kapočiūtė-Dzikienė[1] and Robertas Damaševičius[2(✉)]

[1] Vytautas Magnus University, K. Donelaičio 58, 44248 Kaunas, Lithuania
j.kapociute-dzikiene@vdu.lt

[2] Kaunas University of Technology, K. Donelaičio 73, 44029 Kaunas, Lithuania
robertas.damasevicius@ktu.lt

Abstract. Neural network-based word embeddings –outperforming traditional approaches in the various Natural Language Processing tasks – have gained a lot of interest recently. Despite it, the Lithuanian word embeddings have never been obtained and evaluated before. Here we have used the Lithuanian corpus of ~234 thousand running words and produced several word embedding models: based on the continuous bag-of-words and skip-gram architectures; softmax and negative sampling training algorithms; varied number of dimensions (100, 300, 500, and 1,000). Word embeddings were evaluated using the Lithuanian WordNet as the resource for the synonym search. We have determined the superiority of the continuous bag-of-words over the skip-gram architecture; while the training algorithm and dimensionality showed no significant impact on the results. Better results were achieved with the continuous bag-of-words, negative sampling and 1,000 dimensions.

Keywords: Intrinsic evaluation · Neural word embeddings
The Lithuanian language

1 Introduction

Due to the tremendous progress in the deep learning field, became possible to train more sophisticated models on the huge quantities of data and thus to achieve much higher accuracy levels. Probably, the distributional representations contributed to this progress the most.

When the "traditional representations" – treating linguistic items (i.e., words, phrases, etc.) as isolated atomic units – seems already exhausting its limits, distributional representations (i.e., word embeddings) – trained on large unlabeled corpora and mapping linguistic items to vectors of the real values – demonstrate the state-of-the-art performance and become a dominant paradigm. Neural network-based embeddings are projected in the n-dimensional vectors spaces and the distances between words determine their semantic similarities. Due to this reason, word embeddings are more suitable for various Natural Language

R. Silhavy (Ed.): CSOC 2018, AISC 764, pp. 394–404, 2019.
https://doi.org/10.1007/978-3-319-91189-2_39

Processing (NLP) tasks (for the detailed survey see [1]): text classification [2], sentiment analysis [3], speech recognition [4], machine translation [5], dependency parsing [6], dialect identification [7], etc.

Though the concept of word embeddings is not new, it has become popular only since 2013, when Mikolov et al. [8] introduced the *word2vec* model.[1] This model (presented as one) actually incorporates two architectures (i.e., continuous bag-of-words & continuous skip-gram) and two training algorithms (i.e., hierarchical softmax & negative sampling). The word embeddings are obtained after training two-layer neural network (NN) on the huge representative corpora. The weights in the projection (hidden) layer map words into n-dimensional vector spaces, where n is a number of nodes in this layer. Although n can be arbitrary selected, is usually varies from 50 to 1,000 dimensions. Mikolov et al. report the significantly better results with 300–1,000 dimensions. For instance, the pre-trained Google word embeddings for the English language have 300 dimensions.[2] The vocabulary of this model has ~3 million words and phrases generated from the corpus of ~100 billion words.

Currently *vord2vec* remains the most popular word embeddings approach, but there are many alternatives. One of the most popular is GloVe (Global Vectors), i.e., log-bilinear (LBL) "count-based" model with a weighted least-squares objective which operates directly on the word co-occurrence statistics of the entire corpus [9]. Some embeddings retain not only the word forms, but their morphology. For instance, the LBL model trained on the morphologically annotated German data in the semi-supervised manner arrange words in the vector space in such manner that the closer ones share similar morphological features [10]. The word embeddings can operate not only on the word, but also on the character level (based on the assumption that the meaning of word is related to the meaning of it's composing characters). For the Chinese language, the character-enhanced word embeddings outperformed other baseline methods which ignore internal character information [11]. Word embeddings can also be trained purely on the characters (so called bag-of-character n-grams): vector representations are associated to each character n-gram. Afterwards words can be represented as the sum of these character representations. This approached offered by Bojanowski et al. [12] can solve out-of-vocabulary problems.

Virtually there is plenty of text for any language. However, in reality there is not easy to get billions of words to train the word embeddings: some corpora are proprietary, some have usage restrictions, etc. A promising direction is to learn cross-lingual word representations. Word embeddings learned from the parallel corpora of some languages can be applied to learn word embeddings for the other, e.g., resource-scarce languages (see a recent review of such techniques is in [13]). Since these methods are not sufficiently perfected yet, word embeddings are still a problem for some languages.

---

[1] Describe in detail in https://code.google.com/archive/p/word2vec/.

[2] The Google word embeddings model can be downloaded from: https://drive.google.com/file/d/0B7XkCwpI5KDYNlNUTTlSS21pQmM/edit.

Unfortunately the Lithuanian word embeddings have never been obtained, tested and applied in the downstream NLP tasks. The aim of this paper is: (1) to create the word embeddings using different models and training algorithms; (2) to evaluate and compare their effectiveness intrinsically, i.e., by directly testing syntactic and semantic relations between words.

2 Corpora

For learning word embeddings, we have used the following Lithuanian corpora (the statistical characteristics of datasets is presented in Table 1):

- *The news corpus*. This corpus is composed of the articles crawled from the largest Lithuanian news portal www.delfi.lt in 2001–2015. It covers various topics about the Lithuanian and foreign policy, economy, law, health, style, sports, etc. The corpus consists of 166,498,007 tokens.
- *Wikipedia dump* of October 20th, 2017.[3] After removal of the special hypertext tags and meta data, the corpus contains 29,692,868 tokens of the purely Lithuanian text.
- *Parliamentary transcripts* of the Lithuanian Seimas (Parliament) meetings from 1990 to 2013.[4] It consists of 23,908,302 tokens.
- *Fiction texts*. This corpus as the separate unit was composed of the source fiction texts taken from the whole corpus of the *Contemporary Lithuanian Language*.[5] It contains 9,762,077 tokens.[6]
- *Literary works corpus* composed of the novels and poems that are obligatory for the Lithuanian pupils of the 5th–8th school grade.[7] The corpus contains 2,853,390 tokens.
- *Vytautas Magnus University corpus* (taken from the corpus of the *Contemporary Lithuanian Language*). This corpus contains texts from 9 domains: national newspapers, local newspapers, popular periodicals, fiction, non-fiction, state documents, philosophical literature translations and memoirs. It has 1,259,702 tokens.

[3] Downloaded from https://dumps.wikimedia.org/ltwiktionary/.
[4] This corpus STENOGRAMOS_INDV can be downloaded from http://dangus.vdu.lt/~jkd/eng/?page_id=16.
[5] The whole corpus of the *Contemporary Lithuanian Language* is at http://clarin.vdu.lt:8080/xmlui/handle/20.500.11821/16.
[6] This corpus of the fiction texts GROŽINĖ_INDV can be downloaded from http://dangus.vdu.lt/~jkd/eng/?page_id=16.
[7] Literary works downloaded from http://ebiblioteka.mkp.emokykla.lt/.

Table 1. Statistics about used Lithuanian corpora.

The corpus	Numb. of tokens
The news corpus	166, 498, 007
Wikipedia dump	29, 692, 868
Parliamentary transcripts	23, 908, 302
Fiction texts	9, 762, 077
Literary works	2, 853, 390
Vytautas Magnus University	1, 259, 702
In total:	233, 974, 346

All corpora were merged into one corpus (composed of ~234 million tokens) and used for learning the Lithuanian word embeddings.

3 Word Embeddings Framework

To train the word embeddings we have used *deeplearning4j* [14] – the open-source distributed deep learning library for the Java Virtual Machine. A list of determined options was explored, the rest were set to their default values (for the summary see Table 2):

- *Word2vec models* (both explained in detail by Mikolov et al. [8]):
  - *Continuous bag-of-words* (CBOW). Under this architecture the neural network model predicts a focus word by inputting it's context words from the surrounding window.
  - *Continuous skip-gram* (Skip-gram). Using this architecture the neural network model uses a focus word as an input to predict the context words from the surrounding window.
- *Training algorithms.* With each incoming training sample, the neural weights have to be adjusted to predict that sample as accurate as possible. Having deep architectures with many neurons there is a huge number of weights that need to be updated. To speed up the calculations the following solutions are used:
  - *Hierarchical softmax algorithm* (HS) (the detailed explanation is in [15]), which arranges the output vocabulary of linguistic items in a binary tree (Huffman) manner and uses the softmax to select the child nodes in that tree. Thus the probability of the output is decomposed into probabilities of choosing the correct branch in that tree.
  - *Negative sampling algorithm* (NegS) (presented by Mikolov et al. [16], explained in [17]). It randomly chooses a restricted small number of so-called "negative words" to update the weights. It also updates the weights of the "positive words". Having the (IN, OUT) pair as the training sample, the "negative word" is any word producing 0 at the output (thus not equal to OUT) and the "positive word" is the OUT producing 1 at the output.

Table 2. Options explored in our experiments.

Option	Values
Word2vec model	Continuous bag-of-words
	Continuous skip-gram
Training algorithm	Hierarchical softmax
	Negative sampling
Vector dimensionality	100
	300
	500
	1000
Min. word frequency	5
Window size	5
Numb. of neg. samples	5

- *Minimum word frequency.* By ignoring infrequent words in the corpus this parameter (1) controls the size of the vocabulary, and (2) helps to get rid of noisy (misspelled) words. Here we set minimum word frequency value to 5.
- *Window size* of the context words (used with the CBOW and skip-gram models). As concluded by Goldberg [1], the larger window produces topical similarities, and smaller window – more functional and syntactic similarities. Here (as in many similar research works) this parameter is set to 5, which means that 10 (−5/+5) words around the focus word are considered.
- *Number of negative samples.* This option is used only with the negative sampling training algorithm. As concluded by Mikolov et al. [16] this number should be in the range 5–20 for small training sets and 2–5 – for larger. Usually training of the word embeddings is performed on the billions of words, thus our corpus (described in Sect. 2) is rather small. In our work this parameter is set to 5.

For the Lithuanian language we used word embeddings from the following sources (the statistics is in Table 3):

- *The Wikipedia pre-trained word embeddings.*[8] These 300-dimensional vectors were generated with the character n-gram skip-gram model (described in [12]) using the Lithuanian Wikipedia data. Since these word embeddings contained some distorted data at the end, some minor treatment was needed: the last 5 words were deleted, leaving 306,750 in the embeddings file. The analysis of the remaining data has revealed:
  - Some included words have punctuation symbols and/or white spaces (e.g., "mirčių/", "gyvenvietė,", "m", "ūkiu –", etc.). Moreover, the same words (e.g., "mirčių", "gyvenvietė", "m", "ūkiu") only without punctuation and/or white space are also on the same list. Thus, the list contains

[8] These embeddings were downloaded from https://fasttext.cc/docs/en/pretrained-vectors.html.

Table 3. Statistics about the Wikipedia pre-trained and our generated word embeddings.

Word embeddings	Numb. of words
Wikipedia	219,243
Our embeddings	687,947
The overlap	161,330

word duplicates likely because some punctuation symbols/white spaces were not included into the separators set before the tokenization.
- Foreign language words containing non-Lithuanian diacritic letters (e.g., “naţională”, etc.) or even Cyrillic, Greek, Arabic letters.

After removing all duplicates and non-Lithuanian words (which cannot be used during testing) the list has decreased to 219,243 unique words.

– *Our word embeddings*, which were obtained during the single-epoch training using the methods presented at the beginning of Sect. 3 and the corpus described in Sect. 2. During the pre-processing stage all punctuation marks, digits, and tokens containing non-Lithuanian letters were eliminated. It resulted in 687,947 words in the vocabulary, which is ~3.14 times more compared to a size of the Wikipedia pre-trained embeddings.

Finally, the 161,330 words overlap in both lists of Wikipedia pre-trained and our word embeddings (see Table 3). The overlap coefficient (according to the Eq. 1) is equal to ~0.74.

$$overlap(X, Y) = \frac{|X \bigcap Y|}{min(|X|, |Y|)} \tag{1}$$

4 Intrinsic Evaluation

The word embeddings are generated in the unsupervised manner, therefore it is important to evaluate their quality. There are two types of evaluations (the review can be found in [18]): intrinsic (by testing syntactic or semantic relations between words) and extrinsic (by testing their quality in the downstream NLP tasks). In this research we are focusing on the intrinsic evaluation.

Usually intrinsic evaluation requires human rating [19]: i.e., the human has to choose the most similar word for each candidate-word and the model that gets a majority of votes is considered as the best. Our approach is very similar but does not require human intervention. Unfortunately the benchmark evaluation sets do not exist for the Lithuanian language, therefore the Lithuanian WordNet was used as the reference for the semantic analysis.

4.1 WordNet as the Evaluation Set

The Lithuanian WordNet [20][9] contains ~13 thousand synonym sets (*synsets*). A number of the synonym sets is not large (compared to the English WordNet

[9] Downloaded from http://korpus.juls.savba.sk/ltskwn_en.html.

of ~117 thousand synsets), moreover, some of these sets contain only a single word. For the intrinsic evaluation, we have selected only those synsets that have at least two words in the list of embeddings.

4.2 Experimental Set-Up and Evaluation

Two words are considered similar if they are placed in the similar contexts. Corresponding vectors of these words are close in the vector space, therefore the cosine distance between those vectors has to be smaller.

Let w_1 and w_2 be two words and v_1 and v_2 their corresponding word vectors. The cosine similarity between v_1 and v_2 is represented by dot product and the magnitude as (see Eq. 2):

$$similarity = cos(w_1, w_2) = \frac{v_1 \bullet v_2}{\|v_1\| \times \|v_2\|} \tag{2}$$

The cosine similarity is used to determine k closest (the most similar) words as the output for the given input w_i. In our experiments $k = 10$ (Fig. 1).

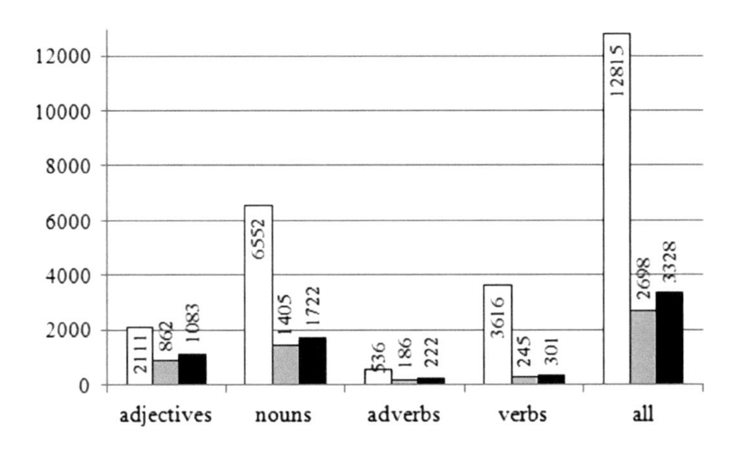

Fig. 1. WordNet statistics. The white column denotes a number of synsets in the WordNet. The gray and black columns denote the number of selected synsets for the pre-trained Wikipedia and our embeddings, respectively.

The evaluation is considered successful if for the given input word w_1 from the synset $\{w_1, w_2, ..., w_n\}$ at least one of $\{w_2, ..., w_n\}$ words is found among 10 closed output words $\{w_1^{w_1}, w_2^{w_1}, w_{10}^{w_1}\}$: i.e., if $\{w_1^{w_1}, w_2^{w_1}, w_{10}^{w_1}\} \cap \{w_2, ..., w_n\} \neq \varnothing$.

The overall accuracy (see Eq. 3) (for all given synsets $count_{all}$) expresses the ratio of successful similarity evaluations, here $count_{all}$ is equal to 2,698 and 3,328 for pre-trained Wikipedia and our embeddings, respectively.

$$accuracy = \frac{count_{successful}}{count_{all}} \tag{3}$$

The research described in [21] claims that the absence of the statistical significance evaluation is one of the major problems related with the evaluation of word embeddings. Moreover, our evaluation set is rather small, therefore there is a risk that differences in the results might be not statistically significant and the conclusions drawn out of it would be worthless. To overcome this problem we have performed the McNemar test [22] with the selected significance level equal to 95%. The calculated *p-value* must be below 0.05 so that the differences in the results would be considered as statistically significant.

4.3 Results

During the experiments we have evaluated the accuracy of pre-trained Wikipedia word embeddings and our word embeddings. Since the *accuracy* on the pre-trained Wikipedia embeddings is zero, Fig. 2 summarizes only the results obtained with our embeddings. Different numbers of dimensions for the same method demonstrated no significant impact of the results.

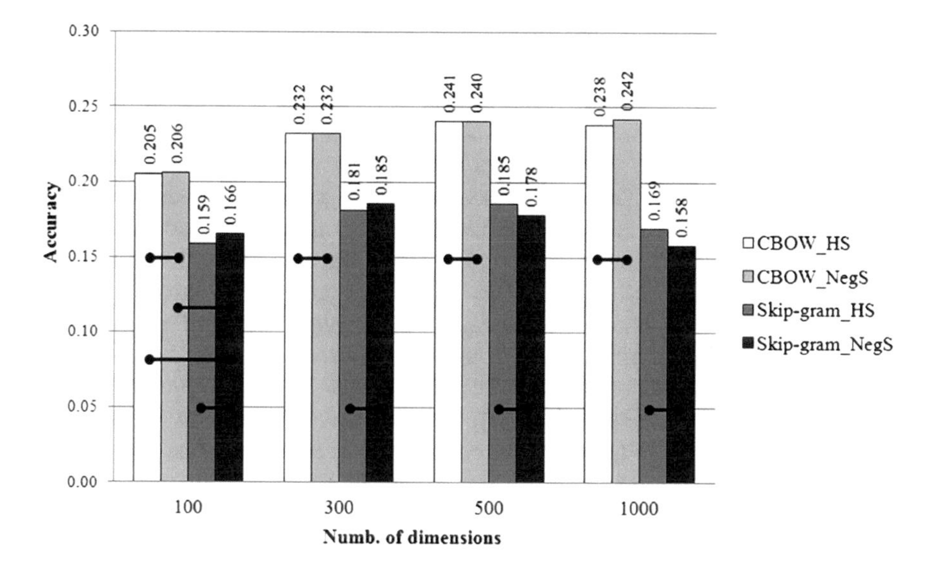

Fig. 2. *Accuracy* evaluated on the different methods. The solid black lines connect columns of those methods for which differences in the results are not statistically significant.

Marginally the best results were obtained with the CBOW_Neg method and 1,000 dimensions, therefore in Fig. 3 we present more detailed results for the separate parts-of-speech.

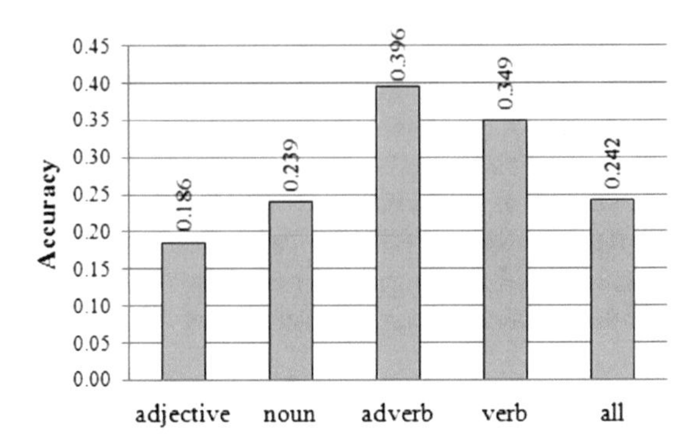

Fig. 3. *Accuracy* for the different parts-of-speech obtained with CBOW_Neg method and 1,000 dimensions.

5 Discussion

CBOW method significantly outperformed skip-gram method, but the negative sampling over hierarchical softmax had insignificant impact on the results. The differences in the results of the same method with a varied number of dimensions (i.e., 100, 300, 500, and 1,000) were not statistically significant. Subsampling method (usually discarding very frequent and meaningless words) was not an issue, because the Lithuanian language does not have articles, auxiliary words, the particles "ne" (not), "nebe" (no any more) also go as the prefixes of words.

The larger part of synsets in the WordNet contains only 2 words (2,333 of 3,328), which means that giving the first word from the synset as the input and searching for it's closest words the overlap is possible only with the second remaining word. The detailed error analysis revealed that among the closed words there are synonyms that differ only slightly from the right answer. These slight differences are due to the morphological forms (e.g., *bėgti* (to run) *bėgu* (I run)), diminutive forms (*vaikas* (child), *vaikelis*, *vaikutis* (child in diminutive)), prefixes (*eiti* (go), *nueiti* (go to)), or reflexiveness (*rengti* (to dress), *rengtis* (to dress up)). If the evaluation procedure would not be so strict, the accuracy would be higher.

The larger corpus (containing billions of words) yield better word embeddings [23]. This is the explanation why results were so bad with the pre-trained Wikipedia dump. Despite our embeddings can serve as the baseline approach for the future research on the Lithuanian language, they are trained only on ~234 thousand words and needs to be refined. The Lithuanian language is highly inflective, derivationally complex and has relatively free word order in the sentences. From this point of view we need several times larger corpus to catch all those similarities between words in their contexts.

Intrinsic evaluation helped to determine more effective word embedding methods in our synonymy detection task. However, the intrinsic evaluation is a bit artificial, therefore in the future research we are planning to plug them first into the downstream NLP applications and then to evaluate.

6 Conclusions and Future Work

This research is the first attempt at finding the accurate method for the Lithuanian word embeddings. Their intrinsic evaluation was performed on the synonym set taken from the Lithuanian WordNet.

The embeddings were generated using continuous bag-of-words/continuous skip-gram word2vec and softmax/negative sampling training algorithms using word vectors of 100, 300, 500, and 1,000 dimensions. Continuous bag-of-words outperform skip-gram method and marginally the best results (evaluating the capability of the method to find similar words based on their cosine distances in the vector space) were achieved with the negative sampling and 1,000 dimensions.

In the future research we are planning to refine the embeddings by adding more data and evaluate them extrinsically.

References

1. Goldberg, Y.: A primer on neural network models for natural language processing. J. Artif. Intell. Res. **57**, 345–420 (2016)
2. Venekoski, V., Puuska, S., Vankka, J.: Vector space representations of documents in classifying finnish social media texts. In: ICIST: Communications in Computer and Information Science, vol. 639, pp. 525–535 (2016)
3. Mandelbaum, A., Shalev, A.: Word embeddings and their use in sentence classification tasks. CoRR, abs/1610.08229 (2016)
4. Bengio, S., Heigold, G.: Word embeddings for speech recognition. In: Proceedings of the 15th Conference of the International Speech Communication Association (Interspeech) (2014)
5. Zou, W.Y., Socher, R., Cer, D.M., Manning, C.D.: Bilingual word embeddings for phrase-based machine translation. In: EMNLP, pp. 1393–1398 (2013)
6. Denis, P., Dehouck, M.: Delexicalized word embeddings for cross-lingual dependency parsing. In: EACL, vol. 1, pp. 241–250 (2017)
7. Tulkens, S., Emmery, C., Daelemans, W.: Evaluating unsupervised dutch word embeddings as a linguistic resource. CoRR abs/1607.00225 (2016)
8. Mikolov, T., Chen, K., Corrado, G., Dean, J.: Efficient estimation of word representations in vector space. CoRR, abs/1301.3781 (2013)
9. Pennington, J., Socher, R., Manning, C.D: Glove: global vectors for word representation. In: EMNLP, vol. 14, pp. 1532–1543 (2014)
10. Cotterell, R., Schütze, H.: Morphological word-embeddings. In: HLT-NAACL, pp. 1287–1292 (2015)
11. Chen, X., Xu, L., Liu, Z., Sun, M., Luan, H.: Joint learning of character and word embeddings. In: IJCAI, pp. 1236–1242 (2015)
12. Bojanowski, P., Grave, E., Joulin, A., Mikolov, T.: Enriching word vectors with subword information. CoRR, abs/1607.04606 (2016)
13. Ruder, S., Vulić, I., Søgaard, A.: A survey of cross-lingual embedding models. CoRR, abs/1706.04902 (2017)
14. Deeplearning4j Development Team: Deeplearning4j: Open-source distributed deep learning for the JVM, Apache Software Foundation License 2.0. http://deeplearning4j.org (2017)

15. Morin, F., Bengio, Y.: Hierarchical probabilistic neural network language model. In: The 10th International Workshop on Artificial Intelligence and Statistics (AISTATS 2005), pp. 246–252 (2005)
16. Mikolov, T., Sutskever, I., Chen, K., Corrado, G.S., Dean, J.: Distributed representations of words and phrases and their compositionality. In: Advances in Neural Information Processing Systems 26: 27th Annual Conference on Neural Information Processing Systems, pp. 3111–3119 (2013)
17. Goldberg, Y., Levy, O.: word2vec Explained: deriving Mikolov et al'.s negative-sampling word-embedding method. CoRR, abs/1402.3722 (2014)
18. Schnabel, T., Labutov, I., Mimno, D.M., Joachims, T.: Evaluation methods for unsupervised word embeddings. In: EMNLP, pp. 298–307 (2015)
19. Gladkova, A., Drozd, A.: Intrinsic evaluations of word embeddings: what can we do better? In: Proceedings of the 1st Workshop on Evaluating Vector-Space Representations for NLP, pp. 36–42 (2016)
20. Garabík, R., Pileckytė, I.: From multilingual dictionary to Lithuanian WordNet. In: Natural Language Processing, Corpus Linguistics, E-Learning, pp. 74–80 (2013)
21. Faruqui, M., Tsvetkov, Y., Rastogi, P., Dyer, C.: Problems with evaluation of word embeddings using word similarity tasks. In: RepEval@ACL, pp. 30–35 (2016)
22. McNemar, Q.: Note on the sampling error of the difference between correlated proportions or percentages. Psychometrika **2**(12), 153–157 (1947)
23. Lai, S., Liu, K., Xu, L., Zhao, J.: How to generate a good word embedding? CoRR, abs/1507.05523 (2015)

Classification of Textures for Autonomous Cleaning Robots Based on the GLCM and Statistical Local Texture Features

Andrzej Seul(✉) and Krzysztof Okarma

Faculty of Electrical Engineering, Department of Signal Processing and Multimedia Engineering, West Pomeranian University of Technology, Szczecin, 26 Kwietnia 10, 71-126 Szczecin, Poland
andrzej.seul@gmail.com, okarma@zut.edu.pl

Abstract. In the paper a texture classification method utilizing the Gray Level Co-occurrence Matrix (GLCM) is proposed which can be applied for autonomous cleaning robots. Our approach is based on the analysis of chosen Haralick features calculated locally together with their selected statistical properties allowing to determine the additional features used for classification purposes. To verify the presented approach a dedicated color image dataset containing textures selected from the Amsterdam Library of Textures (ALOT) representing surfaces typical for the autonomous cleaning robots has been used. The results obtained for various color models and three different classifiers confirm the influence of the color model as well as the advantages of the proposed extended GLCM based approach.

Keywords: Texture analysis · GLCM · Haralick features
Cleaning robots

1 Introduction

Applications of machine vision in autonomous robotics and intelligent self-driving vehicles raises its popularity during recent years [8,9]. Such a situation has been caused by growing availability of relatively cheap high quality cameras and the great number of image processing and analysis algorithms implemented in many programming environments and libraries e.g. OpenCV. A great interest in vision can also be observed in remote sensing where multispectral analysis can be applied e.g. for the classification of textures representing different vegetation areas [7]. Another field of research with growing interest in the use of machine vision is the detection of defects of various surfaces [10] and recently 3D prints [3].

Nevertheless, in most applications images captured by cameras mounted on the robot and/or outside are considered as a source of supplementary information with the use of some classical sensors so further data fusion algorithms are usually implemented. One of the reasons is the sensitivity of cameras to changes of

R. Silhavy (Ed.): CSOC 2018, AISC 764, pp. 405–414, 2019.
https://doi.org/10.1007/978-3-319-91189-2_40

lighting conditions as well as weather issues like fog, rain or snow, if they work outdoor.

Popularity of cameras causes the interest of using the image data applying relatively simple algorithms which can be implemented in embedded systems with limited computational power as well. As they usually require the predictable lighting conditions in many cases the additional illuminators can be applied to ensure the uniform lighting of surfaces and objects. One of the examples of such robotic systems where the additional image data can be used is the wide family of mobile walking or wheel robots where the texture analysis methods can be utilized to gather information about the type of surface under or in front of the robot and e.g. floor detection [6]. A special example considered in this paper is the application of texture classification for autonomous cleaning robots due to the possibility of optimizing the cleaning process by application of appropriate detergents or brushes as well as e.g. changing the movement speed or even omitting some fragments of the floor.

2 Methods

The most popular approach to texture analysis and classification is the use of Haralick features [4] which can be calculated from the normalized symmetrical Gray-Level Co-occurrence Matrix (GLCM) representing the probability of co-existing of some pairs of pixels' values in the specified neighborhood. Although in Haralick's approach 14 features were originally proposed, four of them became the most popular, namely: contrast, correlation, homogeneity and energy. Their implementation is available e.g. in Matlab Image Processing Toolbox (*graycoprops* function). Some interesting details related to FPGA based implementation of the GLCM texture descriptor can also be found in the paper [5].

It should be noticed that the GLCM is typically calculated for grayscale images and therefore the color to grayscale conversion becomes an important issue as well. Nevertheless, an extension of the GLCM for color images can also be directly applied for texture classification and the details can be found in the paper [1].

2.1 Statistical Feature Extraction from Texture

The GLCM Based Approach

The GLCM describes the spatial relations between the pixels and therefore it is also known as the second order histogram. Assuming the offset $(\Delta x, \Delta y)$ it can be computed for the $M \times N$ pixels image as:

$$C(i,j) = \sum_{p=1}^{P} \sum_{q=1}^{Q} \begin{cases} 1 & \text{if } A(p,q) = i \text{ and } A(p+\Delta x, q+\Delta y) = j \\ 0 & \text{otherwise} \end{cases} \tag{1}$$

where $P = M - \Delta x$ and $Q = N - \Delta y$.

Since each element is the number of occurrences of pixels of the luminance level i with pixels of the other chosen luminance level j in the specified neighborhood, defined by the offset $(\Delta x, \Delta y)$, its sum depends on the resolution of the image. To compare the textures illustrated on the images independently on their resolutions, the GLCM can be normalized dividing its values by the sum so that the sum of the normalized GLCM is equal to 1. Typically the symmetrical GLCM is considered so he luminance level i of a pixel left from j is considered equally as i right from j for the horizontal GLCM (and similarly above and below for the vertical matrix). For each image four directions can be considered leading to four symmetrical matrices (as two diagonals can be added). However, changing the offset $(\Delta x, \Delta y)$ even more matrices can be computed for various pixels' distances higher then one pixel. The four features mentioned earlier can be calculated as follows:

- Contrast: $\sum_{i,j} |i-j|^2 C(i,j)$
- Correlation: $\sum_{i,j} \frac{(i-\mu_i)(j-\mu_j)C(i,j)}{\sigma_j \sigma_i}$
- Energy: $\sum_{i,j} C(i,j)^2$
- Homogeneity: $\sum_{i,j} \frac{C(i,j)}{1+|i-j|}$

where: $C(i,j)$ - value of the normalized symmetrical GLCM at position (i,j); i,j - gray levels of the neighboring pixels; μ - average value of p, σ - standard deviation of p (p_i and p_j are obtained summing the i-th row or j-th column of the normalized GLCM).

To calculate the GLCM of color texture images we have treated each channel as a separate monochromatic image. In our experiments five most popular color spaces have been used and therefore we have obtained 15 channels for: RGB (original representation), HSV, CIE LAB, CIE XYZ and YUV. We have calculated Haralick features for each channel obtaining 12 values (4 features $\times$ 3 channels) for each texture.

Extended GLCM Based Approach

As the application of the Haralick features for various color spaces does not guarantee satisfactory results, we have proposed to slit the image into smaller parts and then calculate local Haralick features like described before for every part separately, put the results into a vector and calculate some statistical measures for this vector. Repeating the procedure for each Haralick feature four "new" values for every channel can be obtained.

In our experiments five different statistical measures have been used which can be calculated assuming the vector A of N observations using the formulas listed below.

- Mean - arithmetic average value of vector

$$\mu = \sum_{i=1}^{N} A_i \tag{2}$$

- Variance - the expectation of the squared deviation of a random variable from its mean. Informally, it measures how far a set of (random) numbers are spread out from their average value.

$$Var = \frac{1}{N-1} \sum_{i=1}^{N} |A_i - \mu|^2 \tag{3}$$

- Standard Deviation - square root of variance

$$Std = \sqrt{\frac{1}{N-1} \sum_{i=1}^{N} |A_i - \mu|^2} \tag{4}$$

- Kurtosis - measure of how outlier-prone a distribution is. Kurtosis of normal distribution is 3. If the distribution cone is more "flat" than normal distribution, kurtosis is lower than 3. If the distribution cone is "sharper" than normal distribution, kurtosis is higher than 3.

$$Kurtosis = \frac{\frac{1}{N} \sum_{i=1}^{N} (A_i - \mu)^4}{(\frac{1}{N} \sum_{i=1}^{N} (A_i - \mu)^2)^2} \tag{5}$$

- Skewness - measure of asymmetry of data; if more data is on the left of the mean, skewness is negative and if more on the right, skewness is positive. For perfectly distributed data skewness is 0. In this paper unbiased skewness will be used.

$$Skewness = \frac{\frac{1}{N} \sum_{i=1}^{N} (A_i - \mu)^3}{\sqrt{\frac{1}{N} \sum_{i=1}^{N} (A_i - \mu)^2}^{\,3}} \tag{6}$$

2.2 Classification

Classifiers

To recognize one of the 7 texture three classifiers have been used: linear discriminant analysis (LDA), naive Bayes classifier (BAY), and decision tree learning (TREE). To invert the covariance matrix, which is the same for every class, the pseudo-inversion has been used in the LDA method. Training of the naive Bayes classifier has been done using Gaussian distribution model. As the decision tree method requires the limiting if the number of splits to avoid the tree overgrowing, we have set the limit to 15 splits preventing the possible large number of false-positive matches and too many decisions.

Performance Measures

Classification performance has been measured using three different methods. The first one is the resubstitution loss error (RL) representing the percentage of misclassified observations from training data set. It is reverse to accuracy which is interpreted as the ratio of all correct classifications (true positives and true negatives) to all samples.

Calculating the precision (PR) - defined as true positives to all positives ratio - and recall (RC) being the ratio of true positives to sum of true positives and false negatives, their harmonic mean can be defined which is widely known binary classification metric - F-Measure known also as F1-score expressed as:

$$\mathrm{FM} = 2 \cdot \frac{PR \cdot RC}{PR + RC} \cdot 100\%. \tag{7}$$

The third metric is the K-Fold Loss measuring classification loss of cross-validated model from non-training data. Its value is proportional to number of misclassified observations.

2.3 Texture Database

The verification of the proposed approach has been made using some images chosen from the widely used Amsterdam Library of Textures (ALOT). As the whole dataset contains many images which are not typical to surfaces which can be specific to the assumed area of applications related to cleaning robots, we have decided to choose 700 color images representing patterns of 7 different materials resizing them to the 320×320 pixels resolution. Individual images have been captured with different parameters e.g. viewing angle, illumination color or illumination angle allowing to obtain 100 images for each material [2]. Exemplary images of 7 textures used in our experiments are presented in Fig. 1.

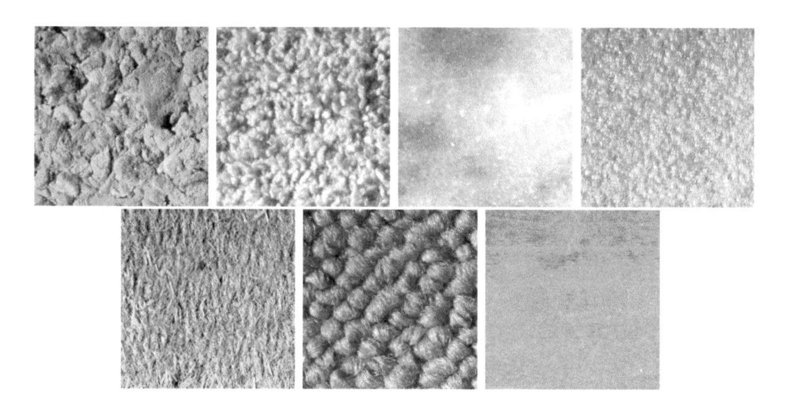

Fig. 1. ALOT texture patterns used in our experiments - upper row: cork (57), terry cloth (74), marble (98) and carpet (184), bottom row: hardboard (189), carpet no.2 (205) and wood plank (206).

For verification and training purposes the database has been divided into two parts. After training the classifiers we have used the second part of the dataset to validate the classifiers using three metrics described earlier.

3 Results

The results obtained during our experiments are shown in Figs. 2, 3, 4, 5, 6, 7, 8, 9 and 10. Analyzing the Figs. 2, 3 and 4 illustrating the F-Measure values for various classifiers, namely naive Bayes, LDA and decision trees, it can be observed that in almost all cases the use of any additional statistical features leads to better classification resulting in higher F-Measure values. For naive Bayes classifier the best results have been obtained for CIE LAB color space using the standard deviation as the additional feature. Surprisingly, the use of variance has lead to worse results for all color models except HSV being the best choice for the combination of variance with local Haralick features.

The use of the LDA approach allows to achieve the highest F-Measure values for the RGB color model with the use of the mean value as the additional feature whereas the best combination for the decision trees is the CIE LAB space with variance as the additional feature. Nevertheless, it is worth noticing that the advantages of using the additional statistical features are not always as high as for the other two classifiers although achieved results are usually worse than for LDA.

Those conclusions can also be confirmed analyzing the plots presented in Figs. 5, 6 and 7 presenting the K-Fold Loss errors. The lowest errors for Bayesian classifier have been obtained using the combination of Haralick features with the standard deviation using CIE LAB color space, similarly as the highest F-Measure. The values of resubstitution loss leading to the same conclusions are presented in Figs. 8, 9 and 10.

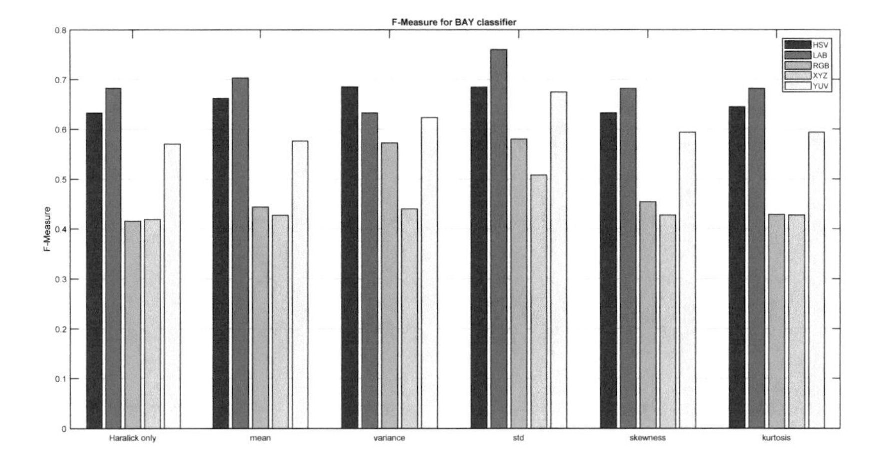

Fig. 2. F-Measure values obtained for Bayesian classifier.

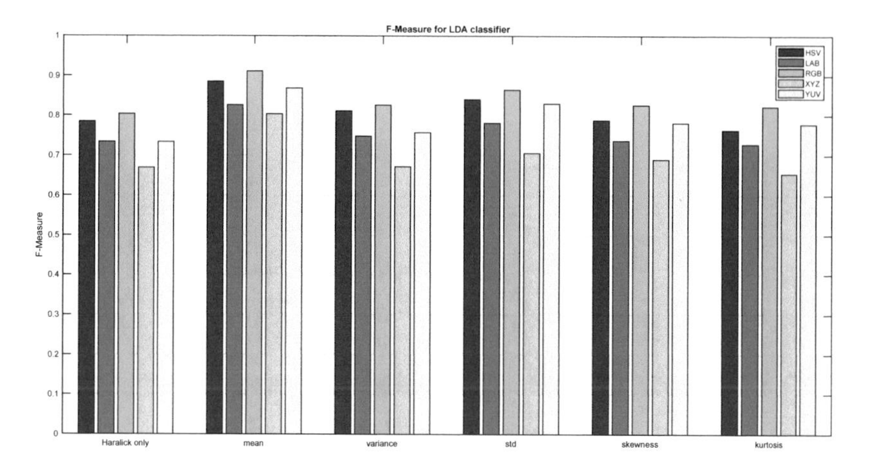

Fig. 3. F-Measure values obtained for LDA classifier.

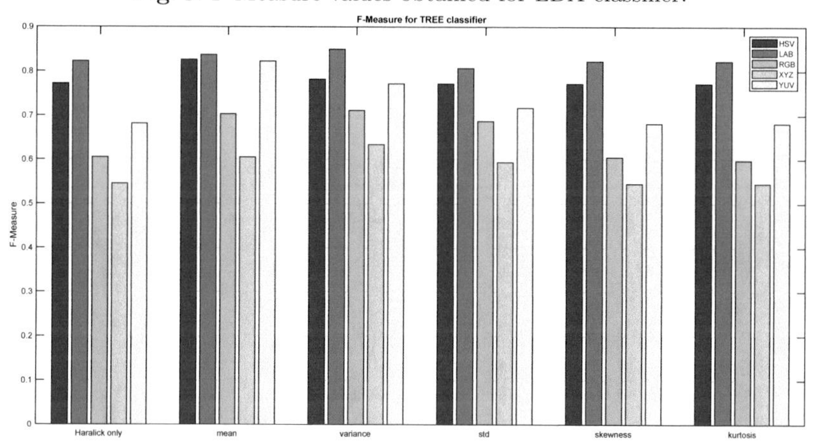

Fig. 4. F-Measure values obtained for decision tree classifier.

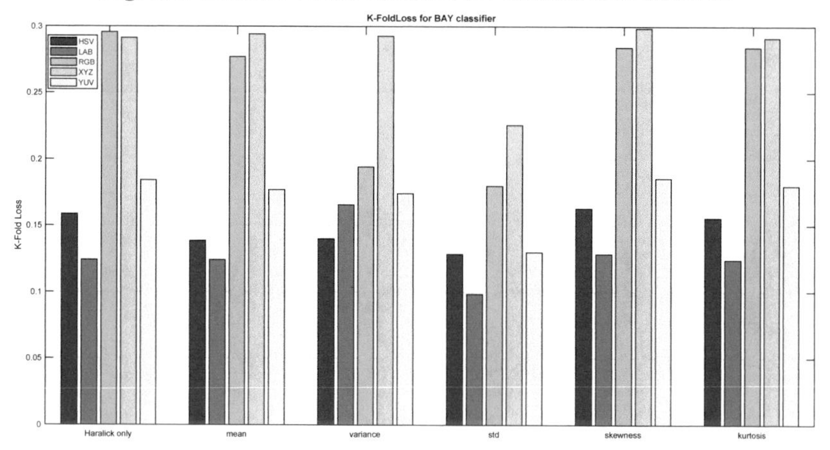

Fig. 5. K-Fold loss values obtained for Bayesian classifier.

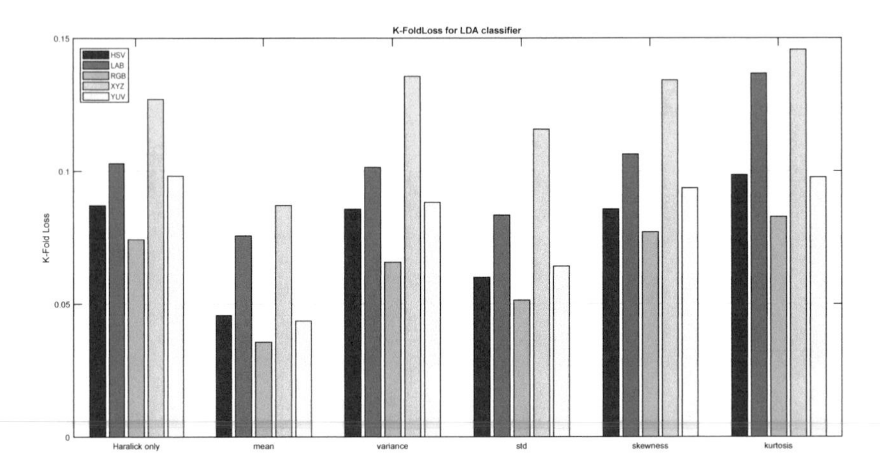

Fig. 6. K-Fold loss values obtained for LDA classifier.

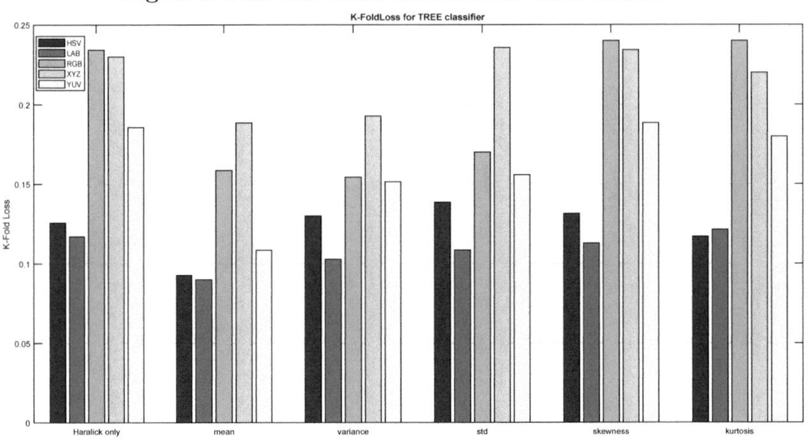

Fig. 7. K-Fold loss values obtained for decision tree classifier.

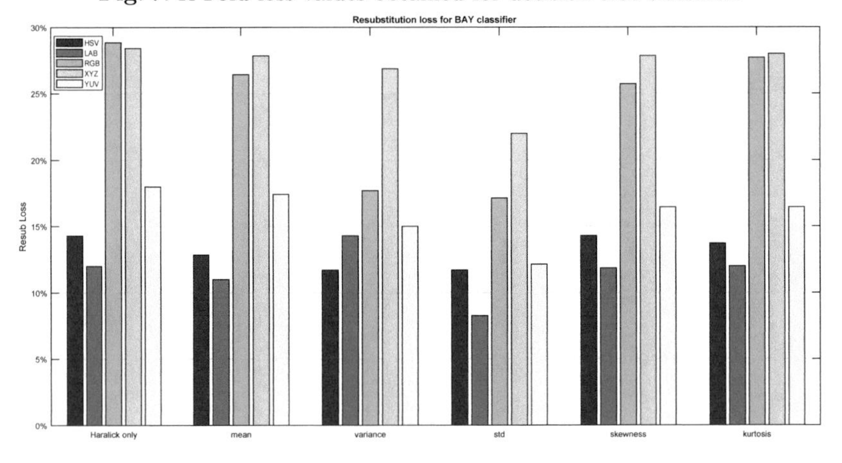

Fig. 8. Resubstitution loss values obtained for Bayesian classifier.

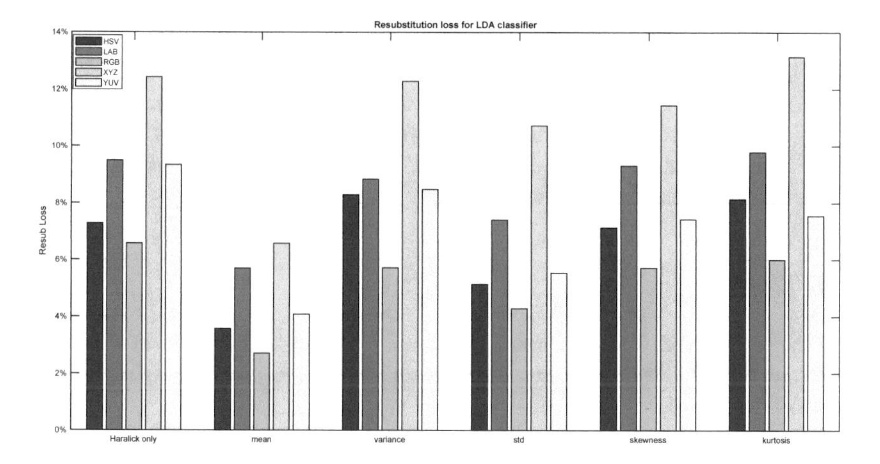

Fig. 9. Resubstitution loss values obtained for LDA classifier.

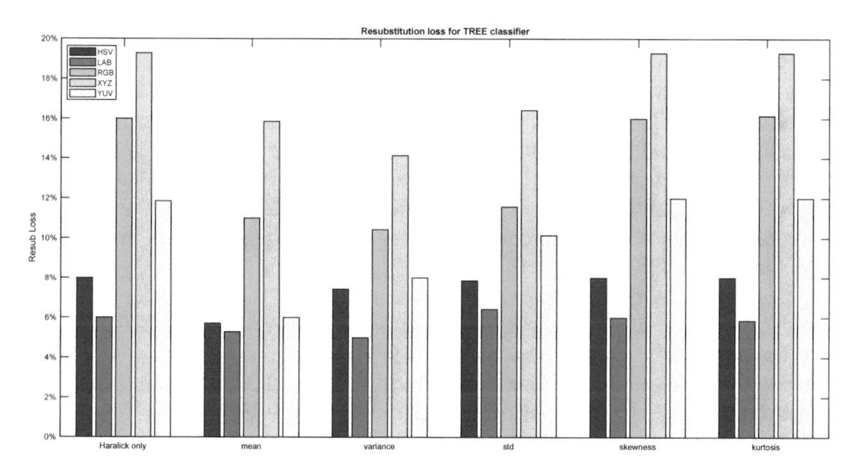

Fig. 10. Resubstitution loss values obtained for decision tree classifier.

4 Conclusions and Future Research

The most important point is that using different color models than the popular RGB, different results of classification can be obtained - better for BAY and TREE methods and a bit surprising results for LDA (it is not a very suitable method for texture classification purposes). Using the extended approach better results can be achieved as well. Some of them are "much better" (e.g. using mean or variance) and some are "less better" or even insignificantly different. It could be concluded which Haralick features are the best in general and which combinations with the additional local features are the best for specific classifiers and color models.

In our future work it is planned to extend the proposed approach further by the analysis of possible combination of local GLCM based features with some other texture descriptors. Another direction of our further experiments will be related to the combination of more statistical features with local Haralick ones as well as the analysis of the influence of the number of blocks used for the calculation of the local features.

References

1. Arvis, V., Debain, C., Berducat, M., Benassi, A.: Generalization of the cooccurrence matrix for colour images: application to colour texture classification. Image Anal. Stereol. **23**(1), 63–72 (2011)
2. Burghouts, G.J., Geusebroek, J.M.: Material-specific adaptation of color invariant features. Pattern Recogn. Lett. **30**(3), 306–313 (2009)
3. Fastowicz, J., Okarma, K.: Texture based quality assessment of 3D prints for different lighting conditions. In: Chmielewski, L.J., Datta, A., Kozera, R., Wojciechowski, K. (eds.) Computer Vision and Graphics: International Conference, ICCVG 2016. Lecture Notes in Computer Science, vol. 9972, pp. 17–28. Springer International Publishing, Cham (2016)
4. Haralick, R.M., Shanmugam, K., Dinstein, I.: Textural features for image classification. IEEE Trans. Syst. Man Cybern. **3**(6), 610–621 (1973)
5. Komorkiewicz, M., Gorgoń, M.: Foreground object features extraction with GLCM texture descriptor in FPGA. In: Proceedings of 2013 Conference on Design and Architectures for Signal and Image Processing (DASIP), Cagliari, Italy, pp. 157–164, October 2013
6. Li, Y., Birchfield, S.T.: Image-based segmentation of indoor corridor floors for a mobile robot. In: 2010 IEEE/RSJ International Conference on Intelligent Robots and Systems (IROS), pp. 837–843. IEEE (2010)
7. Murray, H., Lucieer, A., Williams, R.: Texture-based classification of sub-antarctic vegetation communities on heard island. Int. J. Appl. Earth Obs. Geoinf. **12**(3), 138–149 (2010)
8. Nixon, M.S., Aguado, A.S.: Feature Extraction & Image Processing for Computer Vision. Academic Press, San Diego (2012)
9. Sonka, M., Hlavac, V., Boyle, R.: Image Processing, Analysis, and Machine Vision. Cengage Learning, Stamford (2014)
10. Xie, X.: A review of recent advances in surface defect detection using texture analysis techniques. ELCVIA Electron. Lett. Comput. Vis. Image Anal. **7**(3), 1–22 (2008)

Hybrid Approach to Solving the Problems of Operational Production Planning

L. A. Gladkov(✉), N. V. Gladkova, and S. A. Gromov

Southern Federal University, Taganrog, Russia
{leo_gladkov,nadyusha.gladkova77}@mail.ru

Abstract. The task of operational planning of production is considered. The hierarchy of tasks of production planning is described. The formulation of the problem in terms of scheduling theory is given. A model for solving the problem of operational planning as an adaptive system is proposed. The software agent architecture is chosen, local and global goals of the adaptive system are formed. The use of the apparatus of fuzzy sets in the determination of the agent state is justified. Computational experiments were carried out and the results obtained were analyzed.

Keywords: Operational planning · Scheduling theory
Evolutionary adaptation · Multi-agent systems · Genetic search
Modified genetic algorithm

1 Introduction

The complexity of the problems solved by the means of scheduling theory leads to the need to introduce a number of assumptions. It is assumed that the nature of the work does not depend on the sequence of their implementation [1–5]. In addition, the following assumptions are used.

1. The work to be performed is defined and known in full. All the given jobs must be performed.
2. The devices necessary for performing the specified tasks are clearly defined.
3. A set of all elementary actions that must be done to perform each type of work is given. There are restrictions on the order of execution of each type of work. Each operation from the given list must be provided with at least one unit of equipment.

To solve the task of scheduling, it is necessary: a timetable for completing all the necessary production operations on each unit of equipment is drawn up. Thus, scheduling can be considered as the task of determining the sequence of operations performed by each machine.

The initial data for the problem of scheduling theory are:

- list of works and operations to be performed;
- the number and types of machines performing each operation;
- sequence of operations;
- restrictions on the use of machines;
- criteria for evaluating schedules.

R. Silhavy (Ed.): CSOC 2018, AISC 764, pp. 415–424, 2019.
https://doi.org/10.1007/978-3-319-91189-2_41

2 Formulation of the Problem

We denote by W_{ij} the waiting time for the ij operation, that is, the time interval between the end (j–1) and the beginning of the j operation of the i operation.

Then the total waiting time for i is the sum of the waiting times for all operations [6–8]:

$$W_i = \sum_{j=1}^{g_i} W_{ij}. \tag{1}$$

The result of scheduling is always the set of numbers W_{ij}. The final choice of this or that variant is based on a comparison of the corresponding sets W_{ij}. The most important quantities that depend on W_{ij} are:

T_i – the end of the task; F_i – duration of task i in the system (production cycle); L_i – - displacement of the work i execution time; Z_i – the time lag from the work i schedule; E_i – is the lead time for the task i.

The values of L_i, Z_i, E_i - allow to estimate the actual time of completion of work in comparison with its planned term. The temporary shift of each work can have a different sign. If the sign is positive, then the work is completed after the planned time, which leads to a delay. If the sign is negative, then the work is completed before the planned time, i.e. ahead of schedule.

To assess the importance of different types of work we use weight coefficients. Then the average duration of the task is determined as follows:

$$\bar{F}_u = \frac{1}{n} \sum_{i=1}^{n} u_i F_i. \tag{2}$$

3 Architecture of the Hybrid Operational Planning Model

Traditionally, the search process is carried out using one of the following approaches: mathematical programming, combinatorial approach and heuristic approach.

Heuristic methods are one of the traditional lines of research on artificial intelligence [9, 10]. Their key feature is that there are practically no restrictions on the formulation of the problem being solved. Methods of heuristic programming unites the general principle of adaptation, which was borrowed by technical systems from biology. Adaptation is the ability of a living organism or technical system to adapt to the external environment, changing its state and behavior (parameters, structure, algorithm and functioning), depending on changes in environmental conditions [11, 12]. At the same time, changes occur as information is accumulated and used from outside. Thus, the adaptive system is understood as a system that operates in the presence of a priori uncertainty and changing external conditions, and the information obtained in the process of working on these conditions is used to improve the efficiency of its operation.

In general, management is a targeted impact on the management object (MO), aimed at ensuring its required behavior [11, 12]. As a control organ (device), there can be a certain technical system that translates the object into the required state.

Thus, adaptation can be considered as a way of management in an environment of uncertainty of the external environment and the system itself. The uncertainty of the system is, in turn, associated with its complexity, which prevents the obtaining of an adequate model. Adaptation acts as a means of controlling the system in the absence of its exact model [11, 12].

In our case, the control object is an operational plan, and the adaptive control device is a special adaptation subsystem (Fig. 1), which can be purposefully influenced. The state of the object is changed by the environment in which it is located.

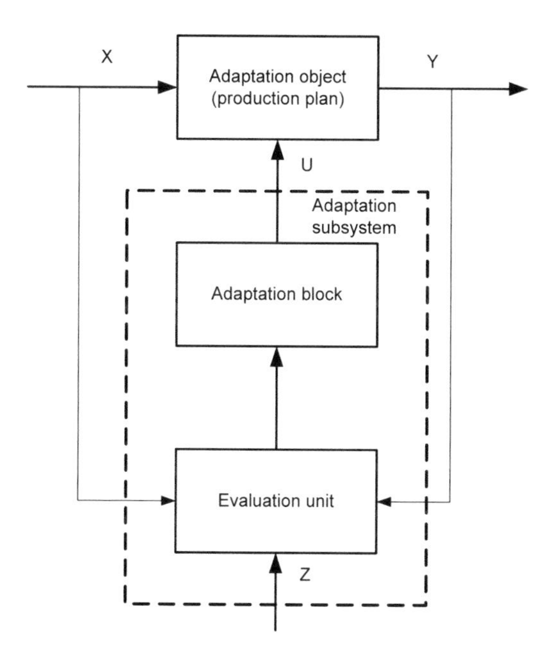

Fig. 1. Production planning as an adaptation object

With reference to the task of operational planning, we have:

X - state of the environment interacting with the adaptation object;
Y - the state of the object, which is determined by the variant of the operational plan, that is, the list of necessary operations.

In addition to the environment, the control action U influences the state of the object. The object can be represented as an operation connecting the input X and the control action U with the output Y: $Y = F\ (X,\ U)$.

The operation F is specified by some algorithm or simulation model. The control commands U are generated by the adaptation unit and have a control effect on the control object.

For the functioning of the evaluation block, you must specify the goal Z, which determines what should be sought in the management process. The resulting Y plan should enable the achievement of the set goals.

Thus, the control process can be described as follows: $U = \varphi\ (X,\ Y,\ Z)$.

Management is primarily related to the goals of Z, which come from outside the management system. These goals are formulated by the subject, who is the consumer of the future management system of the object. As a rule, the subject formulates his goal in the form of a vector $Z = (z_1, \ldots, z_m)$. Each component of this vector is subject to certain requirements. In the tasks of compiling the production schedule, as components of the target vector Z can be:

- Variable attributes of production tasks (line number for the task, the order of the job on the line);
- Target criteria for the optimization task (for example, the number of tasks not completed within the specified time frame);
- boundary conditions (boundaries of service intervals of production lines, intervals for changing lines, etc.).

Models of adaptation and optimization are closely interrelated. The adaptation task arises in the event that there is no information needed to optimize the object. Adaptation provides optimization in an environment of significant interference due to the roughness of the evaluation of the functional [12]. At the same time, adaptation is also very similar to the task of directional search in the state space. At the initial moment of time, an arbitrarily taken decision, obtained at random or with the help of an expert, is assumed. The idea of directional search is to consistently improve the solution at each subsequent iteration. The search process consists of repetitive steps, each of which represents a transition from one solution to another, the best, which forms a procedure for the sequential improvement of the solution

$$Y[0] \rightarrow Y[1] \rightarrow \ldots \rightarrow Y[t] \rightarrow Y[t+1] \rightarrow \ldots, t = (1, 2, \ldots, T), \quad (3)$$

where T is the number of iterations of the search algorithm.

Each subsequent solution is obtained from the previous one by means of some search algorithm A. The algorithm indicates which operations need to be done for $Y[t]$ in order to obtain a more preferable solution $Y[t+1]$: $Y[t+1] = A(Y[t])$.

Since the operational plan model is specified by the dynamic list of production tasks, it will be logical to divide the management of these tasks. To this end, approaches and methods of multi-agent systems are used to implement the adaptation block [13]. As a result, the structure and parameters of the operational plan are changed during the adaptation of the team of software agents. When using the methods of adaptation, it is suggested to be guided by the principles of decentralization. For an example of using the decentralized approach, let's compare the main points characterizing the process of scheduling by a person and program agents (Table 1).

Given the variety of types of adaptation, the specifics of the tasks of constructing production schedules, we can conclude that the most preferred approaches for solving discrete optimization problems are: evolutionary adaptation; alternative adaptation with the collective of objects of adaptation with reflection.

Table 1. Decentralization in scheduling

Symptoms	Human	Team of software agents
The base of rules	One centralized rule base	The rules base distributed between program agents
Set of conditions or situations	Significant, hard to formalize	Limited, usually a small number of basic situations
Experience accumulation, training	Centralized	Decentralized. The motivation for applying the same type of rule in a similar situation for different agents may differ
The nature of the search process	Sequential	Parallelism for unrelated agents is possible

Suppose that each production task is associated with an abstract entity - a software agent, which is treated as an adaptation object. The task of the agent is to develop and implement control signals that change the values of the attribute vector of the corresponding task. The task identifier for the resource and the order of its execution are selected as job attributes. Depending on the specifics of the problem being solved, the list of attributes can be expanded. For example, if the duration of a job is a variable, it must be highlighted in a separate attribute. If you can change the relationships between tasks (which is important in the reconciliation or network planning tasks), you must select individual attributes that specify the link indicators for the individual jobs. Thus, one agent emulates the management of one production application.

With such an organization, each agent acts as a separate sub-level of the lower level, responsible for adapting a single, accountable, production task from the operational plan. We introduce the matrix Y that characterizes the operational production plan (its columns consist of task attribute vectors (Table 2).

Table 2. Solution matrix

Attribute value	Task 1	Task 2	.....	Task i	.....	Task n
Priority of the assignment	y_{11}	y_{21}	.....	$y_{i1} \in \{N\}$	.....	y_{1n}
Line number	y_{12}	y_{22}	.....	$y_{i2} \in \{M\}$	.....	Y_{2n}

The solution space is represented by a set of Y-matrices. The search for the variant of the operational plan is reduced to the search for the matrix Y, namely, to finding a set of values y_{ij}, which optimize the target criterion and satisfy the constraints, i.e. achieve the goal of adapting Z.

Accordingly, each subsystem or agent has its own state and goals, perhaps even different from those of the adaptive top-level system, which is responsible for adapting the entire operational plan. The local purpose of the individual agent is to achieve the fulfillment of the directive requirements for the output and avoid conflict situations involving the use of common production equipment, as well as situations of violation of technological limitations.

For the entire team of agents, the global goal of adaptation is to optimize the value of the criterion that has been put. Thus, in the process of finding a solution, each agent needs to select the values of its attribute vector y_k, which optimize the value of the criterion. In this case, the best requirements will be met and the imposed restrictions will not be violated. Thus, we have:

The local goal of each adaptation object is to achieve a satisfactory state, in which the directive requirements for the production task corresponding to the given agent are met.

The global goal of the collective of adaptation objects is to achieve an environment in which all goals are fulfilled.

The general formulation of the problem is as follows: the set of possible situations of production planning {S} and the actions of agents-planners {A} are finite. It is assumed that in each situation information about the current state of the agent should be reflected: whether it violates any restriction; how much the agent has reached the local goal; whether the policy requirements are met. Sometimes the situation shows how the current state of the agent affected the attainability of the global goal - optimization of the target criterion. It is for the organization of the procedure for a directed transition from one solution to another that each agent is given the ability to perform actions that change its state, that is, the correcting values of the vector of its attributes (y_{k1}, y_{k2}, …, y_{kn}). It remains to determine the algorithm for applying a particular rule, depending on the situation. To do this, you must select the software agent implementation architecture.

As the general architecture of the agent, an animat (animated animal) model based on reinforcement learning is proposed, developed in the context of the study of the adaptive behavior of artificial agents [14]. The general scheme of training with reinforcement is shown in Fig. 2.

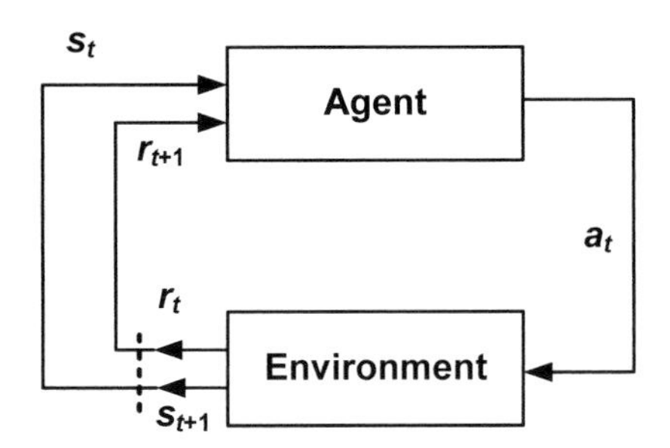

Fig. 2. Training with reinforcement in the animat model

In the current situation, the s_t agent executes the a_t action. Depending on the logic of the selected action, the components of the attribute vector (y_{k1}, y_{k2}, …, y_{kn}), $y_k^{t+1} = a_t(y_k^t)$. Then the agent gets reinforcement r_t and gets into the next situation s_{t+1} (t – is the model time of the adaptive system, which is assumed to be discrete: $t = 1, 2, \ldots$). Reinforcement is regarded as a response signal of the environment (promotion or punishment), assessing the appropriateness of the selected action in the situation in question.

In practice, the fulfillment of requirements in the tasks of production planning is usually determined by the expert, who estimates the amount of delay or advance of the release date relative to the guideline. It is convenient to give such estimates in the form of linguistic variables. Therefore, when solving this problem, it is proposed to use the capabilities of the apparatus of linguistic variables and fuzzy sets. Let's introduce the auxiliary linguistic variable "Deflection of terms". The set of its terms (the set of linguistic meanings) consists of the following values {"Small", "Exactly in time", "Big"}. Also, the linguistic variable "Stock status" with the values {"Early replenishment", "Exactly in time", "Later replenishment"} can be used. As the initial approximation, as the membership function, the simplest symmetric triangular function is chosen.

The fuzzy estimate of the variable "Rejection of time" is carried out for issues according to the operational plan. It is considered that the estimate "Exactly in time" corresponds to the fact of fulfilling the local goal. Otherwise, there is a situation of advance or delay. The use of such judgments makes it possible to increase the adequacy and realism of assessments of the state of agents, which in turn affects the convergence of the search algorithm.

Also, for solving planning problems, bioengineered algorithms that imply the application of natural mechanisms to the modeling of technical systems are successfully used. The hybrid algorithm implies the use of fuzzy logic to change the control parameters of the algorithm and the type of evolution operators used in the course of its operation [15–18].

Changes are subject to the probability of mutation and the probability of crossing-over. Fuzzy rules for adapting these parameters are applied. The parameters change based on the proposed heuristic rules. The type of mutation operator also changes. The fuzzy logic controller implements the changes.

4 Experimental Research

The proposed planning algorithms were implemented programmatically in the form of extensions to the standard functionality of information systems affecting production planning processes. Development of the solution was carried out in two stages: the development of prototypes, the transfer of prototypes to the industrial platform.

The development of the schedule calculation module is performed in the ANSI C/C++ language using the standard STL template library.

The purpose of experimental studies is to analyze the dependence of the rate of convergence and the quality of solutions of the developed algorithms for various cases of their use.

It is necessary to establish the nature of the dependence of the dimension of the problem (the value of the number of jobs N, the number of lines M), and the time spent on the solution.

The study was conducted on twelve points, selected expertly.

The convergence time increases linearly with increasing dimension of the problem. It should also be noted faster convergence of the adaptive search algorithm (ASA) in comparison with the modified genetic algorithm (MGA). The obtained results confirmed the assumption of the expediency of using an approximate algorithm for obtaining a quasi-optimal solution. The nature of the dependencies reflects the almost linear dependence of the counting time on the dimension of the problem, thus confirming the assumption of the polynomial time complexity of the proposed algorithms.

Also, an assessment was made of the improvement in the quality of the solution with respect to the initial value of the objective function. For clarity, the algorithm included a random search for a solution. In carrying out this study, the same initial data was taken for each algorithm: the number of jobs N = 100, the number of lines, of which paired M/LG = 24/8, the planning horizon D = 120; number of rigging RQ = 4. When comparing the results of the algorithms studied, the following considerations are adopted:

- Each generation of MGA corresponds to a hundred iterations of ASA;
- Every hundredth decision of the random search algorithm corresponds to the calculation of one generation of MGA;
- Each cycle of the hybrid algorithm corresponds to the calculation of one generation of MGAs.

The values of the parameters of the MGA and ASA algorithms are given in Tables 3 and 4, respectively.

Table 3. Parameters of the modified genetic algorithm (MGA)

Parameter name	Value
Population size, pcs.	100
Probability of crossing	0.9
Probability of mutation	0.1
Migration scheme	"best" - "worst"
Period of migration, gen.	5
Share of migratory decisions, %	10
Number of crossing points, %	0.0005
Number of mutation points, %	0.001

Table 4. Parameters of the adaptive search algorithm

Parameter name	Value
Number of cycles of algorithm stagnation	20
Initial values in the learning matrix *Q*. *MAX/MIN/ST*	7/1/3
The amount of reinforcement r_t	0.2
The probability of choosing an action for an ε-greedy policy	0.8

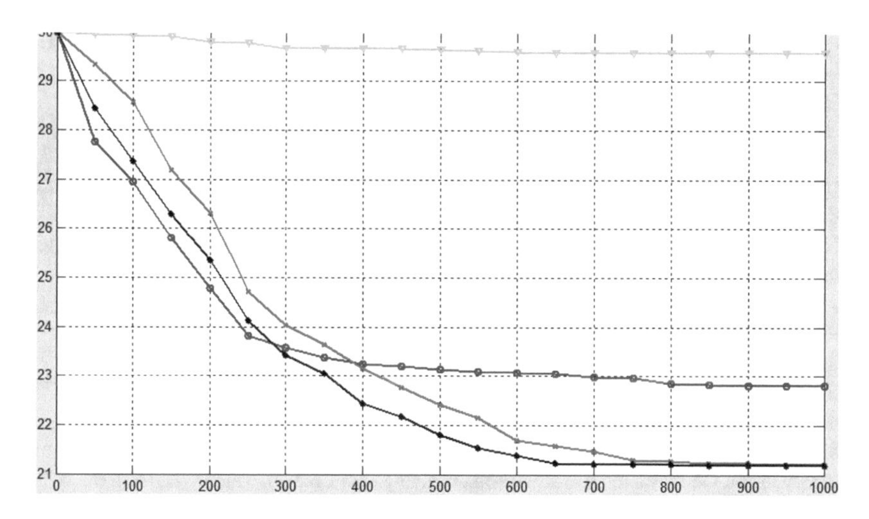

Fig. 3. Absolute effectiveness of algorithms

5 Conclusion

Figure 3 shows the comparative results demonstrated by different search algorithms (random search - green graph, adaptive algorithm - blue chart, modified genetic algorithm - red graph, hybrid algorithm - black chart). Analyzing the obtained dependencies, we can note the fact that all algorithms, with the exception of the random ones, demonstrate the directed character of search and convergence. Moreover, as can be seen from the graphs, an increase in the number of iterations of the algorithm does not lead to a significant improvement in the result. Concerning the rate of convergence, the most rapid is the adaptive search algorithm.

Acknowledgment. This research is supported by the grant from the Russian Foundation for Basic Research (project# 16-01-00715, 17-01-00627).

References

1. Conway, R.M., Maxwell, W.L., Miller, L.W.: Theory of Scheduling, 2nd edn. Dover Publications, Mineola (2004)
2. Pinedo, M.: Scheduling: Theory, Algorithms and Systems, 3rd edn. Springer, New York (2008)
3. Leung, J.Y.T.: Handbook of Scheduling. Chapman & Hall/CRC, Boca Raton (2004)
4. Luger, G.F.: Artificial Intelligence: Structures and Strategies for Complex Problem Solving, 6th edn. Addison Wesley, Boston (2009)
5. Michael, A., Takagi, H.: Dynamic control of genetic algorithms using fuzzy logic techniques. In: Proceedings of the Fifth International Conference on Genetic Algorithms, pp. 76–83. Morgan Kaufmann (1993)

6. Lee, M.A., Takagi, H.: Integrating design stages of fuzzy systems using genetic algorithms. In: Proceedings of the 2nd IEEE International Conference on Fuzzy System, pp. 612–617 (1993)
7. Herrera, F., Lozano, M.: Fuzzy adaptive genetic algorithms: design, taxonomy, and future directions. J. Soft Comput. **7**, 545–562 (2003)
8. Gladkov, L.A., Kureichik, V.V., Kureichik, V.M.: Genetic Algorithms. Phizmatlit, Moscow (2010)
9. Gladkov, L.A., Kureichik, V.V., Kureichik, V.M., Sorokoletov, P.V.: Bioinspirated Methods in Optimization. Phizmatlit, Moscow (2009)
10. Kureichik, V.M., Lebedev, B.K., Lebedev, O.B.: Search Adaptation: Theory and Practice. Phizmatlit, Moscow (2006)
11. Rasstrigin, L.A.: Adaptation of Complex Systems (1981)
12. Redko, V.G.: Evolutionary Cybernetics. Nauka, Moscow (2001)
13. King, R.T.F.A., Radha, B., Rughooputh, H.C.S.: A fuzzy logic controlled genetic algorithm for optimal electrical distribution network reconfiguration. In: Proceedings of 2004 IEEE International Conference on Networking, Sensing and Control, Taipei, Taiwan, pp. 577–582 (2004)
14. Gladkov, L.A., Gladkova, N.V., Leiba, S.N.: Hybrid intelligent approach to solving the problem of service data queues. In: Proceedings of 1st International Scientific Conference Intelligent Information Technologies for Industry (IITI 2016), vol. 1, pp. 421–433 (2016)
15. Gladkov, L.A., Gladkova, N.V., Legebokov, A.A.: Organization of knowledge management based on hybrid intelligent methods. In: Software Engineering in Intelligent Systems, Proceedings of the 4th Computer Science On-line Conference 2015 (CSOC 2015), Vol 3: Software Engineering in Intelligent Systems, pp. 107–113. Springer, Cham (2015)
16. Gladkov, L., Gladkova, N., Leiba, S.: Manufacturing scheduling problem based on fuzzy genetic algorithm. In: Proceedings of IEEE East-West Design & Test Symposium–(EWDTS 2014), Kiev, Ukraine, pp. 209–212 (2014)
17. Gladkov, L.A., Gladkova, N.V., Leiba, S.N.: Electronic computing equipment schemes elements placement based on hybrid intelligence approach. In: Advanced in Intelligent Systems and Computing. Intelligent Systems in Cybernetics and Automation Theory, vol. 348, pp. 35–45. Springer, Cham (2015)
18. Gladkov, L.A., Gladkova, N.V., Gromov, S.A.: Hybrid fuzzy algorithm for solving operational production planning problems. In: Advances in Intelligent Systems and Computing. Proceedings of the 6th Computer Science On-line Conference 2017 (CSOC 2017), Vol 1: Artificial Intelligence Trends in Intelligent Systems, vol. 573, pp. 444–456. Springer, Cham (2017)

Stacked Autoencoder for Segmentation of Bone Marrow Histological Images

Dorota Oszutowska-Mazurek[1(✉)], Przemyslaw Mazurek[2], and Oktawian Knap[3]

[1] Department of Epidemiology and Management, Pomeranian Medical University, Szczecin, Zolnierska 48 Street, 71-210 Szczecin, Poland
adorotta@pum.edu.pl

[2] Department of Signal Processing and Multimedia Engineering, West-Pomeranian University of Technology, Szczecin, 26. Kwietnia 10 Street, 71-126 Szczecin, Poland
przemyslaw.mazurek@zut.edu.pl

[3] Department of Forensic Medicine, Pomeranian Medical University, Szczecin, Powstancow Wielkopolskich 72 Street, 70-111 Szczecin, Poland

Abstract. Stacked autoencoder was used for the segmentation of trabeculas from bone marrow histological images derived from patients after hip joints arthroplasty. Additional filtering of areas smaller than 20000 pixels is necessary. The method has 95% efficiency. Proposed stacked autoencoder processes input images without special intervention automatically that is the main advantage of unsupervised learning over the supervised learning.

Keywords: Deep learning · Autoencoders · Stacked autoencoders Image segmentation · Bone marrow

1 Introduction

The segmentation of medical images is very often very challenging due to the complexity of biological structures. The problem of segmentation is related to low contrast, low resolution of images and lack of well defined boundaries between regions.

Advances in image acquisition devices reduce resolution problem. Contrast improvement could be obtained by biochemical preparation of structures, for example the staining of microscopic slides. Such biochemical segmentation is used over decades for some types of tissue samples, but recent advances in immunohistochemistry add the possibility of segmentation specific to tissue activity. Microscopic slides are problematic due to technique of tissue preparation. Slides observed by microscope are thin 3D structure. The application of microtome provides to separation of thin layers about 3 μm, so some small structures are located as whole object in such layer and are well visible. Some small structures are separated partially if they are at the edge of cutting and they are more

R. Silhavy (Ed.): CSOC 2018, AISC 764, pp. 425–435, 2019.
https://doi.org/10.1007/978-3-319-91189-2_42

transparent. Such stereological effect provides different contrast of object of the same type.

The processing of microscopic slide for automatic or computer assisted diagnosis is challenging due to very large resolutions. Resolutions larger then $10k \times 10k$ are typical. The process of slide preparation influences the flatness of the structure. Shallow focus at high magnification requires focus stating and multiple measurements of slide. Virtual slides that are obtained during the scanning using dedicated scanner or microscope from numerous patients are big data.

The processing of virtual slides requires dedicated algorithms for the segmentation purposes. It is serious problem because there are various different textures corresponding even for single tissue type. The design of the segmentation algorithm using graph of basic image processing algorithms is very difficult. Alternative approach is based on learning algorithms. Supervised learning requires prior segmentation of images manually that is time consuming. The number of training examples with manually segmented structures should be large. Another solution is unsupervised learning with specific segmentation obtained automatically without the requirement of the availability of manually segmented images. Unsupervised learning requires large image database, but for cases when structures are simple augmentation of small subset of images is sufficient.

1.1 Content and Contribution of Paper

The segmentation task is difficult for bone marrow microscopic images, because there are many factors with impact on tissue structure, including. e.g. age of the patient and neoplasms that are important for pathomorphologist assessment, The segmentation is therefore essential for further quantitative analysis [8].

Red bone marrow [2–4, 10] is located between trabeculae, contains three types of cells: osteoblasts (function: bone synthesis) osteocytes in lacunas, osteoclasts (function: bone resorption). Detailed structure of red bone marrow was described in [8].

Stacked autoencoder for the unsupervised learning for further automatic segmentation is proposed in this paper.

1.2 Related Works

The application of Convolutional Neural Networks for the segmentation of bone marrow was described in [8]. The efficacy of this method equaled 92%.

Autoencoders were proposed for segmentation of biological objects in various applications, e.g. computed tomography, macroscopic images, histopathological images in a few papers.

Convolutional autoencoder approach for mining features in cellular electron cryo-tomograms and weakly supervised coarse segmentation was described in [13]. It was described, that the autoencoder is possible to detect non cellular features related to sample preparation and patterns indication of spatial interactions between cellular elements and for weakly supervised semantic segmentation purpose with small amount of manual work.

Deep-Stacked Auto Encoder for Liver Segmentation from CT images was demonstrated in [1]. It was described that this classification supports segmentation of the liver from the abdomen and has high accuracy.

Stacked Autoencoders were also used for skin segmentation [7], algorithm used blocks for learning the representations and detecting skin and background areas.

Stacked denoising autoencoders (sDAE) were used for cells segmentation in histopathological Images [11]. Proposed algorithm was tested on two data sets containing more than 3000 cells derived from brain tumor and lung cancer microscopic images and achieved the best performance compared with other researchers.

Stacked Sparse Autoencoder (SSAE) was used for nuclei detection from breast cancer histopathology images [12]. The SSAE learns high-level features from pixel intensities for identification of nuclei features. A cohort of 500 histopathological images (2200×2200) and approximately 3500 manually segmented individual nuclei serving was used, method based on SSAE showed improved F-measure 84.49% and an average area under Precision–Recall curve (AveP) 78.83%.

Stacked Auto-Encoder method was described in [9] as possible method of medical image analysis, based on learning or discovering of highly non-linear and complicated patterns like the relations among input values.

2 Proposed Segmentation System

2.1 Stacked Autoencoder

Image segmentation using unsupervised learning could be obtained using different algorithms. The stacked autoencoder is possible solution. Autoencoders are obtained by the application of input image as desired output. Autoencoder transforms input data to different data (encoding part) with reduced number of outputs and these outputs are processed by decoding part to desired output (identical to input). Autoencoder uses a vector quantisation of data and compression of data is obtained. Proper autoencoder reduces data due to inherit redundancy of data to simpler form and could reconstruct original data. Image processing of autoencoder requires the application of sliding window approach for the training of the autoencoder.

The simplest autoencoders use the linear transformation of data, and such processing algorithm is the loose linear transform. Better results could be achieved using series of encoders with non linear transformation. Multiple encoders or decoders with linear transformation cannot be used, because linear transformation of previous linearly transformed data could be replaced by single linear transformation. The stack of linear transformations has no advantage over single liner transformation. Different properties are achieved for nonlinear transformations. Stacked autoencoder is achieved using iterative process.

First autoencoder (Fig. 1) is obtained by the training with the use of original data. The reduction of inputs from 961 to assumed 200 is achieved. The encoder is responsible for finding low-level features in images.

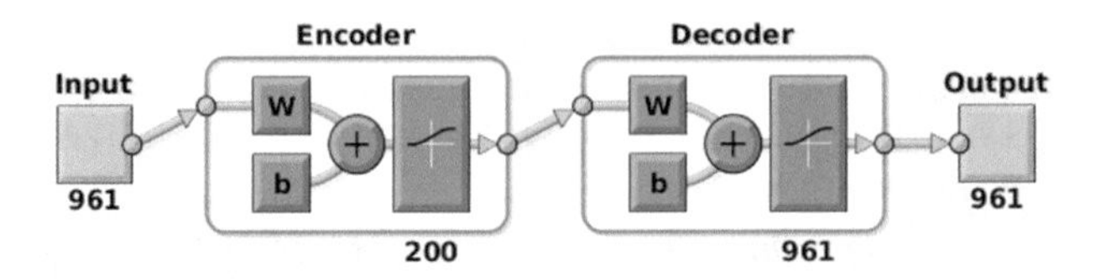

Fig. 1. First autoencoder

Achieved autoencoder is used for the transformation of input data to a new set corresponding to reduced number of data from inner outputs. Such data are applied for training a new autoencoder, and they are input and desired output data for second autoencoder (Fig. 2).

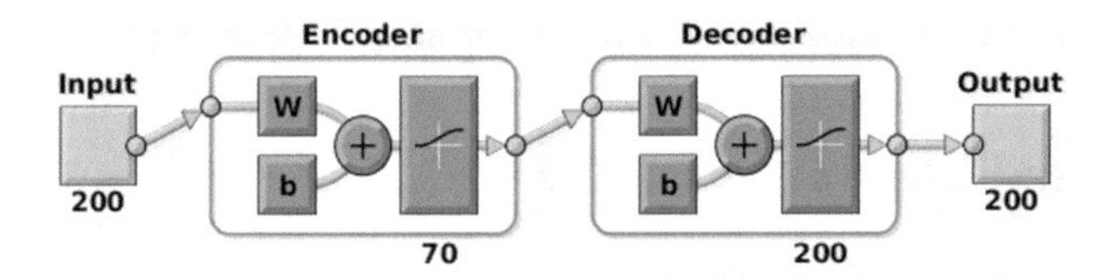

Fig. 2. Second autoencoder

This process is repeated again, and third autoencoder is applied in similar way (Fig. 3).

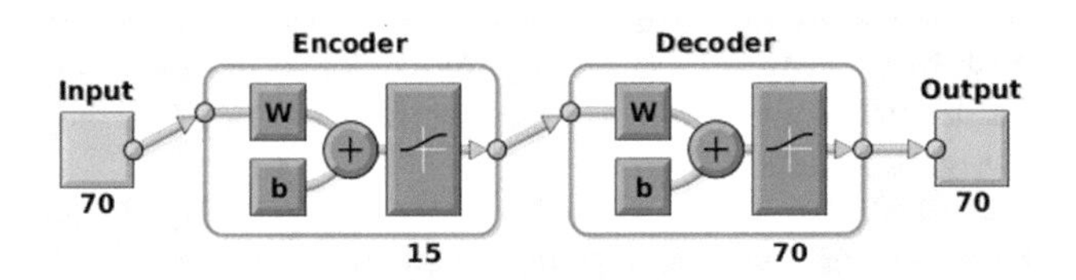

Fig. 3. Third autoencoder

Obtained autoencoders should be merged together (stacked), so three encoders: Encoder1, Encoder2 and Encoder 3 are responsible for maximal assumed data reduction. The restoring of original input images is possible by the processing of reduced data by decoders: Decoder3, Decoder2 and Decoder1.

This autoencoder cannot be used directly for image segmentation purposes. Softmax layer is applied as final layer for the normalisation of output values, so the sum of outputs is 1.0.

Stacked autoencoders (exactly stacked encoders) create deep convolutional neural network [5,6] and the learning process is well controlled, by iterative process of training individual layers. This approach reduces training time to a few minutes (Fig. 4).

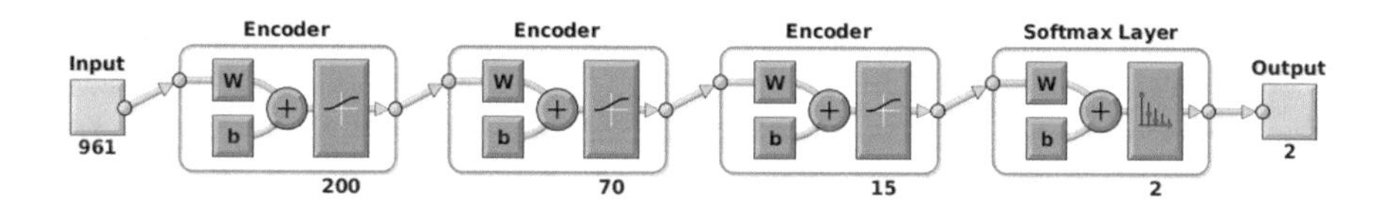

Fig. 4. Stacked autoencoder.

The output has two values. It is possible to use single value (binary), but two values allow the testing of results - assignment of false value to particular pixel.

2.2 Preprocessing of Input Images

Numerous variants of autoencoders were tested and improvement of processing was achieved by the application of additional preprocessing. The database was extended by the application of random rotation.

The second type of preprocessing is related to embedding estimator values into the image. Some areas that are not trabeculas are assigned to them and are related to noisy areas. The reduction of this influence was achieved by the calculation of local standard deviation. This value is populated over four rows of image (31×4).

2.3 Postprocessing of Outputs

The maximal value is applied as criteria of selection of output binary value $Y(x, y)$ for particular pixel (x, y) from two outputs ($OUT_1(x, y)$, $OUT_2(x, y)$) for softmax layer:

$$Y(x, y) = \begin{cases} 0 : & OUT_1(x, y) > OUT_2(x, y) \\ 1 : & otherwise \end{cases} \quad (1)$$

The output of the algorithm uses additional processing for removal of small islands, that are visible as small white spots on segmented images: Figs. 7 and 8. As a removal threshold is used value 30000 (pixels), so numerous small islands are removed.

3 Results

Slides of femoral heads were derived from patients after hip joint arthroplasty and stained via Hematoxylin–eosin method (H&E) Tissue images were acquired with the use of microscope Imager D1 (Carl Zeiss) and Axio–CamMRc5 camera with 2584 × 1936 resolution [8]. Grayscale conversion of colour images was applied for processing purposes.

Training data set uses random 20000 pieces of 31 × 27 subimages from database. Half of them are related to trabeculas and half of them are related to background with different features. This assignment is possible by the application of manual segmentation for testing purposes. The application of stacked autoencoders does not require manual segmentation - it is only for test purposes and segmentation is not ideal, some parts are difficult to assign to particular class: the trabecula or the background.

Numerous deep learning software tools are available and Matlab R2017b with Neural Network Toolbox was chosen. This toolbox supports learning using CPU, but much faster is GPU processing and NVidia Titan X GPGPU card (General–Purpose Graphics Processing Unit) was used.

Interesting property of deep learning approach using neural networks is the possibility analysis of obtained weights. Traditional neural networks are difficult to analysis. The sliding window approach used with autoencoder gives convolutional deep learning network and kernels related to the detection of features are shown in Fig. 5. This image shows that obtained kernels are correct, because low frequency spatial components are visible and practically there is no noise in kernel images.

Obtained kernels are used for random initialisation of weights and pretraining is not used.

Training process is quite fast comparing to typical deep learning algorithm because it consumes time equal to a few minutes. The training of stacked autoencoder is fast because at one time there are only two nonlinear layers (encoder and decoder layers). This approach influences fast convergence that is shown in Fig. 6.

Example results for large, about 4Mpix images are shown in Figs. 7 and 8. Input image is colour due to staining, but particular parts of the image could be recognised without colour. Manual segmentation is shown as the reference. Stacked autoencoder output is shown as binary image. Small lacunas of trabeculas are not assigned to trabeculas. Segmented image is processed by additional filtering and small white areas below 20000 pixels are removed. The efficiency of process equals 95%.

4 Discussion

Results shown in Figs. 7 and 8 and 95% efficiency indicate that segmentation method proposed in this paper is superior to method proposed in [8] with 92% efficacy.

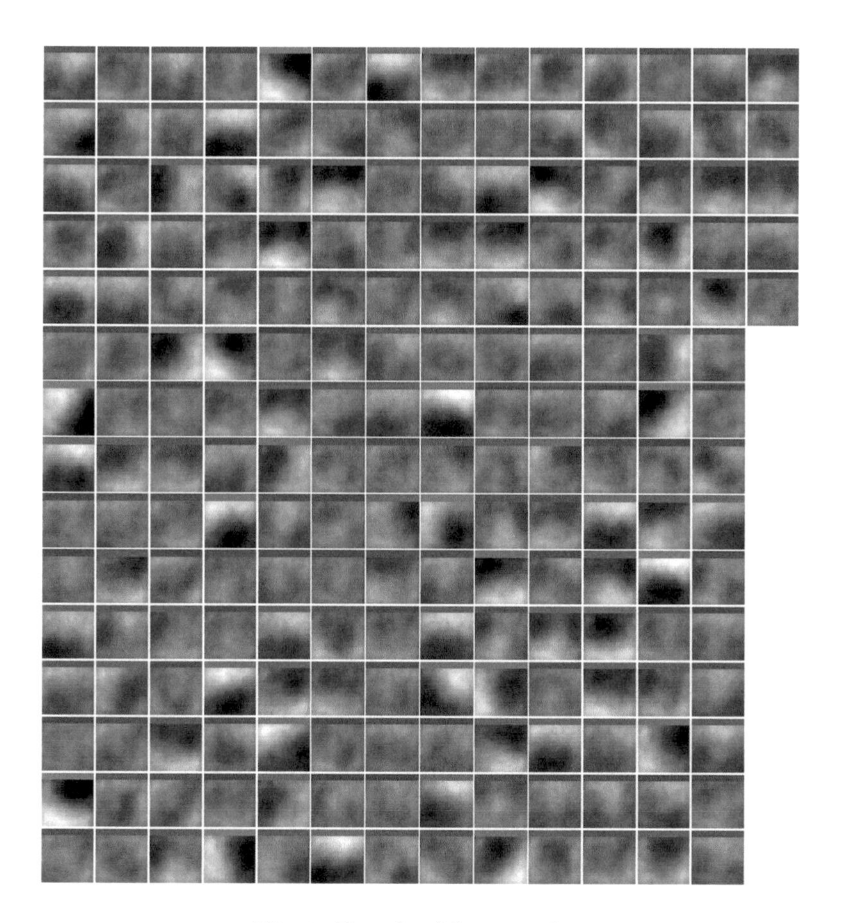

Fig. 5. Kernels of first encoder.

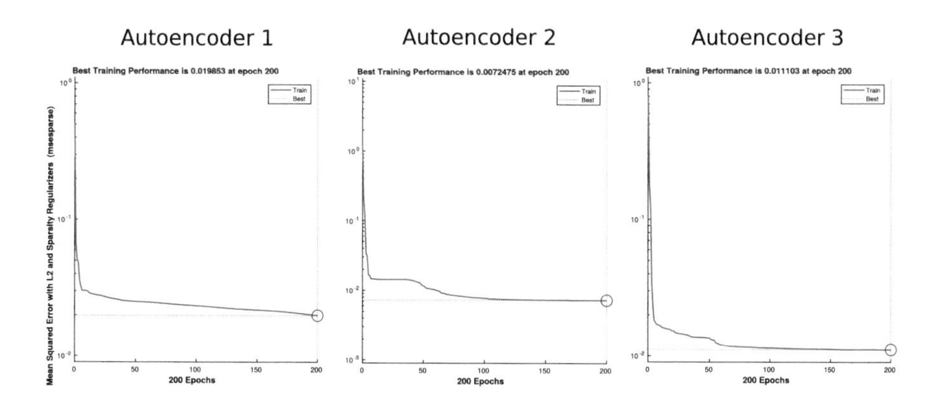

Fig. 6. Example training process.

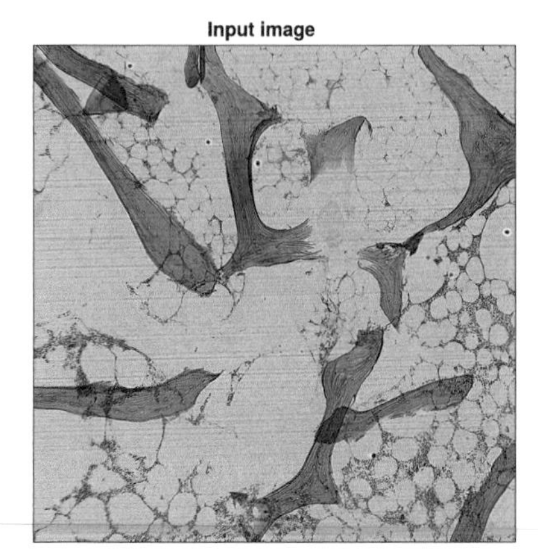

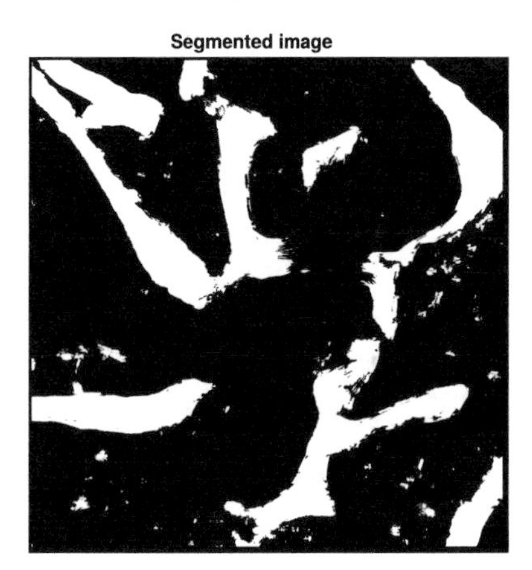

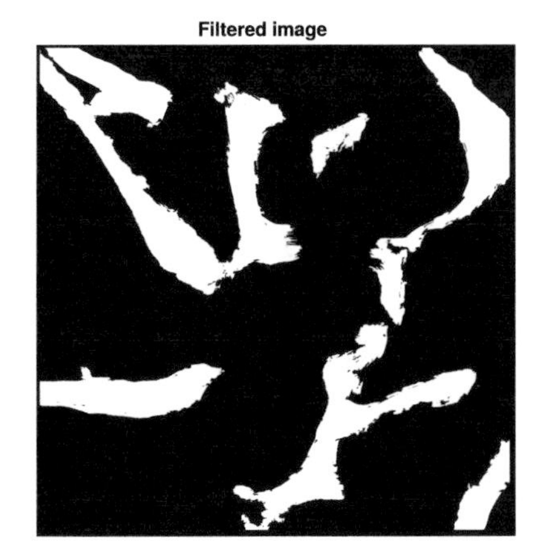

Fig. 7. Example results.

The application of standard deviation as input is motivated by the content of images from database. Kernels consist of spatial low frequency components, so the image is filtered and noise area with high standard deviation cannot be distinguished from low contrast areas with similar values.

The additional removal of small areas is necessary for the improvement of the segmentation. It is necessary because quite small (less then 20000 pixels) sliding window is applied. There are some areas with artefacts that are with similar or larger size, so they are recognised as trabeculas.

Essential problem of obtained autoencoder is the lack of details support, that is especially visible for small lacunas in trabeculas. Training process was based on two databases with random positions of windows. Reduced probability of presentation of such cases influences the results, as well as a small difference between

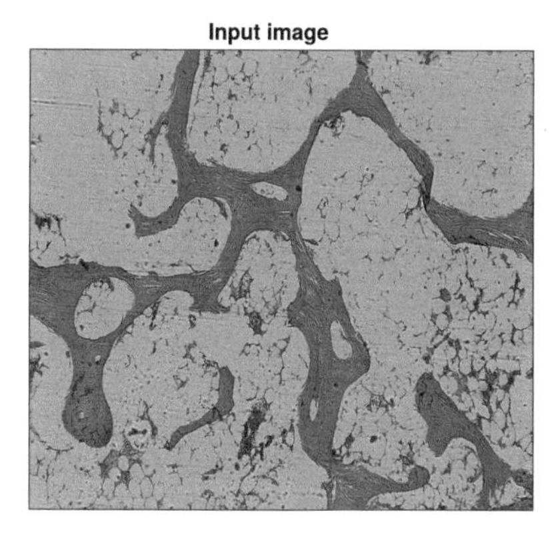

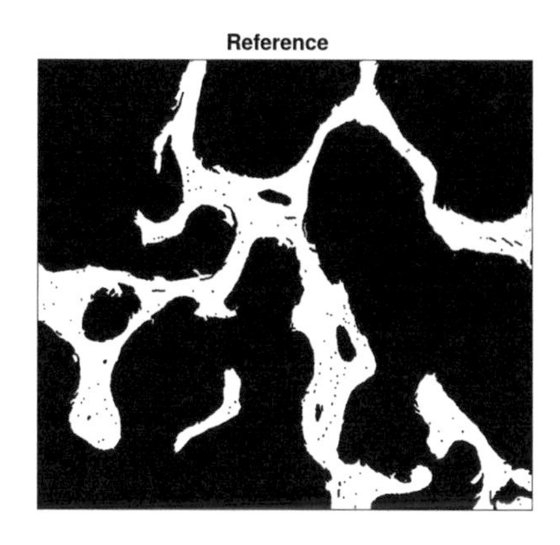

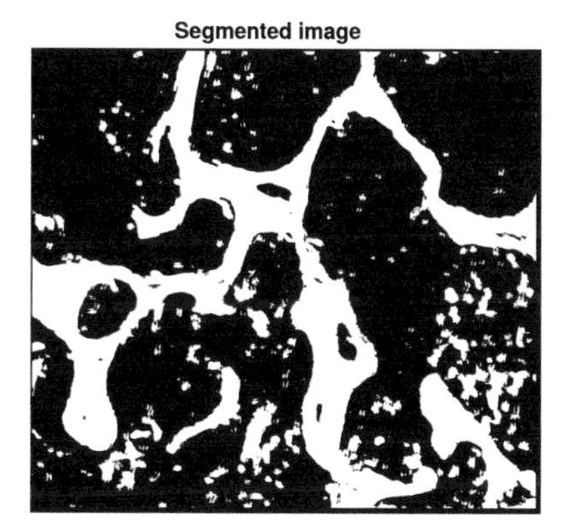

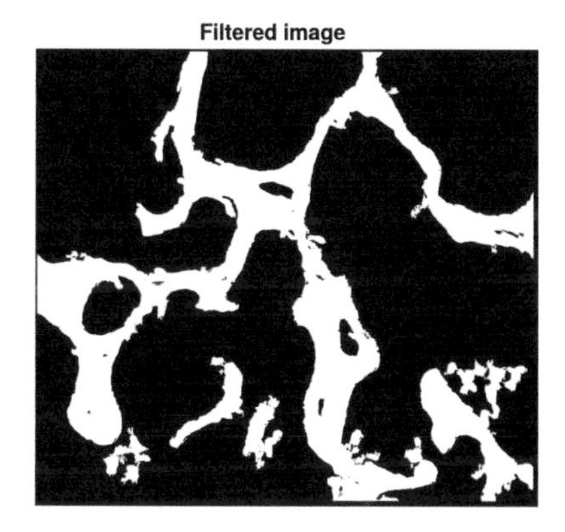

Fig. 8. Example results.

trabeculas subimages with and without lacunas. It is an effect of unsupervised learning also.

5 Conclusions

Proposed stacked autoencoder processes input images without special intervention automatically that is the main advantage of unsupervised learning over the supervised learning, but some removal of artefacts is necessary with the use of morphological filtering.

Further work will be related to the improvement of the stacked autoencoder for removal of artefacts using autoencoder only. The problem of segmentation of lacunas could be addressed using autoencoder and larger sliding window of the autoencoder. The improvement of the lacunas segmentation requires larger data set that supports numerous cases of lacunas for proper training.

Alternative approach is possible for unsupervised learning that is also interesting research area. Assumed output data from softmax layer are binary (after postprocessing) and are related to two outputs. The assumption of the multiple outputs from softmax layer could provide to different segmentation schemes with support of the detection of different features. The support of lacunas using three or more output of softmax layer is very desired, as well as the detection of granular textures that are related to the background. These granular textures are sometimes assigned to bone marrow class and are suppressed from this class by the application of morphological filtering (removal of small areas).

Stacked autoencoders could be processed by the application of CPU or GPU, but GPU is much faster a few times then CPU. Sliding window approach using assumed processing of the row of with about 2000 windows takes about 0.2 s for GPU. The processing of full image with 2000 rows consumes about 7 min, that is very long time. More efficient implementation is necessary for processing of full image at one times. It is required for the processing of medical data that are usually much larger. The problem of processing time could addressed but he application of different deep learning frameworks or more careful implementation of Matlab code.

Acknowledgement. This work is supported by the UE EFRR ZPORR project Z/2.32/I/1.3.1/267/05 "Szczecin University of Technology – Research and Education Center of Modern Multimedia Technologies" (Poland).

We gratefully acknowledge the support of NVIDIA Corporation with the donation of the Titan X GPU used for this research.

References

1. Ahmad, M., Yang, J., Ai, D., Qadri, S.F., Wang, Y.: Deep-stacked auto encoder for liver segmentation. In: Wang, Y., et al. (eds.) Advances in Image and Graphics Technologies, IGTA 2017, vol. 757, pp. 243–251. Springer, Singapore (2018)
2. Domagala, W., Chosia, M., Urasinska, E.: Atlas of histopathology. Wydawnictwo Lekarskie PZWL (2007)
3. Kumar, B., Abbas, A.K., Aster, J.: Robbins Basic Pathology. Elsevier, Philadelphia (2013)
4. Kumar, V., Abbas, A.K., Aster, J.C.: Robbins & Cotran Pathologic Basis of Disease. Elsevier, Philadelphia (2015)
5. LeCun, Y., Kavukcuoglu, K., Farabet, C.: Convolutional networks and applications in vision. In: Proceedings of 2010 IEEE International Symposium on Circuits and Systems, pp. 253–256, May 2010
6. LeCun, Y., Bengio, Y., Hinton, G.: Deep learning. Nature **521**(7553), 436–444 (2015)
7. Lei, Y., Yuan, W., Wang, H., Wenhu, Y., Bo, W.: A skin segmentation algorithm based on stacked autoencoders. IEEE Trans. Multimedia **19**(4), 740–749 (2017)
8. Oszutowska-Mazurek, D., Knap, O.: The use of deep learning for segmentation of bone marrow histological images. In: Silhavy, R., Senkerik, R., Kominkova, O.Z., Prokopova, Z., Silhavy, P. (eds.) Artificial Intelligence Trends in Intelligent Systems, CSOC 2017, vol. 573, pp. 466–473. Springer International Publishing, Cham (2017)

9. Shen, D., Wu, G., Suk, H.I.: Deep learning in medical image analysis. Ann. Rev. Biomed. Eng. **19**(1545–4274 (Electronic)), 221–248 (2017)
10. Sobotta, J.: Histology. Urban & Schwarzenberg, Baltimore (1983)
11. Su, H., Xing, F., Kong, X., Xie, Y., Zhang, S., Yang, L.: Robust cell detection and segmentation in histopathological images using sparse reconstruction and stacked denoising autoencoders. In: Navab, N., Hornegger, J., Wells, W., Frangi, A. (eds.) Medical Image Computing and Computer-Assisted Intervention – MICCAI 2015, vol. 9351, pp. 383–390. Springer International Publishing, Cham (2015)
12. Xu, J., Xiang, L., Liu, Q., Gilmore, H., Wu, J., Tang, J., Madabhushi, A.: Stacked sparse autoencoder (SSAE) for nuclei detection on breast cancer histopathology images. IEEE Trans. Med. Imaging **35**(1), 119–130 (2016)
13. Zeng, X., Ricardo Leung, M., Zeev-Ben-Mordehai, T., Xu, M.: A convolutional autoencoder approach for mining features in cellular electron cryo-tomograms and weakly supervised coarse segmentation, June 2017

Exploiting User Expertise and Willingness of Participation in Building Reputation Model for Scholarly Community-Based Question and Answering (CQA) Platforms

Tauseef Ur Rahman, Shah Khusro(✉), Irfan Ullah, and Zafar Ali

Department of Computer Science, University of Peshawar, Peshawar 25120, Pakistan
tauseef@icp.edu.pk,
{khusro, cs.irfan, zafarali}@uop.edu.pk

Abstract. Several scholarly social networking platforms are available on the Web, which build a collaborative research & development (R&D) environment for academicians and researchers to connect and collaborate with each other in solving real-world problems. The collaboration happens in the form of uploading, sharing and following research outcomes including technical reports, research publications, books, etc.; giving feedback on these research outputs; and community-based question & answering (CQA). In such systems, the reputation of users plays a key role, which acts as a trust indicator for the quality of questions, answers, and in recommending scholars and scholarly data. It is therefore necessary to build the reputation of a scholar in manner that reflects their active participation in the CQA activities. Therefore, the paper contributes a reputation model that besides expertise, considers the willingness of the user to participate in the CQA activities. The proposed reputation model is the first step towards recommending experts and active scholars that can potentially answer a given question. The empirical results show that the user expertise and their willingness to participate in the scholarly social Q&A activities play a major role in building more accurate reputation.

Keywords: Question & answering (Q&A) · Scholars · Reputation model Expertise · Willingness

1 Introduction

Several online collaborative web applications are available to facilitate collaboration among its users in generating and sharing knowledge, asking and answering questions, and requesting recommendations [1] in the form of community-based Q&A or CQA. The quality of the generated and shared data varies and depends greatly on the knowledge, expertise (and active participation) of users. Distinguishing good posts from the others requires the accurate and precise identification of user expertise in the form of reputation models [2]. The reputation model is the source of trust on the

R. Silhavy (Ed.): CSOC 2018, AISC 764, pp. 436–444, 2019.
https://doi.org/10.1007/978-3-319-91189-2_43

user-generated content [3] and therefore, has gained considerable attention in recent years [4], especially in generating fine-tune recommendations e.g., [5], collaborative web search solutions e.g., HeyStaks [6], and in CQA solutions, etc.

Reputation is also important in CQA platforms, where users ask and answer questions. These platforms are effective sources of crowd-generated knowledge [7]. Their wide adaptation is due to the easiest and effective ways of knowledge sharing CQA [8]. Several popular CQA sites including StackOverflow[1], Yahoo! Answers[2], Naver's Knowledge-iN[3], etc., are available on the Web where millions of users post tens of millions of questions and answers, this type of information seeking is far better than a common Web search [9]. Unlike generic CQA, ResearchGate[4] is the emerging scholarly CQA platform that allows scholars to discuss scholarly outputs and share thoughts in the form of social Q&A activities [10]. However, the reputation model used by ResearchGate is proprietary and cannot be reproduced [11], which introduces a research gap, that should be filled by introducing an open reputation model that can be used in CQA platforms for research purposes, which is the aim of the this paper. The paper investigates how user expertise and willingness of participation in CQA affect the user reputation, which can be considered as the first step towards recommending experts in CQA platforms. Section 2 briefly presents related work. Section 3 presents the proposed user reputation model. Section 4 presents the results of our empirical investigation. Section 5 concludes the paper, followed by references.

2 Related Work

Assessing the quality of user-generated content is one of the major issues in CQA platforms. As the main source of information, in these platforms, is the users/experts, therefore, the user reputation can be used as an essential indicator to evaluate the quality and the reliability of the generated content [9]. This is one of the reasons that CQA platforms are using voting and reputation models to provide users with the facility of finding the trustworthy and accurate content [12].

Several reputation models have been proposed for CQA platforms. In generating user reputation, the dynamics of CQA activities are considered, which include questions asked and answered, voting and rating, and users following other users. Reputation gets changed due to the arrival of new users [12] who contribute the user-generated content, which can be exploited in several ways. This is because user reputation has a direct correlation with the diversity of user-generated content, especially tags [13].

Several state-of-the-art research contributions have focused user reputation in CQA platforms. Wei et al. [14] use the user-related data from Stack Exchange to find the effects of virtual rewards and reputation on the quality and quantity of user

[1] http://www.stackoverflow.com.

[2] http://www.anwers.yahoo.com.

[3] http://www.naver.com.

[4] https://www.researchgate.net.

contribution. Their findings show that the willingness of participation gets affected by their relative rankings and not by the user reputation. Bian et al. [9] uses a semi-supervised coupled mutual reinforcement framework to find high-quality questions, answers, and users. Their framework simultaneously identifies high-quality content and user reputation using little labeled example in initiating training process as compared to manual labeling of large data [9]. Jurczyk et al. [8] adapted Hypertext Induced Topic Search (HITS) ranking algorithm in finding experts in the CQA platform. Through an empirical study, they found that link analysis helps in discovering authorities (experts) in topical categories. Chen et al. [15] compute user reputation by combining social network analysis and user ratings and analyze the influence members' relation to corresponding reputations to obtain explicit judgments about user reputation. Alam et al. [3] proposed a dynamic points-based user reputation model by exploiting ratings and social network analysis based on current levels of askers, answerers, and question's difficulty level.

Unlike these related studies, we focus on user reputation in scholarly CQA platforms and aim to investigate the effect of user expertise and their willingness to participate on the reputation score of the user through an empirical investigation like e.g., [8] did in their study. In this regard, an emerging [yet rapidly getting popular] CQA platform is ResearchGate that acts as a virtual community for the scholars and their peers [10]. It computes user reputation in the form of RG Score, which has been criticized by researchers and is therefore, subject to frequent changes. Although the formula for computing RG score has been proprietary/trade secret and is not reproducible, in doing so it exploits several aspects. Researchers have different observations. Ovadia [16] says it is the user-generated content that affects RG score and includes questions asked and answered, user profile, publications, downloads, views, and citations, however, like h-index, the RG score is not an efficient bibliographic measurement [16]. Kraker and Lex [11] enumerates its three major limitations including (i) irreproducibility with no transparency; (ii) using journal impact factor [although not used now]; and (iii) inability to reconstruct changes in RG score [11]. The undisclosed user reputation model of ResearchGate [17] introduces a gap in the literature, where it becomes difficult to identify the factors that actually affect the user reputation in scholarly CQA, which needs to be filled in order to generate open-source and better user reputation models for the said purpose. Therefore, as an initial step, the paper empirically investigates the potential impact of user expertise and willingness to participate on the reputation of the user in scholarly CQA platforms.

3 The User Reputation Model

To model user reputation in scholarly CQA, we use a hybrid approach that combines user expertise and willingness of participation in collaborative activities. The user expertise is based on several factors including the level of qualification, research output and its quality, etc., whereas the willingness is obtained from active participation in CQA activities. These are discussed in the following subsections. The proposed user reputation model combines user expertise E_u and willingness W_u of participation in CQA activities. Equation 1 computes the user reputation R_u.

$$R_u = E_u + W_u \tag{1}$$

The user expertise E_u and willingness W_u are discussed in detail in the following subsections.

3.1 Expertise

The expertise shows academic and professional experience of a CQA user. Equation 2 computes E_u from educational qualification (L_q), normalized projects-based reputation and publications-based reputation of user u.

$$E_u = \log\left[L_q + \left\{\left(\frac{Project_r}{Project_{r-total}} + \frac{Publication_r}{Publication_{r-total}}\right) + 1\right\}\right] \tag{2}$$

For L_q, we define three qualification levels including PhD, MS, and BS, and its values are given in Table 1. Together with qualification level L_q, Eq. 2 also measures the project-based reputation $Project_r$ and publications-based reputation $Publication_r$ of the user u. Equation 3 computes $Project_r$ from the available data about projects the user u is currently working on or has worked before.

$$Project_r = \sum\nolimits_{i=1}^{p} Score(Project_i) \tag{3}$$

Where p is the total number of projects, and $Score(Project_i)$ gives the score of each project, computed using Eq. 4.

$$\begin{aligned} &Score(Project_i) \\ &\quad = \left(Project_{collab.} + Project_{fund.} + Project_{dur.} + Project_{type} + Project_{del.}\right) \end{aligned} \tag{4}$$

Where, $Project_{collab.}$ means number of collaborators, $Project_{fund.}$ means whether the project is funded by a funding agency and has a funding number, $Project_{dur.}$ means duration of the project, $Project_{type}$ means the type of the project (industry-level, academic-level, both), and $Project_{del.}$ means number of deliverables of the project in terms of products, software prototypes, technical reports, etc. To normalize the effect of the $Project_r$ in Eq. 3, it is divided by $Project_{r-total}$, i.e., the projects-based reputation $Project_r$ of all users. Table 2 gives the possible values of the parameters used in Eq. 4. Equation 5 computes $Project_{r-total}$.

$$Project_{r-total} = \sum\nolimits_{i=1}^{n} Project_{r_i} \tag{5}$$

To compute publications-based reputation $Publication_r$ of the user u, which considers the number of publications, quality of publication (journal vs. conference), number of citations, number of self-citations, etc., (here, self-citations reduces the reputation score contributed by number of citations). Equation 6 gives $Publication_r$ of user u.

Table 1. The values of L_q used in Eq. 1.

Qualification level	Value
PhD/higher	1.0
MS/MPhil	0.5
BS	0.25

Table 2. The values for different parameters in Eq. 4.

Parameter		Value
$Project_{collab.}$	≤ 25	0.25
	≤ 50 and > 25	0.50
	≤ 75 and > 50	0.75
	> 75	1.00
$Project_{fund.}$	Yes	1.0
	No	0.0
$Project_{dur.}$	6 months	0.25
	12 months	0.50
	$\geq 24\, months$	1.0
$Project_{type}$	Industry-level	0.50
	Academic-level	0.50
	Academic + Industry	1.0
$Project_{del.}$	Yes	1.0
	No	0.0

$$Publication_r = \sum\nolimits_{i=1}^{p} Score(Publication_i) \tag{6}$$

Where p is the total number of publications, and $Score(Publication_i)$ gives the score of each publication, computed using Eq. 7.

$$Score(Publication_i) = \left(Publication_{type} + Publication_{citations} \times Publication_{self-citations}\right) \tag{7}$$

To normalize the effect of the $Publication_r$ in Eq. 6, it is divided by $Publication_{r-total}$, i.e., the publications-based reputation $Publication_r$ of all users. $Publication_{r-total}$ can be computed using Eq. 8.

$$Publication_{r-total} = \sum\nolimits_{i=1}^{n} Publication_{r_i} \tag{8}$$

Although some self-citations relate closely to the current work of the author, and therefore should not be excluded, this is one of the limitations of publications-based reputation computed by Eq. 6, which needs further consideration. Table 3 gives the possible values of the parameters used in Eq. 7. Note that the values used in Tables 1, 2, and 3, are taken for the sake of empirical investigation and could be adjusted after doing further research or using the reputation model in practical scholarly CQA platform, which we plan as our future work.

3.2 Willingness

The willingness W_u of the user shows their tendency towards the active participation in CQA activities. Equation 9 computes W_u from the CQA activities of the user performed in the last 30 days (subject to change, which needs further experimentation), whereas users whose activities fall behind this threshold are considered as inactive and therefore, their reputation is based only on their expertise. This ensures the active participation of the user, which is one of the desired conditions in CQA, has been considered while computing reputation besides their expertise. In Eq. 9, n represents the index of the last day i.e., 30th day, and m represents the total number of answers given on i th day. A_{ij} shows j th answer given on i th day and $\text{avg}(R_{ij})$ is the average of the ratings given by users to the answer A_{ij}. The value of A_{ij} is kept equal to 1 as each answer is treated separately with the quality of that answer.

$$W_u = \log\left[\sum\nolimits_{i=n-29}^{n} \left\{\sum\nolimits_{j=0}^{m} \left(A_{ij} * \text{avg}\left(R_{ij}\right)\right)\right\} + 1\right] \tag{9}$$

Putting the values of E_u and W_u from Eqs. 2 and 3, Eq. 1 presents the resulting reputation R_u of the user as given in Eq. 10.

$$\begin{aligned} R_u = \log\left[L_q + \left\{\left(\frac{Project_r}{Project_{r-total}} + \frac{Publication_r}{Publication_{r-total}}\right) + 1\right\}\right] \\ + \log\left[\sum\nolimits_{i=n-29}^{n} \left\{\sum\nolimits_{j=0}^{m} \left(A_{ij} * \text{avg}\left(R_{ij}\right)\right)\right\} + 1\right] \end{aligned} \tag{10}$$

A question may get several answers, where users may rate an answer higher. The rating for the answer represents its quality, and therefore, an answer with greater ratings means that it is of high quality. To capture this aspect in the willingness and finally in the reputation model, Table 4 introduces 5-star ratings of answer's quality based on the ratings it receives. The ratings show the satisfaction level of a user from the answer to the question. If an answer is not rated, it gets 0 showing that the irrelevant answer is given no importance and therefore, ignored.

4 Experimental Settings

To evaluate the effect of user expertise and willingness of participation on their reputation, the authors performed an empirical study. Although several datasets including Yahoo! Answers and Stack Exchange etc., are available, these could not be used in our case as we target scholarly CQA platforms like ResearchGate. As such datasets are not available therefore, we created our own dataset comprising of values for the different parameters used in Eqs. 1–10. The reputation model was evaluated by analyzing the

Table 3. Values for different parameters in Eq. 7

Parameter		**Value**
$Publication_{type}$	Journal	1.0
	Conference	0.5
$Publication_{citations}$	≤ 25	0.25
	≤ 50 and > 25	0.50
	≤ 75 and > 50	0.75
	> 75	1.00
$Publication_{self-citations}$	Yes	0.5
	No	1.0

Table 4. Answer levels with their corresponding ratings and values.

Rating	Value
No ratings	0
*	0.2
**	0.4
***	0.6
****	0.8
*****	1.0

data of 20 researchers with varying expertise and willingness values. The numerical computation resulted in Fig. 1, showing expertise (E_u), willingness of participation (W_u) and reputation (R_u), respectively for each of the 20 users.

By looking at Fig. 1, it can be observed that a user gets higher R_u value if he is more active W_u with moderate expertise E_u. In other words, two users having the same E_u but different W_u will result in different reputation, i.e., the expertise does affect the overall reputation of the user, but it is the willingness that boosts it.

5 Conclusion

We investigated how user expertise and willingness of participation in scholarly CQA platforms affect their reputation. For this purpose, we developed a user reputation model and tested it empirically. The reputation model currently exploits user expertise in terms of their qualification and the quality of their research output in terms of quality of publications, research projects, citations, etc., whereas for measuring their willingness to participate, it uses their active involvement in CQA activities. It hypothesizes that a scholar can better answer a question and therefore, has high reputation, if they are expert in the field, have high research output, and are comparatively more active in the CQA activities. The reputation model is an initial step in developing open-source scholarly CQA platform, where exploiting the details about the domain of the collaborators and their matching to the relevant questions are yet to be achieved. Once matched, the proposed reputation model can be exploited either in its current or in some

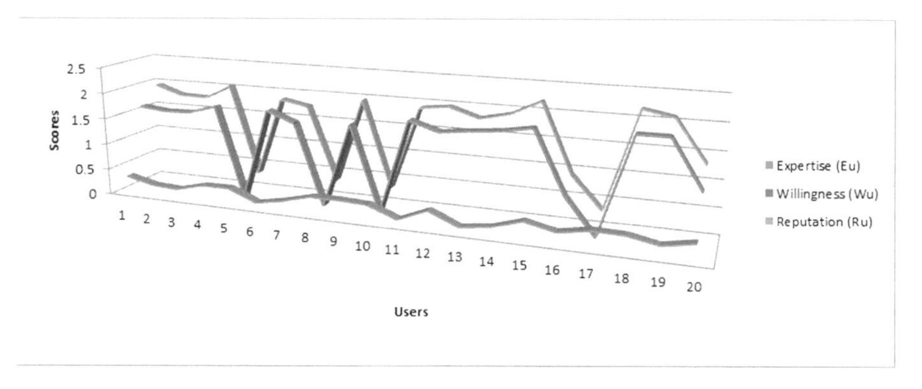

Fig. 1. Users 4 and 5 have almost the same expertise but different willingness of participation, which results in different reputation for both users. Here, user 4 has comparatively high willingness score, and therefore, its reputation is high. The same can be observed for users 8 & 9 as well as 17 & 18. This means that besides expertise, willingness of participation has a significant effect on final reputation score.

modified form to measure the reputation of the answerers so that the askers can trust on the answers received to their question. The model can also be exploited in recommending experts in scholarly CQA platforms so that a user may find experts and contact them personally to get answers that are more relevant and explanatory on the topic under consideration.

We performed an empirical investigation and found that the user reputation gets affected by changing the willingness score and keeping expertise almost same. This complies with the basic requirement of any CQA platform, where the active participation of the users has a key role in its success. The values considered in this empirical study are arbitrary and therefore, needs further research to find their true alternatives, which is possible by implementing the model in a scholarly CQA platform, and is therefore, one of the possible future research avenues to explore. In addition, in computing the expertise and willingness of participation, we exploited only a few factors, where several other factors should be considered and exploited in producing a more concrete and true representative model of user reputation.

References

1. Huo, Q., Palmer, A.: Analysing voluntary contribution to online forums using a proposed critical mass contribution model. J. Appl. Bus. Res. **31**(2), 687 (2015)
2. Abdaoui, A., Azé, J., Bringay, S., Poncelet, P.: Collaborative content-based method for estimating user reputation in online forums. In: Wang, J., Cellary, W., Wang, D., Wang, H., Chen, S.-C., Li, T., Zhang, Y. (eds.) WISE 2015. LNCS, vol. 9419, pp. 292–299. Springer, Cham (2015). https://doi.org/10.1007/978-3-319-26187-4_26
3. Alam, A., et al.: Confluence of social network, social question and answering community, and user reputation model for information seeking and experts generation. J. Inf. Sci. **43**(2), 260–274 (2016). https://doi.org/10.1177/0165551516637322

4. Mui, L., Mohtashemi, M., Halberstadt, A.: Notions of reputation in multi-agents systems: a review. In: Proceedings of the First International Joint Conference on Autonomous Agents and Multiagent Systems: Part 1. ACM (2002)
5. Abdel-Hafez, A., Xu, Y., Tian, N.: Item reputation-aware recommender systems. In: Proceedings of the 16th International Conference on Information Integration and Web-based Applications and Services. ACM (2014)
6. McNally, K., et al.: Towards a reputation-based model of social web search. In: Proceedings of the 15th International Conference on Intelligent User Interfaces. ACM (2010)
7. Fu, M., Zhu, M., Su, Y., Zhu, Q., Li, M.: Modeling temporal behavior to identify potential experts in question answering communities. In: Luo, Y. (ed.) CDVE 2016. LNCS, vol. 9929, pp. 51–58. Springer, Cham (2016). https://doi.org/10.1007/978-3-319-46771-9_7
8. Jurczyk, P., Agichtein, E.: Discovering authorities in question answer communities by using link analysis. In: Proceedings of the Sixteenth ACM Conference on Information and Knowledge Management. ACM (2007)
9. Bian, J., et al.: Learning to recognize reliable users and content in social media with coupled mutual reinforcement. In: Proceedings of the 18th International Conference on World Wide Web. ACM (2009)
10. Li, L., et al.: Answer quality characteristics and prediction on an academic Q&A site: a case study on ResearchGate. In: Proceedings of the 24th International Conference on World Wide Web. ACM (2015)
11. Kraker, P., Lex, E.: A critical look at the ResearchGate score as a measure of scientific reputation. In: Proceedings of the Quantifying and Analysing Scholarly Communication on the Web workshop (ASCW 2015), Web Science Conference (2015)
12. Anderson, A., et al.: Discovering value from community activity on focused question answering sites: a case study of stack overflow. In: Proceedings of the 18th ACM SIGKDD International Conference on Knowledge Discovery and Data Mining. ACM (2012)
13. MacLeod, L.: Reputation on stack exchange: tag, you're it! In: 2014 28th International Conference on Advanced Information Networking and Applications Workshops (WAINA). IEEE (2014)
14. Wei, X., Chen, W., Zhu, K.: Motivating user contributions in online knowledge communities: virtual rewards and reputation. In: 2015 48th Hawaii International Conference on System Sciences (HICSS). IEEE (2015)
15. Chen, W., et al.: A user reputation model for a user-interactive question answering system. Concurrency Comput. Pract. Experience **19**(15), 2091–2103 (2007)
16. Ovadia, S.: ResearchGate and Academia. edu: academic social networks. Behav. Soc. Sci. Librarian **33**(3), 165–169 (2014)
17. Yu, M.-C., et al.: ResearchGate: an effective altmetric indicator for active researchers? Comput. Hum. Behav. **55**, 1001–1006 (2016)

Performance of the Bison Algorithm on Benchmark IEEE CEC 2017

Anezka Kazikova(✉), Michal Pluhacek, and Roman Senkerik

Faculty of Applied Informatics, Tomas Bata University in Zlin, T.G. Masaryka 5555, 760 01 Zlin, Czech Republic
{kazikova, pluhacek, senkerik}@fai.utb.cz

Abstract. This paper studies the performance of a newly developed optimization algorithm inspired by the behavior of bison herds: the Bison Algorithm. The algorithm divides its population into two groups. The exploiting group simulates the swarming behavior of bison herds endangered by predators. The exploring group systematically runs through the search space in order to avoid local optima.

At the beginning of the paper, the Bison Algorithm is introduced. Then the performance of the algorithm is compared to the Particle Swarm Optimization and the Cuckoo Search on the set of 30 benchmark functions of IEEE CEC 2017. Finally, the outcome of the experiments is discussed.

Keywords: Bison Algorithm · Bison · Optimization · Swarm algorithms
Metaheuristic

1 Introduction

Modern optimization often finds inspiration in nature phenomena. So far, various nature inspired principles have been employed for optimization purposes. The Evolutionary Algorithms simulate the Darwinian evolution [1], Genetic Algorithms build artificial genomes [2] and swarm algorithms mimic behavior patterns of species living in swarms [3–5].

Since the source of inspiration is so vast, researchers find themselves simulating the movement patterns from various occasions like mating, searching for a prey or even running away from predators. These new optimization techniques are very successful in solving modern nontrivial optimization problems dealing with both discrete and continuous minimization tasks such as the Travelling Salesman Problem [6, 7], navigating pilot-free air combat vehicles [8] and many others.

The Bison Algorithm was proposed in 2017 by the author et al. [9] as an optimization technique with a simple inner mechanism preventing the population from getting stuck in local optima. The performance of the algorithm was compared to other optimization algorithms on a set of four functions. However, as the objective functions sample in the study was very small, the results could have been taken as preliminary.

This paper investigates the performance of the Bison Algorithm on the complete set of 30 IEEE CEC 2017 benchmark functions [10]. The algorithm is compared to the performance of the Particle Swarm Optimization [3] and the Cuckoo Search [4]

R. Silhavy (Ed.): CSOC 2018, AISC 764, pp. 445–454, 2019.
https://doi.org/10.1007/978-3-319-91189-2_44

implementations from the EvoloPy library [11]. In Sect. 2 the inner mechanisms of the Bison Algorithm are introduced. Section 3 presents the experiment design and results. Finally, Sect. 4 discusses the outcomes.

2 Bison Algorithm

Bison are unusual creatures. Even though their enormous mass and constitution effortlessly cause respect, they have developed rather interesting protection mechanisms. When bison herd is attacked by predators, they form a circle with the strong ones on the outline, while the weaker ones (like calves) try to get into the circle to find a safer location. Bison are also known as good runners, as they can reach the maximum velocity even of 56 km per hour [12].

2.1 Main Loop of the Bison Algorithm

The Bison Algorithm divides the population into two groups, each of which simulates one of the mentioned behavior practices (Algorithm 1). The group of the stronger individuals exploits the search space by the swarming movement, while the weaker individuals explore the search space by running through it. The population is being sorted after every iteration, therefore the groups can swap their members easily.

Algorithm 1. Bison Algorithm Pseudocode

```
1.   Initialization:
       objective function: f(x) = (x_1, x_2, ..., x_d)
       generate the swarming group with random position
       generate the running group around the last bison
       generate the run direction vector (Eq. 8)
2.    for every migration round do
3.       compute the center of the elite bison group
4.       for every bison in the swarming group do
5.          compute a new position candidate x_new (Eq. 2)
6.          place the out-bounded onto the hypersphere
7.          if f(x_new) < f(x_old) then move to the x_new
8.       end
9.       adjust the run direction vector (Eq. 9)
10.      for every bison in the running group do
11.         move to the new position (Eq. 10)
12.         place the out-bounded on the hypersphere
13.      end
14.      sort population by objective function value
15.   end for
```

Swarming. The swarming behavior starts by computing the center of several strongest individuals (Eq. 1). Every member of the swarming group then computes a new possible position closer to the center (Eq. 2) and can overstep the center by the overstep parameter. The movement happens only, if it improves its value.

$$\boldsymbol{direction} = \boldsymbol{center} - \boldsymbol{x} \tag{1}$$

$$\boldsymbol{x_{new}} = \boldsymbol{x_{old}} + \boldsymbol{direction} * \boldsymbol{random}(\boldsymbol{0}, \boldsymbol{overstep})_{\boldsymbol{dim}} \tag{2}$$

Center Computation. There are three ways of computing the center of the strongest individuals: the arithmetic (Eq. 3), weighted (Eqs. 4, 5) and ranked center (Eqs. 6, 7) as shown in Table 1. The first is built from the positional information only, while the latter and the last consider the fitness of the strongest individuals as well.

Table 1. Center computations of the Bison Algorithm

Center computation	Equation	
Arithmetic center	$\boldsymbol{center} = \sum_{i=1}^{s} \frac{\boldsymbol{x_i}}{s}$	(3)
Weighted center	$weight(\boldsymbol{x}) = -f(\boldsymbol{x}) + \sum_{i=1}^{s} f(\boldsymbol{x_i})$	(4)
	$\boldsymbol{center} = \sum_{i=1}^{s} \frac{weight(\boldsymbol{x_i}) * \boldsymbol{x_i}}{\sum_{j=1}^{s} weight(\boldsymbol{x_i})}$	(5)
Ranked center	$\boldsymbol{weight} = (10, 20, \ldots, 10 * s)$	(6)
	$\boldsymbol{center} = \sum_{i=1}^{s} \frac{weight_i * \boldsymbol{x_i}}{\sum_{j=1}^{s} weight_i}$	(7)

Where s is the number of the strongest elite bison (also called the elite group size)

Running. The running group is exploiting the search space throughout the whole optimization process. The movement is based on the *run direction vector*. The vector is generated during the initialization of the algorithm (Eq. 8) and is slightly altered at the beginning of each iteration (Eq. 9). The running movement is formulated in Eq. 10. Each member of the running herd is then shifted in the run direction vector.

$$\boldsymbol{run\,direction} = \boldsymbol{random}\left(\frac{\boldsymbol{up\,bound} - \boldsymbol{low\,bound}}{45}, \frac{\boldsymbol{up\,bound} - \boldsymbol{low\,bound}}{15}\right)_{\boldsymbol{dim}} \tag{8}$$

$$\boldsymbol{run\,direction} = \boldsymbol{run\,direction} * \boldsymbol{random}(\boldsymbol{0.9}, \boldsymbol{1.1})_{\boldsymbol{dim}} \tag{9}$$

$$\boldsymbol{x_{new}} = \boldsymbol{x_{old}} + \boldsymbol{run\,direction} \tag{10}$$

Where *up bound* and *low bound* define the search space limitations. The space has the features of a hyper-sphere, therefore crossing the boundaries means appearing on the other side of the dimensions.

Parameters. Table 2 presents all the configurable parameters with their recommended values based on an early testing.

Table 2. Parameters of the Bison Algorithm and their recommended values

Parameter	Description	Recommended
Population		50
Elite group size	Number of the strongest individuals. Their center is used in the swarming movement	20
Swarm group size	Number of individuals performing the swarming movement	40
Center computation	Arithmetic/Weighted/Ranked	
Overstep	Defines the maximum length of the swarming movement. (0 = no movement; 1 = to the center)	3.5–4.1

Validation of the Bison Algorithm. to prove the functionality of the introduced algorithm, Fig. 1 shows the movement patterns of the Bison Algorithm on the 2-dimensional Rastrigin's function. The swarming group is exploiting the discovered promising landscapes and the running group is exploring the search space to make sure that the population is not stuck in any local optimum.

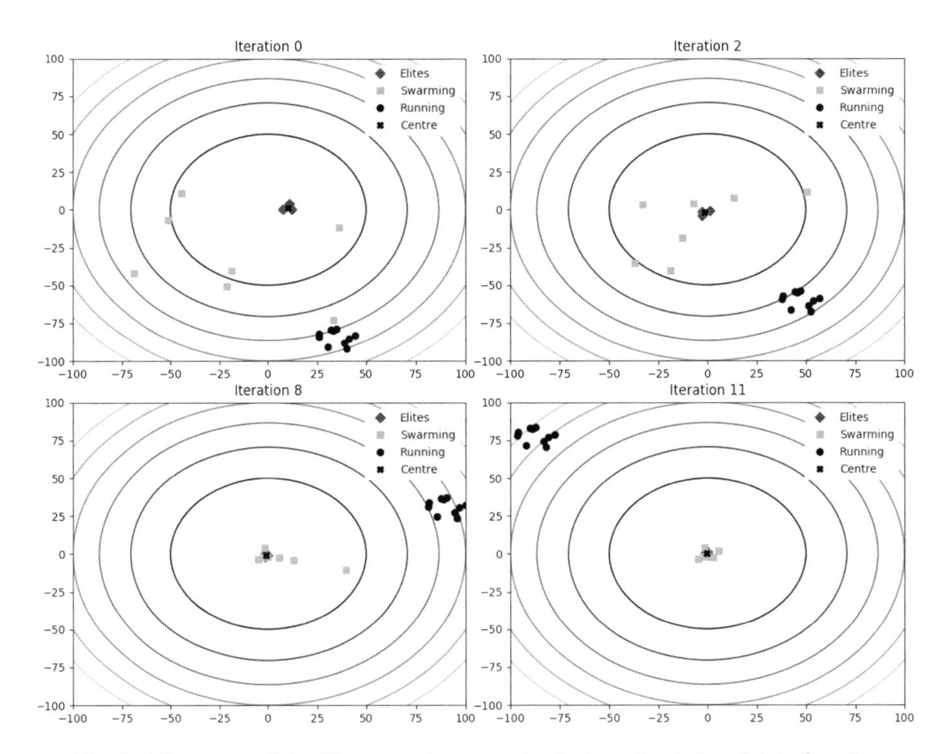

Fig. 1. Movement of the Bison population on the 2-dimensional Rastrigin's function

3 Methods and Results

The Bison Algorithm was tested on 30 nontrivial functions of the IEEE CEC 2017 benchmark [10] and compared to the Particle Swarm Optimization and the Cuckoo Search algorithms with both the implementation and the parameter configuration (Table 3) based on the EvoloPy library default settings [11]. The following shortcuts are used: PSO for Particle Swarm Optimization, CS for Cuckoo Search and BIA for Bison Algorithm.

Table 3. Experimental parameter settings

Bison Algorithm		PSO		Cuckoo Search	
Population	50	Population	50	Population	50
Elite group size	20	Vmax	6	Pa	0.25
Swarm group size	40	wMax	0.9		
Center computation	Ranked	wMin	0.2		
Overstep	3.5	C1	2		
		C2	2		

In accordance with the CEC 2017 benchmark instructions, the algorithms were tested on 51 independent runs, each consisting of $n = 10\,000 \cdot \textit{dimensions}$ evaluations of the objective function. Tables 4 and 5 present the comparison of the average solution quality and standard deviation found by BIA, PSO and CS on 10 and 30 dimensional problems respectively. The last column names the most successful algorithm according to Wilcoxon rank-sum test ($\alpha = 0.05$). The winner is recognized only if it outperforms all the other algorithms with statistical significance according to the rank-sum test.

Table 4. Performance of Bison Algorithm, PSO and CS on 10 dimensional CEC 2017

	Bison Algorithm		PSO		Cuckoo Search		Min
	avg	*std*	*avg*	*std*	*avg*	*std*	*avg*
F1	4.26E+02	5.77E+02	9.75E+02	1.38E+03	2.07E+03	1.00E+03	BIA
F2	7.84E−02	3.37E−01	4.70E−01	3.36E+00	0.00E+00	0.00E+00	–
F3	0.00E+00	0.00E+00	0.00E+00	0.00E+00	6.00E−02	5.00E−02	BIA
F4	4.72E−01	5.60E−01	4.86E+00	9.97E+00	4.20E−01	4.40E−01	–
F5	7.90E+00	6.21E+00	3.16E+01	1.03E+01	1.78E+01	3.52E+00	BIA
F6	1.43E−02	1.02E-01	3.02E+00	4.30E+00	5.79E+00	1.96E+00	BIA
F7	2.48E+01	8.04E+00	2.20E+01	5.59E+00	2.90E+01	4.60E+00	–
F8	8.17E+00	6.99E+00	1.65E+01	7.02E+00	1.80E+01	3.92E+00	BIA
F9	4.63E−02	2.49E−01	8.93E+00	6.38E+01	5.41E+01	3.30E+01	–
F10	1.12E+03	2.97E+02	8.32E+02	2.78E+02	6.35E+02	1.20E+02	CS
F11	2.46E+00	2.39E+00	2.58E+01	1.30E+01	5.06E+00	1.42E+00	BIA
F12	7.53E+03	3.90E+03	1.40E+04	1.00E+04	3.15E+03	1.25E+03	CS

(*continued*)

Table 4. (*continued*)

	Bison Algorithm		PSO		Cuckoo Search		Min
	avg	*std*	*avg*	*std*	*avg*	*std*	*avg*
F13	3.87E+03	2.09E+03	6.94E+03	6.30E+03	2.27E+01	6.26E+00	CS
F14	3.65E+01	4.85E+00	6.06E+01	5.43E+01	2.05E+01	4.05E+00	CS
F15	2.62E+01	1.35E+01	9.72E+01	1.09E+02	8.10E+00	3.13E+00	CS
F16	2.23E+01	4.84E+01	2.27E+02	1.25E+02	5.89E+00	3.72E+00	BIA
F17	3.38E+01	2.21E+01	5.84E+01	3.07E+01	3.15E+01	4.46E+00	–
F18	2.96E+03	2.68E+03	8.08E+03	9.56E+03	1.19E+02	5.34E+01	CS
F19	2.03E+01	8.02E+00	6.94E+02	9.85E+02	5.17E+00	1.02E+00	CS
F20	6.00E+00	1.45E+01	9.78E+01	5.86E+01	2.96E+01	5.47E+00	BIA
F21	1.14E+02	3.19E+01	1.26E+02	4.25E+01	1.02E+02	1.65E+01	–
F22	1.01E+02	7.06E−01	1.23E+02	1.77E+02	7.19E+01	3.12E+01	CS
F23	3.10E+02	4.71E+00	3.68E+02	2.62E+01	3.17E+02	4.12E+00	BIA
F24	3.20E+02	4.02E+01	1.50E+02	8.40E−01	1.34E+02	3.84E+01	CS
F25	4.43E+02	1.15E+01	3.10E+02	3.49E+01	3.00E+02	1.12E+02	–
F26	2.94E+02	2.38E+01	3.84E+02	2.70E+02	2.02E+02	8.59E+01	CS
F27	4.00E+02	4.35E+00	6.88E+02	9.90E+01	3.90E+02	5.70E−01	CS
F28	4.44E+02	1.38E+02	2.50E+02	0.00E+00	2.99E+02	3.67E+01	PSO
F29	2.75E+02	1.71E+01	4.50E+02	1.00E+02	2.81E+02	1.47E+01	–
F30	2.77E+04	1.75E+05	2.35E+05	4.23E+05	5.64E+03	6.26E+03	BIA
Winning algorithms according to the Wilcoxon rank-sum test (α = 0.05):							
	None	**BIA**	**PSO**	**CS**		**Control sum**	
	8	10	1	11		30	

Table 5. Performance of Bison Algorithm, PSO and CS on 30 dimensional CEC 2017

	Bison Algorithm		PSO		Cuckoo Search		Min
	avg	*std*	*avg*	*std*	*avg*	*std*	*avg*
F1	1.67E+03	1.85E+03	3.91E+03	5.11E+03	8.90E+02	7.49E+02	–
F2	1.17E+11	5.87E+11	5.02E+23	3.58E+24	2.85E+11	6.21E+11	–
F3	8.92E+01	8.57E+01	3.42E−04	6.75E−04	3.39E+04	5.76E+03	PSO
F4	1.59E+01	2.51E+01	9.31E+01	2.89E+01	6.82E+01	1.89E+01	BIA
F5	7.22E+01	6.11E+01	1.52E+02	2.82E+01	1.37E+02	2.00E+01	BIA
F6	2.84E−04	1.05E−03	3.06E+01	9.78E+00	4.15E+01	9.07E+00	BIA
F7	1.77E+02	3.17E+01	1.02E+02	2.21E+01	1.63E+02	2.11E+01	PSO
F8	7.97E+01	5.85E+01	1.06E+02	1.90E+01	1.36E+02	1.76E+01	–
F9	5.50E+00	9.01E+00	2.21E+03	9.68E+02	3.72E+03	1.14E+03	BIA
F10	6.97E+03	2.79E+02	3.51E+03	6.38E+02	3.74E+03	2.57E+02	PSO
F11	2.94E+01	2.33E+01	1.00E+02	3.55E+01	8.84E+01	1.82E+01	BIA
F12	2.30E+04	1.17E+04	6.57E+05	2.05E+06	1.50E+05	5.39E+04	BIA

(*continued*)

Table 5. (*continued*)

	Bison Algorithm		PSO		Cuckoo Search		Min
	avg	*std*	*avg*	*std*	*avg*	*std*	*avg*
F13	1.12E+04	7.72E+03	1.03E+05	6.20E+05	5.91E+03	4.01E+03	CS
F14	2.07E+03	1.71E+03	5.08E+03	5.80E+03	1.18E+02	1.84E+01	CS
F15	1.42E+03	2.09E+03	1.06E+04	1.28E+04	9.92E+02	6.81E+02	–
F16	1.01E+03	4.17E+02	8.49E+02	2.54E+02	9.25E+02	1.35E+02	–
F17	1.39E+02	1.52E+02	5.21E+02	1.90E+02	3.01E+02	8.57E+01	BIA
F18	1.37E+05	1.03E+05	1.53E+05	1.55E+05	5.57E+04	1.50E+04	CS
F19	4.30E+03	3.96E+03	4.86E+03	6.42E+03	3.82E+02	2.91E+02	CS
F20	2.09E+02	1.29E+02	1.83E+02	1.30E+02	4.13E+02	8.57E+01	–
F21	2.62E+02	5.87E+01	3.37E+02	2.85E+01	3.25E+02	4.04E+01	BIA
F22	1.00E+02	6.68E−01	1.98E+03	1.96E+03	8.29E+02	1.52E+03	BIA
F23	3.78E+02	1.38E+01	6.58E+02	1.04E+02	4.86E+02	2.73E+01	BIA
F24	4.49E+02	1.17E+01	6.98E+02	6.72E+01	5.44E+02	4.98E+01	BIA
F25	3.90E+02	8.93E+00	3.90E+02	4.03E+00	3.85E+02	1.27E+00	CS
F26	8.69E+02	6.42E+02	2.07E+03	1.60E+03	1.05E+03	4.45E+02	–
F27	5.32E+02	1.11E+01	5.77E+02	5.06E+01	5.29E+02	7.43E+00	–
F28	3.38E+02	5.47E+01	4.24E+02	4.38E+01	3.87E+02	3.43E+01	BIA
F29	5.62E+02	1.28E+02	9.34E+02	2.20E+02	9.28E+02	7.87E+01	BIA
F30	3.68E+03	7.83E+02	5.19E+03	3.11E+03	1.10E+04	3.43E+03	BIA
Winning algorithms according to the Wilcoxon rank-sum test (α = 0.05):							
	None	**BIA**	**PSO**	**CS**		**Control sum**	
	8	14	3	5		30	

Table 6 shows the results of the Friedman rank test with relevance at $p < 0.05$. Figure 2 shows the Nemenyi critical distance between the algorithms in 30 and 50 dimensions. Table 7 sums the results of the Wilcoxon rank-sum test in all dimensions.

Table 6. Average rank according to Friedman test ($p < 0.05$)

Dimension	BIA	PSO	CS	P-Value
10 D	1.80	2.60	1.60	7.29E−05
30 D	1.57	2.50	1.93	7.13E−04
50 D	1.57	2.30	2.13	9.68E−03

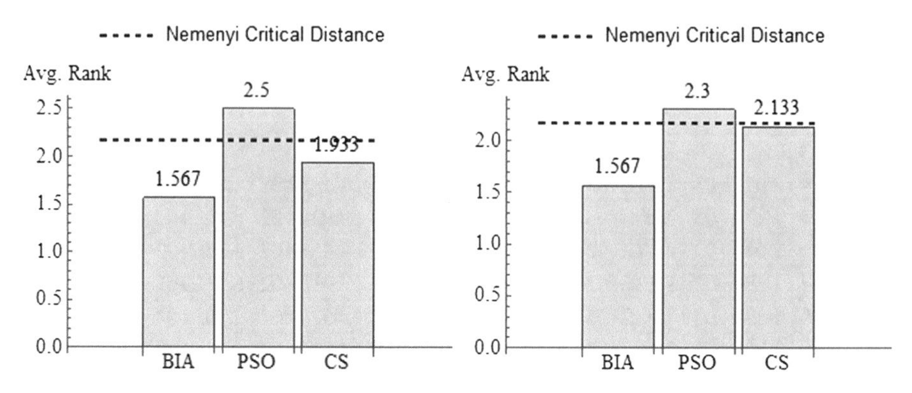

Fig. 2. Ranking with critical distance in 30 (left) and 50 (right) dimensions

Table 7. Winning algorithms according to the Wilcoxon rank-sum test ($\alpha = 0.05$)

Dimension	None	BIA	PSO	CS
10 D	8	10	1	11
30 D	8	14	3	5
50 D	5	15	5	5

Figures 3 and 4 show the average (mean) error of the solution functional value on 10 and 30 dimensional problems using the logarithmic scale.

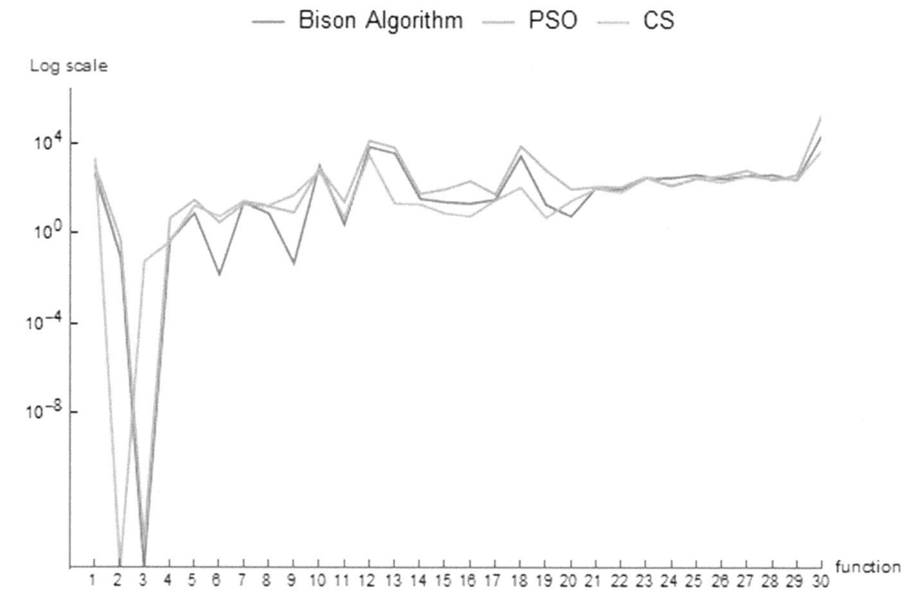

Fig. 3. Mean solution of tested algorithms in 10 dimensions

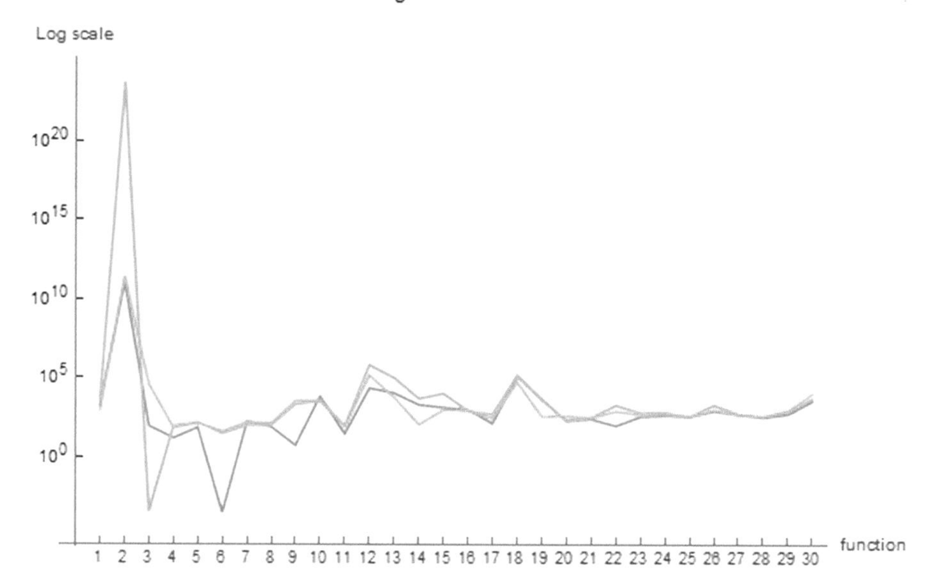

Fig. 4. Mean solution of tested algorithms in 30 dimensions

The experiments made on 10 dimensional functions (Table 4) generated promising results of the Bison Algorithm, which outperformed the other metaheuristics in 10 cases out of 30. Worth to notice are especially solutions of F3, F6, F9 and F20, where the quality difference is visible. The Cuckoo Search algorithm found the best solution in 11 cases out of 30, proving the advancement of this metaheuristic.

In 30 dimensions (Table 5), the Bison Algorithm outperformed the other metaheuristics in 14 cases out of 30. It was particularly successful in the functions F6 and F9, which is also confirmed in Figs. 3 and 4.

When solving F22, the Bison Algorithm seemed to generally find a perspective local optimum with objective function value 100 in all of the tested dimensions with an extremely low standard deviation.

According to Table 6, the Bison Algorithm significantly outperformed the PSO in the Friedman rank tests. The Wilcoxon rank-sum test in Table 7 showed that the BIA outperformed the other algorithms in 30 and 50 dimensions.

4 Discussion

The Bison Algorithm achieved impressive results on IEEE CEC 2017 benchmark. It outperformed the Particle Swarm Optimization and the Cuckoo Search in most of the tested functions in 30 and 50 dimensions. The Cuckoo Search excelled in 10 dimensions, closely followed by the Bison Algorithm.

The Bison Algorithm found admirable solutions especially when solving the F6 and F9 in all the tested dimensions. These functions seem to be conveniently fitting for the algorithm, as the standard deviation of the results is very low.

5 Conclusion

The Bison Algorithm proved to be able to solve complex nontrivial problems even on the large set of 30 CEC 2017 benchmark functions and often found significantly better solutions than the other optimization techniques

However, it is more than possible, that with a different parametric approach, the algorithm could perform better, as the parameters in this paper were based on a preliminary study. The Bison Algorithm could also be enriched by adding the Levy flights implementation, as the real European Bison were observed to hold the Levy movement patterns.

To summarize, the promising results prove, that the Bison Algorithm deserves a further development and could be fairly used to solve future optimization tasks.

References

1. Bäck, T.: Evolutionary Algorithms in Theory and Practice: Evolution Strategies, Evolutionary Programming, Genetic Algorithms. Oxford University Press, Oxford (1996)
2. Goldberg, D.: Genetic Algorithms in Search, Optimization, and Machine Learning. Addison-Wesley, Reading (1989)
3. Kennedy, J., Eberhart, R.: Particle swarm optimization. In: Proceedings of the IEEE International Conference on Neural Networks, vol. 4, 1942–1948 (1995)
4. Yang, X.-S., Deb, S.: Cuckoo search via Levy flights. In: Proceedings of World Congress on Nature & Biologically Inspired Computing (NaBIC 2009), December 2009, India, USA, pp. 210–214. IEEE Publications (2009)
5. Mirjalili, S., Mirjalili, S.M., Lewis, A.: Grey wolf optimizer. Adv. Eng. Softw. **69**, 46–61 (2014)
6. Ouaarab, A., Ahiod, B., Yang, X.S.: Discrete cuckoo search algorithm for the travelling salesman problem. Neural Comput. Appl. **24**(7–8), 1659–1669 (2014)
7. Duan, Y., Ying, S.: A particle swarm optimization algorithm with ant search for solving traveling salesman problem. In: 2009 International Conference on Computational Intelligence and Security, Beijing, pp. 137–141 (2009)
8. Duan, H., Guo, L., Wang, G., Wang, H.: A modified firefly algorithm for UCAV path planning. Int. J. Hybrid Inf. Technol. **5**, 123–144 (2012)
9. Kazikova, A., Pluhacek, M., Senkerik, R., Viktorin, A.: Proposal of a new swarm optimization method inspired in bison behavior. In: Matousek, R. (ed.) Recent Advances in Soft Computing (Mendel 2017). Advances in Intelligent Systems and Computing. Springer (in press)
10. Awad, N.H., Ali, M.Z., Liang, J.J., Qu, B.Y., Suganthan, P.N.: Problem Definitions and Evaluation Criteria for the CEC 2017 Special Session and Competition on Single Objective Bound Constrained Real-Parameter Numerical Optimization, Technical report. Nanyang Technological University, Singapore (2016)
11. Faris, H., Aljarah, I., Mirjalili, S., Castillo, P., Merelo, J.: EvoloPy: an open-source nature-inspired optimization framework in Python. In: Proceedings of the 8th International Joint Conference on Computational Intelligence (IJCCI 2016), vol. 1, pp. 171–177. ECTA (2016)
12. Berman, R.: American Bison (Nature Watch). Lerner Publications, Minneapolis (2008)

Distance vs. Improvement Based Parameter Adaptation in SHADE

Adam Viktorin(✉), Roman Senkerik, Michal Pluhacek, and Tomas Kadavy

Faculty of Applied Informatics, Tomas Bata University in Zlin, T. G. Masaryka 5555, 760 01 Zlin, Czech Republic
{aviktorin, senkerik, pluhacek, kadavy}@utb.cz

Abstract. This work studied a relationship between optimization qualities of Success-History based Adaptive Differential Evolution algorithm (SHADE) and its self-adaptive parameter strategy. Original SHADE with improvement based adaptation is compared to the SHADE with Distance based parameter adaptation (Db_SHADE) on the basis of the CEC2015 benchmark set for continuous optimization and a novel approach combining both distance and improvement adaptation (DIb_SHADE) is presented and tested as a trade-off between both approaches.

Keywords: Differential evolution · SHADE · Db_SHADE · DIb_SHADE Parameter adaptation

1 Introduction

The Differential Evolution (DE) is a heuristic algorithm developed for solving numerical optimization problems. It was created by Storn and Price in 1995 [1] and since then it has formed a basis for many successful heuristic optimization algorithms [2–4]. For example, last four CEC competitions in numerical optimization were dominated by DE-based algorithms: CEC2014 – L-SHADE [5], CEC2015 – SPS-L-SHADE-EIG [6], CEC2016 – LSHADE_EpSin [7], CEC2017 – jSO [8]. A common denominator of these four algorithms is a self-adaptive variant of the DE, which was developed by Tanabe and Fukunaga in 2013 and was titled Success-History based Adaptive Differential Evolution (SHADE) [9]. This algorithm competed in CEC2013 and ended up 3rd as a first participant from the DE family.

One of the biggest problems of SHADE is its premature convergence to local extremes in higher dimensional decision spaces. This issue was addressed in [10] by exchanging the weights in an original adaptive strategy. Weights in SHADE are based on the relative improvement in objective function value between generations and promote the exploitation of the decision space. Weights in SHADE with Distance based parameter adaptation (Db_SHADE) are based on the distance between individuals from successive generations, and therefore promote exploration. In this paper, a combination of both approaches is studied in order to determine whether a trade-off between distance and improvement based adaptations is the appropriate balance between exploration and exploitation abilities. The novel approach presented here was

R. Silhavy (Ed.): CSOC 2018, AISC 764, pp. 455–464, 2019.
https://doi.org/10.1007/978-3-319-91189-2_45

titled Distance/Improvement based parameter adaptation and hence the abbreviation DIb_SHADE.

The structure of the remaining sections is as follows: Sect. 2 describes DE, SHADE, Db_SHADE and DIb_SHADE algorithms, Sect. 3 provides the experimental setting and results and the paper is concluded in Sect. 4.

2 DE, SHADE, Db_SHADE and DIb_SHADE

In order to describe the Success-History based Adaptive Differential Evolution algorithm (SHADE) and its Distance based variants (Db_SHADE and DIb_SHADE), it is important to start from the canonical Differential Evolution (DE) by Storn and Price [1].

The canonical 1995 DE is based on the idea of evolution from a randomly generated set of solutions of the optimization task called population $\boldsymbol{P}$, which has a preset size of NP. Each individual (solution) in the population consists of a vector $\boldsymbol{x}$ of length D (each vector component corresponds to one attribute of the optimized task) and objective function value $f(\boldsymbol{x})$, which mirrors the quality of the solution. The number of optimized attributes D is often referred to as the dimensionality of the problem and such generated population $\boldsymbol{P}$, represent the first generation of solutions.

The individuals in the population are combined in an evolutionary manner in order to create improved offspring for the next generation. This process is repeated until the stopping criterion is met (either the maximum number of generations, or the maximum number of objective function evaluations, or the population diversity lower limit, or overall computational time), creating a chain of subsequent generations, where each following generation consists of better solutions than those in previous generations – a phenomenon called elitism.

The combination of individuals in the population consists of three main steps: Mutation, crossover and selection.

In the mutation, attribute vectors of selected individuals are combined in simple vector operations to produce a mutated vector $\boldsymbol{v}$. This operation uses a control parameter – scaling factor F. In the crossover step, a trial vector $\boldsymbol{u}$ is created by selection of attributes either from mutated vector $\boldsymbol{v}$ or the original vector $\boldsymbol{x}$ based on the crossover probability given by a control parameter – crossover rate CR. And finally, in the selection, the quality $f(\boldsymbol{u})$ of a trial vector is evaluated by an objective function and compared to the quality $f(\boldsymbol{x})$ of the original vector and the better one is placed into the next generation.

From the basic description of the DE algorithm, it can be seen, that there are three control parameters, which have to be set by the user – population size NP, scaling factor F and crossover rate CR. It was shown in [11, 12], that the setting of these parameters is crucial for the performance of DE. Fine-tuning of the control parameter values is a time-consuming task and therefore, many state-of-the-art DE variants use self-adaptation in order to avoid this cumbersome task. Which is also a case of SHADE algorithm proposed by Tanabe and Fukunaga in 2013 [9] and since it is used in this paper, the algorithm is described in more detail in the next section along with the novel distance based parameter adaptation.

2.1 Shade

As aforementioned, SHADE algorithm was proposed with a self-adaptive mechanism of some of its control parameters in order to avoid their fine-tuning. Control parameters in question are scaling factor F and crossover rate CR. It is fair to mention, that SHADE algorithm is based on Zhang and Sanderson's JADE [13] and shares a lot of its mechanisms. The main difference is in the historical memories $\boldsymbol{M}_F$ and $\boldsymbol{M}_{CR}$ for successful scaling factor and crossover rate values with their update mechanism.

Following subsections describe individual steps of the SHADE algorithm: Initialization, mutation, crossover, selection and historical memory update.

Initialization

The initial population $\boldsymbol{P}$ is generated randomly and for that matter, a Pseudo-Random Number Generator (PRNG) with uniform distribution is used. Solution vectors $\boldsymbol{x}$ are generated according to the limits of solution space – *lower* and *upper* bounds (1).

$$\boldsymbol{x}_{j,i} = U\left[lower_j, upper_j\right] \text{ for } j = 1, \ldots, D;\ i = 1, \ldots, NP, \tag{1}$$

where i is the individual index and j is the attribute index. The dimensionality of the problem is represented by D, and NP stands for the population size.

Historical memories are preset to contain only 0.5 values for both, scaling factor and crossover rate parameters (2).

$$M_{CR,i} = M_{F,i} = 0.5 \text{ for } i = 1, \ldots, H, \tag{2}$$

where H is a user-defined size of historical memories.

Also, the external archive of inferior solutions $\boldsymbol{A}$ has to be initialized. Because of no previous inferior solutions, it is initialized empty, $\boldsymbol{A}$ = Ø. And index k for historical memory updates is initialized to 1.

The following steps are repeated over the generations until the stopping criterion is met.

Mutation

Mutation strategy "current-to-*p*best/1" was introduced in [13] and it combines four mutually different vectors in a creation of the mutated vector $\boldsymbol{v}$. Therefore, $\boldsymbol{x}_{pbest} \neq \boldsymbol{x}_{r1} \neq \boldsymbol{x}_{r2} \neq \boldsymbol{x}_i$ (3).

$$\boldsymbol{v}_i = \boldsymbol{x}_i + F_i\left(\boldsymbol{x}_{pbest} - \boldsymbol{x}_i\right) + F_i(\boldsymbol{x}_{r1} - \boldsymbol{x}_{r2}), \tag{3}$$

where $\boldsymbol{x}_{pbest}$ is randomly selected individual from the best $NP \times p$ individuals in the current population. The p value is randomly generated for each mutation by PRNG with uniform distribution from the range [p_{min}, 0.2] and p_{min} = $2/NP$. Vector $\boldsymbol{x}_{r1}$ is randomly selected from the current population $\boldsymbol{P}$. Vector $\boldsymbol{x}_{r2}$ is randomly selected from the union of the current population $\boldsymbol{P}$ and external archive $\boldsymbol{A}$. The scaling factor value F_i is given by (4).

$$F_i = C\left[M_{F,r}, 0.1\right], \tag{4}$$

where $M_{F,r}$ is a randomly selected value (index r is generated by PRNG from the range 1 to H) from $\boldsymbol{M}_F$ memory and C stands for Cauchy distribution. Therefore the F_i value is generated from the Cauchy distribution with location parameter value $M_{F,r}$ and scale parameter value of 0.1. If the generated value F_i higher than 1, it is truncated to 1 and if it is F_i less or equal to 0, it is generated again by (4).

Crossover

In the crossover step, trial vector $\boldsymbol{u}$ is created from the mutated $\boldsymbol{v}$ and original $\boldsymbol{x}$ vectors. For each vector component, a PRNG with uniform distribution is used to generate a random value. If this random value is less or equal to given crossover rate value CR_i, current vector component will be taken from a trial vector, otherwise, it will be taken from the original vector (5). There is also a safety measure, which ensures, that at least one vector component will be taken from the trial vector. This is given by a randomly generated component index j_{rand}.

$$u_{j,i} = \begin{cases} v_{j,i} & \text{if } U[0,1] \le CR_i \text{ or } j = j_{rand} \\ x_{j,i} & \text{otherwise} \end{cases}. \tag{5}$$

The crossover rate value CR_i is generated from a Gaussian distribution with a mean parameter value $M_{CR,r}$ selected from the crossover rate historical memory $\boldsymbol{M}_{CR}$ by the same index r as in the scaling factor case and standard deviation value of 0.1 (6).

$$CR_i = N\left[M_{CR,r}, 0.1\right]. \tag{6}$$

When the generated CR_i value is less than 0, it is replaced by 0 and when it is greater than 1, it is replaced by 1.

Selection

The selection step ensures, that the optimization will progress towards better solutions because it allows only individuals of better or at least equal objective function value to proceed into the next generation *G+1* (7).

$$\boldsymbol{x}_{i,G+1} = \begin{cases} \boldsymbol{u}_{i,G} & \text{if } f(\boldsymbol{u}_{i,G}) \le f(\boldsymbol{x}_{i,G}) \\ \boldsymbol{x}_{i,G} & \text{otherwise} \end{cases}, \tag{7}$$

where G is the index of the current generation.

Historical Memory Updates

Historical memories $\boldsymbol{M}_F$ and $\boldsymbol{M}_{CR}$ are initialized according to (2), but their components change during the evolution. These memories serve to hold successful values of F and CR used in mutation and crossover steps. Successful in terms of producing trial individual better than the original individual. During every single generation, these successful values are stored in their corresponding arrays $\boldsymbol{S}_F$ and $\boldsymbol{S}_{CR}$. After each generation, one cell of $\boldsymbol{M}_F$ and $\boldsymbol{M}_{CR}$ memories is updated. This cell is given by the index k, which starts at 1 and increases by 1 after each generation. When it overflows the memory size H, it is reset to 1. The new value of k-th cell for $\boldsymbol{M}_F$ is calculated by (8) and for $\boldsymbol{M}_{CR}$ by (9).

$$M_{F,k} = \begin{cases} \text{mean}_{WL}(\boldsymbol{S}_F) & \text{if } \boldsymbol{S}_F \neq \emptyset \\ M_{F,k} & \text{otherwise} \end{cases}, \tag{8}$$

$$M_{CR,k} = \begin{cases} \text{mean}_{WL}(\boldsymbol{S}_{CR}) & \text{if } \boldsymbol{S}_{CR} \neq \emptyset \\ M_{CR,k} & \text{otherwise} \end{cases}, \tag{9}$$

where $\text{mean}_{WL}()$ stands for weighted Lehmer (10) mean.

$$\text{mean}_{WL}(\boldsymbol{S}) = \frac{\sum_{k=1}^{|S|} w_k \cdot S_k^2}{\sum_{k=1}^{|S|} w_k \cdot S_k}, \tag{10}$$

where the weight vector $\boldsymbol{w}$ is given by (11) and is based on the improvement in objective function value between trial and original individuals in current generation G.

$$w_k = \frac{\text{abs}\left(f\left(\boldsymbol{u}_{k,G}\right) - f\left(\boldsymbol{x}_{k,G}\right)\right)}{\sum_{m=1}^{|S_{CR}|} \text{abs}\left(f\left(\boldsymbol{u}_{m,G}\right) - f\left(\boldsymbol{x}_{m,G}\right)\right)}. \tag{11}$$

And since both arrays $\boldsymbol{S}_F$ and $\boldsymbol{S}_{CR}$ have the same size, it is arbitrary which size will be used for the upper boundary for m in (11).

The last Eq. (11) is the subject of change in the novel Db_SHADE algorithm, which is described in the next section.

2.2 Db_SHADE

The original adaptation mechanism for scaling factor and crossover rate values uses weighted forms of means (10), where weights are based on the improvement in objective function value (11). This approach promotes exploitation over exploration and therefore might lead to premature convergence, which could be a problem especially in higher dimensions.

Distance approach is based on the Euclidean distance between the trial and the original individual, which slightly increases the complexity of the algorithm by exchanging simple difference for Euclidean distance computation for the price of stronger exploration. In this case, scaling factor and crossover rate values connected with the individual that moved the furthest will have the highest weight (12).

$$w_k = \frac{\sqrt{\sum_{j=1}^{D}\left(u_{k,j,G} - x_{k,j,G}\right)^2}}{\sum_{m=1}^{|S_{CR}|}\sqrt{\sum_{j=1}^{D}\left(u_{m,j,G} - x_{m,j,G}\right)^2}}. \tag{12}$$

Therefore, the exploration ability is rewarded and this should lead to avoidance of the premature convergence in higher dimensional objective spaces. Such approach might be also useful for constrained problems, where constrained areas could be overcome by increased changes of individual's components.

Below is the pseudo-code of the Db_SHADE algorithm for a clear overview.

Algorithm pseudo-code 2: Db_SHADE

1. Set NP, H and stopping criterion;
2. $G = 0$, $x_{best} = \{\}$, $k = 1$, $p_{min} = 2/NP$, $\mathbf{A} = \emptyset$;
3. Randomly initialize (1) population $\mathbf{P} = (\mathbf{x}_{1,G}, \ldots, \mathbf{x}_{NP,G})$;
4. Set $\mathbf{M}_F$ and $\mathbf{M}_{CR}$ according to (2);
5. $\mathbf{P}_{new} = \{\}$, $\mathbf{x}_{best}$ = best from population $\mathbf{P}$;
6. **while** stopping criterion not met
7. $\quad\mathbf{S}_F = \emptyset$, $\mathbf{S}_{CR} = \emptyset$;
8. $\quad$**for** $i = 1$ to NP **do**
9. $\quad\quad\mathbf{x}_{i,G} = \mathbf{P}[i]$;
10. $\quad\quad r = U[1, H]$, $p_i = U[p_{min}, 0.2]$;
11. $\quad\quad$Set F_i by (4) and CR_i by (6);
12. $\quad\quad\mathbf{v}_{i,G}$ by mutation (3);
13. $\quad\quad\mathbf{u}_{i,G}$ by crossover (5);
14. $\quad\quad$**if** $f(\mathbf{u}_{i,G}) < f(\mathbf{x}_{i,G})$ **then**
15. $\quad\quad\quad\mathbf{x}_{i,G+1} = \mathbf{u}_{i,G}$;
16. $\quad\quad\quad\mathbf{x}_{i,G} \rightarrow \mathbf{A}$;
17. $\quad\quad\quad F_i \rightarrow \mathbf{S}_F$, $CR_i \rightarrow \mathbf{S}_{CR}$;
18. $\quad\quad$**else**
19. $\quad\quad\quad\mathbf{x}_{i,G+1} = \mathbf{x}_{i,G}$;
20. $\quad\quad$**end**
21. $\quad\quad$**if** $|\mathbf{A}| > NP$ **then** randomly delete individuals from $\mathbf{A}$ **end**;
22. $\quad\quad\mathbf{x}_{i,G+1} \rightarrow \mathbf{P}_{new}$;
23. $\quad$**end**
24. $\quad$**if** $\mathbf{S}_F \neq \emptyset$ **and** $\mathbf{S}_{CR} \neq \emptyset$ **then**
25. $\quad\quad$Update $\mathbf{M}_{F,k}$ (8) and $\mathbf{M}_{CR,k}$ (9) with distance based weights from (12), k++;
26. $\quad\quad$**if** $k > H$ **then** $k = 1$, **end**;
27. $\quad$**end**
28. $\quad\mathbf{P} = \mathbf{P}_{new}$, $\mathbf{P}_{new} = \{\}$, $\mathbf{x}_{best}$ = best from population $\mathbf{P}$;
29. **end**
30. **return** $\mathbf{x}_{best}$ as the best found solution

2.3 DIb_SHADE

The DIb_SHADE algorithm is a trade-off between simple improvement based SHADE and distance based Db_SHADE. While Db_SHADE neglects the improvement in objective function value and works only with distance between solutions when it calculates parameter weights (12), DIb_SHADE combines both approaches. The resulting weight $\boldsymbol{w}$ for the parameter combination is computed as a sum (13) of distance based weight $\boldsymbol{w}_d$ (11) and improvement based weight $\boldsymbol{w}_i$ (12).

$$\boldsymbol{w} = \boldsymbol{w}_d + \boldsymbol{w}_i. \tag{13}$$

Thus, this approach combines both explorative and exploitative weights in order to balance those two characteristics.

3 Experimental Setting and Results

All three versions of SHADE with different parameter adaptation schemes (SHADE, Db_SHADE and DIb_SHADE) were tested on the CEC2015 benchmark set of 15 test functions in 30*D* because the results from [10] shown significant differences in higher dimensional spaces. Algorithms were run with following parameters:

- Populations size $NP = 100$.
- Historical memory size $H = 10$.
- External archive size $|\boldsymbol{A}| = NP$.
- Stopping criterion – maximum number of function evaluations $MAXFES = 10{,}000 \times D = 300{,}000$.
- Number of runs $r = 51$.

The obtained objective function values were subjected to the basic descriptive statistic and also compared via Wilcoxon rank-sum and Friedman rank tests. Table 1 contains mean results from the 51 runs and a column with a statistically significant winner selected by Wilcoxon rank-sum test with significance level of 5%. Winning algorithm is the one that performed significantly better than all others in pairwise comparisons.

The results presented in Table 1 show that the optimization performance of the Db_SHADE algorithm is still significantly superior to that of SHADE and DIb_-SHADE algorithms on three test functions – *f6*, *f7* and *f11*.

Friedman rank test provides a nice overview of the average ranking of algorithms over the 15 test functions. The ranking is provided in Fig. 1.

While Friedman rank test p-value was 0.089 and therefore, according to this test, there is no significant difference between tested algorithms on the significance level of 5%, there is a visible pattern in ranking which decreases in favor of distance based parameter adaptation. Thus, it can be concluded, that the explorative factor in weight definitions of the adaptive scheme is more beneficial to the overall performance and there is no need for balancing distance and improvement factors. This could be due to the fact, that the exploitative ability is promoted in the mutation step via $\boldsymbol{x}_{pbest}$ individual and therefore there is no need to promote it even further by adapting parameters according to the objective function improvement.

Table 1. SHADE, DIb_SHADE and Db_SHADE on CEC2015 in 30*D*.

f	SHADE	DIb_SHADE	Db_SHADE	Winner
	Mean	Mean	Mean	
1	2.62E+02	2.82E+02	2.42E+02	-
2	0.00E+00	0.00E+00	0.00E+00	-
3	2.01E+01	2.01E+01	2.01E+01	-
4	1.41E+01	1.38E+01	1.31E+01	-
5	1.50E+03	1.52E+03	1.52E+03	-
6	5.73E+02	4.80E+02	3.48E+02	Db_SHADE
7	7.26E+00	7.17E+00	6.74E+00	Db_SHADE
8	1.21E+02	1.05E+02	7.38E+01	-
9	1.03E+02	1.03E+02	1.03E+02	-
10	6.22E+02	5.72E+02	5.32E+02	-
11	4.50E+02	4.31E+02	4.16E+02	Db_SHADE
12	1.05E+02	1.05E+02	1.05E+02	-
13	9.50E+01	9.45E+01	9.50E+01	-
14	3.24E+04	3.23E+04	3.24E+04	-
15	1.00E+02	1.00E+02	1.00E+02	-

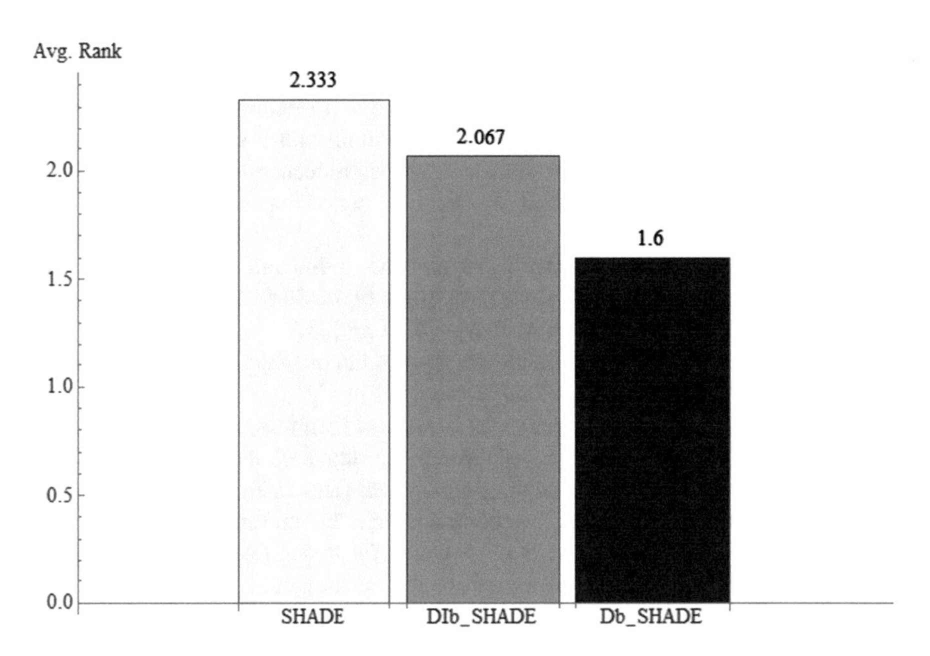

Fig. 1. Friedman ranking of SHADE, DIb_SHADE and Db_SHADE on CEC2015 in 30*D*.

4 Conclusion

This paper studied an influence of distance and improvement based parameter adaptation to the performance of SHADE algorithm on the basis of CEC2015 benchmark set of 15 test functions in $30D$. In addition, a novel adaptation scheme that combines both distance and improvement was presented in order to find the optimal trade-off between explorative and exploitative factors. However, the results showed that the exploitative ability is still very strong in the mutation step and therefore, the distance based parameter adaptation is more beneficial for the overall performance. These findings support the idea, that there is still a lot of room for improvement in balancing between exploitation and exploration of SHADE optimization algorithm. This will also be authors' future research direction in the DE field.

Acknowledgements. This work was supported by the Ministry of Education, Youth and Sports of the Czech Republic within the National Sustainability Programme Project no. LO1303 (MSMT-7778/2014), further by the European Regional Development Fund under the Project CEBIA-Tech no. CZ.1.05/2.1.00/03.0089 and by Internal Grant Agency of Tomas Bata University under the Projects no. IGA/CebiaTech/2018/003. This work is also based upon support by COST (European Cooperation in Science & Technology) under Action CA15140, Improving Applicability of Nature-Inspired Optimisation by Joining Theory and Practice (ImAppNIO), and Action IC1406, High-Performance Modelling and Simulation for Big Data Applications (cHiPSet). The work was further supported by resources of A.I.Lab at the Faculty of Applied Informatics, Tomas Bata University in Zlin (ailab.fai.utb.cz).

References

1. Storn, R., Price, K.: Differential evolution-a simple and efficient adaptive scheme for global optimization over continuous spaces, vol. 3. ICSI, Berkeley (1995)
2. Neri, F., Tirronen, V.: Recent advances in differential evolution: a survey and experimental analysis. Artif. Intell. Rev. **33**(1–2), 61–106 (2010)
3. Das, S., Suganthan, P.N.: Differential evolution: a survey of the state-of-the-art. IEEE Trans. Evol. Comput. **15**(1), 4–31 (2011)
4. Das, S., Mullick, S.S., Suganthan, P.N.: Recent advances in differential evolution–an updated survey. Swarm Evol. Comput. **27**, 1–30 (2016)
5. Tanabe, R., Fukunaga, A.S.: Improving the search performance of SHADE using linear population size reduction. In: 2014 IEEE Congress on Evolutionary Computation (CEC), pp. 1658–1665. IEEE, July 2014
6. Guo, S.M., Tsai, J.S.H., Yang, C.C., Hsu, P.H.: A self-optimization approach for L-SHADE incorporated with eigenvector-based crossover and successful-parent-selecting framework on CEC 2015 benchmark set. In: 2015 IEEE Congress on Evolutionary Computation (CEC), pp. 1003–1010. IEEE, May 2015
7. Awad, N.H., Ali, M.Z., Suganthan, P.N., Reynolds, R.G.: An ensemble sinusoidal parameter adaptation incorporated with L-SHADE for solving CEC2014 benchmark problems. In: 2016 IEEE Congress on Evolutionary Computation (CEC), pp. 2958–2965. IEEE, July 2016
8. Brest, J., Maučec, M.S., Bošković, B.: Single objective real-parameter optimization: algorithm jSO. In: 2017 IEEE Congress on Evolutionary Computation (CEC), pp. 1311–1318. IEEE, June 2017

9. Tanabe, R., Fukunaga, A.: Success-history based parameter adaptation for differential evolution. In: 2013 IEEE Congress on Evolutionary Computation (CEC), pp. 71–78. IEEE, June 2013
10. Viktorin, A., Senkerik, R., Pluhacek, M., Kadavy, T., Zamuda, A.: Distance based parameter adaptation for differential evolution. In: 2017 IEEE Symposium Series on Computational Intelligence (SSCI), pp. 1–7 IEEE (2017)
11. Gämperle, R., Müller, S.D., Koumoutsakos, P.: A parameter study for differential evolution. In: Advances in Intelligent Systems, Fuzzy Systems, Evolutionary Computation, vol. 10, pp. 293–298 (2002)
12. Liu, J., Lampinen, J.: On setting the control parameter of the differential evolution method. In: Proceedings of the 8th International Conference on Soft Computing (MENDEL 2002), pp. 11–18 (2002)
13. Zhang, J., Sanderson, A.C.: JADE: adaptive differential evolution with optional external archive. IEEE Trans. Evol. Comput. **13**(5), 945–958 (2009)

Dogface Detection and Localization of Dogface's Landmarks

Alzbeta Vlachynska(✉), Zuzana Kominkova Oplatkova, and Tomas Turecek

Faculty of Applied Informatics, Tomas Bata University in Zlin, Nam T.G. Masaryka 5555, 760 01 Zlin, Czech Republic
{vlachynska, oplatkova, tturecek}@utb.cz

Abstract. The paper deals with an approach for a reliable dogface detection in an image using the convolutional neural networks. Two detectors were trained on a dataset containing 8351 real-world images of different dog breeds. The first detector achieved the average precision equal to 0.79 while running real-time on single CPU, the second one achieved the average precision equal to 0.98 but more time for processing is necessary. Consequently, the facial landmark detector using the cascade of regressors was proposed based on those, which are commonly used in human face detection. The proposed algorithm is able to detect dog's eyes, a muzzle, a top of the head and inner bases of the ears with the 0.05 median location error normalized by the inter-ocular distance. The proposed two-step technique – a dogface detection with following facial landmark detector - could be utilized for a dog breeds identification and consequent auto-tagging and image searches. The paper demonstrates a real-world application of the proposed technique – a successful supporting system for taking pictures of dogs facing the camera.

Keywords: Convolutional neural networks · Dogface · Landmark detection

1 Introduction

The development of convolutional neural networks (CNNs) boosted massive progress in object detection in recent years [1]. Among other things, the improvement in performance was caused by: (i) the availability of large training sets with millions of labeled examples; (ii) powerful implementations using the graphics processing unit (GPU) computations, enabling the training of very large models to be realistic; and (iii) development of new model regularization strategies, such as dropout [2] or dense-sparse-dense training [3]. One of the main advantages of deep learning techniques is the possibility to use pre-trained checkpoints[1] as a starting point for custom training on new objects.

---

[1] The models are pre-trained on big public datasets and are available on-line. For example, the list of checkpoints provided by Tensorflow Object Detection API can be found at https://github.com/tensorflow/models/blob/master/research/object_detection/g3doc/detection_model_zoo.md.

R. Silhavy (Ed.): CSOC 2018, AISC 764, pp. 465–476, 2019.
https://doi.org/10.1007/978-3-319-91189-2_46

This paper presents the applicability of CNN detection on the special dataset with a rich inter-variability of objects – the dataset containing 8351 real-world images of 133 different dog breeds[2] [4]. As was claimed in [5], the domestic dog (*Canis lupus familiaris*) displays greater levels of morphological and behavioral diversity than have been recorded for any other land mammal. The dog's diversity in its visual appearance might present significant challenges to object detection. On the other hand, dogs are probably the most photographed species apart from humans. Millions of dog images with nearly infinite variety can be easily obtained from image search engines, showing dogs of all sizes, colors, and shapes; in different poses; under different illumination; and in diverse places.

The dogface detection was carried out because the dogface is usually in the camera view. Moreover, the dogface contains some common features which make the detection easier than the detection of a whole dog body. Successful dogface detection and its following analysis, i.e., landmarks detection may show the possible way of an image investigation, which can be consequently used in the broader domain of automatic species identification. Dogface detection may also be useful for a dog breed identification, consequent auto-tagging, and image searches. The effective detection of a dogface might be the crucial step resulting in a successful segmentation of the complete animal body.

Being pet lovers, the authors of this paper do recognize the troubles with making a good photo of a dog. The dogs typically do not look directly into the camera and quite frequently do turn the head away from the camera. Therefore, in order to prevent the frustration with numerous failed attempts, a successful supporting technique for taking pictures of a dog facing the camera is very helpful.

We see the novel contribution of our paper in the following points:

- The paper presents a real-world application of object detection. We demonstrate that when the training is adapted appropriately to the detector purpose, the deep neural network could obtain even higher average precision than was presented in the original papers describing net architectures and testing them on standard datasets with diverse objects categories (i.e., COCO dataset[3] or VOC2012 dataset[4]).
- We also provide evidence that the object detection could be realized in real time with enough reliability using only single CPU computation. With GPU computation, it is possible to obtain results that are even more precise.
- The real world application of the proposed technique is able to take a picture of dogs placed in the camera view at the moment when they are all looking into the camera.

[2] The dataset is available at http://faceserv.cs.columbia.edu/DogData/.

[3] COCO dataset – Common Objects in Context, available online at http://cocodataset.org/.

[4] VOC2012 dataset – Visual Object Classes Challenge 2012, available online at http://host.robots.ox.ac.uk/pascal/VOC/voc2012/index.html.

2 Related Work

The paper [4] is especially relevant to our work. Liu et al. present a dog breed classifier which utilizes the support vector machine based detector for dogfaces, and the consensus of models approach to localize parts of the dogface. From the precision recall curve, we can estimate the average precision of the proposed dogface detection to be 0.79. Unfortunately, these authors did not make a comment about the speed of the detection. We can only estimate that the detection is not done in the real-time.

Regarding the real-time detection, there has been a great deal of interest in methods focused on human faces. A cascaded AdaBoost classifier with Haar and Haar-like wavelets have been mostly used for this task [6, 7]. The paper [4] shows that even the cascade detectors are sufficiently effective and successful in finding human faces, it struggles quite a lot in the case of the dogface detection and produces unwanted false positive detections. This is presumably caused by greater variation in geometry and appearance of dogface in comparison with a human face. The authors of [4] implemented a cascaded AdaBoost model for dogface detection, its average precision, 0.63, can be estimated from the precision recall curve.

Indeed, authors have often focused on cats and dogs as examples of highly deformable objects for which recognition and detection are particularly challenging [8–10]. The authors of [8, 9] extend template-based detector built on the deformable parts model by combining the low-level image features of histogram oriented gradients (capturing shape) and local binary patterns (capturing texture). This model achieves the average precision 0.61 for the dog/cat head detector. The paper [10] presents similar two-step approach as [8] for the cathead detection, which gains advantages from the use of different sets of features based on histogram oriented gradients, and achieves similar average precision 0.63.

3 Dogface Detection

The overall proposed system consists of two standalone parts; the first module serves for the dogface detection (Fig. 5b) and the second one for the dogface's landmarks localization (i.e., eyes, a muzzle, etc., see (Fig. 5c)). This section discusses the dogface detection, whereas the dogface's landmark localization is described in Sect. 4.

We were interested in the object detection based on the deep convolutional networks, specifically dogface detection. For this purpose, two different net designs were utilized: (i) Single Shot Detector (SSD) [11] with MobileNet v1 feature extractor [12], and (ii) faster Region-based convolutional neural network (R-CNN) [14] with feature extractor Resnet 101 [15]. Both models were implemented and trained via Tensorflow Object Detection API (application programming interface) [1]. To speed up the process, the training of both nets was initialized by checkpoints provided by Tensorflow Object Detection API (see footnote 1), concretely by a model trained on COCO dataset (see footnote 3). Following subsections are focused on a particular configuration of each model architecture in detail.

3.1 Single Shot Detector (SSD)

In this paper, the term Single Shot Detector (SSD) is used for the architecture consisting of a single feed-forward convolutional network that directly predicts classes and an anchor offsets without the need of a second stage per-proposal classification operation. The network architecture consists of a feed-forward convolutional network that produces a fixed-size collection of proposals and classification scores for these proposals, followed by a non-maximum suppression step that outputs the final detections [11]. For an illustration of an SSD network architecture, please refer to Fig. 1.

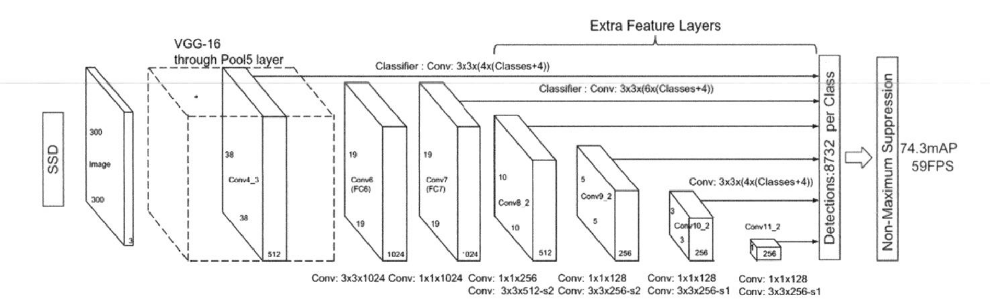

Fig. 1. The architecture of a Single Shot Detector (SSD) network. For more information about the model, please refer to [11].

The original architecture of SSD described in the main paper [11] was followed closely in the Tensorflow's implementation. Anchors were generated in the same way, selecting the topmost convolutional feature map and a higher resolution feature map at a net lower level. A sequence of convolutional layers with spatial resolution decaying by a factor of two follows each of these additional layers is used for prediction. In addition to the original implementation, we used a dropout. A principle of this technique is described in [2]. The setting of the dropout probability was kept at 0.8 during the training. The batch size was set up to 20 images. The data augmentation schemes were utilized according to the description in the most recent version of [11] on arXiv[5] to obtain better results; concretely the random horizontal flip and the SSD random cop were used. The specifics of Mobile Net feature extractor [12] follow.

MobileNet [12]. Following the paper, Tensorflow's implementation Mobile Net feature extractor use depthwise separable convolution layers, appending four additional convolutional layers with decaying resolution with depths 512, 256, 256, 128 respectively. Rectified linear 6 (ReLU6)[6] was utilized as the non-linear activation function for each convolution layer. During training, a base-learning rate was set up to 0.004, followed by learning rate decay of 0.95 every 800 thousand steps.

[5] arXiv is an e-print service operated by Cornell University. It can be reached via website https://arxiv.org/.

[6] The ReLU6 activation function counts min(max(features, 0), 6), for more information about this technique please see the manuscript [13].

3.2 Faster R-CNN

The object detection system called Faster R-CNN (Region-based convolutional neural net) combines two modules. The first module is a deep fully convolutional network for region proposing and the second module is the Fast R-CNN detector that process and classifies these regions into object categories or background [14]. The schematic view of the combined system can be seen in Fig. 2.

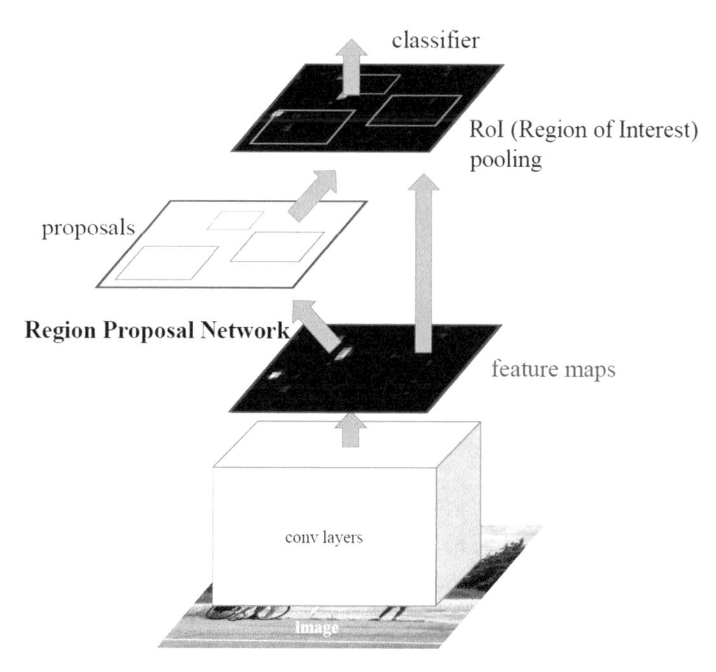

Fig. 2. Faster R-CNN unifies two modules into one network for object detection; the first module serves for predicting proposals, the second for classification. For more information about the model, please refer to [14].

The Tensorflow's implementation follows the original description of Faster R-CNN [14] closely, but it uses Tensorflow's "crop_and_resize" operation instead of standard RoI (Region of Interest) pooling. The batch normalization was used after each convolution layer. Because of the memory requirements, the batch size during training was set to 1. As in the case of SSD, in addition to the original Faster R-CNN implementation dropout was used and its probability was kept at 0.8 during the training. The initial learning rate was 0.0003, and the learning rate was reduced by 10 times after 900 thousand iterations and another 10 times after 1.2 million iterations. The utilized feature extractor Resnet 101 [15] is explained below.

Resnet 101 [15]. Tensorflow's implementation follows the description in the original paper and extracts features from the last layer of the convolutional block, in the manuscript [15] named as "conv4 block". The stride size is set to 8 pixels in atrous-mode[7] and to 16 pixels in all other cases. Position-sensitive score maps are cropped and resized to 14×14 then max-pooled to 7×7. The initial learning rate was 0.0003, and this value was kept during the whole training.

3.3 Dataset Preparation for Dogface Detection

The used dataset for training is a copy of the Columbia Dogs Dataset (see footnote 2), which was introduced in [4]. The dataset was revised, and some additional bounding boxes were added as well as some very loose ones were fixed. Afterwards, we split the dataset into two parts for the purpose of experiments. The dataset contains 8351 images with 9240 dogfaces marked in all of them. Randomly chosen 6700 images ($\approx$ 80.2%) was used for training, and the rest 1651 images was left for later evaluation. This way, the training dataset contains 7421 dogfaces ($\approx$ 80.3%) and the second part of images, left for evaluation (testing), contains 1819 dogfaces. Samples of different dog breeds images from the dataset are shown in Fig. 3.

Fig. 3. Samples of images from the dataset.

3.4 Evaluation of Dogface Detectors

We have trained two different versions of dogface detector, each of them using different feature extractor and net design, each of them meant to represent different

[7] atrous convolution, also known as convolution with holes or dilated convolution, based on the French word "trous" meaning holes in English. For a description of atrous convolution and how it can be used for dense feature extraction, please see [16].

computational complexity. The implementation of the detectors is described in Sects. 3.1 and 3.2. For the purpose of the evaluation both dogface detectors, outputs were computed on images from the testing dataset, and the precision recall curve was recorded (Fig. 4). Precision P (1) is defined as the number of true positives (T_p) over the number of true positives plus the number of false positives (F_p):

$$P = \frac{T_p}{T_p + F_p} \tag{1}$$

Recall R (2) is determined as the number of true positives (T_p) divided by the number of true positives plus the number of false negatives (F_n):

$$R = \frac{T_p}{T_p + F_n} \tag{2}$$

As a true positive detection, we consider the bounding box whose intersection over union (IoU) ratio (3) is at least 0.5 and which is not a duplicate. IoU ratio can be computed as a proportion between intersection and union of areas of a predicted bounding box B_p and a ground truth bounding box B_{gt}, as presents the formula:

$$IoU = \frac{area\left(B_p \cap B_{gt}\right)}{area\left(B_p \cup B_{gt}\right)} \tag{3}$$

Precision recall curve presents the relation between precision and recall recorded for different detector thresholds. Consequently, the average precision of the detector was determined as the area under the precision recall curve (indicated as AP@0.5 in this paper, where a value 0.5 refers to IoU > 0.5). The average precision of both dogface detectors trained in this paper are listed in Table 1.

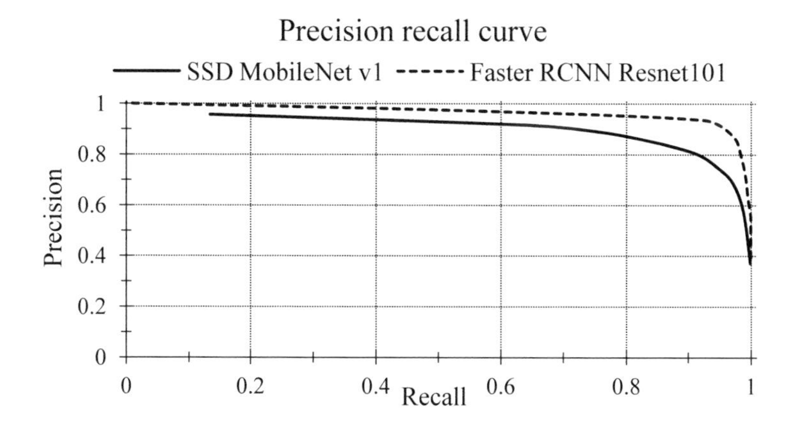

Fig. 4. Precision recall curve for dogface detectors.

An average timing needed for evaluation of one image was recorded too. The timing includes only time consumed to the dogface detection, not for the pre/post-processing of the image. Time evaluation was performed out on a computer with AMD A8-7600 Radeon R7, 10 Compute Cores 4C + 6G 3.10 GHz and RAM memory 4.00 GB using only single CPU computation. Timing results of both dogface detectors can be seen in Table 1.

Table 1. Average precision and timing comparison of dogface detectors.

Net design	AP@0.5	Timing [ms]
SSD MobileNet v1	0.790	225
Faster RCNN Resnet 101	0.980	231

4 Localization of Dogface's Landmarks

One of the reasons we chose the dog head for the detection is its relative rigidity – even though the dogs might be quite visually different, they all have common features: eyes, ears, and a muzzle (nose). The logical next step is to detect the position of these dogface's parts (landmarks).

4.1 Cascade Landmark Detector

A face land-marking model presented in the paper [17] was utilized for the purpose of dogface's landmark detection. Although this model is often used for human face landmarking, it is quite general and can be used for a variety of shape prediction tasks. The implementation of Dlib library[8] was adopted. The most significant advantage of this algorithm is its speed. The runtime complexity of the algorithm on a single image depends linearly on the number of strong and weak regressors, and on the depth of the trees [17]. Practically, one estimation of face landmarks computed on single CPU takes only a few milliseconds.

The method utilizes a cascade of regressors. In the beginning, the rough guess of the desired landmark position is made. Afterwards, each regressor in the cascade predicts an update vector from the image and the current estimate of the landmark position. This updated vector is consequently added to the current landmark position estimate to improve the estimation. If we designate $x_i \in \mathbb{R}^2$ to be the x, y-coordinates of the i-th landmark in an image marked as I, then the vector $S = \left(x_1^T, x_2^T, \ldots, x_p^T\right)^T \in \mathbb{R}^{2p}$ denotes the coordinates of all the p landmarks in the image I. The vector S represents the desired shape and $\widehat{S}^{(t)}$ is the current estimate of S. Let $r_t(.,.)$ be the regressor, then the update of the vector made in each regressor step can be written as:

$$\widehat{S}^{(t+1)} = \widehat{S}^t + r_t\left(I, \widehat{S}^{(t)}\right) \tag{4}$$

[8] For more info about Dlib library visit http://blog.dlib.net/.

The regressor r_t makes its predictions based on image features relative to the current shape estimate $\widehat{S}^{(t)}$, such as pixel intensity values. The initial landmark positions guess is usually chosen simply as the mean landmark positions of the training data, centered and scaled according to the bounding box output. Training of each regressor r_t is accomplished by the gradient tree boosting algorithm with a sum of square error loss as is described in [18].

A facial landmark detector for detection of dog's eyes, a muzzle, and three points at the top of the dog's head were trained. The example of landmarks positions can be seen in the Fig. 5. We did not implement the detection of the tips of dog's ears because its position may vary a lot between different dog breeds. Moreover, the tips of the ears are often far outside the bounding box predicted by a dogface detector. During the training process, explicit details and tips for training described by authors of human face landmark detector [17] were adopted, with an exception of the following adjustments: (i) the number of strong regressors was set to 20 and (ii) the region where the algorithm should search to compare the relative intensity of pixel pairs in the input image was expanded by the 0.2 padding, effectively multiplying the area by 1.44.

4.2 Dataset Preparation for Landmark Detection

The dataset used for training is the same copy of the Columbia Dogs Dataset (see footnote 2), which was introduced in the paper [4] and used for the dogface detection described in Sect. 3. The dataset contains 8351 images with 9240 dogfaces marked in all of them. Three human specialists [4] marked both eyes, a muzzle, tips of both ears, top of the head, and inner bases of ears in each dogface in the dataset. The same dataset splitting to training and testing part was used as in the case of dogface detector training (for more information, please refer the Sect. 3.3).

4.3 Evaluation of Landmark Detector

To evaluate the dogface's landmark detector, the localization error was calculated as the Euclidean distance between the landmark position predicted by the detector and the ground truth position of the landmark. For the purpose of the normalization,

Table 2. Dogface's landmark detector average and median location error normalized by the inter-ocular distance.

Dogface's landmark	Average Loc. Error/Eye Dist.	Median Loc. Error/Eye Dist.
Left eye	0.047	0.032
Right eye	0.047	0.034
Muzzle	0.079	0.039
Inner base of left ear	0.116	0.088
Inner base of right ear	0.116	0.088
Top of the head	0.083	0.058
All dogface's landmarks	0.082	0.051

we divided the error distance by the inter-ocular distance. The average and median localization error are shown in Table 2. The average timing needed to detect dogface's landmarks was also recorded. The timing includes only time consumed to the landmark detection inside an already found dogface bounding box. Time evaluation was carried out on the same computer described in Sect. 3.3. The average time spent by landmark detection was 4.338 ms.

5 Supportive System for Taking Dog Photos

Simple supportive system for taking pictures of dogs facing the camera was created by the combination of proposed dogface detector and consequent landmark localization. The system processes the preview video from the camera (Fig. 5a), detects all dogfaces in the camera view (Fig. 5b) and consequently localizes their facial landmarks (Fig. 5c). The dog gaze direction could be estimated from the detected position of eyes and muzzle. The system compares distances between left and right eye from a muzzle. If the difference does not exceed the defined threshold, we assume that the dog is looking in the direction of the camera lens, so the system takes a photograph. The preliminary results of the system are promising. The development of a more precise method for gaze estimation will be the aim of the future research.

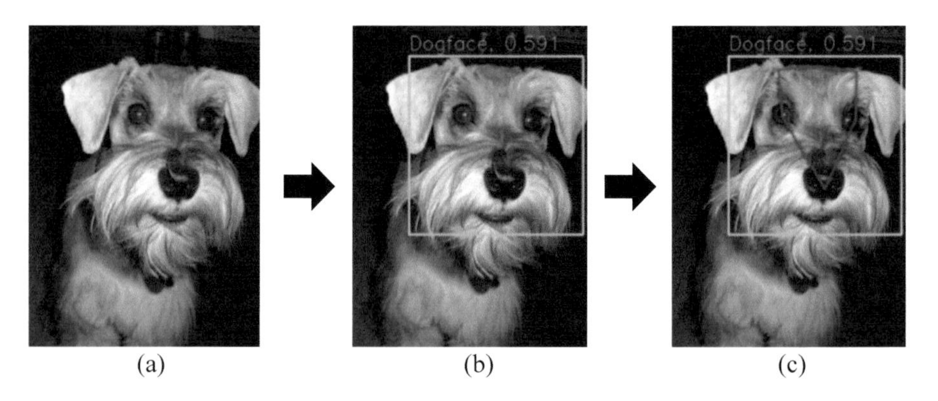

Fig. 5. Sample image evaluation by the proposed system: (a) input image, (b) dogface detection, (c) dogface's landmark detection.

6 Conclusion

This paper dealt with the dogface detection and dogface's landmark localization. Compared to human face detection, the area of the animal face and animal facial landmark detection is not studied so deeply. The proposed technique presents the reliable and fast dogface detector followed by the facial landmark detector finding eyes, a muzzle, a top of the head and inner bases of the ears. We show that the appropriately tuned deep convolutional network trained on the special dataset could be a powerful tool providing enough reliability that could serve in a real-world application. The dogface variability is huge and standard algorithms normally used for a human detection cannot cope with it. Nevertheless, the more robust implementation of convolutional neural net dogface detector, based on the Faster RCNN Resnet101, achieved the average precision equal to 0.98. The land-marking model achieves the median localization error (normalized by the inter-ocular distance) equal to 0.05. The proposed models can find application in an automatic system capable of photographing the dogs facing the camera (as we demonstrated) or could be used for dog breeds identification and consequent auto-tagging and image searches.

Acknowledgements. This work was supported by the Ministry of Education, Youth and Sports of the Czech Republic within the National Sustainability Programme Project no. LO1303 (MSMT-7778/2014), further by the European Regional Development Fund under the Project CEBIA-Tech no. CZ.1.05/2.1.00/03.0089 and by Internal Grant Agency of Tomas Bata University under the Projects no. IGA/CebiaTech/2018/003. This work is also based upon support by COST (European Cooperation in Science & Technology) under Action CA15140, Improving Applicability of Nature-Inspired Optimisation by Joining Theory and Practice (ImAppNIO), and Action IC1406, High-Performance Modelling and Simulation for Big Data Applications (cHiPSet). The work was further supported by resources of A.I. Lab at the Faculty of Applied Informatics, Tomas Bata University in Zlin (ailab.fai.utb.cz).

References

1. Huang, J., Rathod, V., Sun, C., Zhu, M., Korattikara, A., Fathi, A., et al.: Speed/Accuracy trade-offs for modern convolutional object detectors. In: CVPR, vol. abs/1611.10012, pp. 3296–3297. IEEE (2017). https://doi.org/10.1109/cvpr.2017.351
2. Srivastava, N., Hinton, G., Krizhevsky, A., Sutskever, I., Salakhutdinov, R.: Dropout: a simple way to prevent neural networks from overfitting. J. Mach. Learn. Res. **15**, 1929–1958 (2014)
3. Han, S., Pool, J., Narang, S., Mao, H., Gong, E., Tang, S., et al.: DSD: dense-sparse-dense training for deep neural networks. CoRR, vol. abs/1607.04381 (2016)
4. Liu, J., Kanazawa, A., Jacobs, D., Belhumeur, P.: Dog breed classification using part localization. In: Fitzgibbon, A., Lazebnik, S., Perona, P., Sato, Y., Schmid, C. (eds.) ECCV 2012. LNCS, vol. 7572, pp. 172–185. Springer, Heidelberg (2012). https://doi.org/10.1007/978-3-642-33718-5_13
5. Spady, T.C., Ostrander, E.A.: Canine behavioral genetics: pointing out the phenotypes and herding up the genes. Am. J. Hum. Genet. **82**, 10–18 (2008). https://doi.org/10.1016/j.ajhg.2007.12.001
6. Dalal, N., Triggs, B.: Histograms of oriented gradients for human detection. In: 2005 IEEE Computer Society Conference on Computer Vision and Pattern Recognition (CVPR 2005), vol. 1, p. 893, June 2005. https://doi.org/10.1109/cvpr.2005.177
7. Cuimei, L., Zhiliang, Q., Nan, J., Jianhua, W.: Human face detection algorithm via Haar cascade classifier combined with three additional classifiers. In: ICEMI, pp. 483–487. IEEE, (2017). https://doi.org/10.1109/icemi.2017.8265863
8. Parkhi, O.M., Vedaldi, A., Jawahar, C.V., Zisserman, A.: The truth about cats and dogs. In: 2011 International Conference on Computer Vision, pp. 1427–1434, November 2011. https://doi.org/10.1109/iccv.2011.6126398

9. Parkhi, O.M., Vedaldi, A., Zisserman, A., Jawahar, C.V.: Cats and dogs. In: 2012 IEEE Conference on Computer Vision and Pattern Recognition, pp. 3498–3505, June 2012. https://doi.org/10.1109/cvpr.2012.6248092
10. Zhang, W., Sun, J., Tang, X.: Cat head detection - how to effectively exploit shape and texture features. In: Forsyth, D., Torr, P., Zisserman, A. (eds.) ECCV 2008. LNCS, vol. 5305, pp. 802–816. Springer, Heidelberg (2008). https://doi.org/10.1007/978-3-540-88693-8_59
11. Liu, W., Anguelov, D., Erhan, D., Szegedy, C., Reed, S., Fu, C.-Y., Berg, A.C.: SSD: single shot multibox detector. In: Leibe, B., Matas, J., Sebe, N., Welling, M. (eds.) ECCV 2016. LNCS, vol. 9905, pp. 21–37. Springer, Cham (2016). https://doi.org/10.1007/978-3-319-46448-0_2
12. Howard, A.G., Zhu, M., Chen, B., Kalenichenko, D., Wang, W., Weyand, T., et al.: MobileNets: efficient convolutional neural networks for mobile vision applications (2017)
13. Ranzato, M., Krizhevsky, A., Hinton, G.E.: Convolutional Deep Belief Networks on CIFAR-10, Toronto (2010). Unpublished manuscript
14. Ren, S., He, K., Girshick, R., Sun, J.: Faster R-CNN: towards real-time object detection with region proposal networks. IEEE Trans. Pattern Anal. Mach. Intell. **39**, 1 (2016). https://doi.org/10.1109/tpami.2016.2577031
15. He, K., Zhang, X., Ren, S., Sun, J.: Deep residual learning for image recognition. In: 2016 IEEE Conference on Computer Vision and Pattern Recognition (CVPR), pp. 770–778, June 2016. https://doi.org/10.1109/cvpr.2016.90
16. Chen, L., Papandreou, G., Kokkinos, I., Murphy, K., Yuille, A.L.: Semantic image segmentation with deep convolutional nets and fully connected CRFs. In: Computer Vision and Pattern Recognition, vol. abs/1412.7062 (2014)
17. Kazemi, V., Sullivan, J.: One millisecond face alignment with an ensemble of regression trees. In: 2014 IEEE Conference on Computer Vision and Pattern Recognition, pp. 1867–1874, June 2014. https://doi.org/10.1109/cvpr.2014.241
18. Hastie, T., Tibshirani, R., Friedman, J.: The Elements of Statistical Learning. Springer, New York (2009). https://doi.org/10.1007/978-0-387-84858-7. ISBN 978-0-387-84858-7

Firefly Algorithm Enhanced by Orthogonal Learning

Kadavy Tomas(✉), Pluhacek Michal, Viktorin Adam, and Senkerik Roman

Tomas Bata University in Zlin,
T. G. Masaryka 5555, 760 01 Zlin, Czech Republic
{kadavy,pluhacek,aviktorin,senkerik}@utb.cz

Abstract. Orthogonal learning strategy, a proven technique, is combined with hybrid optimization metaheuristic, which is based on Firefly Algorithm and Particle Swarm Optimization. The hybrid algorithm Firefly Particle Swarm Optimization is then compared, together with canonical Firefly Algorithm, with the newly created Orthogonal Learning Firefly Algorithm. Comparisons have been conducted on five selected basic benchmark functions, and the results have been evaluated for statistical significance using Wilcoxon rank-sum test.

Keywords: Firefly algorithm · Particle Swarm Optimization
Orthogonal learning

1 Introduction

In general, the swarm-based metaheuristic optimization algorithms are still quite popular amongst researchers. Nowadays [1], instead of developing a new swarm algorithm, the utilization and smart improvement of existing ones are more favorable. Various techniques are used to achieve this improvement. For example, the ensemble method, which was already successfully adapted [2, 3]. Another possible method could be the hybridization of the already existing algorithms or their inner dynamics to achieve the better results. Both methods typically work with two or more dynamics borrowed from different optimization algorithms. The basic idea is that the hybrid version should combine the advantages of both and eliminate their disadvantages. However, achieving this ideal state is a difficult task, and this is a clear motivation behind the presented research.

The most typical representative of an optimization algorithm with a long history and many modifications [4–6] is Particle Swarm Optimization (PSO) [7]. One of the possible modification is based on the orthogonal learning [8]. The basic idea behind orthogonal learning lies in the identification of combination or rather parts of solutions in particular dimensions offering the best results and using them in the next optimization steps. The principle of the orthogonal learning will be more explained in the relevant section.

In recent years, another swarm-based algorithm proves its usefulness and is becoming quite popular. This novel optimization technique is a Firefly Algorithm

R. Silhavy (Ed.): CSOC 2018, AISC 764, pp. 477–488, 2019.
https://doi.org/10.1007/978-3-319-91189-2_47

(FA) which was introduced in 2010 [9]. Since then, like PSO, many extensions and modifications were proposed for this powerful optimization algorithm [10–12]. Thanks to versatility and popularity of both optimization algorithms, the creation of their hybrid was only a matter of time. The outcome hybrid Firefly Particle Swarm Optimization (FFPSO) [13] combines both, the canonical FA and PSO just as the name suggests. Due to the fundamental principles of orthogonal learning, its hybridization with FA would be quite tricky. However, with FFPSO, the situation is much more simplified.

The proposed optimization algorithm is tested and statistically evaluated on selected well-known benchmark functions. Since the primary aim of this research study was to precisely track the performance differences of the novel hybridization scheme on different recognized objective functions surfaces, we did not used extremely complex IEEE CEC benchmark sets with composite, rotated and shifted functions. We also include the comparisons with original hybridized algorithm FFPSO and with the original version of the FA.

The rest of the paper is structured as follows. Brief descriptions of PSO, FA, and FFPSO, are in Sects. 2 and 3. Section 3 also covers a description of the proposed algorithm based on the orthogonal learning. In Sect. 4, the benchmark functions are defined and the parameter settings of tested algorithms are shown as well. The results and conclusion sections follow afterwards.

2 Particle Swarm Optimization

The PSO is one of the current leading representatives on a field of swarm intelligence based algorithms. It was first published by Eberhart and Kennedy in 1995 [7]. This algorithm mimics the social behavior of swarming animals in nature. Despite the fact that its quite long time from its first appearance, its still plenty used across many optimization problems.

Every particle has a position in n-dimensional solution space, and this position represents the input parameters of the optimized problem. This position of particles changes over the time due to two factors. One of them is the current position of a particle. The second one is the velocity of a particle, labeled as v. Each particle also remembers its best position (solution of the problem) obtained so far. This solution is tagged as the *pBest*, personal best solution. Also, each particle has access to the global best solution, *gBest*, which is selected from all *pBest*s. These variables set the direction for every particle and a new position in the next iteration. PSO usually stops after a number of iterations or a number of FEs (objective function evaluations).

In every iteration of the algorithm, the new positions of particles are calculated based on the previous positions and velocities. The new position of a particle is checked if it still lies in the space of possible solutions.

The position of particle x is calculated according to the formula (1).

$$x_i' = x_i + v_i \tag{1}$$

Where x_i' is a new position of particle *i*, x_i is the previous old position of a particle and *v* is the velocity of a particle. The velocity of a particle *v* is calculated according to (2).

$$v_i' = w \cdot v_i + c_1 \cdot r_1 \cdot (pBest_i - x_i) + c_2 \cdot r_2 \cdot (gBest - x_i) \tag{2}$$

Where *w* is inertia weight [14], c_1 and c_2 are learning factors, and r_1 and r_2 are random numbers of unimodal distribution in the range <0, 1> .

3 Firefly Algorithm

This optimization nature-based algorithm was developed and introduced by Yang in 2008 [9]. The fundamental principle of this algorithm lies in simulating the mating behavior of fireflies at night when fireflies emit light to attract a suitable partner. The main idea of FA is that the objective function value that is optimized is associated with the flashing light of these fireflies. The author for simplicity set a couple of rules to describe the algorithm itself:

- The brightness of each firefly is based on the objective function value.
- The attractiveness of a firefly is proportional to its brightness. This means that the less bright firefly is lured towards, the brighter firefly. The brightness depends on the environment or the medium in which fireflies are moving and decreases with the distance between each of them.
- All fireflies are sexless, and it means that each firefly can attract or be lured by any of the remaining ones.

The movement of one firefly towards another one is then defined by Eq. (3). Where x_i' is a new position of a firefly *i*, x_i is the current position of firefly *i* and x_j is a selected brighter firefly (with better objective function value). The α is a randomization parameter and *sign* simply provides random direction -1 or 1.

$$x_i' = x_i + \beta \cdot \left(x_j - x_i\right) + \alpha \cdot sign \tag{3}$$

The brightness *I* of a firefly is computed by the Eq. (4). This equation of brightness consists of three factors mentioned in the rules above. On the objective function value, the distance between two fireflies and the last factor is the absorption factor of a media in which fireflies are.

$$I = \frac{I_0}{1 + \gamma r^m} \tag{4}$$

Where I_o is the objective function value, the γ stands for the light absorption parameter of a media in which fireflies are and the *m* is another user-defined coefficient and it should be set $m \geq 1$. The variable *r* is the Euclidian distance (5) between the two compared fireflies.

$$r_{ij} = \sqrt{\sum\nolimits_{k=1}^{d} \left(x_{i,k} - x_{j,k}\right)^2} \tag{5}$$

Where r_{ij} is the Euclidian distance between fireflies x_i and x_j. The d is current dimension size of the optimized problem.

The attractiveness β (6) is proportional to brightness I as mentioned in rules above and so these equations are quite similar to each other. The β_0 is the initial attractiveness defined by the user, the γ is again the light absorption parameter and the r is once more the Euclidian distance. The m is also the same as in Eq. (4).

$$\beta = \frac{\beta_0}{1 + \gamma r^m} \tag{6}$$

The pseudocode below shows the fundamentals of FA operations.

```
1.  FA initialization
2.  while(terminal condition not met)
3.     for i = 1 to all fireflies
4.       for j = 1 to all fireflies
5.         if(Ij < Ii) then
6.           move xi to xj
7.           evaluate xi
8.         end if
9.       end for j
10.    end for i
11.    record the best firefly
12. end while
```

Another more advanced versions of the FA could be represented by its modifications, or more likely hybridizations, with others successful metaheuristic algorithms. The basic idea behind such an approach is that the new hybrid strategy can share advantages from both algorithms and hopefully eliminate their disadvantages.

The typical example is a hybrid of the FA and PSO algorithms, the FFPSO [13] introduced in late 2015 by Padmavathi Kora and K. Sri Rama Krishna. The main principle remains the same as in the standard FA, but the equation for firefly motion (3) is slightly changed according to PSO movement and is newly computed as (7).

$$x_i' = wx_i + c_1 e^{-r_{px}^2}(pBest_i - x_i) + c_2 e^{-r_{gx}^2}(gBest - x_i) + \alpha \cdot sign \tag{7}$$

Where w, c_1, and c_2 are control parameters transferred from PSO and their values often depends on the user. Also, the *pBest* and *gBest* are variables originally belonging to PSO algorithm. They both represent the memory of the best position where *pBest* is best position of each particle and *gBest* is globally achieved best position so far. The remaining variables r_{px} (8) and r_{gx} (9) are distances between particle x_i and both $pBest_i$ and *gBest*.

$$r_{px} = \sqrt{\sum_{k=1}^{d} \left(pBest_{i,k} - x_{i,k}\right)^2} \tag{8}$$

$$r_{gx} = \sqrt{\sum_{k=1}^{d} \left(gBest_{k} - x_{i,k}\right)^2} \tag{9}$$

3.1 Orthogonal Learning Firefly Algorithm

Our proposed algorithm, OLFA, is based on FFPSO mentioned above and it uses the orthogonal learning technique. Our application of orthogonal learning is similar as in the Orthogonal Learning Particle Swarm Optimization (OLPSO). The OLFA also generates the promising learning exemplar by adopting an orthogonal learning strategy for each particle to learn from. This means that the equation of firefly moves (7) is slightly changed and does not contain the *pBest* and *gBest* any longer. The new Eq. (10) contains only the trial exemplar $gVector$.

$$x_i' = wx_i + c \cdot e^{-r^2}(gVector_i - x_i) + \alpha \cdot sign \tag{10}$$

Where Euclidian distance r is computed as (11) between firefly x_i and trial $gVector$.

$$r = \sqrt{\sum_{k=1}^{d} \left(gVector_{i,k} - x_{i,k}\right)^2} \tag{11}$$

The $gVector$ is used as the guide for each particle until it cannot improve the solution quality for a certain number of generations, which is called refreshing gap G. When the number of non-improved generations reaches the refreshing gap limit, the learning $gVector$ is reconstructed. For the construction of the $gVector$ and more details about orthogonal learning technique, please refer to [8].

4 Experimental Setup

The experiments were performed on a set of five well-known benchmark functions. As already explained in the introduction section, we have not use any complex IEEE CEC benchmark set, since the aim was to precisely track the performance differences on the simpler unimodal and multimodal function types:

- Sphere function (f_1) (12),
- Rosenbrock function (f_2) (13),
- Rastrigin function (f_3) (14),
- Schwefel function (f_4) (15),
- Ackley function (f_5) (16).

$$f(x)_1 = \sum_{i=1}^{d} x_1^2 \tag{12}$$

$$f(x)_2 = \sum_{i=1}^{d-1} \left[100 \cdot \left(x_{i+1} - x_i^2\right)^2 + (1 - x_i)^2\right] \tag{13}$$

$$f(x)_3 = 10 \cdot d + \sum_{i=1}^{d} \left[x_i^2 - 10 \cdot \cos(2\pi x_i)\right] \tag{14}$$

$$f(x)_4 = 418.9829 \cdot d - \sum_{i=1}^{d} \left(x_i \cdot \sin\sqrt{|x_i|}\right) \tag{15}$$

$$f(x)_5 = -20 \cdot e^{-0.2 \cdot \sqrt{d^{-1} \cdot \sum_{i=1}^{d} x_i^2}} - e^{d^{-1} \cdot \sum_{i=1}^{d} \cos(2 \cdot \pi \cdot x_i)} + 20 + e^1 \tag{16}$$

The selected and tested dimensions were 2, 5, 10 and 15. The maximal number of the evaluation was set as 2000 · d (dimension size). The number of particles was set to 40 for all dimension sizes. Every test function was repeated for 30 independent runs, and the results were statistically evaluated.

The control parameters settings for all tested and compared algorithms are given in Table 1. Parameters were set to optimal values according to literature.

Table 1. Parameters of tested algorithms

Name	Description	Parameters
FA	Original Firefly Algorithm	$\alpha = 0.5, \beta_0 = 0.2, \gamma = 1$
FFPSO	A hybrid of the Firefly Algorithm and Particle Swarm Optimization	$\alpha = 0.5, \beta_0 = 0.2, \gamma = 1, w = 0.729,$ $c_1 = c_2 = 1.49445$
OLFA	A new prosed Orthogonal Learning Firefly Algorithm	$\alpha = 0.5, \beta_0 = 0.2, \gamma = 1,$ $w = 0.729, c = 1.49445,$ $G = 5$

5 Results

The results of all performed experiments are reported here in details. The results overview and simple statistical comparisons are presented in Table 3, depicting *mean, std. dev., min.* and *max.* final objective function values. Examples of convergence behavior of the compared methods are given in Figs. 1, 2, 3, 4, 5, 6 and 7. Further, the Wilcoxon rank-sum test pairwise comparisons are presented in Table 2 with statistical significance 0.05. The Wilcoxon rank-sum performs a hypothesis between two data sets with null hypothesis H_O that the true median difference $\mu_1 - \mu_2 = \mu_0$ against H_a that $\mu_1 - \mu_2 \neq \mu_0$.

Table 2 contains test results for three tested algorithms on five benchmark functions and four compared dimension sizes. Each cell of Table 2 contains a simple matrix of compared results. The number of columns and rows are equal to a number of compared algorithms (in this order: FA, FFPSO, and OLFA). The "<" means that the algorithm outperformed the other compared algorithm from the test couple, the ">" means that the algorithm performs worse than the compared one. The "=" stand for the equivalent performance. The "0" symbol is present in Table 2 since the matrix is symmetrical by the main diagonal. For example, how to read the results in Table 2, for function f_1 in dimension 2, the first algorithm achieved better results than the other two and the second algorithm had a better result than the third algorithm. To conclude this example, the best performing algorithm for this test and dimension is the first algorithm FA, the second-best performing one is FFPSO, and the worst is the OLFA.

From the all given results, it is noticeable, that the proposed OLFA achieved significantly better results only for the Rastrigin test function f_3. For the most cases, the best results were obtained by original FA, and the performance of OLFA lies between the FA and FFPSO. With increasing dimension size, the OLFA results are less significant, and the convergence speed is slower, which is supported in selected Figs. 1, 2, 3, 4, 5, 6 and 7. This kind of behavior could be caused by the orthogonal learning

Table 2. Wilcoxon rank-sum test

F	Dimension											
	2			*5*			*10*			*15*		
f_1	0	<	<	0	<	<	0	<	<	0	<	<
	0	0	>	0	0	>	0	0	>	0	0	>
	0	0	0	0	0	0	0	0	0	0	0	0
f_2	0	<	<	0	<	<	0	<	<	0	<	<
	0	0	>	0	0	>	0	0	>	0	0	<
	0	0	0	0	0	0	0	0	0	0	0	0
f_3	0	<	<	0	=	>	0	>	>	0	>	>
	0	0	>	0	0	>	0	0	<	0	0	<
	0	0	0	0	0	0	0	0	0	0	0	0
f_4	0	<	<	0	<	<	0	<	<	0	<	<
	0	0	=	0	0	=	0	0	=	0	0	=
	0	0	0	0	0	0	0	0	0	0	0	0
f_5	0	<	<	0	<	<	0	<	<	0	<	<
	0	0	>	0	0	>	0	0	<	0	0	<
	0	0	0	0	0	0	0	0	0	0	0	0

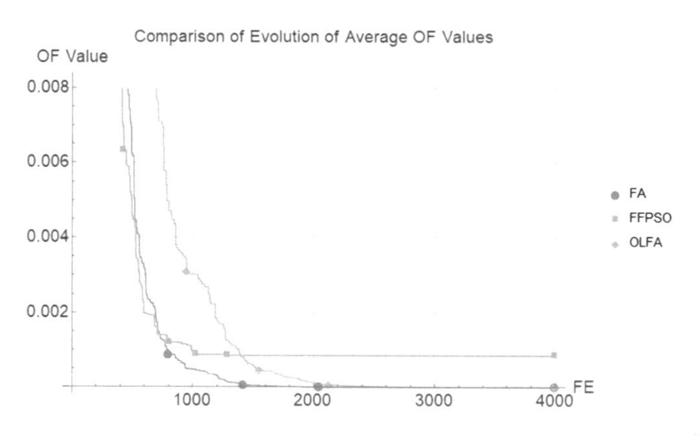

Fig. 1. Convergence graph for f_1 dimension 2.

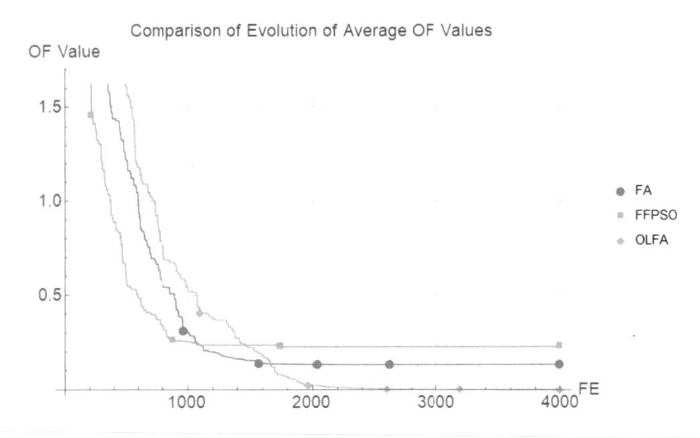

Fig. 2. Convergence graph for f_3 dimension 2.

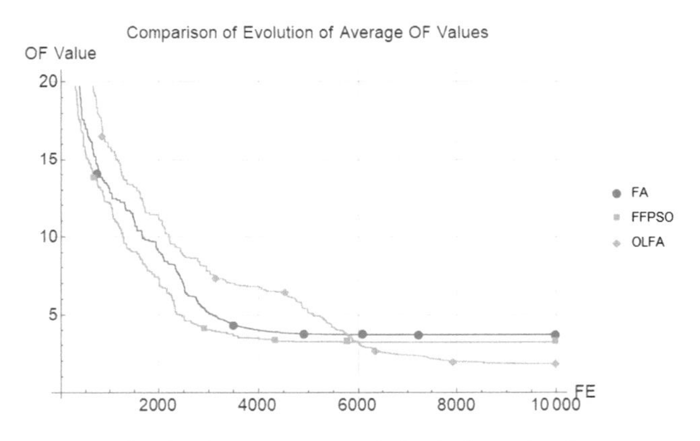

Fig. 3. Convergence graph for f_3 dimension 5.

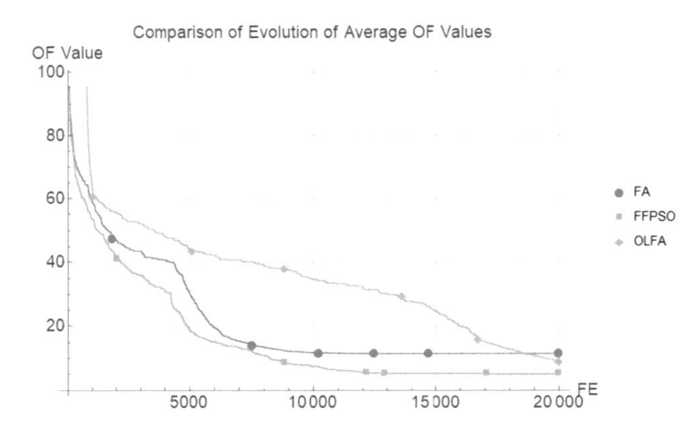

Fig. 4. Convergence graph for f_5 dimension 5.

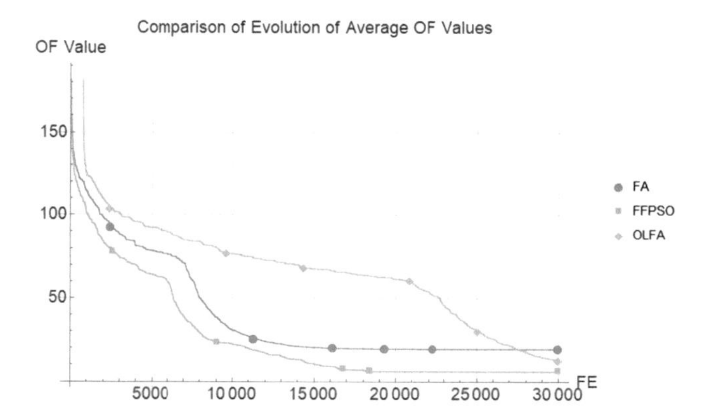

Fig. 5. Convergence graph for f_3 dimension 10.

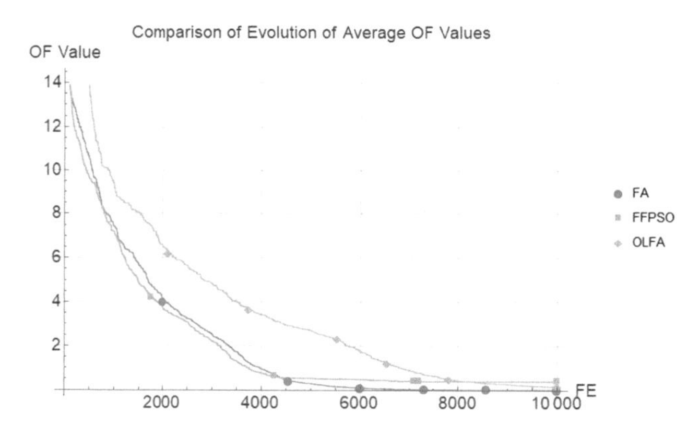

Fig. 6. Convergence graph for f_3 dimension 15.

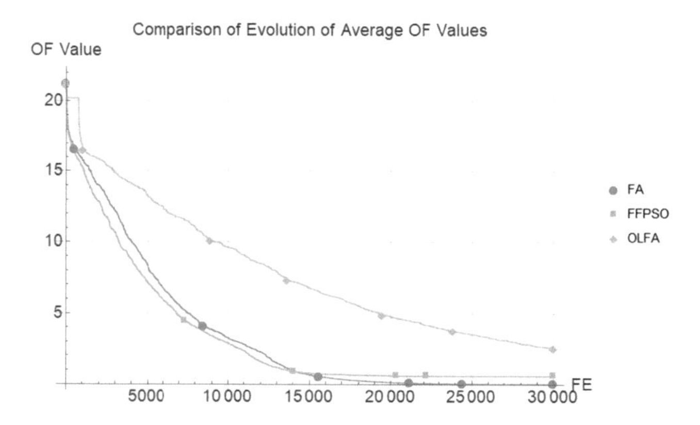

Fig. 7. Convergence graph for f_5 dimension 15.

Table 3. Statistical results (mean, min, max, std. dev.)

F	*FA*				*FFPSO*				*OLFA*			
Dimension 2												
f_1	6.3E−10	2.3E−11	2.2E−09	6.2E−10	8.4E−04	1.0E−06	3.9E−03	9.2E−04	1.8E−07	2.7E−09	7.1E−07	1.9E−07
f_2	6.2E−08	1.1E−09	7.3E−07	1.3E−07	2.7E−04	1.1E−04	7.6E−02	2.1E−01	7.9E−03	2.0E−04	2.7E−02	8.0E−03
f_3	1.3E−01	6.3E−9	9.9E−01	3.4E−01	2.2E−01	1.7E−01	9.9E−01	2.4E−01	4.9E−05	2.4E−07	1.7E−04	4.7E−05
f_4	7.3E+01	2.5E−05	2.3E+02	7.2E+01	2.3E+02	9.6E−01	4.5E+02	1.0E+02	2.3E+02	2.9E−01	4.1E+02	1.1E+02
f_5	4.3E−04	2.5E−05	9.6E−04	2.3E−04	1.0E−01	1.5E−02	2.0E−01	6.0E−02	6.0E−03	5.7E−04	1.4E−02	3.5E−03
Dimension 5												
f_1	3.1E−08	1.1E−08	5.2E−08	1.1E−08	2.2E−02	4.4E−03	4.7E−02	1.0E−02	1.0E−04	2.6E−05	2.1E−04	4.7E−05
f_2	1.1E+00	1.7E−01	7.6E+00	1.8E+00	3.9E+00	2.6E+00	5.3E+00	6.7E−01	2.4E+00	1.4E+00	3.0E+00	4.0E−01
f_3	3.6E+00	9.9E−01	7.9E+00	2.1E+00	3.2E+00	2.0E−01	6.8E+00	2.1E+00	1.8E+00	2.5E−02	7.1E+00	1.7E+00
f_4	6.5E+02	2.1E+02	9.2E+02	1.4E+02	1.1E+03	8.5E+02	1.3E+03	1.1E+02	1.1E+03	7.7E+02	1.3E+03	1.6E+02
f_5	1.8E−03	9.8E−04	3.0E−03	4.4E−04	3.8E−01	1.1E−01	7.9E−01	1.9E−01	1.5E−01	7.4E−02	2.5E−01	3.9E−02
Dimension 10												
f_1	2.0E−07	6.1E−08	2.9E−07	5.7E−08	7.0E−02	1.8E−02	1.1E−01	2.3E−02	1.2E−02	6.1E−03	1.7E−02	2.7E−03
f_2	9.9E+00	4.4E−01	1.3E+02	2.3E+01	1.1E+01	9.0E+00	1.7E+01	1.6E+01	1.0E+01	8.5E+00	1.2E+01	7.3E−01
f_3	1.1E+01	5.9E+00	2.4E+01	4.4E+00	5.0E+00	1.7E−03	1.3E+01	4.1E+00	8.8E+00	2.4E+00	2.1E+01	5.5E+00
f_4	1.6E+03	8.6E+02	2.2E+03	3.1E+02	2.6E+03	2.0E+03	3.0E+03	2.4E+02	2.7E+03	2.1E+03	3.0E+03	2.3E+02
f_5	3.6E−03	2.5E-03	5.3E−03	8.2E-04	5.7E−01	2.6E−02	9.0E−01	2.1E−01	1.8E+00	1.0E+00	2.3E+00	3.2E−01
Dimension 15												
f_1	5.4E−07	2.4E−07	8.6E−07	1.4E−07	9.0E−01	5.1E−02	1.2E−01	1.8E−02	3.1E−02	2.2E−02	5.1E−02	6.2E−03
f_2	1.6E+01	1.0E+01	8.0E+01	1.6E+01	1.6E+01	1.4E+01	2.3E+01	2.2E+00	1.8E+01	1.6E+01	2.1E+01	1.2E+00
f_3	1.9E+01	7.9E+00	3.5E+01	6.7E+00	6.0E+00	5.9E−01	2.7E+01	5.9E+00	1.2E+01	6.9E+00	2.5E+01	3.7E+00
f_4	2.7E+03	1.6E+03	3.5E+03	4.0E+02	4.5E+03	3.5E+03	4.8E+03	2.8E+02	4.5E+03	3.9E+03	4.9E+03	2.6E+02
f_5	4.7E−03	2.5E−03	6.5E−03	9.2E−04	5.7E−01	1.4E−01	7.9E−01	1.6E−01	2.5E+00	1.9E+00	2.9E+00	2.0E−01

technique, which in general needs more objective function evaluations to achieve similar results as the original FA/FFPSO algorithms. Nevertheless, all the presented results supported by the convergence plots, lend weight to the argument, that the proposed OLFA offers a noticeable potential.

6 Conclusion

In this study, a new hybrid algorithm based on interesting strategy of orthogonal learning is proposed and tested. Our proposed algorithm was compared to the original version of FA and the other hybrid algorithm FFPSO. According to analyzed statistical data, the proposed algorithm outperformed the FFPSO. However, with raising dimension size of the solved optimization problem, the performance of OLFA is proportionally decreasing, apparently caused by the slower convergence speed induced by orthogonal learning strategy.

All the presented results show that the proposed hybrid OLFA offers a noticeable potential. Nevertheless, more research including larger test bench is needed to increase the understandability of the algorithm behavior and its performance. Design of the proposed algorithm offers several possible changes that could affect its performance.

Acknowledgements. This work was supported by the Ministry of Education, Youth and Sports of the Czech Republic within the National Sustainability Programme Project no. LO1303 (MSMT-7778/2014), further by the European Regional Development Fund under the Project CEBIA-Tech no. CZ.1.05/2.1.00/03.0089 and by Internal Grant Agency of Tomas Bata University under the Projects no. IGA/CebiaTech/2018/003. This work is also based upon support by COST (European Cooperation in Science & Technology) under Action CA15140, Improving Applicability of Nature-Inspired Optimisation by Joining Theory and Practice (ImAppNIO), and Action IC1406, High-Performance Modelling and Simulation for Big Data Applications (cHiPSet). The work was further supported by resources of A.I. Lab at the Faculty of Applied Informatics, Tomas Bata University in Zlin (ailab.fai.utb.cz)

References

1. Fister Jr, I., Mlakar, U., Brest, J., Fister, I.: A new population-based nature-inspired algorithm every month: is the current era coming to the end. In Proceedings of the 3rd Student Computer Science Research Conference, pp. 33–37. University of Primorska Press (2016)
2. Du, W., Li, B.: Multi-strategy ensemble particle swarm optimization for dynamic optimization. Inf. Sci. **178**(15), 3096–3109 (2008)
3. Wang, H., Wu, Z., Rahnamayan, S., Sun, H., Liu, Y., Pan, J.: Multi-strategy ensemble artificial bee colony algorithm. Inf. Sci. **279**(20), 587–603 (2014)
4. Lynn, N., Suganthan, P.N.: Heterogeneous comprehensive learning particle swarm optimization with enhanced exploration and exploitation. Swarm Evol. Comput. **24**, 11–24 (2015)
5. Nepomuceno, F.V., Engelbrecht, A.P.: A self-adaptive heterogeneous PSO for real-parameter optimization. IEEE (2013)

6. Zhan, Z.-H., Zhang, J., Li, Y., Shi, Y.-H.: Orthogonal learning particle swarm optimization. TEVC **15**(6), 832–847 (2011)
7. Eberhart, R., Kennedy, J.: A new optimizer using particle swarm theory (1995)
8. Zhan, Z.H., Zhang, J., Li, Y., Shi, Y.H.: Orthogonal learning particle swarm optimization. IEEE Trans. Evol. Comput. **15**(6), 832–847 (2011)
9. Yang, X.: Nature-Inspired Metaheuristic Algorithms. Luniver Press, Bristol (2010)
10. Gandomi, A.H., Yang, X., Talatahari, S., Alavi, A.H.: Firefly algorithm with chaos. Commun. Nonlinear Sci. Numer. Simul. **18**(1), 89–98 (2013)
11. Yang, X.: Firefly algorithm, lévy flights and global optimization. In: Research and Development in Intelligent Systems XXVI, pp. 209–218 (2010)
12. Farahani, S.M., Abshouri, A.A., Nasiri, B., Meybodi, M.R.: A gaussian firefly algorithm. Int. J. Mach. Learn. Comput. **1**(5), 448 (2011)
13. Kora, P., Rama Krishna, K.S.: Hybrid firefly and particle swarm optimization algorithm for the detection of bundle branch block. Int. J. Cardiovasc. Acad. **2**(1), 44–48 (2016)
14. Kennedy, J.: The particle swarm: social adaptation of knowledge. In: Proceedings of the IEEE International Conference on Evolutionary Computation, pp. 303–308 (1997)

On the Applicability of Random and the Best Solution Driven Metaheuristics for Analytic Programming and Time Series Regression

Roman Senkerik(✉), Adam Viktorin, Michal Pluhacek, Tomas Kadavy, and Zuzana Kominkova Oplatkova

Faculty of Applied Informatics, Tomas Bata University in Zlin, Nam T.G. Masaryka 5555, 760 01 Zlin, Czech Republic
{senkerik, aviktorin, pluhacek, kadavy, oplatkova}@utb.cz

Abstract. This paper provides a closer insight into applicability and performance of the hybridization of symbolic regression open framework, which is Analytical Programming (AP) and Differential Evolution (DE) algorithm in the task of time series regression. AP can be considered as a robust open framework for symbolic regression thanks to its usability in any programming language with arbitrary driving metaheuristic. The motivation behind this research is to explore and investigate the applicability and differences in performance of AP driven by basic canonical entirely random or best solution driven mutation strategies of DE. An experiment with four case studies has been carried out here with the several time series consisting of GBP/USD exchange rate. The differences between regression/prediction models synthesized using AP as a direct consequence of different DE strategies performances are statistically compared and briefly discussed in conclusion section of this paper.

Keywords: Analytic programming · Differential Evolution
Time series regression

1 Introduction

This paper provides a deeper insight into hybridization of symbolic regression open framework, which is Analytical Programming (AP) [1], and Differential Evolution (DE) [2] algorithm. The task was a time series regression with the research emphasis on the selection schemes of individuals inside driving heuristics to the performance of AP.

The most current intelligent methods are mostly based on the artificial intelligence paradigm. The most popular of these methods are machine learning, fuzzy logic, (deep) neural networks, evolutionary algorithms (EA's) and symbolic regression approaches like genetic programming (GP). Currently, EA's together with symbolic regression techniques are known as a robust set of tools for almost any difficult and complex optimization problems. One of such a challenging problem is naturally the regression/prediction of data/time series. In recent years, it attracts the researches' attention, and it has been solved by GP or hybrid mutual connection of EA's, Genetic Programming (GP), fuzzy systems, neural networks and more complex models [3–5].

R. Silhavy (Ed.): CSOC 2018, AISC 764, pp. 489–498, 2019.
https://doi.org/10.1007/978-3-319-91189-2_48

Currently, AP is a novel approach to symbolic structure synthesis which uses EA for its computation. Since it can utilize any metaheuristic (i.e., evolutionary/ swarm-based algorithm) and it can be easily applied in any programming language, it can be considered as an open symbolic regression framework. AP was introduced by I. Zelinka in 2001 and since its introduction; it has been proven on numerous problems to be as suitable for symbolic structure synthesis as GP or Grammatical Evolution (GE) [6–10]. AP is based on the set of functions and terminals called General Functional Set. The individual of an EA is translated from individual domain to program domain using this set (more in Sect. 3).

The chosen metaheuristic driving AP was DE. More precisely, this research is focused on the two original strategies based either on the full randomized selection of individuals for mutation or based on both influences of stochastics and the current global best-obtained result within the population.

Currently, DE [11–14] is a well-known evolutionary optimization technique for continuous optimization domain. Many DE strategies have been recently developed with the emphasis on learning or control parameter adjustment and adaptation. Over recent years, DE has won most of the evolutionary algorithm competitions in major scientific conferences [15–22], as well as being applied to many complex optimization applications.

This research is an extension and continuation of the previous successful experiments [23–25] as well as experiments with the connection of state of the art Success-History based Adaptive Differential Evolution (SHADE) algorithm and AP on regression of simple functions [26].

Our clear motivation behind this research is to provide a closer insight into the connection between AP and two different original mutation schemes of DE. The motivation and the difference from the above mentioned previous research can be summarized in following points:

- To show the results of DE driven AP for time series regression/prediction problems. This research encompasses four case studies with different dataset/snapshots of data and longer prediction data-part.
- To investigate the differences in performances of AP driven by basic canonical entirely random or best solution driven mutation strategies of DE.
- To give better statistical validation than in the previous research.

The organization of this paper is following: The next sections are focused on the description of the used metaheuristic strategies and the concept of AP. Experiment design, results and conclusion with discussion follow afterward.

2 Differential Evolution

This section describes the basics of canonical DE strategy. The original DE [1] has four static control parameters – a number of generations *G*, population size *NP*, scaling factor *F* and crossover rate *CR*. In the evolutionary process of DE, these four parameters remain unchanged and depend on the initial user setting. The concept of essential operations in DE algorithm is shown in the following subsections, for a detailed description of the canonical DE refer to [1]. In this research, we have used canonical DE "rand/1/bin" (1) and DE "best/1/bin" (2) mutation strategies and binomial crossover (3).

Mutation Strategies and Parent Selection

In canonical forms of DE, parent indices (vectors) are selected by classic PRNG with uniform distribution. Mutation strategy "rand/1/bin" uses three random parent vectors with indexes *r1*, *r2* and *r3*, where $r1 = U[1, NP]$, $\mathrm{r2} = U[1, NP]$, $r3 = U[1, NP]$ and $r1 \neq r2 \neq r3$. Mutated vector $\boldsymbol{v}_{i,G}$ is obtained from three different vectors $\boldsymbol{x}_{r1}$, $\boldsymbol{x}_{r2}$, $\boldsymbol{x}_{r3}$ from current generation G with the help of static scaling factor F as follows (1):

$$\boldsymbol{v}_{i,G} = \boldsymbol{x}_{r1,G} + F\left(\boldsymbol{x}_{r2,G} - \boldsymbol{x}_{r3,G}\right) \tag{1}$$

Whereas mutation strategy "best/1/bin" uses only two random parent vectors with indexes *r1*, and *r2*, and best individual solution in the current generation. The selection respects the very same rules and features as in the previous case. Mutated vector $\boldsymbol{v}_{i,G}$ is obtained as follows (2):

$$\boldsymbol{v}_{i,G} = \boldsymbol{x}_{best,G} + F\left(\boldsymbol{x}_{r1,G} - \boldsymbol{x}_{r2,G}\right) \tag{2}$$

Crossover and Elitism

The trial vector $\boldsymbol{u}_{i,G}$ which is compared with original vector $\boldsymbol{x}_{i,G}$ is completed by crossover operation (3). CR_i value in canonical DE algorithm is static, i.e., $CR_i = CR$.

$$\boldsymbol{u}_{j,i,G} = \begin{cases} \boldsymbol{v}_{j,i,G} & \text{if } U[0,1] \leq CR_i \text{ or } j = j_{rand} \\ \boldsymbol{x}_{j,i,G} & \text{otherwise} \end{cases} \tag{3}$$

Where j_{rand} is randomly selected index of a feature, which has to be updated ($j_{rand} = U[1, D]$), D is the dimensionality of the problem.

The vector which will be placed into the next generation $G + 1$ is selected by elitism. When the objective function value of the trial vector $\boldsymbol{u}_{i,G}$ is better or equal than that of the original vector $\boldsymbol{x}_{i,G}$, the trial vector will be selected for the next population. Otherwise, the original will survive (4).

$$\boldsymbol{x}_{i,G+1} = \begin{cases} \boldsymbol{u}_{i,G} & \text{if } f\left(\boldsymbol{u}_{i,G}\right) \leq f\left(\boldsymbol{x}_{i,G}\right) \\ \boldsymbol{x}_{i,G} & \text{otherwise} \end{cases} \tag{4}$$

3 Analytic Programming

The basic functionality of AP is formed by three parts – General Functional Set (GFS), Discrete Set Handling (DSH) and Security Procedures (SPs). GFS contains all elementary objects which can be used to form a program, DSH carries out the mapping of individuals to programs and SPs are implemented into mapping process to avoid mapping to pathological programs and into cost function to avoid critical situations.

3.1 General Function Set

AP uses sets of functions and terminals. The synthesized program is branched by functions requiring two and more arguments, and the length of it is extended by functions which require one argument. Terminals do not contribute to the complexity of the synthesized program (length) but are needed to synthesize a non-pathological program (a program that can be evaluated by cost function). Therefore, each non-pathological program must contain at least one terminal.

The combined set of functions and terminals forms GFS (See Fig. 1), which is used for mapping from individual domain to program domain. The content of GFS is dependent on user choice. GFS is nested and can be divided into subsets according to the number of arguments that the subset requires. GFS_{0arg} is a subset which requires zero arguments, thus contains only terminals. GFS_{1arg} contains all terminals and functions requiring one argument, GFS_{2arg} contains all objects from GFS_{1arg} and functions requiring two arguments, and so on, GFS_{all} is a complete set of all elementary objects. For mapping from individual to the program, it is important to note that objects in GFS are ordered by a number of arguments they require in descending order.

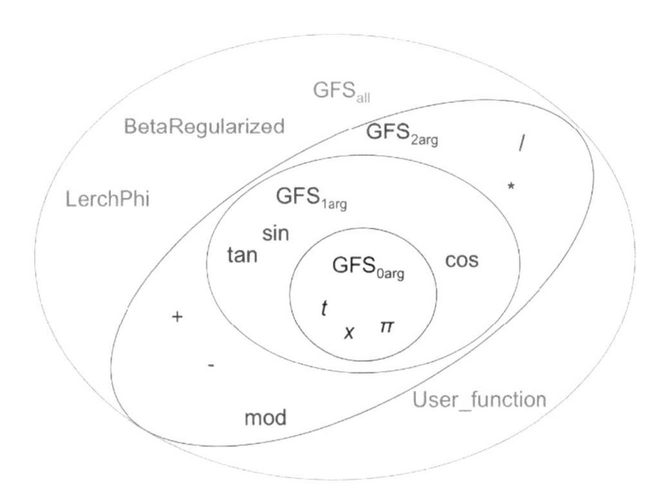

Fig. 1. DSH in AP

3.2 Discrete Set Handling

DSH is used for mapping the individual to the synthesized program. Most of the EAs use individuals with real number encoded individuals. The first significant step for DSH to work is to get individual with integer components which are done by rounding real number values. The integer values of an individual are indexes into the discrete set, in this case, GFS and its subsets. If the index value is greater than the size of used GFS, modulo operation with the size of the discrete set is performed. An illustrative example of mapping is given in (5).

$$\begin{gathered}
Individual = \left\{ \begin{array}{c} 0.12, 4.29, 6.92, 6.12, 2.45, \\ 6.33, 5.78, 0.22, 1.94, 7.32 \end{array} \right\} \\
Rounded\ individual = \{0, 4, 7, 6, 2, 6, 6, 0, 2, 7\} \\
GFS_{all} = \{+, -, *, /, sin, cos, x, k\} \\
\boldsymbol{Program}: \mathbf{sin}\ \boldsymbol{x} + \boldsymbol{k}
\end{gathered} \tag{5}$$

The objects in GFS_{all} are indexed from 0, and the mapping is as follows: The first rounded individual feature is 0 which represents a + function in GFS_{all}. This function requires two arguments, and those are represented by next two indexes – 4 and 7, which are mapped to function *sin* and constant *k*. The *sin* function requires one argument which is given by next index (rounded feature) – 6, and it is mapped to variable *x*. Since there is no possible way of branching the program further, other features are ignored and synthesized program is *sin x* + *k*.

3.3 Security Procedures

SPs are used in AP to avoid critical situations. Some of the SPs are implemented into the AP itself, and some have to be implemented into the cost function evaluation. The typical representatives of the later are checking synthesized programs for loops, infinity, and imaginary numbers if not expected (dividing by 0, the square root of negative numbers, etc.).

3.4 Constants Handling

The constant values in synthesized programs are usually estimated by second EA (meta-heuristic) or by non-linear fitting, which can be very time-demanding. Alternatively, it is possible to use the extended part of the individual in EA for the evolution of constant values. Details can be found here [23].

4 Experiment Design

For performance comparison of AP driven by two different mutation schemes of DE strategies, an experiment with time series regression has been carried out. Time series consisting of 300 data-points of GBP/USD exchange rate has been utilized. To support the robustness of the performance comparisons, the experiment encompasses of totally four case studies:

- Case study 1: The entire dataset was used: The first 2/3 of data (200 points) were used for regression process, and the last 1/3 of the data (100 points) were used as a verification for prediction process.
- Case study 2: Snapshot No. 1 of the time series (data points 1–150). The first 100 points for regression and 50 points for prediction.
- Case study 3: Snapshot No. 2 of the time series (data points 101–250). The first 100 points for regression and remaining 50 points for prediction.
- Case study 4: Snapshot No. 3 of the time series (data points 151–300). The first 100 points for regression and remaining 50 points for prediction.

The parameter setting for both strategies of original DE was following: Population size of 75, canonical DE parameters $F = 0.5$ and $Cr = 0.8$. The maximum number of generations was fixed at 2000 generations. The cost function (CF) was defined as a simple difference between given time series and the synthesized model given using AP (6) on regression interval.

$$CF = \sum_{i=1}^{nreg} |dataTS_i - dataAP_i| \tag{6}$$

Where *nreg* represents the length of the time series regression part (*nreg* data points), *dataTS* given time series data; and *dataAP* synthesized model given by AP.

The setting for AP was following:

- Max length of the individual (max D) = 150, where all max. 150 positions were used for functions, no constants were synthesized here.
- GFS_{All} = {+ , −, * , /abs, cos, x^3, exp, ln, log10, mod, x^2, sin, sigmoid, sqrt, tan, a^b, x}

Experiments were performed in the environment of *Java* and *Wolfram Mathematica*. Overall, 50 independent runs for each DE strategy were performed.

5 Results

Simple statistical results of the experiments are shown in comprehensive Tables 1, 2, 3 and 4 for all 50 repeated runs of both DE mutation strategies. These tables contain basic statistical characteristics for the cost function values like *minimum, maximum, mean, median and standard deviation*. The bold values depict the best-obtained results (except the last attribute).

Table 1. Simple statistical comparisons for both DE strategies and 50 runs, case study 1.

DE strategy	Min	Max	Median	Mean	Std. Dev.
DE “rand/1/bin”	**0.8052**	**1.3922**	**1.0577**	**1.1306**	**0.2171**
DE “best/1/bin”	0.8399	65.1916	1.3922	5.9035	12.2741

Table 2. Simple statistical comparisons for both DE strategies and 50 runs, case study 2.

DE strategy	Min	Max	Median	Mean	Std. Dev.
DE “rand/1/bin”	**0.3304**	**0.5246**	**0.4182**	**0.4308**	**0.0602**
DE “best/1/bin”	0.4084	10.5937	0.5246	0.9740	1.8423

Table 3. Simple statistical comparisons for both DE strategies and 50 runs, case study 3.

DE strategy	Min	Max	Median	Mean	Std. Dev.
DE “rand/1/bin”	0.3710	**0.5639**	**0.4699**	**0.4846**	**0.0625**
DE “best/1/bin”	**0.3186**	2.1188	0.5639	0.6024	0.3655

Table 4. Simple statistical comparisons for both DE strategies and 50 runs, case study 4.

DE strategy	Min	Max	Median	Mean	Std. Dev.
DE "rand/1/bin"	0.4569	**0.7904**	**0.6635**	**0.6645**	**0.0988**
DE "best/1/bin"	**0.4440**	7.7090	0.7868	1.1181	1.3882

Statistical non-parametric tests supporting the conclusions have been performed for a compared pair of DE strategies. Wilcoxon Sum Rank test with the significance level of 0.05 has been used. The p-values for alternative hypotheses "unequal," "greater," and "lower" are depicted in Table 5.

Table 5. p-values for Wilcoxon non-parametric tests, All case studies.

DE "rand" vs. DE "best"	Case study 1	Case study 2	Case study 3	Case study 4
"unequal"	0.00005	0.00002	0.1155	0.0104
"lower"	0.00002	0.00001	0.0578	0.0052
"greater"	0.9999	0.9999	0.9449	0.9951

Convergence plots for the time evolution of average cost function values are in Fig. 2. The best results (synthesized models) for the case studies 1–4 are depicted in Fig. 3. Possible missing/discrete areas in plots of complex symbolic formulas are present due to the numerical instabilities occurring in Plot function in *Wolfram Mathematica SW.*

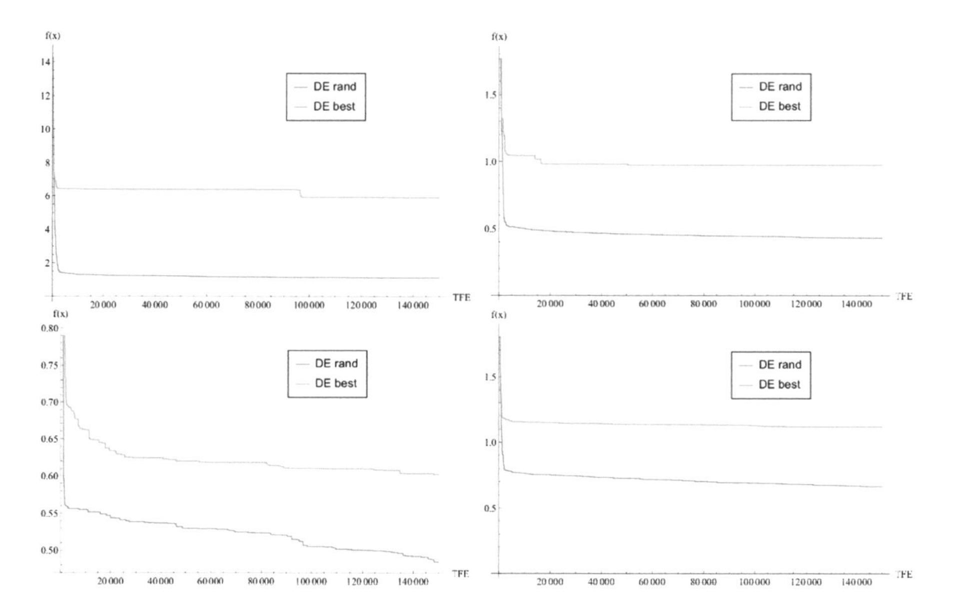

Fig. 2. Convergence plots from all 50 repeated runs; Case study 1 (upper left); Case study (upper right); Case study 3 – (bottom left); Case study 4 – (bottom right).

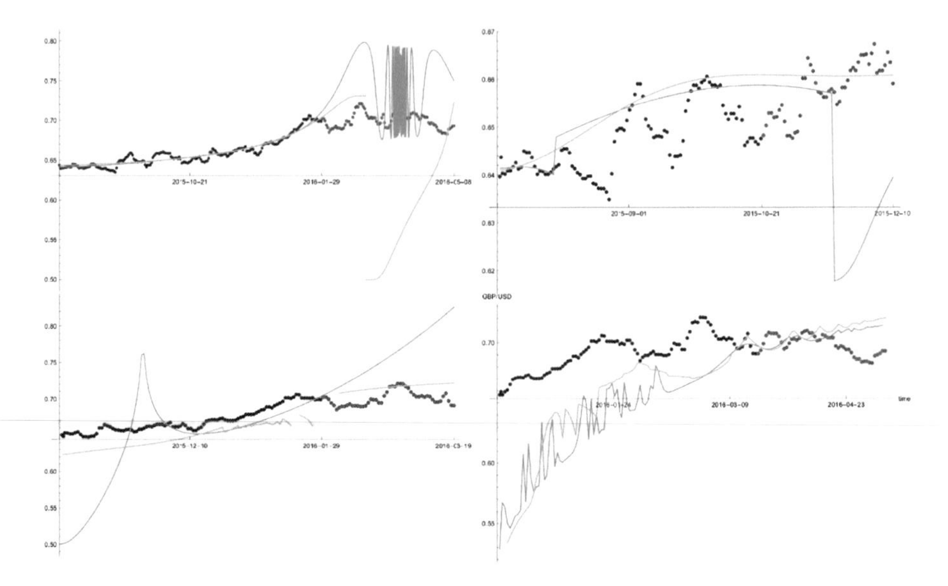

Fig. 3. Comparison of the best results given by two different DE strategies and AP framework for time series prediction problem of GBP/USD exchange rate. Black points (100) used for regression, red points (50) as reference for prediction; Case study 1 (upper left); Case study (upper right); Case study 3 – (bottom left); Case study 4 – (bottom right)

6 Conclusion and Results Analysis

This paper presented a deeper insight into the performance of hybridization between AP and different mutation strategies of DE. The motivation behind this research was to explore the level of applicability and to investigate the differences in performance of AP hybridized with either basic fully randomized or global best solution driven mutation schemes of DE. Therefore, no comprehensive comparison to other methods (GP, EP, NNs, etc.) was performed. The findings can be summarized as follows:

- Obtained graphical comparisons depicted in Figs. 2 and 3 together with statistical data in Tables 1, 2, 3, 4 and 5 support the claim that there are significant performance differences between particular DE mutation strategies in the task of synthesizing time series regression/prediction models using AP.
- The complexity of the best results (synthesized models) was higher than of the worse results.
- The primary logical assumption that the global best result driven mutation variant will outperform the most straightforward fully randomized DE strategy has not been confirmed. Nevertheless, the graphical outputs of synthesized models, statistical data, and paired tests revealed slightly different aspects in performance comparisons. The "best/1/bin" was able to find the best min result regarding CF value for case studies 3 and 4. Moreover, there is no statistically significant difference between both strategies in case study 3.

- The "best/1/bin" strategy shows the tendency of very fast and premature convergence to local extremes in high dimensional complex search space, whereas randomized driven strategy shows the long searching process before stagnation/not-updating of the best result.
- An interesting phenomenon was discovered within this experimental research. It seems to be a very good choice to hybridize the AP and simpler random-based DE strategy, to obtain a perfect synthesized model structure fitting and predicting the data with higher accuracy. The mutual connection of AP and "rand/1/bin" strategy was able to quickly search in very complex high dimensional space for fine individual (solution) structure for discrete set handling process inside AP resulting in good synthesized model structure.

Acknowledgments. This work was supported by the Ministry of Education, Youth and Sports of the Czech Republic within the National Sustainability Programme Project no. LO1303 (MSMT-7778/2014), further by the European Regional Development Fund under the Project CEBIA-Tech no. CZ.1.05/2.1.00/03.0089 and by Internal Grant Agency of Tomas Bata University under the Projects no. IGA/CebiaTech/2018/003. This work is also based upon support by COST (European Cooperation in Science & Technology) under Action CA15140, Improving Applicability of Nature-Inspired Optimisation by Joining Theory and Practice (ImAppNIO), and Action IC406, High-Performance Modelling and Simulation for Big Data Applications (cHiPSet).

References

1. Zelinka, I., Davendra, D., Senkerik, R., Jasek, R., Oplatkova, Z.: Analytical programming - a novel approach for evolutionary synthesis of symbolic structures. In: Kita, E. (ed.) Evolutionary Algorithms. InTech (2011)
2. Storn, R., Price, K.: Differential evolution – a simple and efficient heuristic for global optimization over continuous spaces. J. Global Optim. **11**(4), 341–359 (1997)
3. Wang, W.-C., Chau, K.-W., Cheng, C.-T., Qiu, L.: A comparison of performance of several artificial intelligence methods for forecasting monthly discharge time series. J. Hydrol. **374** (3), 294–306 (2009)
4. Santini, M., Tettamanzi, A.: Genetic programming for financial time series prediction. In: European Conference on Genetic Programming, pp. 361–370. Springer (2001)
5. Pallit, A., Popovic, D.: Computational Intelligence in Time Series Forecasting. Springer, London (2005)
6. Koza, J.R.: Genetic programming: on the programming of computers by means of natural selection, vol. 1. MIT press, Cambridge (1992)
7. Zelinka, I., Oplatková, Z., Nolle, L.: Boolean symmetry function synthesis by means of arbitrary evolutionary algorithms-comparative study. Int. J. Simul. Syst. Sci. Technol. **6**(9), 44–56 (2005)
8. Oplatková, Z., Zelinka, I.: Investigation on artificial ant using analytic programming. In: Proceedings of the 8th Annual Conference on Genetic and Evolutionary Computation, pp. 949–950. ACM (2006)
9. Zelinka, I., Chen, G., Celikovsky, S.: Chaos synthesis by means of evolutionary algorithms. Int. J. Bifurcat. Chaos **18**(04), 911–942 (2008)

10. Senkerik, R., Oplatkova, Z., Zelinka, I., Davendra, D.: Synthesis of feedback controller for three selected chaotic systems by means of evolutionary techniques: analytic programming. Math. Comput. Model. **57**(1–2), 57–67 (2013)
11. Price, K.V., Storn, R.M., Lampinen, J.A.: Differential Evolution - A Practical Approach to Global Optimization. Natural Computing Series. Springer, Heidelberg (2005)
12. Neri, F., Tirronen, V.: Recent advances in differential evolution: a survey and experimental analysis. Artif. Intell. Rev. **33**(1–2), 61–106 (2010)
13. Das, S., Suganthan, P.N.: Differential evolution: a survey of the state-of-the-art. IEEE Trans. Evol. Comput. **15**(1), 4–31 (2011)
14. Das, S., Mullick, S.S., Suganthan, P.N.: Recent advances in differential evolution–an updated survey. Swarm Evol. Comput. **27**, 1–30 (2016)
15. Brest, J., Greiner, S., Boskovic, B., Mernik, M., Zumer, V.: Self-adapting control parameters in differential evolution: a comparative study on numerical benchmark problems. IEEE Trans. Evol. Comput. **10**(6), 646–657 (2006)
16. Qin, A.K., Huang, V.L., Suganthan, P.N.: Differential evolution algorithm with strategy adaptation for global numerical optimization. IEEE Trans. Evol. Comput. **13**(2), 398–417 (2009)
17. Zhang, J., Sanderson, A.C.: JADE: adaptive differential evolution with optional external archive. IEEE Trans. Evol. Comput. **13**(5), 945–958 (2009)
18. Das, S., Abraham, A., Chakraborty, U.K., Konar, A.: Differential evolution using a neighborhood-based mutation operator. IEEE Trans. Evol. Comput. **13**(3), 526–553 (2009)
19. Mininno, E., Neri, F., Cupertino, F., Naso, D.: Compact differential evolution. IEEE Trans. Evol. Comput. **15**(1), 32–54 (2011)
20. Mallipeddi, R., Suganthan, P.N., Pan, Q.-K., Tasgetiren, M.F.: Differential evolution algorithm with ensemble of parameters and mutation strategies. Appl. Soft Comput. **11**(2), 1679–1696 (2011)
21. Brest, J., Korošec, P., Šilc, J., Zamuda, A., Bošković, B., Maučec, M.S.: Differential evolution and differential ant-stigmergy on dynamic optimisation problems. Int. J. Syst. Sci. **44**(4), 663–679 (2013)
22. Tanabe, R., Fukunaga, A.S.: Improving the search performance of SHADE using linear population size reduction. In: 2014 IEEE Congress on Evolutionary Computation (CEC), pp. 1658–1665. IEEE (2014)
23. Senkerik, R., Viktorin, A., Pluhacek, M., Kadavy, T., Zelinka, I.: Differential evolution driven analytic programming for prediction. In: Rutkowski, L., Korytkowski, M., Scherer, R., Tadeusiewicz, R., Zadeh, L., Zurada, J. (eds.) International Conference on Artificial Intelligence and Soft Computing, pp. 676–687. Springer, Cham (2017)
24. Senkerik, R., Viktorin, A., Pluhacek, M., Kadavy, T., Zelinka, I.: Hybridization of analytic programming and differential evolution for time series prediction. In: Martínez de Pisón, F., Urraca, R., Quintián, H., Corchado, E. (eds.) International Conference on Hybrid Artificial Intelligence Systems, pp. 686–698. Springer, Cham (2017)
25. Senkerik, R., Viktorin, A., Pluhacek, M., Kadavy, T., Oplatkova, Z.K.: Performance comparison of differential evolution driving analytic programming for regression. In: 2017 IEEE Symposium Series on Computational Intelligence (SSCI), pp. 1–8. IEEE, November 2017
26. Viktorin, A., Pluhacek, M., Oplatkova, Z.K., Senkerik, R.: Analytical programming with extended individuals. In: European Conference on Modelling and Simulation, pp. 237–244 (2016). ISBN: 978-0-9932440-2-5

Author Index

R. Silhavy (Ed.): CSOC 2018, AISC 764, pp. 499–501, 2019.
https://doi.org/10.1007/978-3-319-91189-2

Printed in the United States
By Bookmasters